FOLLOWING

THE EQUATOR

and Anti-imperialist Essays

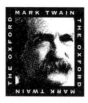

THE OXFORD MARK TWAIN

Shelley Fisher Fishkin, Editor

The Celebrated Jumping Frog of Calaveras County, and Other Sketches
  Introduction: Roy Blount Jr.
  Afterword: Richard Bucci

The Innocents Abroad
  Introduction: Mordecai Richler
  Afterword: David E. E. Sloane

Roughing It
  Introduction: George Plimpton
  Afterword: Henry B. Wonham

The Gilded Age
  Introduction: Ward Just
  Afterword: Gregg Camfield

Sketches, New and Old
  Introduction: Lee Smith
  Afterword: Sherwood Cummings

The Adventures of Tom Sawyer
  Introduction: E. L. Doctorow
  Afterword: Albert E. Stone

A Tramp Abroad
  Introduction: Russell Banks
  Afterword: James S. Leonard

The Prince and the Pauper
  Introduction: Judith Martin
  Afterword: Everett Emerson

Life on the Mississippi
  Introduction: Willie Morris
  Afterword: Lawrence Howe

Adventures of Huckleberry Finn
  Introduction: Toni Morrison
  Afterword: Victor A. Doyno

A Connecticut Yankee in King Arthur's Court
  Introduction: Kurt Vonnegut, Jr.
  Afterword: Louis J. Budd

Merry Tales
  Introduction: Anne Bernays
  Afterword: Forrest G. Robinson

The American Claimant
  Introduction: Bobbie Ann Mason
  Afterword: Peter Messent

The £1,000,000 Bank-Note and Other New Stories
  Introduction: Malcolm Bradbury
  Afterword: James D. Wilson

Tom Sawyer Abroad
  Introduction: Nat Hentoff
  Afterword: M. Thomas Inge

The Tragedy of Pudd'nhead Wilson and the Comedy
Those Extraordinary Twins
  Introduction: Sherley Anne Williams
  Afterword: David Lionel Smith

Following the Equator and Anti-imperialist Essays

Mark Twain

FOREWORD

SHELLEY FISHER FISHKIN

INTRODUCTION

GORE VIDAL

AFTERWORD

FRED KAPLAN

New York   Oxford

OXFORD UNIVERSITY PRESS

1996

OXFORD UNIVERSITY PRESS

Oxford   New York

Athens, Auckland, Bangkok, Bogotá, Bombay
Buenos Aires, Calcutta, Cape Town, Dar es Salaam
Delhi, Florence, Hong Kong, Istanbul, Karachi
Kuala Lumpur, Madras, Madrid, Melbourne
Mexico City, Nairobi, Paris, Singapore
Taipei, Tokyo, Toronto
and associated companies in
Berlin, Ibadan

Published by
Oxford University Press, Inc.
198 Madison Avenue, New York,
New York 10016

Oxford is a registered trademark of
Oxford University Press

Library of Congress
Cataloging-in-Publication Data

Twain, Mark, 1835–1910.
Following the equator & anti-imperialist essays / by
Mark Twain; with an introduction by Gore Vidal
and an afterword by Fred Kaplan.
p. cm. — (The Oxford Mark Twain)
Includes bibliographical references.
Contents: Following the equator — To the person
sitting in darkness — King Leopold's soliloquy.
1. Voyages around the world. 2. Imperialism. I. Title.
II. Series: Twain, Mark, 1835–1910. Works. 1996.
PS1310.A1 1996
813'.403—dc20
[B]
96-16584
CIP
ISBN 0-19-510151-0 (trade ed.)
ISBN 0-19-511419-1 (lib. ed.)
ISBN 0-19-509088-8 (trade ed. set)
ISBN 0-19-511345-4 (lib. ed. set)

9 8 7 6 5 4 3 2 1

Printed in the United States of America
on acid-free paper

FRONTISPIECE
Samuel L. Clemens is photographed here in 1901,
the year he published "To the Person Sitting in
Darkness." (The Mark Twain House, Hartford,
Connecticut)

CONTENTS

EDITOR'S NOTE

The Oxford Mark Twain consists of twenty-nine volumes of facsimiles of the first American editions of Mark Twain's works, with an editor's foreword, new introductions, afterwords, notes on the texts, and essays on the illustrations in volumes with artwork. The facsimiles have been reproduced from the originals unaltered, except that blank pages in the front and back of the books have been omitted, and any seriously damaged or missing pages have been replaced by pages from other first editions (as indicated in the notes on the texts).

In the foreword, introduction, afterword, and essays on the illustrations, the titles of Mark Twain's works have been capitalized according to modern conventions, as have the names of characters (except where otherwise indicated). In the case of discrepancies between the title of a short story, essay, or sketch as it appears in the original table of contents and as it appears on its own title page, the title page has been followed. The parenthetical numbers in the introduction, afterwords, and illustration essays are page references to the facsimiles.

FOREWORD

Shelley Fisher Fishkin

Samuel Clemens entered the world and left it with Halley's Comet, little dreaming that generations hence Halley's Comet would be less famous than Mark Twain. He has been called the American Cervantes, our Homer, our Tolstoy, our Shakespeare, our Rabelais. Ernest Hemingway maintained that "all modern American literature comes from one book by Mark Twain called *Huckleberry Finn*." President Franklin Delano Roosevelt got the phrase "New Deal" from *A Connecticut Yankee in King Arthur's Court*. *The Gilded Age* gave an entire era its name. "The future historian of America," wrote George Bernard Shaw to Samuel Clemens, "will find your works as indispensable to him as a French historian finds the political tracts of Voltaire."[1]

There is a Mark Twain Bank in St. Louis, a Mark Twain Diner in Jackson Heights, New York, a Mark Twain Smoke Shop in Lakeland, Florida. There are Mark Twain Elementary Schools in Albuquerque, Dayton, Seattle, and Sioux Falls. Mark Twain's image peers at us from advertisements for Bass Ale (his drink of choice was Scotch), for a gas company in Tennessee, a hotel in the nation's capital, a cemetery in California.

Ubiquitous though his name and image may be, Mark Twain is in no danger of becoming a petrified icon. On the contrary: Mark Twain lives. *Huckleberry Finn* is "the most taught novel, most taught long work, and most taught piece of American literature" in American schools from junior high to the graduate level.[2] Hundreds of Twain impersonators appear in theaters, trade shows, and shopping centers in every region of the country.[3] Scholars publish hundreds of articles as well as books about Twain every year, and he is the subject of daily exchanges on the Internet. A journalist somewhere in the world finds a reason to quote Twain just about every day. Television series such as *Bonanza, Star Trek: The Next Generation,* and *Cheers* broadcast episodes that feature Mark Twain as a character. Hollywood screenwriters regularly produce movies inspired by his works, and writers of mysteries and science fiction continue to weave him into their plots.[4]

A century after the American Revolution sent shock waves throughout Europe, it took Mark Twain to explain to Europeans and to his countrymen alike what that revolution had wrought. He probed the significance of this new land and its new citizens, and identified what it was in the Old World that America abolished and rejected. The founding fathers had thought through the political dimensions of making a new society; Mark Twain took on the challenge of interpreting the social and cultural life of the United States for those outside its borders as well as for those who were living the changes he discerned.

Americans may have constructed a new society in the eighteenth century, but they articulated what they had done in voices that were largely inter-changeable with those of Englishmen until well into the nineteenth century. Mark Twain became the voice of the new land, the leading translator of what and who the "American" was — and, to a large extent, is. Frances Trollope's *Domestic Manners of the Americans,* a best-seller in England, Hector St. John de Crèvecoeur's *Letters from an American Farmer,* and Tocqueville's *Democracy in America* all tried to explain America to Europeans. But Twain did more than that: he allowed European readers to *experience* this strange "new world." And he gave his countrymen the tools to do two things they had not quite had the confidence to do before. He helped them stand before the cultural icons of the Old World unembarrassed, unashamed of America's lack of palaces and shrines, proud of its brash practicality and bold inventiveness, unafraid to reject European models of "civilization" as tainted or corrupt. And he also helped them recognize their own insularity, boorishness, arrogance, or ignorance, and laugh at it — the first step toward transcending it and becoming more "civilized," in the best European sense of the word.

Twain often strikes us as more a creature of our time than of his. He appreciated the importance and the complexity of mass tourism and public relations, fields that would come into their own in the twentieth century but were only fledgling enterprises in the nineteenth. He explored the liberating potential of humor and the dynamics of friendship, parenting, and marriage. He narrowed the gap between "popular" and "high" culture, and he meditated on the enigmas of personal and national identity. Indeed, it would be difficult to find an issue on the horizon today that Twain did not touch on somewhere in his work. Heredity versus environment? Animal rights? The boundaries of gender? The place of black voices in the cultural heritage of the United States? Twain was there.

With startling prescience and characteristic grace and wit, he zeroed in on many of the key challenges — political, social, and technological — that would face his country and the world for the next hundred years: the challenge of race relations in a society founded on both chattel slavery and ideals of equality, and the intractable problem of racism in American life; the potential of new technologies to transform our lives in ways that can be both exhilarating and terrifying — as well as unpredictable; the problem of imperialism and the difficulties entailed in getting rid of it. But he never lost sight of the most basic challenge of all: each man or woman's struggle for integrity in the face of the seductions of power, status, and material things.

Mark Twain's unerring sense of the right word and not its second cousin taught people to pay attention when he spoke, in person or in print. He said things that were smart and things that were wise, and he said them incomparably well. He defined the rhythms of our prose and the contours of our moral map. He saw our best and our worst, our extravagant promise and our stunning failures, our comic foibles and our tragic flaws. Throughout the world he is viewed as the most distinctively American of American authors — and as one of the most universal. He is assigned in classrooms in Naples, Riyadh, Belfast, and Beijing, and has been a major influence on twentieth-century writers from Argentina to Nigeria to Japan. The Oxford Mark Twain celebrates the versatility and vitality of this remarkable writer.

The Oxford Mark Twain reproduces the first American editions of Mark Twain's books published during his lifetime.[5] By encountering Twain's works in their original format — typography, layout, order of contents, and illustrations — readers today can come a few steps closer to the literary artifacts that entranced and excited readers when the books first appeared. Twain approved of and to a greater or lesser degree supervised the publication of all of this material.[6] The Mark Twain House in Hartford, Connecticut, generously loaned us its originals.[7] When more than one copy of a first American edition was available, Robert H. Hirst, general editor of the Mark Twain Project, in cooperation with Marianne Curling, curator of the Mark Twain House (and Jeffrey Kaimowitz, head of Rare Books for the Watkinson Library of Trinity College, Hartford, where the Mark Twain House collection is kept), guided our decision about which one to use.[8] As a set, the volumes also contain more than eighty essays commissioned especially for The Oxford Mark Twain, in which distinguished contributors reassess Twain's achievement as a writer and his place in the cultural conversation that he did so much to shape.

Each volume of The Oxford Mark Twain is introduced by a leading American, Canadian, or British writer who responds to Twain — often in a very personal way — as a fellow writer. Novelists, journalists, humorists, columnists, fabulists, poets, playwrights — these writers tell us what Twain taught them and what in his work continues to speak to them. Reading Twain's books, both famous and obscure, they reflect on the genesis of his art and the characteristics of his style, the themes he illuminated, and the aesthetic strategies he pioneered. Individually and collectively their contributions testify to the place Mark Twain holds in the hearts of readers of all kinds and temperaments.

Scholars whose work has shaped our view of Twain in the academy today have written afterwords to each volume, with suggestions for further reading. Their essays give us a sense of what was going on in Twain's life when he wrote the book at hand, and of how that book fits into his career. They explore how each book reflects and refracts contemporary events, and they show Twain responding to literary and social currents of the day, variously accepting, amplifying, modifying, and challenging prevailing paradigms. Sometimes they argue that works previously dismissed as quirky or eccentric departures actually address themes at the heart of Twain's work from the start. And as they bring new perspectives to Twain's composition strategies in familiar texts, several scholars see experiments in form where others saw only form-lessness, method where prior critics saw only madness. In addition to eluci-dating the work's historical and cultural context, the afterwords provide an overview of responses to each book from its first appearance to the present.

Most of Mark Twain's books involved more than Mark Twain's words: unique illustrations. The parodic visual send-ups of "high culture" that Twain himself drew for *A Tramp Abroad*, the sketch of financial manipulator Jay Gould as a greedy and sadistic "Slave Driver" in *A Connecticut Yankee in King Arthur's Court*, and the memorable drawings of Eve in *Eve's Diary* all helped Twain's books to be sold, read, discussed, and preserved. In their es-says for each volume that contains artwork, Beverly R. David and Ray Sapirstein highlight the significance of the sketches, engravings, and pho-tographs in the first American editions of Mark Twain's works, and tell us what is known about the public response to them.

The Oxford Mark Twain invites us to read some relatively neglected works by Twain in the company of some of the most engaging literary figures of our time. Roy Blount Jr., for example, riffs in a deliciously Twain-like manner on "An Item Which the Editor Himself Could Not Understand," which may well rank as one of the least-known pieces Twain ever published. Bobbie Ann Mason celebrates the "mad energy" of Twain's most obscure comic novel, *The American Claimant*, in which the humor "hurtles beyond tall tale into simon-pure absurdity."[9] Garry Wills finds that *Christian Science* "gets us very close to the heart of American culture." Lee Smith reads "Political Economy" as a sharp and funny essay on language. Walter Mosley sees "The Stolen White Elephant," a story "reduced to a series of ridiculous telegrams related by an untrustworthy narrator caught up in an adventure that is as impossible as it is ludicrous," as a stunningly compact and economical satire of a world we still recognize as our own. Anne Bernays returns to "The Private History of a Campaign That Failed" and finds "an antiwar manifesto that is also confession, dramatic monologue, a plea for understanding and absolution, and a romp that gradually turns into atrocity even as we watch." After revisiting Captain Stormfield's heaven, Frederik Pohl finds that there "is no imaginable place more pleasant to spend eternity." Indeed, Pohl writes, "one would almost be willing to die to enter it."

While less familiar works receive fresh attention in The Oxford Mark Twain, new light is cast on the best-known works as well. Judith Martin ("Miss Manners") points out that it is by reading a court etiquette book that Twain's pauper learns how to behave as a proper prince. As important as etiquette may be in the palace, Martin notes, it is even more important in the slums.

> That etiquette is a sorer point with the ruffians in the street than with the proud dignitaries of the prince's court may surprise some readers. As in our own streets, etiquette is always a more volatile subject among those who cannot count on being treated with respect than among those who have the power to command deference.

And taking a fresh look at *Adventures of Huckleberry Finn,* Toni Morrison writes,

> much of the novel's genius lies in its quiescence, the silences that pervade it and give it a porous quality that is by turns brooding and soothing. It lies in ... the subdued images in which the repetition of a simple word, such as "lonesome," tolls like an evening bell; the moments when nothing is said, when scenes and incidents swell the heart unbearably precisely because unarticulated, and force an act of imagination almost against the will.

Engaging Mark Twain as one writer to another, several contributors to The Oxford Mark Twain offer new insights into the processes by which his books came to be. Russell Banks, for example, reads *A Tramp Abroad* as "an important revision of Twain's incomplete first draft of *Huckleberry Finn*, a second draft, if you will, which in turn made possible the third and final draft." Erica Jong suggests that *1601*, a freewheeling parody of Elizabethan manners and mores, written during the same summer Twain began *Huckleberry Finn*, served as "a warm-up for his creative process" and "primed the pump for other sorts of freedom of expression." And Justin Kaplan suggests that "one of the transcendent figures standing behind and shaping" *Joan of Arc* was Ulysses S. Grant, whose memoirs Twain had recently published, and who, like Joan, had risen unpredictably "from humble and obscure origins" to become a "military genius" endowed with "the gift of command, a natural eloquence, and an equally natural reserve."

As a number of contributors note, Twain was a man ahead of his times. *The Gilded Age* was the first "Washington novel," Ward Just tells us, because "Twain was the first to see the possibilities that had eluded so many others." Commenting on *The Tragedy of Pudd'nhead Wilson*, Sherley Anne Williams observes that "Twain's argument about the power of environment in shaping character runs directly counter to prevailing sentiment where the negro was concerned." Twain's fictional technology, wildly fanciful by the standards of his day, predicts developments we take for granted in ours. DNA cloning, fax machines, and photocopiers are all prefigured, Bobbie Ann Mason tells us, in *The American Claimant*. Cynthia Ozick points out that the "telelectrophonoscope" we meet in "From the 'London Times' of 1904" is suspiciously like what we know as "television." And Malcolm Bradbury suggests that in the "phrenophones" of "Mental Telegraphy" "the Internet was born."

Twain turns out to have been remarkably prescient about political affairs as well. Kurt Vonnegut sees in *A Connecticut Yankee* a chilling foreshadowing (or perhaps a projection from the Civil War) of "all the high-tech atrocities which followed, and which follow still." Cynthia Ozick suggests that "The Man That Corrupted Hadleyburg," along with some of the other pieces collected under that title — many of them written when Twain lived in a Vienna ruled by Karl Lueger, a demagogue Adolf Hitler would later idolize — shoot up moral flares that shed an eerie light on the insidious corruption, prejudice, and hatred that reached bitter fruition under the Third Reich. And Twain's portrait in this book of "the dissolving Austria-Hungary of the 1890s," in Ozick's view, presages not only the Sarajevo that would erupt in 1914 but also "the disintegrated components of the former Yugoslavia" and "the *fin-de-siècle* Sarajevo of our own moment."

Despite their admiration for Twain's ambitious reach and scope, contributors to The Oxford Mark Twain also recognize his limitations. Mordecai Richler, for example, thinks that "the early pages of *Innocents Abroad* suffer from being a tad broad, proffering more burlesque than inspired satire," perhaps because Twain was "trying too hard for knee-slappers." Charles Johnson notes that the Young Man in Twain's philosophical dialogue about free will and determinism (*What Is Man?*) "caves in far too soon," failing to challenge what through late-twentieth-century eyes looks like "pseudoscience" and suspect essentialism in the Old Man's arguments.

Some contributors revisit their first encounters with Twain's works, recalling what surprised or intrigued them. When David Bradley came across "Fenimore Cooper's Literary Offences" in his college library, he "did not at first realize that Twain was being his usual ironic self with all this business about the 'nineteen rules governing literary art in the domain of romantic fiction,' but by the time I figured out there was no such list outside Twain's own head, I had decided that the rules made *sense*. . . . It seemed to me they were a pretty good blueprint for writing — Negro writing included." Sherley Anne Williams remembers that part of what attracted her to *Pudd'nhead Wilson* when she first read it thirty years ago was "that Twain, writing at the end of the nineteenth century, could imagine negroes as characters, albeit white ones, who actually thought for and of themselves, whose actions were the product of their thinking rather than the spontaneous ephemera of physical instincts that stereotype assigned to blacks." Frederik Pohl recalls his first reading of *Huckleberry Finn* as "a watershed event" in his life, the first book he read as a child in which "bad people" ceased to exercise a monopoly on doing "bad things." In *Huckleberry Finn* "some seriously bad things — things like the possession and mistreatment of black slaves, like stealing and lying, even like killing other people in duels — were quite often done by people who not only thought of themselves as exemplarily moral but, by any other standards I knew how to apply, actually *were* admirable citizens." The world that

Tom and Huck lived in, Pohl writes, "was filled with complexities and con-tradictions," and resembled "the world I appeared to be living in myself."

Other contributors explore their more recent encounters with Twain, ex-plaining why they have revised their initial responses to his work. For Toni Morrison, parts of *Huckleberry Finn* that she "once took to be deliberate eva-sions, stumbles even, or a writer's impatience with his or her material," now strike her "as otherwise: as entrances, crevices, gaps, seductive invitations flashing the possibility of meaning. Unarticulated eddies that encourage div-ing into the novel's undertow — the real place where writer captures reader." One such "eddy" is the imprisonment of Jim on the Phelps farm. Instead of dismissing this portion of the book as authorial bungling, as she once did, Morrison now reads it as Twain's commentary on the 1880s, a period that "saw the collapse of civil rights for blacks," a time when "the nation, as well as Tom Sawyer, was deferring Jim's freedom in agonizing play." Morrison be-lieves that Americans in the 1880s were attempting "to bury the combustible issues Twain raised in his novel," and that those who try to kick Huck Finn out of school in the 1990s are doing the same: "The cyclical attempts to re-move the novel from classrooms extend Jim's captivity on into each genera-tion of readers."

Although imitation-Hemingway and imitation-Faulkner writing contests draw hundreds of entries annually, no one has ever tried to mount a faux-Twain competition. Why? Perhaps because Mark Twain's voice is too much a part of who we are and how we speak even today. Roy Blount Jr. suggests that it is impossible, "at least for an American writer, to parody Mark Twain. It would be like doing an impression of your father or mother: he or she is al-ready there in your voice."

Twain's style is examined and celebrated in The Oxford Mark Twain by fellow writers who themselves have struggled with the nuances of words, the structure of sentences, the subtleties of point of view, and the trickiness of opening lines. Bobbie Ann Mason observes, for example, that "Twain loved the sound of words and he knew how to string them by sound, like different shades of one color: 'The earl's barbaric eye,' 'the Usurping Earl,' 'a double-dyed humbug.'" Twain "relied on the punch of plain words" to show writers how to move beyond the "wordy romantic rubbish" so prevalent in nine-teenth-century fiction, Mason says; he "was one of the first writers in America to deflower literary language." Lee Smith believes that "American writers have benefited as much from the way Mark Twain opened up the possibilities of first-person narration as we have from his use of vernacular language." (She feels that "the ghost of Mark Twain was hovering someplace in the back-ground" when she decided to write her novel *Oral History* from the stand-point of multiple first-person narrators.) Frederick Busch maintains that "A Dog's Tale" "boasts one of the great opening sentences" of all time: "My fa-ther was a St. Bernard, my mother was a collie, but I am a Presbyterian." And Ursula Le Guin marvels at the ingenuity of the following sentence that she en-counters in *Extracts from Adam's Diary*.

> . . . This made her sorry for the creatures which live in there, which she calls fish, for she continues to fasten names on to things that don't need them and don't come when they are called by them, which is a matter of no consequence to her, as she is such a numskull anyway; so she got a lot of them out and brought them in last night and put them in my bed to keep warm, but I have noticed them now and then all day, and I don't see that they are any happier there than they were before, only quieter.[10]

Le Guin responds,

> Now, that is a pure Mark-Twain-tour-de-force sentence, covering an im-mense amount of territory in an effortless, aimless ramble that seems to be heading nowhere in particular and ends up with breathtaking accuracy at the gold mine. Any sensible child would find that funny, perhaps not fol-lowing all its divagations but delighted by the swing of it, by the word "numskull," by the idea of putting fish in the bed; and as that child grew older and reread it, its reward would only grow; and if that grown-up child had to write an essay on the piece and therefore earnestly studied and pored over this sentence, she would end up in unmitigated admiration of its vocabulary, syntax, pacing, sense, and rhythm, above all the beautiful

timing of the last two words; and she would, and she does, still find it funny.

The fish surface again in a passage that Gore Vidal calls to our attention, from *Following the Equator*: "'The Whites always mean well when they take human fish out of the ocean and try to make them dry and warm and happy and comfortable in a chicken coop,' which is how, through civilization, they did away with many of the original inhabitants. Lack of empathy is a principal theme in Twain's meditations on race and empire."

Indeed, empathy — and its lack — is a principal theme in virtually all of Twain's work, as contributors frequently note. Nat Hentoff quotes the following thoughts from Huck in *Tom Sawyer Abroad*:

> I see a bird setting on a dead limb of a high tree, singing with its head tilt-
> ed back and its mouth open, and before I thought I fired, and his song
> stopped and he fell straight down from the limb, all limp like a rag, and I
> run and picked him up and he was dead, and his body was warm in my
> hand, and his head rolled about this way and that, like his neck was broke,
> and there was a little white skin over his eyes, and one little drop of blood
> on the side of his head; and laws! I could n't see nothing more for the tears;
> and I hain't never murdered no creature since that war n't doing me no
> harm, and I ain't going to.[11]

"The Humane Society," Hentoff writes, "has yet to say anything as powerful — and lasting."

Readers of The Oxford Mark Twain will have the pleasure of revisiting Twain's Mississippi landmarks alongside Willie Morris, whose own lower Mississippi Valley boyhood gives him a special sense of connection to Twain. Morris knows firsthand the mosquitoes described in *Life on the Mississippi* — so colossal that "two of them could whip a dog" and "four of them could hold a man down"; in Morris's own hometown they were so large during the flood season that "local wags said they wore wristwatches." Morris's Yazoo City and Twain's Hannibal shared a "rough-hewn democracy . . . complicated by all the visible textures of caste and class, . . . harmless boyhood fun and mischief right along with ... rank hypocrisies, churchgoing sanctimonies, racial hatred, entrenched and unrepentant greed."

For the West of Mark Twain's *Roughing It*, readers will have George Plimpton as their guide. "What a group these newspapermen were!" Plimpton writes about Twain and his friends Dan De Quille and Joe Goodman in Virginia City, Nevada. "Their roisterous carryings-on bring to mind the kind of frat-house enthusiasm one associates with college humor magazines like the *Harvard Lampoon*." Malcolm Bradbury examines Twain as "a living example of what made the American so different from the European." And Hal Holbrook, who has interpreted Mark Twain on stage for some forty years, describes how Twain "played" during the civil rights movement, during the Vietnam War, during the Gulf War, and in Prague on the eve of the demise of Communism.

Why do we continue to read Mark Twain? What draws us to him? His wit? His compassion? His humor? His bravura? His humility? His understanding of who and what we are in those parts of our being that we rarely open to view? Our sense that he knows we can do better than we do? Our sense that he knows we can't? E. L. Doctorow tells us that children are attracted to *Tom Sawyer* because in this book "the young reader confirms his own hope that no matter how troubled his relations with his elders may be, beneath all their disapproval is their underlying love for him, constant and steadfast." Readers in general, Arthur Miller writes, value Twain's "insights into America's always uncertain moral life and its shifting but everlasting hypocrisies"; we appreciate the fact that he "is not using his alienation from the public illusions of his hour in order to reject the country implicitly as though he could live without it, but manifestly in order to correct it." Perhaps we keep reading Mark Twain because, in Miller's words, he "wrote much more like a father than a son. He doesn't seem to be sitting in class taunting the teacher but standing at the head of it challenging his students to acknowledge their own humanity, that is, their immemorial attraction to the untrue."

Mark Twain entered the public eye at a time when many of his countrymen considered "American culture" an oxymoron; he died four years before a world conflagration that would lead many to question whether the contradiction in terms was not "European civilization" instead. In between he worked in journalism, printing, steamboating, mining, lecturing, publishing, and editing, in virtually every region of the country. He tried his hand at humorous sketches, social satire, historical novels, children's books, poetry, drama, science fiction, mysteries, romance, philosophy, travelogue, memoir, polemic, and several genres no one had ever seen before or has ever seen since. He invented a self-pasting scrapbook, a history game, a vest strap, and a gizmo for keeping bed sheets tucked in; he invested in machines and processes designed to revolutionize typesetting and engraving, and in a food supplement called "Plasmon." Along the way he cheerfully impersonated himself and prior versions of himself for doting publics on five continents while playing out a charming rags-to-riches story followed by a devastating riches-to-rags story followed by yet another great American comeback. He had a long-running real-life engagement in a sumptuous comedy of manners, and then in a real-life tragedy not of his own design: during the last fourteen years of his life almost everyone he ever loved was taken from him by disease and death.

Mark Twain has indelibly shaped our views of who and what the United States is as a nation and of who and what we might become. He understood the nostalgia for a "simpler" past that increased as that past receded — and he saw through the nostalgia to a past that was just as complex as the present. He recognized better than we did ourselves our potential for greatness and our potential for disaster. His fictions brilliantly illuminated the world in which he lived, changing it — and us — in the process. He knew that our feet often danced to tunes that had somehow remained beyond our hearing; with perfect pitch he played them back to us.

My mother read *Tom Sawyer* to me as a bedtime story when I was eleven. I thought Huck and Tom could be a lot of fun, but I dismissed Becky Thatcher as a bore. When I was twelve I invested a nickel at a local garage sale in a book that contained short pieces by Mark Twain. That was where I met Twain's Eve. Now, *that's* more like it, I decided, pleased to meet a female character I could identify *with* instead of against. Eve had spunk. Even if she got a lot wrong, you had to give her credit for trying. "The Man That Corrupted

Hadleyburg" left me giddy with satisfaction: none of my adolescent reveries of getting even with my enemies were half as neat as the plot of the man who got back at that town. "How I Edited an Agricultural Paper" set me off in uncontrollable giggles.

People sometimes told me that I looked like Huck Finn. "It's the freckles," they'd explain — not explaining anything at all. I didn't read *Huckleberry Finn* until junior year in high school in my English class. It was the fall of 1965. I was living in a small town in Connecticut. I expected a sequel to *Tom Sawyer*. So when the teacher handed out the books and announced our assignment, my jaw dropped: "Write a paper on how Mark Twain used irony to attack racism in *Huckleberry Finn*."

The year before, the bodies of three young men who had gone to Mississippi to help blacks register to vote — James Chaney, Andrew Goodman, and Michael Schwerner — had been found in a shallow grave; a group of white segregationists (the county sheriff among them) had been arrested in connection with the murders. America's inner cities were simmering with pent-up rage that began to explode in the summer of 1965, when riots in Watts left thirty-four people dead. None of this made any sense to me. I was confused, angry, certain that there was something missing from the news stories I read each day: the why. Then I met Pap Finn. And the Phelpses.

Pap Finn, Huck tells us, "had been drunk over in town" and "was just all mud." He erupts into a drunken tirade about "a free nigger . . . from Ohio — a mulatter, most as white as a white man," with "the whitest shirt on you ever see, too, and the shiniest hat; and there ain't a man in town that's got as fine clothes as what he had."

> . . . they said he was a p'fessor in a college, and could talk all kinds of languages, and knowed everything. And that ain't the wust. They said he could *vote*, when he was at home. Well, that let me out. Thinks I, what is the country a-coming to? It was 'lection day, and I was just about to go and vote, myself, if I warn't too drunk to get there; but when they told me there was a State in this country where they'd let that nigger vote, I drawed out. I says I'll never vote agin. Them's the very words I said. . . . And to see the

cool way of that nigger — why, he wouldn't a give me the road if I hadn't shoved him out o' the way.[12]

Later on in the novel, when the runaway slave Jim gives up his freedom to nurse a wounded Tom Sawyer, a white doctor testifies to the stunning altruism of his actions. The Phelpses and their neighbors, all fine, upstanding, well-meaning, churchgoing folk,

> agreed that Jim had acted very well, and was deserving to have some notice took of it, and reward. So every one of them promised, right out and hearty, that they wouldn't curse him no more.
>
> Then they come out and locked him up. I hoped they was going to say he could have one or two of the chains took off, because they was rotten heavy, or could have meat and greens with his bread and water, but they didn't think of it.[13]

Why did the behavior of these people tell me more about why Watts burned than anything I had read in the daily paper? And why did a drunk Pap Finn railing against a black college professor from Ohio whose vote was as good as his own tell me more about white anxiety over black political power than anything I had seen on the evening news?

Mark Twain knew that there was nothing, absolutely *nothing*, a black man could do — including selflessly sacrificing his freedom, the only thing of value he had — that would make white society see beyond the color of his skin. And Mark Twain knew that depicting racists with chilling accuracy would expose the viciousness of their world view like nothing else could. It was an insight echoed some eighty years after Mark Twain penned Pap Finn's rantings about the black professor, when Malcolm X famously asked, "Do you know what white racists call black Ph.D.'s?" and answered, " '*Nigger!*' "[14]

Mark Twain taught me things I needed to know. He taught me to understand the raw racism that lay behind what I saw on the evening news. He taught me that the most well-meaning people can be hurtful and myopic. He taught me to recognize the supreme irony of a country founded in freedom that continued to deny freedom to so many of its citizens. Every time I hear of another effort to kick Huck Finn out of school somewhere, I recall everything that Mark Twain taught *this* high school junior, and I find myself jumping into the fray.[15] I remember the black high school student who called CNN during the phone-in portion of a 1985 debate between Dr. John Wallace, a black educator spearheading efforts to ban the book, and myself. She accused Dr. Wallace of insulting her and all black high school students by suggesting they weren't smart enough to understand Mark Twain's irony. And I recall the black cameraman on the *CBS Morning News* who came up to me after he finished shooting another debate between Dr. Wallace and myself. He said he had never read the book by Mark Twain that we had been arguing about — but now he really wanted to. One thing that puzzled him, though, was why a white woman was defending it and a black man was attacking it, because as far as he could see from what we'd been saying, the book made whites look pretty bad.

As I came to understand *Huckleberry Finn* and *Pudd'nhead Wilson* as commentaries on the era now known as the nadir of American race relations, those books pointed me toward the world recorded in nineteenth-century black newspapers and periodicals and in fiction by Mark Twain's black contemporaries. My investigation of the role black voices and traditions played in shaping Mark Twain's art helped make me aware of their role in shaping all of American culture.[16] My research underlined for me the importance of changing the stories we tell about who we are to reflect the realities of what we've been.[17]

Ever since our encounter in high school English, Mark Twain has shown me the potential of American literature and American history to illuminate each other. Rarely have I found a contradiction or complexity we grapple with as a nation that Mark Twain had not puzzled over as well. He insisted on taking America seriously. And he insisted on *not* taking America seriously: "I think that there is but a single specialty with us, only one thing that can be called by the wide name 'American,'" he once wrote. "That is the national devotion to ice-water."[18]

Mark Twain threw back at us our dreams and our denial of those dreams, our greed, our goodness, our ambition, and our laziness, all rattling around together in that vast echo chamber of our talk — that sharp, spunky American talk that Mark Twain figured out how to write down without robbing it of its energy and immediacy. Talk shaped by voices that the official arbiters of "culture" deemed of no importance — voices of children, voices of slaves, voices of servants, voices of ordinary people. Mark Twain listened. And he made us listen. To the stories he told us, and to the truths they conveyed. He still has a lot to say that we need to hear.

Mark Twain lives — in our libraries, classrooms, homes, theaters, movie houses, streets, and most of all in our speech. His optimism energizes us, his despair sobers us, and his willingness to keep wrestling with the hilarious and horrendous complexities of it all keeps us coming back for more. As the twenty-first century approaches, may he continue to goad us, chasten us, delight us, berate us, and cause us to erupt in unrestrained laughter in unexpected places.

NOTES

1. Ernest Hemingway, *Green Hills of Africa* (New York: Charles Scribner's Sons, 1935), 22. George Bernard Shaw to Samuel L. Clemens, July 3, 1907, quoted in Albert Bigelow Paine, *Mark Twain: A Biography* (New York: Harper and Brothers, 1912), 3:1398.

2. Allen Carey-Webb, "Racism and *Huckleberry Finn*: Censorship, Dialogue and Change," *English Journal* 82, no. 7 (November 1993):22.

3. See Louis J. Budd, "Impersonators," in J. R. LeMaster and James D. Wilson, eds., *The Mark Twain Encyclopedia* (New York: Garland Publishing Company, 1993), 389–91.

4. See Shelley Fisher Fishkin, "Ripples and Reverberations," part 3 of *Lighting Out for the Territory: Reflections on Mark Twain and American Culture* (New York: Oxford University Press, 1996).

5. There are two exceptions. Twain published chapters from his autobiography in the *North American Review* in 1906 and 1907, but this material was not published in book form in Twain's lifetime; our volume reproduces the material as it appeared in the *North American Review*. The other exception is our final volume, *Mark Twain's Speeches*, which appeared two months after Twain's death in 1910.

An unauthorized handful of copies of *1601* was privately printed by an Alexander Gunn of Cleveland at the instigation of Twain's friend John Hay in 1880. The first American edition authorized by Mark Twain, however, was printed at the United States Military Academy at West Point in 1882; that is the edition reproduced here.

It should further be noted that four volumes — *The Stolen White Elephant and Other Detective Stories, Following the Equator and Anti-imperialist Essays, The Diaries of Adam and Eve,* and *1601, and Is Shakespeare Dead?* — bind together material originally published separately. In each case the first American edition of the material is the version that has been reproduced, always in its entirety. Because Twain constantly recycled and repackaged previously published works in his collections of short pieces, a certain amount of duplication is unavoidable. We have selected volumes with an eye toward keeping this duplication to a minimum.

Even the twenty-nine-volume Oxford Mark Twain has had to leave much out. No edition of Twain can ever claim to be "complete," for the man was too prolix, and the file drawers of both ephemera and as yet unpublished texts are deep.

6. With the possible exception of *Mark Twain's Speeches.* Some scholars suspect Twain knew about this book and may have helped shape it, although no hard evidence to that effect has yet surfaced. Twain's involvement in the production process varied greatly from book to book. For a fuller sense of authorial intention, scholars will continue to rely on the superb definitive editions of Twain's works produced by the Mark Twain Project at the University of California at Berkeley as they become available. Dense with annotation documenting textual emendation and related issues, these editions add immeasurably to our understanding of Mark Twain and the genesis of his works.

7. Except for a few titles that were not in its collection. The American Antiquarian Society in Worcester, Massachusetts, provided the first edition of *King Leopold's Soliloquy*; the Elmer Holmes Bobst Library of New York University furnished the 1906–7 volumes of the *North American Review* in which *Chapters from My Autobiography* first appeared; the Harry Ransom Humanities Research Center at the University of Texas at Austin made their copy of the West Point edition of *1601* available; and the Mark Twain Project provided the first edition of *Extract from Captain Stormfield's Visit to Heaven.*

8. The specific copy photographed for Oxford's facsimile edition is indicated in a note on the text at the end of each volume.

9. All quotations from contemporary writers in this essay are taken from their introductions to the volumes of The Oxford Mark Twain, and the quotations from Mark Twain's works are taken from the texts reproduced in The Oxford Mark Twain.

10. *The Diaries of Adam and Eve*, The Oxford Mark Twain [hereafter OMT] (New York: Oxford University Press, 1996), p. 33.

11. *Tom Sawyer Abroad*, OMT, p. 74.

12. *Adventures of Huckleberry Finn*, OMT, p. 49–50.

13. Ibid., p. 358.

14. Malcolm X, *The Autobiography of Malcolm X*, with the assistance of Alex Haley (New York: Grove Press, 1965), p. 284.

15. I do not mean to minimize the challenge of teaching this difficult novel, a challenge for which all teachers may not feel themselves prepared. Elsewhere I have developed some concrete strategies for approaching the book in the classroom, including teaching it in the context of the history of American race relations and alongside books by black writers. See Shelley Fisher Fishkin, "Teaching *Huckleberry Finn*," in James S. Leonard, ed., *Making Mark Twain Work in the Classroom* (Durham: Duke University Press, forthcoming). See also Shelley Fisher Fishkin, *Was Huck Black? Mark Twain and African-American Voices* (New York: Oxford University Press, 1993), pp. 106–8, and a curriculum kit in preparation at the Mark Twain House in Hartford, containing teaching suggestions from myself, David Bradley, Jocelyn Chadwick-Joshua, James Miller, and David E. E. Sloane.

16. See Fishkin, *Was Huck Black?* See also Fishkin, "Interrogating 'Whiteness,' Complicating 'Blackness': Remapping American Culture," in Henry Wonham, ed., *Criticism and the Color Line: Desegregating American Literary Studies* (New Brunswick: Rutgers UP, 1996, pp. 251–90 and in shortened form in *American Quarterly* 47, no. 3 (September 1995):428–66.

17. I explore the roots of my interest in Mark Twain and race at greater length in an essay entitled "Changing the Story," in Jeffrey Rubin-Dorsky and Shelley Fisher Fishkin, eds., *People of the Book: Thirty Scholars Reflect on Their Jewish Identity* (Madison: U of Wisconsin Press, 1996), pp. 47–63.

18. "What Paul Bourget Thinks of Us," *How to Tell a Story and Other Essays*, OMT, p. 197.

INTRODUCTION
Gore Vidal

Both Mark Twain and his inventor, Samuel Clemens, continue to give trouble to those guardians of the national mythology to which Twain added so much in his day, often deliberately. The Freudians are still on his case even though Dr. Freud and his followers are themselves somewhat occluded these days. Yet as recently as 1991, an academic critic tells us that Clemens was sexually infantile, burnt out at fifty (if not before), and given to degenerate reveries about little girls, all the while exhibiting an unnatural interest in outhouse humor and other excremental vilenesses. It is hard to believe that at century's end, academics of this degraded sort are still doing business, as Twain would put it, at the same old stand.

As is so often the case, this particular critic is a professor emeritus, and emerituses often grow reckless once free of the daily grind of dispensing received opinion. Mr. Guy Cardwell, for reasons never quite clear, wants to convince us that Twain (we'll drop the Clemens because he's very much dead while Twain will be with us as long as there are English-speakers in the United States) "suffered from erectile dysfunction at about the age of fifty. . . . Evidence that he became impotent ranges from the filmy to the relatively firm." This is a fair example of the good professor's style. "Filmy" evidence suggests a slightly blurred photograph of an erection gone south, while "relatively firm" is a condition experienced by many men over fifty who drink as much Scotch whiskey as Twain did. But filmy — or flimsy? — as the evidence is, the professor wants to demolish its owner, who, sickeningly, married above his station in order to advance himself socially as well as to acquire a surrogate mother; as his own mother was — yes! — a strong figure while his father was — what else? — cold and uncaring.

No Freudian cliché is left unstroked. To what end? To establish that Twain hated women as well as blacks, Jews, foreigners, American imperialists, Christian missionaries, and Mary Baker Eddy. Since I join him in detesting the last three, I see no need to find a Freudian root to our shared loathing of, say, that imperialist jingo Theodore Roosevelt. Actually, Twain was no more neurotic or dysfunctional than most people and, on evidence, rather less out of psychic kilter than other major figures in the American literary canon.

Twain was born on November 30, 1835, in Missouri. He spent his boyhood, famously, in the Mississippi River town of Hannibal. When he was eleven, his father died, becoming *truly* absent as Dr. Freud might sagely have observed, and Twain went to work as a printer's apprentice. Inevitably, he started writing the copy that was to be printed, and in essence, he was a journalist to the end of his days. Literature as such did not really engage him. *Don Quixote* was his favorite novel (as it was Flaubert's). He could not read Henry James, who returned the compliment by referring to him only once in his own voluminous bookchat, recently collected and published by the Library of America.

Exactly where and how the "Western Storyteller," as such, was born is unknown. He could have evolved from Homer or, later, from the Greek Milesian tales of run-on anecdote. In any case, an American master of the often scabrous tall story, Twain himself was predated by, among others, Abraham Lincoln, many of whose stories were particularly noisome as well as worse — worse! — *politically incorrect*. Our stern Freudian critic finds Twain's smutty stories full of "slurs" on blacks and women and so on. But so are those of Rabelais and Ariosto and Swift, Rochester and Pope and. . . Whatever the "true" motivation for telling such stories, Twain was a master in this line both in print and on the lecture circuit.

Primarily, of course, he was a popular journalist, and with the best-seller *Innocents Abroad* (1869) he made the hicks back home laugh and Henry James, quite rightly, shudder. Yet when the heavy-handed joky letters, written from the first cruise liner, *Quaker City*, became a text, it turned out to be an unusually fine-meshed net in which Twain caught up old Europe and an even older Holy Land and then, as he arranged his catch on the — well — deck of his art, he Americanized the precedent civilization and vulgarized it in the most satisfactory way ("Lump the whole thing! Say that the Creator made Italy from designs by Michael Angelo!"), and made it possible for an American idea to flourish someday.

But Twain was far too ambitious to be just a professional hick, as opposed to occasional hack. He had social ambitions; he also lusted for money (in a "banal anal" way, according to the Freudian emeritus — as opposed to "floral oral"?).

In the great tradition of men on the make, Twain married above his station to one Olivia Langdon of the first family of Elmira, New York. He got her to polish him socially. He also became a friend of that currently underestimated novelist-editor William Dean Howells, a lad from the Western Reserve who had superbly made it in Boston as editor of the *Atlantic Monthly*. Howells encouraged Twain to celebrate the American "West" as the sort of romanticized Arcadia that Rousseau might have wanted his chainless noble savage to roam.

While knocking about the West and Southwest, Twain worked as a pilot on Mississippi steamboats from 1857 to 1861; he joined the Civil War, briefly, on the Confederate side. When he saw how dangerous war might be, he moved on to the Nevada Territory, where his brother had been made secretary to the governor. He wrote for newspapers. In 1863, he started to use the pseudonym "Mark Twain," a river pilot's measurement of depth, called out on approaching landfall — some twelve feet, a bit on the shallow side.

After the war, Twain began to use life on the river and the river's bank as a background for stories that were to place him permanently at the center of American literature: *The Adventures of Tom Sawyer* (1876); *Life on the Mississippi* (1883); *Adventures of Huckleberry Finn* (1885). He liked fame and money, the last perhaps too much since he was forever going broke speculating on experimental typesetting machines and underfinanced publishing houses. He lived in considerable bourgeois splendor at Hartford, Connecticut; oddly for someone who had made his fortune out of being the American writer, as he once described himself, Twain spent more than a decade in Europe. One reason, other than *douceur de la vie*, was that he was admired on the Continent in a way that he never was, or so he felt, by the Eastern seaboard gentry, who were offended by his jokes, his profanity, his irreligion, and all those Scotch sours he drank. Fortunately, no one then suspected his erectile dysfunction.

Whenever cash was needed and a new book not ready to be sold to the public, Twain took to the lecture circuit. An interesting, if unanswerable question: Was Mark Twain a great actor who wrote, or a great writer who could act? Or was he an even balance like Charles Dickens or George Bernard Shaw? Much of what Twain writes is conversation — dialogue — with different voices thrown in to delight the ear of an audience. But, whichever he was, he was always, literally, a journalist, constantly describing daily things while recollecting old things. In the process, he made, from time to time, essential literature, including the darkest of American novels *Pudd'nhead Wilson* (1894).

Mark Twain's view of the human race was not sanguine, and much has been made of the Calvinism out of which he came. Also, his great river, for all its fine amplitude, kept rolling along, passing villages filled with fierce monotheistic folk in thrall to slavery, while at river's end there were the slave markets of New Orleans. Calvinist could easily become Manichean if he brooded too much on the river world of the mid-1800s. *Pudd'nhead Wilson* contains the seeds of Twain's as yet unarticulated notion that if there is a God (*What is Man?*, 1906) he is, if not evil in the Manichean sense, irrelevant, since man, finally, is simply a machine acted upon by a universe "frankly and hysterically insane" (*No. 44, The Mysterious Stranger*): "Nothing exists but You. And You are but a *thought*."

The agony of the two boys in *Pudd'nhead Wilson*, one brought up white, the other black, becomes exquisite for the "white" one, who is found to be black and gets shipped downriver, his question to an empty Heaven unanswered: "What crime did the uncreated first nigger commit that the curse of birth was decreed for him?" All this, then, is what is going on in Mark Twain's mind as he gets ready for a second luxury tour, this time around the world.

When one contemplates the anti-imperialism of Mark Twain, it is hard to tell just where it came from. During his lifetime the whole country was — like himself — on the make, in every sense. But Mark Twain was a flawed materialist. As a Southerner he should have had some liking for the peculiar institution of slavery; yet when he came to write of antebellum days, it is Miss Watson's "nigger," Jim, who represents what little good Twain ever found in man. Lynchings shocked him. But then, pace Hemingway, so did Spanish bullfights. Despite the various neuroses ascribed to him by our current political correctionists, he never seemed in any doubt that he was a man, and therefore never felt, like so many sissies of the Hemingway sort, a need to swagger about, bullying those not able to bully him.

In 1898, the United States provoked a war with Spain (a war with England over Venezuela was contemplated but abandoned as there was a good chance that we would have lost). The Spanish empire collapsed more from dry rot than from our military skills. Cuba was made "free," and Puerto Rico was attached to us while the Spanish Philippines became our first Asian real estate and the inspiration for close to a century now of disastrous American adventures in that part of the world.

Mark Twain would have had a good time with the current demise of that empire, which he greeted, with some horror, in the first of his meditations on imperialism. *To the Person Sitting in Darkness* was published as a pamphlet in 1901, a year in which we were busy telling the Filipinos that although we had, at considerable selfless expense, freed them from Spain they were not yet ready for the higher democracy, as exemplified by Tammany Hall, to use Henry James' bitter analogy. Strictly for their own good, we would have to kill one or two hundred thousand men, women, and children in order to make their country into an American-style democracy. Most Americans were happy to follow the exuberant lead of the prime architect of empire, Theodore Roosevelt — known to the sour Henry Adams as "our Dutch-American Napoleon." But then, suddenly, Mark Twain quite forgot that he was *the* American writer and erupted, all fire and lava.

The people who sit in darkness are Kipling's "lesser breeds," waiting for the white man to take up his burden and "civilize" them. Ironically, Twain compares our bloody imperialism favorably with that of the white European

powers then abroad in the "unlit" world, busy assembling those colonial empires that now comprise today's desperate third world. Twain, succinctly for
him, lists who was stealing what from whom and when, and all in the name of
the "Blessings-of-Civilization Trust." But now the American writer is so
shocked at what his countrymen are capable of doing in the imperial line that
he proposes a suitable flag for the "Philippine Province": "We can have just
our usual flag, with the white stripes painted black and the stars replaced by
the skull and cross-bones."

In 1905, Twain published a second pamphlet (for the Congo Reform
Association), *King Leopold's Soliloquy*, subtitled "A Defense of his Congo
Rule." On the cover there is a crucifix crossed by a machete and bearing the
cheery inscription "By this sign we prosper."

The soliloquy is just that. The King of the Belgians is distressed by reports
of his bloody rule over a large section of black Africa. Leopold, an absolute
ruler in Africa if not in Belgium, is there "to root out slavery and stop the
slave-raids, and lift up those twenty-five millions of gentle and harmless
blacks out of darkness into light . . ." He is in rather the same business as
Presidents McKinley and Roosevelt in the earlier pamphlet.

Leopold free-associates, noting happily that Americans were the first to
recognize his rule. As he defends himself, his night-mind (as the surrealists
used to say) gets the better of him and he keeps listing his crimes as he defends them. He notes that his enemies "concede — reluctantly — that I have
one match in history, but only one — the *Flood*. This is intemperate." He
blames his current "crash" on "the incorruptible *kodak*," the "only witness I
have encountered in my long experience that I couldn't bribe." Twain provides us with a page of nine snapshots of men and women, each lacking a
hand, the King's usual punishment. Twain's intervention was not unlike
those of Voltaire and Zola or, closer to home, Howells' denunciation of the
American legal system — and press — that had found guilty the non-perpetra-
tors of the Haymarket riots. Imperialism and tyranny for Twain were great
evils, but the more he understood — or thought he understood — the human
race, the darker his view of the whole lot became, as he had begun to demon-

strate in the epigraphs from Pudd'nhead Wilson's New Calendar at the head of each chapter of his travel book *Following the Equator* (1897).

In 1895, Twain, his wife, Olivia, and their daughter Clara started on a round-the-world lecture tour. They crossed the Atlantic; then the United States; then, on August 23, they set sail from Vancouver bound for Sydney, Australia. For several years Twain had undergone a series of financial setbacks. Now the lecture tour would make him some money while a look at the whole world would provide him with a great deal of copy, most of which he was to use in *Following the Equator*. At the start of the tour, Twain seems not to have been his usual resilient self. "Mr. Clemens," wrote Olivia to a friend, "has not as much courage as I wish he had, but, poor old darling, he has been pursued with colds and inabilities of various sorts. Then he is so impressed with the fact that he is sixty years old." Definitely a filmy time for someone Olivia had nicknamed "Youth."

The pleasures of travel have not been known for two generations now; even so, it is comforting to read again about the soothing boredom of life at sea and the people that one meets aboard ship as well as on shore in exotic lands. One also notes that it was Twain in Australia, and not an English official recently testifying in an Australian court, who first observed that someone "was economical of the truth."

In his travel journal, Twain muses about his past; contemplates General Grant, whose memoirs he had published and, presumably, edited a decade earlier. One would like to know more about that relationship, since Gertrude Stein, among others, thought Grant our finest prose writer. When the ship stops in Honolulu, Twain notes that the bicycle is now in vogue and "the riding horse is retiring from business everywhere in the world." Twain is not pleased by the combined influences of Christian missionaries and American soldiers upon what had once been a happy and independent Pacific kingdom.

They pass the Fiji Islands, ceded to England in 1858. Twain tells the story that when the English commissioner remarked to the Fiji king that it was merely "a sort of hermit-crab formality," the king pointed out that "the crab moves into an unoccupied shell, but mine isn't."

A great comfort to Twain aboard ship is *The Sentimental Song Book* of the Sweet Singer of Michigan, one Mrs. Julia A. Moore, who has, for every human occasion, numerous sublimely inapt verses that never, even by accident, scan. As one reads Twain's own prose, written in his own character, one is constanty reminded that he is very much a stand-up comedian whose laugh-lines are carefully deployed at the end of every observation, thus reducing possible tension with laughter. Of the colonists sent out to Australia by England, Twain observes that they came from the jails and from the army. "The colonists trembled. It was feared that next there would be an importation of the nobility."

In general, Australia gets high marks. Twain and family travel widely; he lectures to large crowds: "The welcome which an American lecturer gets from a British colonial audience is a thing which will move him to his deepest deeps, and veil his sight and break his voice." He is treated as what he was, a Great Celebrity, and "I was conscious of a pervading atmosphere of envy which gave me deep satisfaction."

Twain continually adverts to the white man's crimes against the original inhabitants of the Pacific islands, noting that "there are many humorous things in the world; among them the white man's notion that he is less savage than the other savages." The Freudian critic cannot quite fathom how the Twain who in his youth made jokes about "Negroes" now, in his filmy years, has turned anti-white and speaks for the enslaved and the dispossessed. Dr. Freud apparently had no formula to explain this sort of sea-change.

New Zealand appeals to Twain; at least they did not slaughter the native population though they did something almost as bad: "The Whites always mean well when they take human fish out of the ocean and try to make them dry and warm and happy and comfortable in a chicken coop," which is how, through civilization, they did away with many of the original inhabitants. Lack of empathy is a principal theme in Twain's meditations on race and empire. Twain notes with approval that New Zealand's women have been able to vote since 1893. At sixty, he seems to have overcome his misogyny; our Freudian critic passes over this breakthrough in dark silence.

Ceylon delights. "Utterly Oriental," though plagued by missionaries who dress the young in western style, rendering them as hideous on the outside as they are making them cruelly superstitious on the inside. Twain broods on slavery as he remembered it a half century before in Missouri. He observes its equivalent in Ceylon and India. He meets a Mohammedan "deity," who discusses Huck Finn in perfect English. Twain now prefers brown or black skin to "white," which betrays the inner state rather too accurately, making "no concealments." Although he prefers dogs to cats, he does meet a dog that he cannot identify, which is odd since it is plainly a dachshund. He tries to get used to pajamas but goes back to the old-fashioned nightshirt. Idly, he wonders why western men's clothes are so ugly and uncomfortable. He imagines himself in flowing robes of every possible color. Heaven knows what *this* means. Heaven and a certain critic. . . .

Benares has its usual grim effect. Here, beside the Ganges, bodies are burned; and people bathe to become pure while drinking the polluted waters of the holiest of holy rivers. It is interesting that Twain never mentions the Buddha, who became enlightened at Benares, but he does go into some detail when he describes the Hindu religion. In fact, he finds the city of Benares "just a big church" to that religion in all its aspects. In Calcutta, he broods on the Black Hole, already filled in. The Taj Mahal induces an interesting reverie. Twain notes that when one has read so many descriptions of a famous place, one can never actually *see* it because of all the descriptions that crowd one's mind. In this perception, Twain anticipates the latest — if not the last — theory of how memory works. He also broods on the phenomenon of Helen Keller, born deaf, dumb, and blind; yet able to learn to speak and think. How *does* the mind work?

From India, Twain and company cross the Indian Ocean to Mauritius. Although he often alludes to his lecturing, he never tells us what he talks about. He does note, "I never could tell a lie that anybody would doubt, nor a truth that anybody would believe." We learn that he dislikes Oliver Goldsmith and Jane Austen. As a prose writer, the imperialist Kipling beguiles him even though Twain likens empires to thieves who take clothes off other people's clotheslines. "In 800 years an obscure tribe of Muscovite savages has risen to the dazzling position of Land-Robber-in-Chief." He is more tolerant of the English. But then he is a confessed Anglophile.

Meanwhile, the ship is taking Twain and family down the east coast of Africa. South Africa is in ferment—Boers against English settlers, white against black. Cecil Rhodes is revealed as a scoundrel. But Twain is now writing as of May 1897, one year after his visit to South Africa, and so the outcome of all this is still unclear to him. He sides with the English, despite reservations about Rhodes and company. "I have always been especially fond of war. No, I mean fond of discussing war; and fond of giving military advice." As for that new territorial entity, Rhodesia, Twain remarks that it is "a happy name for that land of piracy and pillage, and puts the right stain upon it"; and he also has Pudd'nhead Wilson observe: "The very ink with which all history is written is merely fluid prejudice."

Finally, "Our trip around the earth ended at the Southampton pier, where we embarked thirteen months before. . . . I seemed to have been lecturing a thousand years. . . ." But he had now seen the whole world, more or less at the equator and, perhaps more to the point, quite a few people got to see Mark Twain in action, in itself something of a phenomenon, never to be repeated on earth unless, of course, his nemesis, Mary Baker Eddy, were to allow him to exchange her scientific deathless darkness for his limelight, our light.

FOLLOWING

THE EQUATOR:

A JOURNEY AROUND

THE WORLD

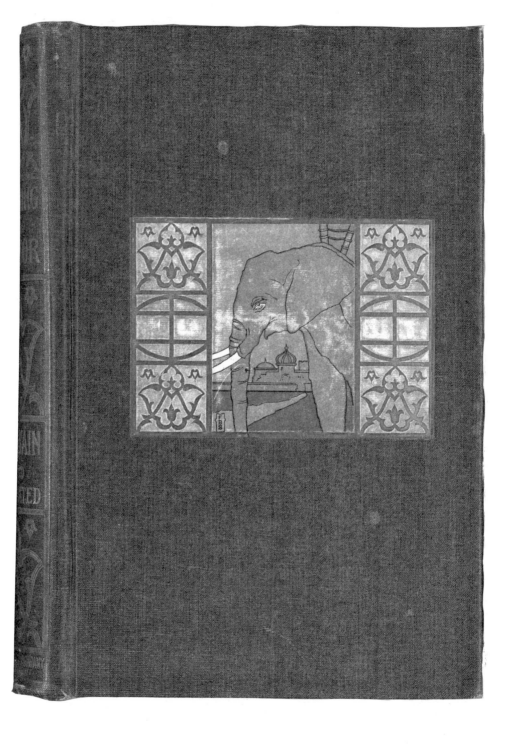

FOLLOWING THE EQUATOR

A JOURNEY AROUND THE WORLD

BY

MARK TWAIN

SAMUEL L. CLEMENS

HARTFORD, CONNECTICUT

THE AMERICAN PUBLISHING COMPANY

MDCCCXCVII

THIS BOOK

Is Affectionately Inscribed to

MY YOUNG FRIEND

HARRY ROGERS,

WITH RECOGNITION

OF WHAT HE IS, AND APPREHENSION OF WHAT HE MAY BECOME
UNLESS HE FORM HIMSELF A LITTLE MORE CLOSELY

UPON THE MODEL OF

THE AUTHOR.

THE PUDD'NHEAD MAXIMS.

THESE WISDOMS ARE FOR THE LURING OF YOUTH TOWARD
HIGH MORAL ALTITUDES. THE AUTHOR DID NOT
GATHER THEM FROM PRACTICE, BUT FROM
OBSERVATION. TO BE GOOD IS NOBLE;
BUT TO SHOW OTHERS HOW
TO BE GOOD IS NOBLER
AND NO TROUBLE.

CONTENTS

CHAPTER XIX.

CHAPTER XX.

CHAPTER XXI.

CHAPTER XXII.

CHAPTER XXIII.

CHAPTER XXIV.

CONTENTS.

2

ILLUSTRATIONS

BY

Dan Beard, A. B. Frost, B. W. Clinedinst, Frederick Dielman, Peter Newell,
F. M. Seinor, T. J. Fogarty, C. H. Warren, A. G. Reinhart,
F. Berkeley Smith, C. Allan Gilbert.

The publishers acknowledge the courtesy extended by Walter G. Chase, Boston, Major J. B. Pond, New York, and F. R. Reynolds, Manchester, England, in furnishing many of the photographs reproduced in this volume.

THEY PASSED IN REVIEW.

FOLLOWING THE EQUATOR

CHAPTER I.

A man may have no bad habits and have worse.
— *Pudd'nhead Wilson's New Calendar.*

THE starting point of this lecturing-trip around the world was Paris, where we had been living a year or two.

We sailed for America, and there made certain preparations. This took but little time. Two members of my family elected to go with me. Also a carbuncle. The dictionary says a carbuncle is a kind of jewel. Humor is out of place in a dictionary.

We started westward from New York in midsummer, with Major Pond to manage the platform-business as far as the Pacific. It was warm work, all the way, and the last fortnight of it was suffocatingly smoky, for in Oregon and British Columbia the forest fires were raging. We had an added week of smoke at the seaboard, where we were obliged to wait awhile for our ship. She had been getting herself ashore in the smoke, and she had to be docked and repaired. We sailed at last; and so ended a snail-paced march across the continent, which had lasted forty days.

We moved westward about mid-afternoon over a rippled and sparkling summer sea; an enticing sea, a clean and cool sea, and apparently a welcome sea to all on board; it certainly was to me, after the distressful dustings and smokings and swelterings

of the past weeks. The voyage would furnish a three-weeks
holiday, with hardly a break in it. We had the whole Pacific
Ocean in front of us, with nothing to do but do nothing and
be comfortable. The
city of Victoria was
twinkling dim in the
deep heart
o f h e r
s m o k e -
cloud, and

EVEN THE GULLS SMILED.

getting ready to
vanish; and now
we closed the field-
glasses a n d s a t
d o w n o n o u r
steamer chairs contented and at peace. But they went to
wreck and ruin under us and brought us to shame before all
the passengers. They had been furnished by the largest fur-
niture-dealing house in Victoria, and were worth a couple of
farthings a dozen, though they had cost us the price of honest
chairs. In the Pacific and Indian Oceans one must still bring
his own deck-chair on board or go without, just as in the old
forgotten Atlantic times — those Dark Ages of sea travel.

Ours was a reasonably comfortable ship, with the customary sea-going fare — plenty of good food furnished by the
Deity and cooked by the devil. The discipline observable on
board was perhaps as good as it is anywhere in the Pacific and
Indian Oceans. The ship was not very well arranged for tropi-

cal service; but that is nothing, for this is the rule for ships which ply in the tropics. She had an over-supply of cock-roaches, but this is also the rule with ships doing business in the summer seas — at least such as have been long in service.

Our young captain was a very handsome man, tall and per-fectly formed, the very figure to show up a smart uniform's finest effects. He was a man of the best intentions, and was polite and courteous even to courtliness. There was a soft grace and finish about his manners which made whatever place he happened to be in seem for the moment a drawing-room. He avoided the smoking-room. He had no vices. He did not smoke or chew tobacco or take snuff; he did not swear, or use slang, or rude, or coarse, or indelicate language, or make puns, or tell anecdotes, or laugh intemperately, or raise his voice above the moderate pitch enjoined by the canons of good form. When he gave an order, his manner modified it into a request. After dinner he and his officers joined the ladies and gentle-men in the ladies' saloon, and shared in the singing and piano playing, and helped turn the music. He had a sweet and sympathetic tenor voice, and used it with taste and effect. After the music he played whist there, always with the same partner and opponents, until the ladies' bedtime. The electric lights burned there as late as the ladies and their friends might desire, but they were not allowed to burn in the smoking-room after eleven. There were many laws on the ship's statute book, of course; but so far as I could see, this and one other were the only ones that were rigidly enforced. The captain explained that he enforced this one because his own cabin adjoined the smoking-room, and the smell of tobacco smoke made him sick. I did not see how our smoke could reach him, for the smoking-room and his cabin were on the upper deck, targets for all the winds that blew; and besides there was no crack of communication between them, no opening of any sort in the solid intervening bulkhead. Still, to a delicate stomach even imaginary smoke can convey damage.

The captain, with his gentle nature, his polish, his sweetness, his moral and verbal purity, seemed pathetically out of place in his rude and autocratic vocation. It seemed another instance of the irony of fate.

He was going home under a cloud. The passengers knew about his trouble, and were sorry for him. Approaching Vancouver through a narrow and difficult passage densely befogged with smoke from the forest fires, he had had the ill-luck to lose his bearings and get his ship on the rocks. A matter like this would rank merely as an error with you and me; it ranks as a crime with the directors of steamship companies. The captain had been tried by the Admiralty Court at Vancouver, and its verdict had acquitted him of blame. But that was insufficient comfort. A sterner court would examine the case in Sydney — the Court of Directors, the lords of a company in whose ships the captain had served as mate a number of years. This was his first voyage as captain.

The officers of our ship were hearty and companionable young men, and they entered into the general amusements and helped the passengers pass the time. Voyages in the Pacific and Indian Oceans are but pleasure excursions for all hands. Our purser was a young Scotchman who was equipped with a grit that was remarkable. He was an invalid, and looked it, as far as his body was concerned, but illness could not subdue his spirit. He was full of life, and had a gay and capable tongue. To all appearances he was a sick man without being aware of it, for he did not talk about his ailments, and his bearing and conduct were those of a person in robust health; yet he was the prey, at intervals, of ghastly sieges of pain in his heart. These lasted many hours, and while the attack continued he could neither sit nor lie. In one instance he stood on his feet twenty-four hours fighting for his life with these sharp agonies, and yet was as full of life and cheer and activity the next day as if nothing had happened.

The brightest passenger in the ship, and the most interesting and felicitous talker, was a young Canadian who was not able to let the whisky bottle alone. He was of a rich and powerful family, and could have had a distinguished career and abundance of effective help toward it if he could have conquered his appetite for drink; but he could not do it, so his great equipment of talent was of no use to him. He had often taken the pledge to drink no more, and was a good sample of what that sort of unwisdom can do for a man — for a man with anything short of an iron will. The system is wrong in two ways: it does not strike at the root of the trouble, for one thing, and to make a *pledge* of any kind is to declare war against nature; for a pledge is a chain that is always clanking and reminding the wearer of it that he is not a free man.

I have said that the system does not strike at the root of the trouble, and I venture to repeat that. The root is not the *drinking*, but the *desire* to drink. These are very different things. The one merely requires will — and a great deal of it, both as to bulk and staying capacity — the other merely requires watchfulness — and for no long time. The desire of course precedes the act, and should have one's first attention; it can do but little good to refuse the act over and over again, always leaving the desire unmolested, unconquered; the desire will continue to assert itself, and will be almost sure to win in the long run. When the desire intrudes, it should be at once banished out of the mind. One should be on the watch for it all the time — otherwise it will get *in*. It must be taken in time and not allowed to get a lodgment. A desire constantly repulsed for a fortnight should die, then. That should cure the drinking habit. The system of refusing the mere *act* of drinking, and leaving the *desire* in full force, is unintelligent war tactics, it seems to me.

I used to take pledges — and soon violate them. My will was not strong, and I could not help it. And then, to be tied in any way naturally irks an otherwise free person and makes him chafe in his bonds and want to get his liberty. But when I finally ceased from taking definite pledges, and merely resolved that I would kill an injurious desire, but leave myself free to resume the desire and the habit whenever I should choose to do so, I had no more trouble. In five days I drove out the desire to smoke and was not obliged to keep watch after that; and I never experienced any strong desire to smoke again. At the end of a year and a quarter of idleness I began to write a book, and presently found that the pen was strangely reluctant to go. I tried a smoke to see if that would help me out of the difficulty. It did. I smoked eight or ten cigars and as many pipes a day for five months; finished the book, and did not smoke again until a year had gone by and another book had to be begun.

I can quit any of my nineteen injurious habits at any time, and without discomfort or inconvenience. I think that the Dr. Tanners and those others who go forty days without eating do it by resolutely keeping out the desire to eat, in the beginning; and that after a few hours the desire is discouraged and comes no more.

Once I tried my scheme in a large medical way. I had been confined to my bed several days with lumbago. My case refused to improve. Finally the doctor said, —

"My remedies have no fair chance. Consider what they have to fight, besides the lumbago. You smoke extravagantly, don't you?"

"Yes."

"You take coffee immoderately?"

" Yes."

" And some tea ? "

" Yes."

" You eat all kinds of things that are dissatisfied with each other's company ? "

" Yes."

" You drink two hot Scotches every night ? "

" Yes."

" Very well, there you see what I have to contend against. We can't make progress the way the matter stands. You must make a reduction in these things; you must cut down your consumption of them considerably for some days."

" I can't, doctor."

" Why can't you."

" I lack the will-power. I can cut them off entirely, but I can't merely moderate them."

He said that that would answer, and said he would come around in twenty-four hours and begin work again. He was taken ill himself and could not come; but I did not need him. I cut off all those things for two days and nights; in fact, I cut off all kinds of food, too, and all drinks except water, and at the end of the forty-eight hours the lumbago was discouraged and left me. I was a well man; so I gave thanks and took to those delicacies again.

It seemed a valuable medical course, and I recommended it to a lady. She had run down and down and down, and had at last reached a point where medicines no longer had any helpful effect upon her. I said I knew I could put her upon her feet in a week. It brightened her up, it filled her with hope, and she said she would do everything I told her to do. So I said she must stop swearing and drinking, and smoking and eating for four days, and then she would be all right again. And it would have happened just so, I know it; but she said she could not stop swearing, and smoking, and drinking, because she had never done those things. So there it was. She had neglected her habits, and hadn't any. Now that they would have come good, there were none in stock. She had nothing to fall back on. She was a sinking vessel, with no freight in her to throw over-

board and lighten ship withal. Why, even one or two little bad habits c o u l d have s a v e d her, but she was j u s t a moral pauper. W h e n she could have ac-quired them she was dis-

"WHEN I WAS A YOUTH."

suaded by her parents, who were ignorant people though reared in the best society, and it was too late to begin now. It seemed such a pity; but there was no help for it. These things ought to be attended to while a person is young; otherwise, when age and disease come, there is nothing effectual to fight them with.

When I was a youth I used to take all kinds of pledges, and do my best to keep them, but I never could, because I didn't strike at the root of the habit — the *desire ;* I generally broke down within the month. Once I tried limiting a habit. That worked tolerably well for a while. I pledged myself to smoke but one cigar a day. I kept the cigar waiting until bedtime, then I had a luxurious time with it. But desire persecuted me every day and all day long; so, within the week I found myself hunting for larger cigars than I had been used to smoke; then larger ones still, and still larger ones. Within the fortnight I was getting cigars *made* for me — on a yet larger pattern. They still grew and grew in size. Within the month my cigar had grown to such proportions that I could have used it as a crutch. It now seemed to me that a one-cigar limit was no real protection to a person, so I knocked my pledge on the head and resumed my liberty.

To go back to that young Canadian. He was a " remittance man," the first one I had ever seen or heard of. Passengers explained the term to me. They said that dissipated ne'er-do-weels belonging to important families in England and Canada were not cast off by their people while there was any hope of reforming them, but when that last hope perished at last, the ne'er-do-weel was sent abroad to get him out of the way. He was shipped off with just enough money in his pocket — no, in the purser's pocket — for the needs of the voyage — and when he reached his destined port he would find a remittance awaiting him there. Not a large one, but just enough to keep him a month. A similar remittance would come monthly thereafter. It was the remittance-man's custom to pay his month's board and lodging straightway — a duty which his landlord did not allow him to forget — then spree away the rest of his money in a single night, then brood and mope and grieve in idleness till the next remittance came. It is a pathetic life.

3

We had other remittance-men on board, it was said. At least *they* said they were R. M.'s. There were two. But they did not resemble the Canadian; they lacked his tidiness, and his brains, and his gentlemanly ways, and his resolute spirit, and his humanities and generosities. One of them was a lad of nineteen or twenty, and he was a good deal of a ruin, as to clothes, and morals, and general aspect. He said he was a scion of a ducal house in England, and had been shipped to Canada for the house's relief, that he had fallen into trouble there, and was now being shipped to Australia. He said he had no title. Beyond this remark he was economical of the truth. The first thing he did in Australia was to get into the lockup, and the next thing he did was to proclaim himself an earl in the police court in the morning and fail to prove it.

CHAPTER II.

ABOUT four days out from Victoria we plunged into hot weather, and all the male passengers put on white linen clothes. One or two days later we crossed the 25th parallel of north latitude, and then, by order, the officers of the ship laid away their blue uniforms and came out in white linen ones. All the ladies were in white by this time. This prevalence of snowy costumes gave the promenade deck an invitingly cool and cheerful and picnicky aspect.

From my diary:

There are several sorts of ills in the world from which a person can never escape altogether, let him journey as far as he will. One escapes from one breed of an ill only to encounter another breed of it. We have come far from the snake liar and the fish liar, and there was rest and peace in the thought; but now we have reached the realm of the boomerang liar, and sorrow is with us once more. The first officer has seen a man try to escape from his enemy by getting behind a tree; but the enemy sent his boomerang sailing into the sky far above and beyond the tree; then it turned, descended, and killed the man. The Australian passenger has seen this thing done to two men, behind two trees — and by the one arrow. This being received with a large silence that suggested doubt, he buttressed it with the statement that his brother once saw the boomerang kill a bird away off a hundred yards *and bring it to the thrower.* But these are ills which must be borne. There is no other way.

The talk passed from the boomerang to dreams — usually a fruitful subject, afloat or ashore — but this time the output was poor. Then it passed to instances of extraordinary memory — with better results. Blind Tom, the negro pianist, was spoken of, and it was said that he could accurately play any piece of music, howsoever long and difficult, after hearing it once; and that six months later he could accurately play it again, without having touched it in the interval. One of the most striking of the stories told was furnished by a gentleman who had served on the staff of the Viceroy of India. He read the details from his note-book, and explained that he had written them down, right after the consummation of the incident which they described, because he thought that if he did not put them down in black and white he might presently come to think he had dreamed them or invented them.

The Viceroy was making a progress, and among the shows offered by the Maharajah of Mysore for his entertainment was a memory-exhibition. The Viceroy and thirty gentlemen of his suite sat in a row, and the memory-expert, a high-caste Brahmin, was brought in and seated on the floor in front of them. He said he knew but two languages, the English and his own, but would not exclude any foreign tongue from the tests to be applied to his memory. Then he laid before the assemblage his program — a sufficiently extraordinary one. He proposed that one gentleman should give him one word of a foreign sentence, and tell him its place in the sentence. He was furnished with the French word *est*, and was told it was second in a sentence of three words. The next gentleman gave him the German word *verloren* and said it was the third in a sentence of four words. He asked the next gentleman for one detail in a sum in addition; another for one detail in a sum of subtraction; others for single details in mathematical problems of various kinds; he got them. Intermediates gave him single words from sentences in Greek, Latin, Spanish, Portuguese, Italian, and other languages, and told him their places in the sentences. When at last everybody had furnished him a single rag from a foreign sentence or a figure from a problem, he went over the ground again, and got a second word and a second figure and was told their places in the sentences and the sums; and so on and so on. He went over the ground again and again until he had collected all the parts of the sums and all the parts of the sentences — and all in disorder, of course, not in their proper rotation. This had occupied two hours.

The Brahmin now sat silent and thinking, a while, then began and repeated all the sentences, placing the words in their proper order, and untangled the disordered arithmetical problems and gave accurate answers to them all.

In the beginning he had asked the company to throw almonds at him during the two hours, he to remember how many each gentleman had thrown; but none were thrown, for the Viceroy said that the test would be a sufficiently severe strain without adding that burden to it.

General Grant had a fine memory for all kinds of things, including even names and faces, and I could have furnished an instance of it if I had thought of it. The first time I ever saw him was early in his first term as President. I had just arrived in Washington from the Pacific coast, a stranger and wholly unknown to the public, and was passing the White House one morning when I met a friend, a Senator from Nevada. He asked me if I would like to see the President. I said I should be very glad; so we entered. I supposed that the President would be in the midst of a crowd, and that I could look at him in peace and security from a distance, as another stray cat might look at another king. But it was in the morning, and the Senator was using a privilege of his office which I had not heard of—the privilege of intruding upon the Chief Magistrate's working hours. Before I knew it, the Senator and I were in the presence, and there was none there but we three. General Grant got slowly up from his table, put his pen down, and stood before me with the iron expression of a man who had not smiled for seven years, and was not intending to smile for another seven. He looked me steadily in the eyes—mine lost confidence and fell. I had never confronted a great man before, and was in a miserable state of funk and inefficiency. The Senator said:—

"Mr. President, may I have the privilege of introducing Mr. Clemens?"

The President gave my hand an unsympathetic wag and dropped it. He did not say a word but just stood. In my trouble I could not think of anything to say, I merely wanted to resign. There was an awkward pause, a dreary pause, a horrible pause. Then I thought of something, and looked up into that unyielding face, and said timidly : —

"Mr. President, I — I am embarrassed. Are you?"

His face broke — just a little — a wee glimmer, the momentary flicker of a summer-lightning smile, seven years ahead of time — and I was out and gone as soon as *it* was.

Ten years passed away before I saw him the second time. Meantime I was become better known; and was one of the people appointed to respond to toasts at the banquet given to General Grant in Chicago by the Army of the Tennessee when he came back from his tour around the world. I arrived late at night and got up late in the morning. All the corridors of the hotel were crowded with people waiting to get a glimpse of General Grant when he should pass to the place whence he was to review the great procession. I worked my way by the suite of packed drawing-rooms, and at the corner of the house I found a window open where there was a roomy

AN AWKWARD PAUSE.

platform decorated with flags, and carpeted.   I stepped out on it, and saw below me millions of people blocking all the streets, and other millions caked together in all the windows and on all the house-tops around.   These masses took me for General Grant, and broke into volcanic explosions and cheers; but it was a good place to see the procession, and I stayed. Presently I heard the distant blare of military music, and far up the street I saw the procession come in sight, cleaving its way through the huzzaing multitudes, with Sheridan, the most martial figure of the War, riding at its head in the dress uniform of a Lieutenant-General.

And now General Grant, arm-in-arm with Major Carter Harrison, stepped out on the platform, followed two and two by the badged and uniformed reception committee.   General Grant was looking exactly as he had looked upon that trying occasion of ten years before — all iron and bronze self-posses- sion.   Mr. Harrison came over and led me to the General and formally introduced me.   Before I could put together the proper remark, General Grant said —

"Mr. Clemens, I am not embarrassed.   Are you?"— and that little seven-year smile twinkled across his face again.

Seventeen years have gone by since then, and to-day, in New York, the streets are a crush of people who are there to honor the remains of the great soldier as they pass to their final resting-place under the monument; and the air is heavy with dirges and the boom of artillery, and all the millions of America are thinking of the man who restored the Union and the flag, and gave to democratic government a new lease of life, and, as we may hope and do believe, a permanent place among the beneficent institutions of men.

We had one game in the ship which was a good time- passer — at least it was at night in the smoking-room when the men were getting freshened up from the day's monotonies

and dullnesses. It was the completing of non-complete stories.
That is to say, a man would tell all of a story except the
finish, then the others would try to supply the ending out
of their own invention. When every one who wanted a
chance had had it, the man who had introduced the story
would give it its original ending — then you could take your
choice. Sometimes the new endings turned out to be better
than the old one. But the story which called out the most
persistent and determined and ambitious effort was one which
had no ending, and so there was nothing to compare the new-
made endings with. The man who told it said he could
furnish the particulars up to a certain point only, because that
was as much of the tale as he knew. He had read it in a
volume of sketches twenty-five years ago, and was interrupted
before the end was reached. He would give any one fifty
dollars who would finish the story to the satisfaction of a jury
to be appointed by ourselves. We appointed a jury and
wrestled with the tale. We invented plenty of endings, but
the jury voted them all down. The jury was right. It was a
tale which the author of it may possibly have completed satisfactorily, and if he really had that good fortune I would like
to know what the ending was. Any ordinary man will find
that the story's strength is in its middle, and that there is
apparently no way to transfer it to the close, where of course
it ought to be. In substance the storiette was as follows:

John Brown, aged thirty-one, good, gentle, bashful, timid, lived in a quiet
viliage in Missouri. He was superintendent of the Presbyterian Sunday-
school. It was but a humble distinction ; still, it was his only official one,
and he was modestly proud of it and was devoted to its work and its interests.
The extreme kindliness of his nature was recognized by all ; in fact, people
said that he was made entirely out of good impulses and bashfulness ; that he
could always be counted upon for help when it was needed, and for bashfulness both when it was needed and when it wasn't.

Mary Taylor, twenty-three, modest, sweet, winning, and in character and
person beautiful, was all in all to him. And he was very nearly all in all to
her. She was wavering, his hopes were high. Her mother had been in

opposition from the first. But she was wavering, too ; he could see it. She was being touched by his warm interest in her two charity-protegés and by his contributions toward their support. These were two forlorn and aged sisters who lived in a log hut in a lonely place up a cross road four miles from Mrs Taylor's farm. One of the sisters was crazy, and sometimes a little violent, but not often.

At last the time seemed ripe for a final advance, and Brown gathered his courage together and resolved to make it. He would take along a contribution of double the usual size, and win the mother over ; with her opposition annulled, the rest of the conquest would be sure and prompt.

He took to the road in the middle of a placid Sunday afternoon in the soft Missourian summer, and he was equipped properly for his mission. He was clothed all in white linen, with a blue ribbon for a necktie, and he had on dressy tight boots. His horse and buggy were the finest that the livery stable could furnish. The lap robe was of white linen, it was new, and it had a hand-worked border that could not be rivaled in that region for beauty and elaboration.

When he was four miles out on the lonely road and was walking his horse over a wooden bridge, his straw hat blew off and fell in the creek, and floated down and lodged against a bar. He did not quite know what to do. He must have the hat, that was manifest ; but how was he to get it ?

Then he had an idea. The roads were empty, nobody was stirring. Yes, he would risk it. He led the horse to the roadside and set it to cropping the grass ; then he undressed and put his clothes in the buggy, petted the horse a moment to secure its compassion and its loyalty, then hurried to the stream. He swam out and soon had the hat. When he got to the top of the bank the horse was gone !

His legs almost gave way under him. The horse was walking leisurely along the road. Brown trotted after it, say-ing, 'Whoa, whoa, there's a good fel-low ; " but whenever he got near enough to chance a jump for the buggy, the horse quickened its pace a little and defeated him. And so this went on, the naked man perishing with anxiety, and expect-ing every moment to see people come in sight. He tagged on and on, imploring the horse, beseeching the horse, till he had left a mile behind him, and was clos-ing up on the Taylor premises ; then at last he was successful, and got into the buggy. He flung on his shirt, his neck-tie, and his coat ; then reached for — but he was too late ; he sat suddenly down and pulled up the lap-robe, for he saw some one coming out of the

THE CLIMAX.

gate — a woman, he thought. He wheeled the horse to the left, and struck briskly up the cross-road. It was perfectly straight, and exposed on both sides ; but there were woods and a sharp turn three miles ahead, and he was very grateful when he got there. As he passed around the turn he slowed down to a walk, and reached for his tr — too late again.

He had come upon Mrs. Enderby, Mrs. Glossop, Mrs. Taylor, and Mary. They were on foot, and seemed tired and excited. They came at once to the buggy and shook hands, and all spoke at once, and said eagerly and earnestly, how glad they were that he was come, and how fortunate it was. And Mrs. Enderby said, impressively :

"It *looks* like an accident, his coming at such a time ; but let no one profane it with such a name ; he was sent — sent from on high."

They were all moved, and Mrs. Glossop said in an awed voice :

"Sarah Enderby, you never said a truer word in your life. This is no accident, it is a special Providence. He *was* sent. He is an angel — an angel as truly as ever angel was — an angel of deliverance. *I* say *angel*, Sarah Enderby, and will have no other word. Don't let any one ever say to me again, that there's no such thing as special Providences ; for if this isn't one, let them account for it that can."

"I *know* it's so," said Mrs. Taylor, fervently. "John Brown, I could worship you ; I could go down on my knees to you. Didn't something tell you ? — didn't you *feel* that you were sent ? I could kiss the hem of your lap-robe."

He was not able to speak ; he was helpless with shame and fright. Mrs. Taylor went on :

"Why, just look at it all around, Julia Glossop. *Any* person can see the hand of Providence in it. Here at noon what do we see ? We see the smoke rising. I speak up and say, 'That's the Old People's cabin afire.' Didn't I, Julia Glossop ? "

"The very words you said, Nancy Taylor. I was as close to you as I am now, and I heard them. You may have said hut instead of cabin, but in substance it's the same. And you were looking pale, too."

"Pale ? I was that pale that if — why, you just compare it with this lap-robe. Then the next thing I said was, 'Mary Taylor, tell the hired man to rig up the team — we'll go to the rescue.' And she said, 'Mother, don't you know you told him he could drive to see his people, and stay over Sunday ? ' And it was just so. I declare for it, I had forgotten it. 'Then,' said I, 'we'll go afoot.' And go we did. And found Sarah Enderby on the road."

"And we all went together," said Mrs. Enderby. "And found the cabin set fire to and burnt down by the crazy one, and the poor old things so old and feeble that they couldn't go afoot. And we got them to a shady place and made them as comfortable as we could, and began to wonder which way to turn to find some way to get them conveyed to Nancy Taylor's house. And I spoke up and said — now what did I say ? Didn't I say, 'Providence will provide ' ? "

"Why sure as you live, so you did ! I had forgotten it."

"So had I," said Mrs. Glossop and Mrs. Taylor ; "but you certainly *said* it. Now wasn't that remarkable ? "

"Yes, I said it. And then we went to Mr. Moseley's, two miles, and all of them were gone to the camp meeting over on Stony Fork; and then we came all the way back, two miles, and then here, another mile — and Providence *has* provided. You see it yourselves."

They gazed at each other awe-struck, and lifted their hands and said in unison:

"It's per-fectly wonderful."

"And then," said Mrs. Glossop, "what do you think we had better do — let Mr. Brown drive the Old People to Nancy Taylor's one at a time, or put both of them in the buggy, and him lead the horse?"

Brown gasped.

"Now, then, that's a question," said Mrs. Enderby. "You see, we are all tired out, and any way we fix it it's going to be difficult. For if Mr. Brown takes both of them, at least one of us must go back to help him, for he can't load them into the buggy by himself, and they so helpless."

"That is so," said Mrs. Taylor. "It doesn't look — oh, how would this do? — one of us drive there *with* Mr. Brown, and the rest of you go along to my house and get things ready. I'll go with him. He and I together can lift one of the Old People into the buggy; then drive her to my house and —"

"But who will take care of the other one?" said Mrs. Enderby. "We musn't leave her there in the woods alone, you know — especially the crazy one. There and back is eight miles, you see."

They had all been sitting on the grass beside the buggy for a while, now, trying to rest their weary bodies. They fell silent a moment or two, and struggled in thought over the baffling situation; then Mrs. Enderby brightened and said:

"I think I've got the idea, now. You see, we can't *walk* any more. Think what we've done: four miles there, two to Moseley's, is six, then back to here — nine miles since noon, and not a bite to eat; I declare I don't see how we've done it; and as for me, I am just famishing. Now, somebody's got to go back, to help Mr. Brown — there's no getting around that; but whoever goes has got to ride, not walk. So my idea is this: one of us to ride back with Mr. Brown, then ride to Nancy Taylor's house with one of the Old People, leaving Mr. Brown to keep the other old one company, you all to go now to Nancy's and rest and wait; then one of you drive back and get the other one and drive *her* to Nancy's, and Mr. Brown walk."

"Splendid!" they all cried. "Oh, that will do — that will answer perfectly." And they all said that Mrs. Enderby had the best head for planning, in the company; and they said that they wondered that they hadn't thought of this simple plan themselves. They hadn't meant to take back the compliment, good simple souls, and didn't know they had done it. After a consultation it was decided that Mrs. Enderby should drive back with Brown, she being entitled to the distinction because she had invented the plan. Everything now being satisfactorily arranged and settled, the ladies rose, relieved and happy, and brushed down their gowns, and three of them started homeward; Mrs. Enderby set her foot on the buggy-step and was about to climb in, when Brown found a remnant of his voice and gasped out —

"Please Mrs. Enderby, call them back — I am very weak ; I can't walk, I can't, indeed."

"Why, dear Mr. Brown ! You *do* look pale ; I am ashamed of myself that I didn't notice it sooner. Come back — all of you ! Mr. Brown is not well. Is there anything I can do for you, Mr. Brown ? — I'm real sorry. Are you in pain ? "

"No, madam, only weak ; I am not sick, but only just weak — lately ; not long, but just lately."

The others came back, and poured out their sympathies and commiserations, and were full of self-reproaches for not having noticed how pale he was. And they at once struck out a new plan, and soon agreed that it was by far the best of all. They would all go to Nancy Taylor's house and see to Brown's needs first. He could lie on the sofa in the parlor, and while Mrs. Taylor and Mary took care of him the other two ladies would take the buggy and go and get one of the Old People, and leave one of themselves with the other one, and —

By this time, without any solicitation, they were at the horse's head and were beginning to turn him around. The danger was imminent, but Brown found his voice again and saved himself. He said —

"But ladies, you are overlooking something which makes the plan impracticable. You see, if you bring *one* of them home, and one remains behind with the other, there will be three persons there when one of you comes back for that other, for some one must drive the buggy back, and *three* can't come home in it."

They all exclaimed, "Why, sure-ly, that is so !" and they were all perplexed again.

"Dear, dear, what *can* we do ? " said Mrs. Glossop ; "it is the most mixed-up thing that ever was. The fox and the goose and the corn and things — oh, dear, they are nothing to it."

They sat wearily down once more, to further torture their tormented heads for a plan that would work. Presently Mary offered a plan ; it was her first effort. She said :

"I am young and strong, and am refreshed, now. Take Mr. Brown to our house, and give him help — you see how plainly he needs it. I will go back and take care of the Old People ; I can be there in twenty minutes. You can go on and do what you first started to do — wait on the main road at our house until somebody comes along with a wagon ; then send and bring away the three of us. You won't have to wait long ; the farmers will soon be coming back from town, now. I will keep old Polly patient and cheered up — the crazy one doesn't need it."

This plan was discussed and accepted ; it seemed the best that could be done, in the circumstances, and the Old People must be getting discouraged by this time.

Brown felt relieved, and was deeply thankful. Let him once get to the main road and he would find a way to escape.

Then Mrs. Taylor said :

" The evening chill will be coming on, pretty soon, and those poor old burnt-out things will need some kind of covering. Take the lap-robe with you, dear."

"Very well, Mother, I will."

She stepped to the buggy and put out her hand to take it —

That was the end of the tale. The passenger who told it said that when he read the story twenty-five years ago in a train he was interrupted at that point — the train jumped off a bridge.

At first we thought we could finish the story quite easily, and we set to work with confidence ; but it soon began to appear that it was not a simple thing, but difficult and baffling. This was on account of Brown's character— great generosity and kindliness, but complicated with unusual shyness and diffidence, particularly in the presence of ladies. There was his love for Mary, in a hopeful state but not yet secure — just in a condition, indeed, where its affair must be handled with great tact, and no mistakes made, no offense given. And there was the mother — wavering, half willing — by adroit and flawless diplomacy to be won over, now, or perhaps never at all. Also, there were the helpless Old People yonder in the woods waiting — their fate and Brown's happiness to be determined by what Brown should do within the next two seconds. Mary was reaching for the lap-robe ; Brown must decide — there was no time to be lost.

Of course none but a happy ending of the story would be accepted by

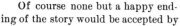
"WE WORKED UNTIL THREE."

the jury ; the finish must find Brown in high credit with the ladies, his behavior without blemish, his modesty unwounded, his character for self-sacrifice maintained, the Old People rescued through him, their benefactor, all the party proud of him, happy in him, his praises on all their tongues.

We tried to arrange this, but it was beset with persistent and irreconcilable difficulties. We saw that Brown's shyness would not allow him to give up the lap-robe. This would offend Mary and her mother ; and it would surprise the other ladies, partly because this stinginess toward the suffering Old People would be out of character with Brown, and partly because he was a special Providence and could not properly act so. If asked to explain his conduct, his shyness would not allow him to tell the truth, and lack of invention and practice would find him incapable of contriving a lie that would wash. We worked at the troublesome problem until three in the morning.

Meantime Mary was still reaching for the lap-robe. We gave it up, and decided to let her continue to reach. It is the reader's privilege to determine for himself how the thing came out.

CHAPTER III.

It is more trouble to make a maxim than it is to do right.
— Pudd'nhead Wilson's New Calendar.

ON the seventh day out we saw a dim vast bulk standing up out of the wastes of the Pacific and knew that that spectral promontory was Diamond Head, a piece of this world which I had not seen before for twenty-nine years. So we were nearing Honolulu, the capital city of the Sandwich Islands — those islands which to me were Paradise; a Paradise which I had been longing all those years to see again. Not any other thing in the world could have stirred me as the sight of that great rock did.

In the night we anchored a mile from shore. Through my port I could see the twinkling lights of Honolulu and the dark bulk of the mountain-range that stretched away right and left. I could not make out the beautiful Nuuana valley, but I knew where it lay, and remembered how it used to look in the old times. We used to ride up it on horseback in those days — we young people — and branch off and gather bones in a sandy region where one of the first Kamehameha's battles was fought. He was a remarkable man, for a king; and he was also a remarkable man for a savage. He was a mere kinglet and of little or no consequence at the time of Captain Cook's arrival in 1788; but about four years afterward he conceived the idea of enlarging his sphere of influence. That is a courteous modern phrase which means robbing your neighbor — for your neighbor's benefit; and the great theater of its benevolences is Africa. Kamehameha went to war, and in the

(48)

From note-book.

The man that invented the cuckoo
clock is ~~dead~~. *no more.* ~~It is old news but~~
~~good.~~

~~As news, this is a little stale, but~~ *old,*
~~some news is better old than not at all.~~
~~As news, this is a little old, but~~
~~better late than never.~~

~~As news this is a little old, for it~~
~~happened 64 years ago, but it~~ *always*
~~is not the newest news that is the best.~~ *most interesting.*
The man that invented the cuckoo
clock is no more. It is old news,
but there is nothing else the
matter with it.

~~Occasionally~~

It is more ~~diffi~~ trouble to ~~coin;~~ *make*
~~struck~~ a maxim than it is to
do right.

FACSIMILE PAGE FROM THE AUTHOR'S NOTE BOOK.

course of ten years he whipped out all the other kings and made himself master of every one of the nine or ten islands that form the group. But he did more than that. He bought ships, freighted them with sandal wood and other native products, and sent them as far as South America and China; he sold to his savages the foreign stuffs and tools and utensils which came back in these ships, and started the march of civilization. It is doubtful if the match to this extraordinary thing is to be found in the history of any other savage. Savages are eager to learn from the white man any new way to kill each other, but it is not their habit to seize with avidity and apply with energy the larger and nobler ideas which he offers them. The details of Kamehameha's history show that he was always hospitably ready to examine the white man's ideas, and that he exercised a tidy discrimination in making his selections from the samples placed on view.

A shrewder discrimination than was exhibited by his son and successor, Liholiho, I think. Liholiho could have qualified as a reformer, perhaps, but as a king he was a mistake. A mistake because he tried to be both king *and* reformer. This is mixing fire and gunpowder together. A king has no proper business with reforming. His best policy is to keep things as they are; and if he can't do that, he ought to try to make them worse than they are. This is not guesswork; I have thought over this matter a good deal, so that if I should ever have a chance to become a king I would know how to conduct the business in the best way.

When Liholiho succeeded his father he found himself possessed of an equipment of royal tools and safeguards which a wiser king would have known how to husband, and judiciously employ, and make profitable. The entire country was under the one scepter, and his was that scepter. There was an Established Church, and he was the head of it. There was a

Standing Army, and he was the head of that; an Army of
114 privates under command of 27 Generals and a Field Mar-
shal. There was a proud and ancient Hereditary Nobility.
There was still one other asset. This was the *tabu* — an agent
endowed with a mysterious and stupendous power, an agent
not found among the properties of any European monarch, a
tool of inestimable value in the business. Liholiho was head-
master of the tabu. The tabu was the most ingenious and
effective of all the inventions that has ever been devised for
keeping a people's privileges satisfactorily restricted.

It required the sexes to live in separate houses. It did not
allow people to eat in either house; they must eat in another
place. It did not allow a man's woman-folk to enter his
house. It did not allow the sexes to eat together; the men
must eat first, and the women must wait on them. Then the
women could eat what was left — if anything was left — and
wait on themselves. I mean, if anything of a coarse or unpalatable sort was left, the women could have it. But not the
good things, the fine things, the choice things, such as pork,
poultry, bananas, coçoanuts, the choicer varieties of fish, and
so on. By the tabu, all these were sacred to the men; the
women spent their lives longing for them and wondering what
they might taste like; and they died without finding out.

These rules, as you see, were quite simple and clear. It
was easy to remember them; and useful. For the penalty for
infringing any rule in the whole list was *death*. Those women
easily learned to put up with shark and taro and dog for a diet
when the other things were so expensive.

It was death for any one to walk upon tabu'd ground; or
defile a tabu'd thing with his touch; or fail in due servility to
a chief; or step upon the king's shadow. The nobles and the
King and the priests were always suspending little rags here
and there and yonder, to give notice to the people that the

decorated spot or thing was tabu, and death lurking near. The struggle for life was difficult and chancy in the islands in those days.

Thus advantageously was the new king situated. Will it be believed that the first thing he did was to destroy his Established Church, root and branch? He did indeed do that. To state the case figuratively, he was a prosperous sailor who burnt his ship and took to a raft. This Church was a horrid thing. It heavily oppressed the people; it kept them always trembling

ROYAL EQUIPMENTS.

in the gloom of mysterious threatenings; it slaughtered them in sacrifice before its grotesque idols of wood and stone; it cowed them, it terrorized them, it made them slaves to its priests, and through the priests to the king. It was the best friend a king could have, and the most dependable. To a professional reformer who should annihilate so frightful and so devastating a power as this Church, reverence and praise would be due; but to a king who should do it, could properly be due nothing but reproach; reproach softened by sorrow; sorrow for his unfitness for his position.

He destroyed his Established Church, and his kingdom is a republic to-day, in consequence of that act.

When he destroyed the Church and burned the idols he did a mighty thing for civilization and for his people's weal — but it was not "business." It was unkingly, it was inartistic. It made trouble for his line. The American missionaries arrived

while the burned idols were still smoking. They found the
nation without a religion, and they repaired the defect. They
offered their own religion and it was gladly received. But it
was no support to arbitrary kingship, and so the kingly power
began to weaken from that day. Forty-seven years later,
when I was in the islands, Kamehameha V. was trying to
repair Liholiho's blunder, and not succeeding. He had set up
an Established Church and made himself the head of it. But
it was only a pinchbeck thing, an imitation, a bauble, an empty
show. It had no power, no value for a king. It could not
harry or burn or slay, it in no way resembled the admirable
machine which Liholiho destroyed. It was an Established
Church without an Establishment; all the people were Dis-
senters.

Long before that, the kingship had itself become but a
name, a show. At an early day the missionaries had turned it
into something very much like a republic; and here lately the
business whites have turned it into something exactly like it.

In Captain Cook's time (1778), the native population of the
islands was estimated at 400,000; in 1836 at something short of
200,000, in 1866 at 50,000; it is to-day, per census, 25,000. All
intelligent people praise Kamehameha I. and Liholiho for con-
ferring upon their people the great boon of civilization.   I
would do it myself, but my intelligence is out of repair, now,
from over-work.

When I was in the islands nearly a generation ago, I was
acquainted with a young American couple who had among
their belongings an attractive little son of the age of seven —
attractive but not practicably companionable with me, because
he knew no English. He had played from his birth with the
little Kanakas on his father's plantation, and had preferred
their language and would learn no other. The family removed
to America a month after I arrived in the islands, and straight-

SOMETHING TOUCHED HIS SHOULDER.

way the boy began to lose his Kanaka and pick up English. By the time he was twelve he hadn't a word of Kanaka left; the language had wholly departed from his tongue and from his comprehension. Nine years later, when he was twenty-one, I came upon the family in one of the lake towns of New York, and the mother told me about an adventure which her son had been having. By trade he was now a professional diver. A passenger boat had been caught in a storm on the lake, and had gone down, carrying her people with her. A few days later the young diver descended, with his armor on, and entered the berth-saloon of the boat, and stood at the foot of the companionway, with his hand on the rail, peering through the dim water. Presently something touched him on the shoulder, and he turned and found a dead man swaying and bobbing about him and seemingly inspecting him inquiringly. He was paralyzed with fright. His entry had disturbed the water, and now he discerned a number of dim corpses making for him and wagging their heads and swaying their bodies like sleepy people trying to dance. His senses forsook him, and in that condition he was drawn to the surface. He was put to bed at home, and was soon very ill. During some days he had seasons of delirium which lasted several hours at a time; and while they lasted he talked *Kanaka* incessantly and glibly; and Kanaka only. He was still very ill, and he talked to me in that tongue; but I did not understand it, of course. The doctor-books tell us that cases like this are not uncommon. Then the doctors ought to study the cases and find out how to multiply them. Many languages and things get mislaid in a person's head, and stay mislaid for lack of this remedy.

Many memories of my former visit to the islands came up in my mind while we lay at anchor in front of Honolulu that night. And pictures — pictures — pictures — an enchanting procession of them! I was impatient for the morning to come

When it came it brought disappointment, of course. Cholera had broken out in the town, and we were not allowed to have any communication with the shore. Thus suddenly did my dream of twenty-nine years go to ruin. Messages came from friends, but the friends themselves I was not to have any sight of. My lecture-hall was ready, but I was not to see that, either.

Several of our passengers belonged in Honolulu, and these were sent ashore; but nobody could go ashore and return. There were people on shore who were booked to go with us to Australia, but we could not receive them; to do it would cost us a quarantine-term in Sydney. They could have escaped the day before, by ship to San Francisco; but the bars had been put up, now, and they might have to wait weeks before any ship could venture to give them a passage any whither. And there were hardships for others. An elderly lady and her son, recreation-seekers from Massachusetts, had wandered westward, further and further from home, always intending to take the return track, but always concluding to go still a little further; and now here they were at anchor before Honolulu — positively their last westward-bound indulgence — they had made up their minds to that — but where is the use in making up your mind in this world? It is usually a waste of time to do it. These two would have to stay with us as far as Australia. Then they could go on around the world, or go back the way they had come; the distance and the accommodations and out-lay of time would be just the same, whichever of the two routes they might elect to take. Think of it: a projected ex-cursion of five hundred miles gradually enlarged, without any elaborate degree of intention, to a possible twenty-four thou-sand. However, they were used to extentions by this time, and did not mind this new one much.

And we had with us a lawyer from Victoria, who had been

sent out by the Government on an international matter, and he had brought his wife with him and left the children at home with the servants — and now what was to be done? Go ashore amongst the cholera and take the risks? Most certainly not. They decided to go on, to the Fiji islands, wait there a fortnight for the next ship, and then sail for home. They couldn't foresee that they wouldn't see a homeward-bound ship again for six weeks, and that no word could come to them from the children, and no word go from them to the children in all that time. It is easy to make plans in this world; even a cat can do it; and when one is out in those remote oceans it is noticeable that a cat's plans and a man's are worth about the same. There is much the same shrinkage in both, in the matter of values.

There was nothing for us to do but sit about the decks in the shade of the awnings and look at the distant shore. We lay in luminous blue water; shoreward the water was green — green and brilliant; at the shore itself it broke in a long white ruffle, and with no crash, no sound that we could hear. The town was buried under a mat of foliage that looked like a cushion of moss. The silky mountains were clothed in soft, rich splendors of melting color, and some of the cliffs were veiled in slanting mists. I recognized it all. It was just as I had seen it long before, with nothing of its beauty lost, nothing of its charm wanting.

A change had come, but that was political, and not visible from the ship. The monarchy of my day was gone, and a republic was sitting in its seat. It was not a material change. The old imitation pomps, the fuss and feathers, have departed, and the royal trademark — that is about all that one could miss, I suppose. That imitation monarchy was grotesque enough, in my time; if it had held on another thirty years it would have been a monarchy without subjects of the king's race.

We had a sunset of a very fine sort. The vast plain of the sea was marked off in bands of sharply-contrasted colors: great stretches of dark blue, others of purple, others of polished bronze; the billowy mountains showed all sorts of dainty browns and greens, blues and purples and blacks, and the rounded velvety backs of certain of them made one want to stroke them, as one would the sleek back of a cat. The long, sloping promontory projecting into the sea at the west turned dim and leaden and spectral, then became suffused with pink — dissolved itself in a pink dream, so to speak, it seemed so airy and unreal. Presently the cloud-rack was flooded with fiery splendors, and these were copied on the surface of the sea, and it made one drunk with delight to look upon it.

From talks with certain of our passengers whose home was Honolulu, and from a sketch by Mrs. Mary H. Krout, I was able to perceive what the Honolulu of to-day is, as compared with the Honolulu of my time. In my time it was a beautiful little town, made up of snow-white wooden cottages deliciously smothered in tropical vines and flowers and trees and shrubs; and its coral roads and streets were hard and smooth, and as white as the houses. The outside aspects of the place suggested the presence of a modest and comfortable prosperity — a general prosperity — perhaps one might strengthen the term and say universal. There were no fine houses, no fine furniture. There were no decorations. Tallow candles furnished the light for the bedrooms, a whale-oil lamp furnished it for the parlor. Native matting served as carpeting. In the parlor one would find two or three lithographs on the walls — portraits as a rule: Kamehameha IV., Louis Kossuth, Jenny Lind; and may be an engraving or two: Rebecca at the Well, Moses smiting the rock, Joseph's servants finding the cup in Benjamin's sack. There would be a center table, with books of a tranquil sort on it: The Whole Duty of Man, Baxter's Saints'

Rest, Fox's Martyrs, Tupper's Proverbial Philosophy, bound copies of The Missionary Herald and of Father Damon's Seaman's Friend. A melodeon; a music stand, with Willie, We have Missed You, Star of the Evening, Roll on Silver Moon, Are We Most There, I Would not Live Alway, and other songs of love and sentiment, together with an assortment of hymns. A what-not with semi-globular glass paperweights, enclosing miniature pictures of ships, New England rural snowstorms, and the like; sea-shells with Bible texts carved on them in cameo style; native curios; whale's tooth with full-rigged ship carved on it. There was nothing reminiscent of foreign parts, for nobody had been abroad. Trips were made to San Francisco, but that could not be called going abroad. Comprehensively speaking, nobody traveled.

But Honolulu has grown wealthy since then, and of course wealth has introduced changes; some of the old simplicities have disappeared. Here is a modern house, as pictured by Mrs. Krout:

"Almost every house is surrounded by extensive lawns and gardens enclosed by walls of volcanic stone or by thick hedges of the brilliant hibiscus.

"The houses are most tastefully and comfortably furnished ; the floors are either of hard wood covered with rugs or with fine Indian matting, while there is a preference, as in most warm countries, for rattan or bamboo furniture ; there are the usual accessories of bric-a-brac, pictures, books, and curios from all parts of the world, for these island dwellers are indefatigable travelers.

"Nearly every house has what is called a *lanai*. It is a large apartment, roofed, floored, open on three sides, with a door or a draped archway opening into the drawing-room. Frequently the roof is formed by the thick interlacing boughs of the *hou* tree, impervious to the sun and even to the rain, except in violent storms. Vines are trained about the sides — the stephanotis or some one of the countless fragrant and blossoming trailers which abound in the islands. There are also curtains of matting that may be drawn to exclude the sun or rain. The floor is bare for coolness, or partially covered with rugs, and the *lanai* is prettily furnished with comfortable chairs, sofas, and tables loaded with flowers, or wonderful ferns in pots.

"The *lanai* is the favorite reception room, and here at any social function the musical program is given and cakes and ices are served ; here morning callers are received, or gay riding parties, the ladies in pretty divided skirts, worn for convenience in riding astride,— the universal mode adopted by Europeans and Americans, as well as by the natives.

"The comfort and luxury of such an apartment, especially at a seashore villa, can hardly be imagined. The soft breezes sweep across it, heavy with the fragrance of jasmine and gardenia, and through the swaying boughs of palm and mimosa there are glimpses of rugged mountains, their summits veiled in clouds, of purple sea with the white surf beating eternally against the reefs,— whiter still in the yellow sunlight or the magical moonlight of the tropics."

There: rugs, ices, pictures, lanais, worldly books, sinful bric-a-brac fetched from everywhere. And the ladies riding astride. These are changes, indeed. In my time the native women rode astride, but the white ones lacked the courage to adopt their wise custom. In my time ice was seldom seen in Honolulu. It sometimes came in sailing vessels from New England as ballast; and then, if there happened to be a man-of-war in port and balls and suppers raging by consequence, the ballast was worth six hundred dollars a ton, as is evidenced by reputable tradition. But the ice-machine has traveled all over the world, now, and brought ice within everybody's reach. In Lapland and Spitzbergen no one uses native ice in our day, except the bears and the walruses.

The bicycle is not mentioned. It was not necessary. We know that it is there, without inquiring. It is everywhere. But for it, people could never have had summer homes on the summit of Mont Blanc; before its day, property up there had but a nominal value. The ladies of the Hawaiian capital learned too late the right way to occupy a horse — too late to get much benefit from it. The riding-horse is retiring from business everywhere in the world. In Honolulu a few years from now he will be only a tradition.

We all know about Father Damien, the French priest who voluntarily forsook the world and went to the leper island of Molokai to labor among its population of sorrowful exiles who wait there, in slow-consuming misery, for death to come and release them from their troubles ; and we know that the thing which he knew beforehand would happen, did happen : that he became a leper himself, and died of that horrible disease. There was still another case of self-sacrifice, it appears. I asked after " Billy " Ragsdale, interpreter to the Parliament in my time — a half-white. He was a brilliant young fellow, and very popular. As an interpreter he would have been hard to match anywhere. He used to stand up in the Parliament and turn the English speeches into Hawaiian and the Hawaiian speeches into English with a readiness and a volubility that were astonishing. I asked after him, and was told that his prosperous career was cut short in a sudden and unexpected way, just as he was about to marry a beautiful half-caste girl. He discovered, by some nearly invisible sign about his skin, that the poison of leprosy was in him. The secret was his own, and might be kept concealed for years; but he would not be treacherous to the girl that loved him; he would not marry her to a doom like his. And so he put his affairs in order, and went around to all his friends and bade them good-bye, and sailed in the leper ship to Molokai. There he died the loathsome and lingering death that all lepers die.

In this place let me insert a paragraph or two from " The Paradise of the Pacific " (Rev. H. H. Gowen) :

"Poor lepers ! It is easy for those who have no relatives or friends among them to enforce the decree of segregation to the letter, but who can write of the terrible, the heart-breaking scenes which that enforcement has brought about ?

"A man upon Hawaii was suddenly taken away after a summary arrest, leaving behind him a helpless wife about to give birth to a babe. The devoted wife with great pain and risk came the whole journey to Honolulu, and pleaded until the authorities were unable to resist her entreaty that she might go and live like a leper with her leper husband.

"A woman in the prime of life and activity is condemned as an incipient leper, suddenly removed from her home, and her husband returns to find his two helpless babes moaning for their lost mother.

"Imagine it ! The case of the babies is hard, but its bitterness is a trifle — less than a trifle — less than nothing — compared to what the mother must suffer ; and suffer minute by minute, hour by hour, day by day, month by month, year by year, without respite, relief, or any abatement of her pain till she dies.

"One woman, Luka Kaaukau, has been living with her leper husband in the settlement for twelve years. The man has scarcely a joint left, his limbs are only distorted ulcerated stumps, for four years his wife has put every particle of food into his mouth. He wanted his wife to abandon his wretched carcass long ago, as she herself was sound and well, but Luka said that she was content to remain and wait on the man she loved till the spirit should be freed from its burden.

"I myself have known hard cases enough : — of a girl, apparently in full health, decorating the church with me at Easter, who before Christmas is taken away as a confirmed leper ; of a mother hiding her child in the mountains for years so that not even her dearest friends knew that she had a child alive, that he might not be taken away ; of a respectable white man taken away from his wife and family, and compelled to become a dweller in the Leper Settlement, where he is counted dead, *even by the insurance companies.*"

And one great pity of it all is, that these poor sufferers are innocent. The leprosy does not come of sins which they committed, but of sins committed by their ancestors, who *escaped* the curse of leprosy !

Mr. Gowan has made record of a certain very striking circumstance. Would you expect to find in that awful Leper Settlement a custom worthy to be transplanted to your own country ? They have one such, and it is inexpressibly touching and beautiful. When death sets open the prison-door of life there, the band salutes the freed soul with a burst of glad music !

CHAPTER IV.

SAILED from Honolulu. From diary:

Sept. 2. Flocks of flying fish — slim, shapely, graceful, and intensely white. With the sun on them they look like a flight of silver fruit-knives. They are able to fly a hundred yards.

Sept. 3. In 9° 50' north latitude, at breakfast. Approaching the equator on a long slant. Those of us who have never seen the equator are a good deal excited. I think I would rather see it than any other thing in the world. We entered the "doldrums" last night — variable winds, bursts of rain, intervals of calm, with chopping seas and a wobbly and drunken motion to the ship — a condition of things findable in other regions sometimes, but present in the doldrums always. The globe-girdling belt called the doldrums is 20 degrees wide, and the thread called the equator lies along the middle of it.

Sept. 4. Total eclipse of the moon last night. At 7.30 it began to go off. At total — or about that — it was like a rich rosy cloud with a tumbled surface framed in the circle and projecting from it — a bulge of strawberry-ice, so to speak. At half-eclipse the moon was like a gilded acorn in its cup.

Sept. 5. Closing in on the equator this noon. A sailor explained to a young girl that the ship's speed is poor because we are climbing up the bulge toward the center of the globe; but that when we should once get over, at the equator, and start

5

down-hill, we should fly.   When she asked him the other day
what the fore-yard was, he said it was the front yard, the open
area in the front end of the ship.   That man has a good deal of
learning stored up, and the girl is likely to get it all.

Afternoon.   Crossed the equator.   In the distance it looked
like a blue ribbon stretched across the ocean.   Several passengers

kodak'd it.   We had no fool ceremonies, no fantastics, no horse-
play.   All that sort of thing
has gone out.   In old times a
sailor, dressed as Neptune, used
to come in over the bows, with
his suite, and lather up and
shave everybody who was crossing the equator for the first
time, and then cleanse these

WATCHING FOR THE BLUE RIBBON.   unfortunates by swinging them
from the yard-arm and ducking them three times in the sea.
This was considered funny.   Nobody knows why.   No, that
is not true.   We do know why.   Such a thing could never be
funny on land; no part of the old-time grotesque performances
gotten up on shipboard to celebrate the passage of the line
could ever be funny on shore — they would seem dreary and
witless to shore people.   But the shore people would change
their minds about it at sea, on a long voyage.   On such a
voyage, with its eternal monotonies, people's intellects dete-
riorate; the owners of the intellects soon reach a point where
they almost seem to prefer childish things to things of a ma-
turer degree.   One is often surprised at the juvenilities which
grown people indulge in at sea, and the interest they take in
them, and the consuming enjoyment they get out of them.
This is on long voyages only.   The mind gradually becomes
inert, dull, blunted; it loses its accustomed interest in intellec-

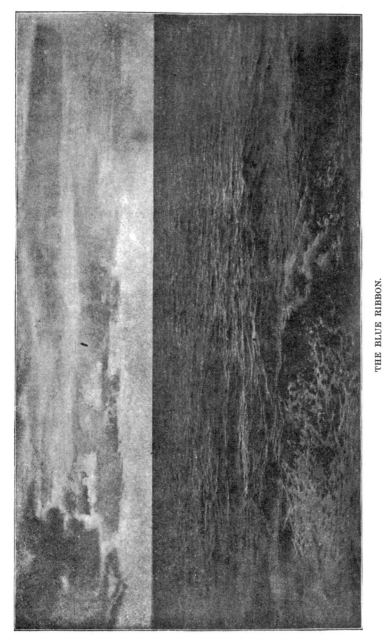

THE BLUE RIBBON.

" HORSE BILLIARDS."

	10	
8	I	6
3	5	7
4	9	2
	10 off	

DIAGRAM.

tual things; nothing but horse-play can rouse it, nothing but wild and foolish grotesqueries can entertain it. On short voyages it makes no such exposure of itself; it hasn't time to slump down to this sorrowful level.

The short-voyage passenger gets his chief physical exercise out of "horse-billiards"—shovel-board. It is a good game. We play it in this ship. A quartermaster chalks off a diagram like this—on the deck.

The player uses a cue that is like a broom-handle with a quarter-moon of wood fastened to the end of it. With this he shoves wooden disks the size of a saucer—he gives the disk a vigorous shove and sends it fifteen or twenty feet along the deck and lands it in one of the squares if he can. If it stays there till the inning is played out, it will count as many points in the game as the figure in the square it has stopped in represents. The adversary plays to knock that disk out and leave his own in its place—particularly if it rests upon the 9 or 10 or some other of the high numbers; but if it rests in the "10-off" he backs it up—lands his disk behind it a foot or two, to make it difficult for its owner to knock it out of that damaging place and improve his record. When the inning is played out it may be found that each adversary has placed his four disks where they count; it may be found that some of them are touching chalk lines and not counting; and very often it will be found that there has been a general wreckage, and that not a disk has been left within the diagram. Anyway, the result is recorded, whatever it is, and the game goes on. The game is 100 points, and it takes from twenty minutes to forty to play it, according to luck and the condition of the sea. It is an exciting game, and the crowd of spectators furnish abundance of applause for fortunate shots and plenty of laughter for the other kind. It is a game of skill, but at the same time the uneasy motion of the ship is constantly interfering with skill; this makes it a chancy game, and the element of luck comes largely in.

We had a couple of grand tournaments, to determine who should be "Champion of the Pacific"; they included among the participants nearly all the passengers, of both sexes, and the officers of the ship, and they afforded many days of stupendous interest and excitement, and murderous exercise — for horse-billiards is a physically violent game.

The figures in the following record of some of the closing games in the first tournament will show, better than any description, how very chancy the game is. The losers here represented had all been winners in the previous games of the series, some of them by fine majorities:

Chase,	102	Mrs. D.,	57	Mortimer,	105	The Surgeon,	92
Miss C.,	105	Mrs. T.,	9	Clemens,	101	Taylor,	92
Taylor,	109	Davies,	95	Miss C.,	108	Mortimer,	55
Thomas,	102	Roper,	76	Clemens,	111	Miss C.,	89
Coomber,	106	Chase,	98				

And so on; until but three couples of winners were left. Then I beat my man, young Smith beat his man, and Thomas beat his. This reduced the combatants to three. Smith and I took the deck, and I led off. At the close of the first inning I was 10 worse than nothing and Smith had scored 7. The luck continued against me. When I was 57, Smith was 97 — within 3 of out. The luck changed then. He picked up a 10-off or so, and couldn't recover. I beat him.

The next game would end tournament No. 1.

Mr. Thomas and I were the contestants. He won the lead and went to the bat — so to speak. And there he stood, with the crotch of his cue resting against his disk while the ship rose slowly up, sank slowly down, rose again, sank again. She never seemed to rise to suit him exactly. She started up once more; and when she was nearly ready for the turn, he let drive and landed his disk just within the left-hand end of the

10. (Applause). The umpire proclaimed "a good 10," and the game-keeper set it down. I played: my disk grazed the edge of Mr. Thomas's disk, and went out of the diagram. (No applause.)

Mr. Thomas played again — and landed his second disk alongside of the first, and almost touching its right-hand side. "Good 10." (Great applause.)

I played, and missed both of them. (No applause.)

Mr. Thomas delivered his third shot and landed his disk just at the right of the other two. "Good 10." (Immense applause.)

There they lay, side by side, the three in a row. It did not seem possible that anybody could miss them. Still I did it. (Immense silence.)

Mr. Thomas played his last disk. It seems incredible, but he actually landed that disk alongside of the others, and just to the right of them — a straight solid row of 4 disks. (Tumultuous and long-continued applause.)

Then I played my last disk. Again it did not seem possible that anybody could miss that row — a row which would have been 14 inches long if the disks had been clamped together; whereas, with the spaces separating them they made a longer row than that. But I did it. It may be that I was getting nervous.

I think it unlikely that that innings has ever had its parallel in the history of horse-billiards. To place the four disks side by side in the 10 was an extraordinary feat; indeed, it was a kind of miracle. To miss them was another miracle. It will take a century to produce another man who can place the four disks in the 10; and longer than that to find a man who can't knock them out. I was ashamed of my performance at the time, but now that I reflect upon it I see that it was rather fine and difficult.

Mr. Thomas kept his luck, and won the game, and later the championship.

In a minor tournament I won the prize, which was a Waterbury watch. I put it in my trunk. In Pretoria, South Africa, nine months afterward, my proper watch broke down and I took the Waterbury out, wound it, set it by the great clock on the Parliament House (8.05), then went back to my room and went to bed, tired from a long railway journey. The parliamentary clock had a peculiarity which I was not aware of at the time — a peculiarity which exists in no other clock, and would not exist in that one if it had been made by a sane person; on the half-hour it strikes the succeeding *hour*, then strikes the hour *again* at the proper time. I lay reading and smoking awhile; then, when I could hold my eyes open no longer and was about to put out the light, the great clock began to boom, and I counted —

I BEAT HER BRAINS OUT.

ten. I reached for the Waterbury to see how it was getting along. It was marking 9.30. It seemed rather poor speed for a three-dollar watch, but I supposed that the climate was affecting it. I shoved it half an hour ahead, and took to my book and waited to see what would happen. At 10 the great clock struck ten *again*. I looked — the Waterbury was marking half-past 10. This was too much speed for the money, and it troubled me. I pushed the hands back a half hour, and waited once more; I had to, for I was vexed and restless now, and my sleepiness was gone. By and by the great clock struck 11. The Waterbury was marking 10.30. I pushed it ahead half an hour, with some show of temper. By and by the great clock struck 11 again. The Waterbury showed up 11.30, now, and I beat her brains out against the bedstead. I was sorry next day, when I found out.

To return to the ship.

The average human being is a perverse creature; and when he isn't that, he is a practical joker. The result to the other person concerned is about the same: that is, he is made to suffer. The washing down of the decks begins at a very early hour in all ships; in but few ships are any measures taken to protect the passengers, either by waking or warning them, or by sending a steward to close their ports. And so the deck-washers have their opportunity, and they use it. They send a bucket of water slashing along the side of the ship and into the ports, drenching the passenger's clothes, and often the passenger himself. This good old custom prevailed in this ship, and under unusually favorable circumstances, for in the blazing tropical regions a removable zinc thing like a sugar-shovel projects from the port to catch the wind and bring it in; this thing catches the wash-water and brings it in, too — and in flooding abundance. Mrs. I., an invalid, had to sleep on the locker-sofa under her port, and every time she over-slept and thus failed to take care of herself, the deck-washers drowned her out.

And the painters, what a good time they had! This ship

would be going into dock for a month in Sydney for repairs; but no matter, painting was going on all the time somewhere or other. The ladies' dresses were constantly getting ruined, nevertheless protests and supplications went for nothing. Sometimes a lady, taking an afternoon nap on deck near a ventilator or some other thing that didn't need painting, would wake up by and by and find that the humorous painter had been noiselessly daubing that thing and had splattered her white gown all over with little greasy yellow spots.

The blame for this untimely painting did not lie with the ship's officers, but with custom. As far back as Noah's time it became law that ships must be constantly painted and fussed at when at sea; custom grew out of the law, and at sea custom knows no death; this custom will continue until the sea goes dry.

A DAY OFF.

Sept. 8.—Sunday. We are moving so nearly south that we cross only about two meridians of longitude a day. This morning we were in longitude 178 west from Greenwich, and 57 degrees west from San Francisco. To-morrow we shall be close to the center of the globe — the 180th degree of west longitude and 180th degree of east longitude.

And then we must drop out a day — lose a day out of our lives, a day never to be found again. We shall all die one day earlier than from the beginning of time we were foreordained to die. We shall be a day behindhand all through eternity. We shall always be saying to the other angels, " Fine day to-day," and they will be always retorting, " But it isn't to-day, it's to-morrow." We shall be in a state of confusion all the time and shall never know what true happiness is.

Next Day. Sure enough, it has happened. Yesterday it was September 8, *Sunday;* to-day, per the bulletin-board at the head of the companionway, it is September 10, *Tuesday.* There is something uncanny about it. And uncomfortable. In fact, nearly unthinkable, and wholly unrealizable, when one comes to consider it. While we were crossing the 180th meridian it was *Sunday* in the stern of the ship where my family were, and *Tuesday* in the bow where I was. They were there eating the half of a fresh apple on the 8th, and I was at the same time eating the other half of it on the 10th — and I could notice how stale it was, already. The family were the same age that they were when I had left them five minutes before, but I was a day older now than I was then. The day they were living in stretched behind them half way round the globe, across the Pacific Ocean and America and Europe; the day I was living in stretched in front of me around the other half to meet it. They were stupendous days for bulk and stretch; apparently much larger days than we had ever been in before. All previous days had been but shrunk-up little things by comparison. The difference in temperature between the two days was very marked, their day being hotter than mine because it was closer to the equator.

Along about the moment that we were crossing the Great Meridian a child was born in the steerage, and now there is no way to tell which day it was born on. The nurse thinks it was Sunday, the surgeon thinks it was Tuesday. The child will never know its own birthday. It will always be choosing first one and then the other, and will never be able to make up its mind permanently. This will breed vacillation and uncertainty in its opinions about religion, and politics, and business, and sweethearts, and everything, and will undermine its principles, and rot them away, and make the poor thing characterless, and its success in life impossible. Every one in the ship says so. And this is not all — in fact, not the worst. For there is an enormously rich brewer in the ship who said as much as ten days ago, that if the child was born on his birthday he would give it ten thousand dollars to start its little life with. His birthday was Monday, the 9th of September.

If the ships all moved in the one direction — westward, I mean — the world would suffer a prodigious loss in the matter of valuable time, through the dumping overboard on the Great Meridian of such multitudes of days by ships' crews and passengers. But fortunately the ships do not all sail west, half of them sail east. So there is no real loss. These latter pick up all the discarded days and add them to the world's stock again; and about as good as new, too; for of course the salt water preserves them.

CHAPTER V.

Noise proves nothing. Often a hen who has merely laid an egg cackles as if she had laid an asteroid.— *Pudd'nhead Wilson's New Calendar.*

WEDNESDAY, *Sept. 11.* In this world we often make mistakes of judgment. We do not as a rule get out of them sound and whole, but sometimes we do. At dinner yesterday evening — present, a mixture of Scotch, English, American, Canadian, and Australasian folk — a discussion broke out about the pronunciation of certain Scottish words. This was private ground, and the non-Scotch nationalities, with one exception, discreetly kept still. But I am not discreet, and I took a hand. I didn't know anything about the subject, but I took a hand just to have something to do. At that moment the word in dispute was the word *three.* One Scotchman was claiming that the peasantry of Scotland pronounced it *three*, his adversaries claimed that they didn't — that they pronounced it *thraw.* The solitary Scot was having a sultry time of it, so I thought I would enrich him with my help. In my position I was necessarily quite impartial, and was equally as well and as ill equipped to fight on the one side as on the other. So I spoke up and said the peasantry pronounced the word *three*, not *thraw.* It was an error of judgment. There was a moment of astonished and ominous silence, then weather ensued. The storm rose and spread in a surprising way, and I was snowed under in a very few minutes. It was a bad defeat for me — a kind of Waterloo. It promised to remain so, and I wished I had had better sense than to enter upon such a forlorn enterprise. But just then I had a saving thought — at least a thought that offered a chance. While the storm was still raging, I made up a Scotch couplet, and then spoke up and said : —

"Very well, don't say any more. I confess defeat. I thought I knew, but I see my mistake. I was deceived by one of your Scotch poets."

"A *Scotch* poet ! O come ! Name him."

"*Robert Burns.*"

It is wonderful the power of that name. These men looked doubtful — but paralyzed, all the same. They were quite silent for a moment; then one of them said — with the reverence in his voice which is always present in a Scotchman's tone when he utters the name :

" Does Robbie Burns say —*what* does he say ? "

" This is what he says :

   ' " There were nae bairns but only three —
    Ane at the breast, twa at the knee." '

It ended the discussion. There was no man there profane enough, disloyal enough, to say any word against a thing which Robert Burns had settled. I shall always honor that great name for the salvation it brought me in this time of my sore need.

It is my belief that nearly any invented quotation, played with confidence, stands a good chance to deceive. There are people who think that honesty is always the best policy. This is a superstition ; there are times when the appearance of it is worth six of it.

We are moving steadily southward — getting further and further down under the projecting paunch of the globe. Yesterday evening we saw the Big Dipper and the north star sink below the horizon and disappear from our world. No, not " we," but they. They saw it — somebody saw it — and told me about it. But it is no matter, I was not caring for those

things, I am tired of them, any way. I think they are well
enough, but one doesn't want them always hanging around.
My interest was all in the Southern Cross. I had never seen
that. I had heard about it all my life, and it was but natural
that I should be burning to see it. No other constellation
makes so much talk. I had nothing against the Big Dipper —
and naturally couldn't have anything against it, since it is a
citizen of our own sky, and the property of the United States —
but I did want it to move out of the way and give this
foreigner a chance. Judging by the size of the talk which the
Southern Cross had made, I supposed it would need a sky all
to itself.

But that was a mistake. We saw the Cross to-night, and it
is not large. Not large, and not strikingly bright. But it was
low down toward the horizon, and it may improve when it gets
up higher in the sky. It is ingeniously named, for it looks just
as a cross would look if it looked like something else. But
that description does not describe; it is too vague, too general,
too indefinite. It does after a fashion suggest a cross — a cross
that is out of repair — or out of drawing; not correctly shaped.
It is long, with a short cross-bar, and the cross-bar is canted
out of the straight line.

It consists of four large stars and
one little one. The little one is out
of line and further damages the shape.
It should have been placed at the inter-
section of the stem and the cross-bar.

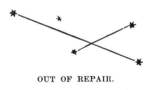

OUT OF REPAIR.

If you do not draw an imaginary line from star to star it does
not suggest a cross — nor anything in particular.

One must ignore the little star, and leave it out of the combination — it confuses everything. If you leave it out, then you
can make out of the four stars a sort of cross — out of true; or
a sort of kite — out of true; or a sort of coffin — out of true.

Constellations have always been troublesome things to name. If you give one of them a fanciful name, it will always refuse to live up to it; it will always persist in not resembling  the thing it has been named for. Ultimately, to satisfy the public, the fanciful name has to be discarded for a common-sense one, a manifestly descriptive one. The Great Bear remained the Great Bear —and unrecognizable as such — for thousands of years; and people complained about it all the time, and quite properly; but as soon as it became the property of the United States, Congress changed it to the Big Dipper, and now every-

SOUTHERN CROSS.

body is satisfied, and there is no more talk about riots. I would not change the Southern Cross to the Southern Coffin, I would change it to the Southern Kite; for up there in the general emptiness is the proper home of a kite, but not for coffins and crosses and dippers. In a little while, now — I cannot tell exactly how long it will be — the globe will belong to the English-speaking race; and of course the skies also. Then the constellations will be re-organized, and polished up, and re-named — the most of them " Victoria," I reckon, but this one will sail thereafter as the Southern Kite, or go out of business. Several towns and things, here and there, have been named for Her Majesty already.

In these past few days we are plowing through a mighty Milky Way of islands. They are so thick on the map that one would hardly expect to find room between them for a canoe;

yet we seldom glimpse one. Once we saw the dim bulk of a couple of them, far away, spectral and dreamy things; members of the Horne — Alofa and Fortuna. On the larger one are two rival native kings — and they have a time together. They are Catholics; so are their people. The missionaries there are French priests.

From the multitudinous islands in these regions the "recruits" for the Queensland plantations were formerly drawn; are still drawn from them, I believe. Vessels fitted up like old-time slavers came here and carried off the natives to serve as laborers in the great Australian province. In the beginning it was plain, simple manstealing, as per testimony of the missionaries. This has been denied, but not disproven. Afterward it was forbidden by law to "recruit" a native without his consent, and governmental agents were sent in all recruiting vessels to see that the law was obeyed — which they did, according to the recruiting people; and which they sometimes didn't, according to the missionaries. A man could be lawfully recruited for a three-years term of service; he could volunteer for another term if he so chose; when his time was up he could return to his island. And would also have the means to do it; for the government required the employer to put money in its hands for this purpose before the recruit was delivered to him.

Captain Wawn was a recruiting shipmaster during many years. From his pleasant book one gets the idea that the recruiting business was quite popular with the islanders, as a rule. And yet that did not make the business wholly dull and uninteresting; for one finds rather frequent little breaks in the monotony of it — like this, for instance:

"The afternoon of our arrival at Leper Island the schooner was lying almost becalmed under the lee of the lofty central portion of the island, about three-quarters of a mile from the shore. The boats were in sight at some dis tance. The recruiter-boat had run into a small nook on the rocky coast, under a high bank, above which stood a solitary hut backed by dense forest. The government agent and mate in the second boat lay about 400 yards to the westward.

"Suddenly we heard the sound of firing, followed by yells from the natives on shore, and then we saw the recruiter-boat push out with a seemingly diminished crew. The mate's boat pulled quickly up, took her in tow, and presently brought her alongside, all her own crew being more or less hurt. It seems the natives had called them into the place on pretence of friendship. A crowd gathered about the stern of the boat, and several fellows even got into her. All of a sudden our men were attacked with clubs and tomahawks. The recruiter escaped the first blows aimed at him, making play with his fists until he had an opportunity to draw his revolver. 'Tom Sayers,' a Maré man, received a tomahawk blow on the head which laid the scalp open but did not penetrate his skull, fortunately. 'Bobby Towns,' another Maré boatman, had both his thumbs cut in warding off blows, one of them being so nearly severed from the hand that the doctors had to finish the operation. Lihu, a Lifu boy, the recruiter's special attendant, was cut and pricked in various places, but nowhere seriously. Jack, an unlucky Tanna recruit, who had been engaged to act as boatman, received an arrow through his forearm, the head of which — a piece of bone seven or eight inches long — was still in the limb, protruding from both sides, when the boats returned. The recruiter himself would have got off scot-free had not an arrow pinned one of his fingers to the loom of the steering-oar just as they were getting off. The fight had been short but sharp. The enemy lost two men, both shot dead."

The truth is, Captain Wawn furnishes such a crowd of instances of fatal encounters between natives and French and English recruiting-crews (for the French are in the business for the plantations of New Caledonia), that one is almost persuaded that recruiting is not thoroughly popular among the islanders; else why this bristling string of attacks and bloodcurdling slaughter? The captain lays it all to "Exeter Hall influence." But for the meddling philanthropists, the native fathers and mothers would be fond of seeing their children carted into exile and now and then the grave, instead of weeping about it and trying to kill the kind recruiters.

CHAPTER VI.

He was as shy as a newspaper is when referring to its own merits.
— Puddn'head Wilson's New Calendar.

CAPTAIN Wawn is crystal-clear on one point: He does not approve of missionaries. They obstruct his business. They make "Recruiting," as he calls it ("Slave-Catching," as *they* call it in their frank way) a trouble when it ought to be just a picnic and a pleasure excursion. The missionaries have their opinion about the manner in which the Labor Traffic is conducted, and about the recruiter's evasions of the law of the Traffic, and about the traffic itself: and it is distinctly uncomplimentary to the Traffic and to everything connected with it, including the law for its regulation. Captain Wawn's book is of very recent date; I have by me a pamphlet of still later date — hot from the press, in fact — by Rev. Wm. Gray, a missionary; and the book and the pamphlet taken together make exceedingly interesting reading, to my mind.

Interesting, and easy to understand — except in one detail, which I will mention presently. It is easy to understand why the Queensland sugar planter should want the Kanaka recruit: he is cheap. Very cheap, in fact. These are the figures paid by the planter: £20 to the recruiter for getting the Kanaka — or "catching" him, as the missionary phrase goes; £3 to the Queensland government for "superintending" the importation; £5 deposited with the Government for the Kanaka's passage home when his three years are up, in case he shall live that long; about £25 to the Kanaka himself for three years' wages and clothing; total payment for the use of a man three years, £53; or, including diet, £60. Altogether, a hundred dollars a year. One can understand why the recruiter is fond of the business; the recruit cheap presents cruit's relatives, not self), and the re- costs him a few (given to the re- to the recruit him- cruit is worth £20 to the recruiter when delivered in Queensland. All this is clear enough; but the thing that is not clear is, what there is about it all to persuade the re- cruit. He is young and brisk; life at home in his beauti- ful island is one lazy, long holiday to him; or if he wants to work he

THE KANAKA'S DEPARTURE.

can turn out a couple of bags of copra per week and sell it for four or five shillings a bag. In Queensland he must get up at dawn and work from eight to twelve hours a day in the canefields — in a much hotter climate than he is used to — and get less than four shillings a week for it.

I cannot understand his willingness to go to Queensland. It is a deep puzzle to me. Here is the explanation, from the planter's point of view; at least I gather from the missionary's pamphlet that it is the planter's:

"When he comes from his home he is a savage, pure and simple. He feels no shame at his nakedness and want of adornment. When he returns home he does so well dressed, sporting a Waterbury watch, collars, cuffs, boots, and jewelry. He takes with him one or more boxes* well filled with clothing, a musical instrument or two, and perfumery and other articles of luxury he has learned to appreciate."

For just one moment we have a seeming flash of com-

prehension o f t h e Kanaka's reason for exiling himself: he goes away to acquire *civilization.* Yes, he was naked and not ashamed, now he is clothed and knows how to be ashamed; he was unenlightened, now he has a Water-bury watch; he was unrefined, now he has jewelry, and some-thing to make him smell good; he was a

THE KANAKA'S RETURN.

nobody, a provincial, now he has been to far countries and can show off.

It all looks plausible — for a moment. Then the mission-ary takes hold of this explanation and pulls it to pieces, and dances on it, and damages it beyond recognition.

"Admitting that the foregoing description is the average one, the average sequel is this : The cuffs and collars, if used at all, are carried off by youngsters, who fasten them round the leg, just below the knee, as ornaments. The Waterbury, broken and dirty, finds its way to the trader, who gives a trifle for it ; or the inside is taken out, the wheels strung on a thread and hung round the neck. Knives, axes, calico, and handker-chiefs are divided among friends, and there is hardly one of these apiece. The boxes, the keys often lost on the road home, can be bought for 2s. 6d. They are to be seen rotting outside in almost any shore village on Tanna. (I speak of what I have seen.) A returned Kanaka has been furiously angry

* "Box " is English for trunk.

with me because I would not buy his trousers, which he declared were just
my fit. He sold them afterwards to one of my Aniwan teachers for 9d. worth
of tobacco — a pair of trousers that probably cost him 8s. or 10s. in Queensland. A coat or shirt is handy for cold weather. The white handkerchiefs,
the 'senet' (perfumery), the umbrella, and perhaps the hat, are kept. The
boots have to take their chance, if they do not happen to fit the copra trader.
'Senet' on the hair, streaks of paint on the face, a dirty white handkerchief
round the neck, strips of turtle shell in the ears, a belt, a sheath and knife,
and an umbrella constitute the rig of returned Kanaka at home the day after
landing."

A hat, an umbrella, a belt, a neckerchief. Otherwise
stark naked. All in a day the hard-earned " civilization "
has melted away to this. And even these perishable things
must presently go. Indeed, there is but a single detail of his
civilization that can be depended on to stay by him : according
to the missionary, he has learned to swear. This is art, and
art is long, as the poet says.

In all countries the laws throw light upon the past. The
Queensland law for the regulation of the Labor Traffic is a
confession. It is a confession that the evils charged by the
missionaries upon the traffic had existed in the past, and that
they still existed when the law was made. The missionaries
make a further charge : that the law is evaded by the recruiters, and that the Government Agent sometimes helps them
to do it. Regulation 31 reveals two things : that sometimes a
young fool of a recruit gets his senses back, after being persuaded to sign away his liberty for three years, and dearly
wants to get out of the engagement and stay at home with his
own people ; and that threats, intimidation, and force are used
to keep him on board the recruiting-ship, and to hold him to
his contract. Regulation 31 forbids these coercions. The law
requires that he shall be allowed to go free ; and another
clause of it requires the recruiter to set him ashore — per boat,
because of the prevalence of sharks. Testimony from Rev.
Mr. Gray :

"There are 'wrinkles' for taking the penitent Kanaka. My first experience of the Traffic was a case of this kind in 1884. A vessel anchored just out of sight of our station, word was brought to me that some boys were stolen, and the relatives wished me to go and get them back. The facts were, as I found, that six boys had recruited, had *rushed* into the boat, the Government Agent informed me. They had all 'signed'; and, said the Government Agent, 'on board they shall remain.' I was assured that the six boys were of age and willing to go. Yet on getting ready to leave the ship I found four of the lads ready to come ashore in the boat! This I forbade. One of them jumped into the water and persisted in coming ashore in my boat. When appealed to, the Government Agent suggested that we go and leave him to be picked up by the ship's boat, a quarter mile distant at the time!"

The law and the missionaries feel for the repentant recruit — and properly, one may be permitted to think, for he is only a youth and ignorant and persuadable to his hurt — but sympathy for him is not kept in stock by the recruiter. Rev. Mr. Gray says:

"A captain many years in the traffic explained to me how a penitent could be taken. 'When a boy jumps overboard we just take a boat and pull ahead of him, then lie between him and the shore. If he has not tired himself swimming, and passes the boat, keep on heading him in this way. The dodge rarely fails. The boy generally tires of swimming, gets into the boat of his own accord, and goes quietly on board."

Yes, exhaustion is likely to make a boy quiet. If the distressed boy had been the speaker's son, and the captors savages, the speaker would have been surprised to see how differently the thing looked from the new point of view; however, it is not our custom to put ourselves in the other person's place. Somehow there is something pathetic about that disappointed young savage's resignation. I must explain, here, that in the traffic dialect, "boy" does not always mean boy; it means a youth above sixteen years of age. That is by Queensland law the age of consent, though it is held that recruiters allow themselves some latitude in guessing at ages.

Captain Wawn of the free spirit chafes under the annoyance of "cast-iron regulations." They and the missionaries have poisoned his life. He grieves for the good old days, vanished to come no more. See him weep; hear him cuss between the lines!

"For a long time we were allowed to apprehend and detain all deserters who had signed the agreement on board ship, but the 'cast-iron' regulations of the Act of 1884 put a stop to that, allowing the Kanaka to sign the agreement for three years' service, travel about in the ship in receipt of the regular rations, cadge all he could, and leave when he thought fit, so long as he did not extend his pleasure trip to Queensland."

Rev. Mr. Gray calls this same restrictive cast-iron law a "farce." "There is as much cruelty and injustice done to natives by acts that are legal as by deeds unlawful. The regulations that exist are unjust and inadequate — unjust and inadequate they must ever be." He furnishes his reasons for his position, but they are too long for reproduction here.

However, if the most a Kanaka advantages himself by a three-years course in civilization in Queensland, is a necklace and an umbrella and a showy imperfection in the art of swearing, it must be that *all* the profit of the traffic goes to the white man. This could be twisted into a plausible argument that the traffic ought to be squarely abolished.

However, there is reason for hope that that can be left alone to achieve itself. It is claimed that the traffic will depopulate its sources of supply within the next twenty or thirty years. Queensland is a very healthy place for white people — death-rate 12 in 1,000 of the population — but the Kanaka death-rate is away above that. The vital statistics for 1893 place it at 52; for 1894 (Mackay district), 68. The first six months of the Kanaka's exile are peculiarly perilous for him because of the rigors of the new climate. The death-rate among the new men has reached as high as 180 in the 1,000. In the Kanaka's native home his death-rate is 12 in time of peace, and 15 in time of war. Thus exile to Queensland — with the opportunity to acquire civilization, an umbrella, and a pretty poor quality of profanity — is twelve times as deadly for him as war. Common Christian charity, common humanity, does seem to require, not only that these people be returned to their homes, but that war, pestilence, and famine be introduced among them for their preservation.

Concerning these Pacific isles and their peoples an eloquent prophet spoke long years ago — five and fifty years ago. In fact, he spoke a little too early. Prophecy is a good line of business, but it is full of risks. This prophet was the Right Rev. M. Russell, LL.D., D.C.L., of Edinburgh:

" Is the tide of civilization to roll only to the foot of the Rocky Mountains, and is the sun of knowledge to set at last in the waves of the Pacific ? No ; the mighty day of four thousand years is drawing to its close ; the sun of humanity has performed its destined course ; but long ere its setting rays are extinguished in the west, its ascending beams have glittered on the isles of the eastern seas. . . . And now we see the race of Japhet setting forth to people the isles, and the seeds of another Europe and a second England sown in the regions of the sun. But mark the words of the prophecy : ' He shall dwell in the tents of Shem, and Canaan shall be his servant.' It is not said Canaan shall be his *slave*. To the Anglo-Saxon race is given the scepter of the globe, but there is not given either the lash of the slave-driver or the rack of the executioner. The East will not be stained with the same atrocities as the West ; the frightful gangrene of an enthralled race is not to mar the destinies of the family of Japhet in the Oriental world ; humanizing, not destroying, as they advance ; uniting, not enslaving, the inhabitants with whom they dwell, the British race may," etc., etc.

And he closes his vision with an invocation from Thomson :

" Come, bright Improvement ! on the car of Time,
    And rule the spacious world from clime to clime."

Very well, Bright Improvement has arrived, you see, with her civilization, and her Waterbury, and her umbrella, and her third-quality profanity, and her humanizing-not-destroying machinery, and her hundred-and-eighty-death-rate, and everything is going along just as handsome !

But the prophet that speaks last has an advantage over the pioneer in the business. Rev. Mr. Gray says :

" What I am concerned about is that we as a Christian nation should wipe out these races to enrich ourselves."

And he closes his pamphlet with a grim Indictment which is as eloquent in its flowerless straightforward English as is the hand-painted rhapsody of the early prophet :

" My indictment of the Queensland-Kanaka Labor Traffic is this :

"1. It generally demoralizes and always impoverishes the Kanaka, deprives him of his citizenship, and depopulates the islands fitted to his home.

"2. It is felt to lower the dignity of the white agricultural laborer in Queensland, and beyond a doubt it lowers his wages there.

"3. The whole system is fraught with danger to Australia and the islands on the score of health.

"4. On social and political grounds the continuance of the Queensland-Kanaka Labor Traffic must be a barrier to the true federation of the Australian colonies.

"5. The Regulations under which the Traffic exists in Queensland are inadequate to prevent abuses, and in the nature of things they must remain so.

"6. The whole system is contrary to the spirit and doctrine of the Gospel of Jesus Christ. The Gospel requires us to help the weak, but the Kanaka is fleeced and trodden down.

"7. The bed-rock of this Traffic is that the life and liberty of a black man are of less value than those of a white man. And a Traffic that has grown out of 'slave-hunting' will certainly remain to the end not unlike its origin."

CHAPTER VII.

Truth is the most valuable thing we have. Let us economize it.

—Pudd'nhead Wilson's New Calendar.

FROM DIARY:—For a day or two we have been plowing among an invisible vast wilderness of islands, catching now and then a shadowy glimpse of a member of it. There does seem to be a prodigious lot of islands this year; the map of this region is freckled and fly-specked all over with them. Their number would seem to be uncountable. We are moving among the Fijis now—224 islands and islets in the group. In front of us, to the west, the wilderness stretches toward Australia, then curves upward to New Guinea, and still up and up to Japan; behind us, to the east, the wilderness stretches sixty degrees across the wastes of the Pacific; south of us is New Zealand. Somewhere or other among these myriads Samoa is concealed, and not discoverable on the map. Still, if you wish to go there, you will have no trouble about finding it if you follow the directions given by Robert Louis Stevenson to Dr. Conan Doyle and to Mr. J. M. Barrie. "You go to America, cross the continent to San Francisco, and then it's the second turning to the left." To get the full flavor of the joke one must take a glance at the map.

Wednesday, September 11.—Yesterday we passed close to an island or so, and recognized the published Fiji characteristics: a broad belt of clean white coral sand around the island; back of it a graceful fringe of leaning palms, with native huts nestling cosily among the shrubbery at their bases; back of these a stretch of level land clothed in tropic vegetation; back of that, rugged and picturesque mountains. A detail of the immediate foreground: a mouldering ship perched high up on a reef-bench. This completes the composition, and makes the picture artistically perfect.

In the afternoon we sighted Suva, the capital of the group, and threaded our way into the secluded little harbor—a placid basin of brilliant blue and green water tucked snugly in among the sheltering hills. A few ships rode at anchor in it — one of them a sailing vessel flying the American flag;

SUVA.

and they said she came from Duluth! There's a journey! Duluth is several thousand miles from the sea, and yet she is entitled to the proud name of Mistress of the Commercial Marine of the United States of America. There is only one free, independent, unsubsidized American ship sailing the foreign seas, and Duluth owns it. All by itself that ship is the American fleet. All by itself it causes the American name and power to be respected in the far regions of the globe. All by itself it certifies to the world that the most populous civilized nation in the earth has a just pride in her stupendous stretch of sea-front, and is determined to assert and maintain her rightful place as one of the Great Maritime Powers of the Planet. All by itself it is making foreign eyes familiar with a Flag which they have not seen before for forty years, outside of the museum. For what Duluth has done, in building, equipping, and maintaining at her sole expense the American Foreign Commercial Fleet, and in thus rescuing the American name from shame and lifting it high for the homage of the nations, we owe her a debt of gratitude which our hearts shall confess with quickened beats whenever her name is named henceforth. Many national toasts will die in the lapse of time, but while the flag flies and the Republic survives, they who live under their shelter will still drink this one, standing and uncovered: Health and prosperity to Thee, O Duluth, American Queen of the Alien Seas!

Row-boats began to flock from the shore; their crews were the first natives we had seen. These men carried no overplus of clothing, and this was wise, for the weather was hot. Handsome, great dusky men they were, muscular, clean-limbed, and with faces full of

BOATS CAME FROM THE SHORE.

character and intelligence. It would be hard to find their superiors anywhere among the dark races, I should think.

Everybody went ashore to look around, and spy out the

land, and have that luxury of luxuries to sea-voyagers—a

land-dinner. And there we saw more natives: Wrinkled old women, with their flat mammals flung over their shoulders, or hanging down in front like the cold-weather drip from the molasses-faucet; plump and smily young girls, blithe and content, easy and graceful, a pleasure to look at; young matrons, tall, straight, comely, nobly built, sweeping by with chin up, and a gait incomparable for

IN TOWN.

unconscious stateliness and dignity; majestic young men— athletes for build and muscle— clothed in a loose arrange-ment of daz-

NATIVES.

zling white, with bronze breast and bronze legs naked, and the head a cannon-swab of solid hair combed straight out from the skull and dyed a rich brick-red. Only sixty years ago they were sunk in darkness; now they have the bicycle.

OUT
OF TOWN.

We strolled about the streets of the white folks' little town, and around over the hills by paths and roads among European dwellings and gardens and plantations, and past clumps of hibiscus that made a body blink, the great blossoms were so intensely red; and by and by we stopped to ask an elderly English colonist a question or two, and to sympathize with him con-concerning the torrid weather; but he was surprised, and said:

"This? This is not hot. You ought to be here in the summer time once."

"We supposed that this was summer; it has the ear-marks of it. You could take it to almost any country and deceive people with it. But if it isn't summer, what does it lack?"

"It lacks half a year. This is mid-winter."

I had been suffering from colds for several months, and a sudden change of season, like this, could hardly fail to do me hurt. It brought on an-other cold. It is odd, these sudden jumps from season to season. A fortnight ago we left America in mid-summer, now it is mid-winter; about a week hence we shall arrive in Australia in the spring.

After dinner I found in the billiard-room a resident whom I had known some-where else in the world, and presently made some new friends and drove with them out into the country to visit

THE RIGORS OF WINTER.

his Excellency the head of the State, who was occupying his country residence, to escape the rigors of the winter weather, I suppose, for it was on breezy high ground and much more comfortable than the lower regions, where the town is, and where the winter has full swing, and often sets a person's hair afire when he takes off his hat to bow. There is a noble and beautiful view of ocean and islands and castellated peaks from the governor's high-placed house, and its immediate surroundings lie drowsing in that dreamy repose and serenity which are the charm of life in the Pacific Islands.

One of the new friends who went out there with me was a large man, and I had been admiring his size all the way. I was still admiring it as he stood by the governor on the veranda, talking; then the Fijian butler stepped out there to announce tea, and dwarfed him. Maybe he did not quite dwarf him, but at any rate the contrast was quite striking. Perhaps that dark giant was a king in a condition of political suspension. I think that in the talk there on the veranda it was said that in Fiji, as in the Sandwich Islands, native kings and chiefs are of much grander size and build than the commoners. This man was clothed in flowing white vestments, and they were just the thing for him; they comported well with his great stature and his kingly port and dignity. European clothes would have degraded him and made him commonplace. I know that, because they do that with everybody that wears them.

It was said that the old-time devotion to chiefs and reverence for their persons still survive in the native commoner, and in great force. The educated young gentleman who is chief of the tribe that live in the region about the capital dresses in the fashion of high-class European gentlemen, but even his clothes cannot damn him in the reverence of his people. Their pride in his lofty rank and ancient lineage lives on, in spite of his lost authority and the evil magic of his tailor. He has no need to defile himself with work, or trouble his heart with the sordid cares of life; the tribe will see to it that he shall not want, and that he shall hold up his head and live like a gentleman. I had a glimpse of him down in the town. Perhaps he is a descendant of the last king — the king with the difficult name whose memory is preserved by a notable monument of cut-stone which one sees in the enclosure in the middle of the town.

Thakombau — I remember, now; that is the name. It is easier to preserve it on a granite block than in your head.

Fiji was ceded to England by this king in 1858. One of the gentlemen present at the governor's quoted a remark made by the king at the time of the session — a neat retort, and with a touch of pathos in it, too. The English Commissioner had offered a crumb of comfort to Thakombau by saying that the transfer of the kingdom to Great Britain was merely " a sort of hermit-crab formality, you know." " Yes," said poor Thakombau, " but with this difference — the crab moves into an unoccupied shell, but mine isn't."

However, as far as I can make out from the books, the King was between the devil and the deep sea at the time, and hadn't much choice. He owed the United States a large debt — a debt which he could pay if allowed time, but time was denied him. He must pay up right away or the warships would be upon him. To protect his people from this disaster he ceded his country to Britain, with a clause in the contract providing for the ultimate payment of the American debt.

In old times the Fijians were fierce fighters; they were very religious, and worshiped idols; the big chiefs were proud and haughty, and they were men of great style in many ways; all chiefs had several wives, the biggest chiefs sometimes had as many as fifty; when a chief was dead and ready for burial, four or five of his wives were strangled and put into the grave with him. In 1804 twenty-seven British convicts escaped from Australia to Fiji, and brought guns and ammunition with them. Consider what a power they were, armed like that, and what an opportunity they had. If they had been energetic men and sober, and had had brains and known how to use them, they could have achieved the sovereignty of the archipelago — twenty-seven kings and each with eight or nine islands under

7

his scepter. But nothing came of this chance. They lived worthless lives of sin and luxury, and died without honor — in most cases by violence. Only one of them had any ambition; he was an Irishman named Connor. He tried to raise a family of fifty children, and scored forty-eight. He died lamenting his failure. It was a foolish sort of avarice. Many a father would have been rich enough with forty.

It is a fine race, the Fijians, with brains in their heads, and an inquiring turn of mind. It appears that their savage ancestors had a doctrine of immortality in their scheme of religion — with limitations. That is to say, their dead friend would go to a happy hereafter if he could be accumulated, but not otherwise. They drew the line; they thought that the missionary's doctrine was too sweeping, too comprehensive. They called his attention to certain facts. For instance, many of their friends had been devoured by sharks; the sharks, in their turn, were caught and eaten by other men; later, these men were captured in war, and eaten by the enemy. The original persons had entered into the composition of the sharks; next, they and the sharks had become part of the flesh and blood and bone of the cannibals. How, then, could the particles of the original men be searched out from the final conglomerate and put together again? The inquirers were full of doubts, and considered that the missionary had not examined the matter with the gravity and attention which so serious a thing deserved.

The missionary taught these exacting savages many valuable things, and got from them one — a very dainty and poetical idea: Those wild and ignorant poor children of Nature believed that the flowers, after they perish, rise on the winds and float away to the fair fields of heaven, and flourish there forever in immortal beauty!

CHAPTER VIII.

It could probably be shown by facts and figures that there is no distinctly native American criminal class except Congress.
— *Pudd'nhead Wilson's New Calendar.*

WHEN one glances at the map the members of the stupendous island wilderness of the Pacific seem to crowd upon each other; but no, there is no crowding, even in the center of a group; and between groups there are lonely wide deserts of sea. Not everything is known about the islands, their peoples and their languages. A startling reminder of this is furnished by the fact that in Fiji, twenty years ago, were living two strange and solitary beings who came from an unknown country and spoke an unknown language. "They were picked up by a passing vessel *many hundreds of miles from any known land*, floating in the same tiny canoe in which they had been blown out to sea. When found they were but skin and bone. No one could understand what they said, and they have never named their country; or, if they have, the name does not correspond with that of any island on any chart. They are now fat and sleek, and as happy as the day is long. In the ship's log there is an entry of the latitude and longitude in which they were found, and this is probably all the clue they will ever have to their lost homes." *

What a strange and romantic episode it is; and how one is tortured with curiosity to know whence those mysterious creatures came, those Men Without a Country, errant waifs

---

* Forbes's "Two Years in Fiji."

who cannot name their lost home, wandering Children of Nowhere.

Indeed, the Island Wilderness is the very home of romance and dreams and mystery. The loneliness, the solemnity, the beauty, and the deep repose of this wilderness have a charm which is all their own for the bruised spirit of men who have fought and failed in the struggle for life in the great world; and for men who have been hunted out of the great world for crime; and for other men who love an easy and indolent existence; and for others who love a roving free life, and stir and change and adventure; and for yet others who love an easy and comfortable career of trading and money-getting, mixed with plenty of loose matrimony by purchase, divorce without trial or expense, and limitless spreeing thrown in to make life ideally perfect.

We sailed again, refreshed.

The most cultivated person in the ship was a young English, man whose home was in New Zealand. He was a naturalist. His learning in his specialty was deep and thorough, his interest in his subject amounted to a passion, he had an easy gift of speech; and so, when he talked about animals it was a pleasure to listen to him. And profitable, too, though he was sometimes difficult to understand because now and then he used scientific technicalities which were above the reach of some of us. They were pretty sure to be above my reach, but as he was quite willing to explain them I always made it a point to get him to do it. I had a fair knowledge of his subject — layman's knowledge — to begin with, but it was his teachings which crystalized it into scientific form and clarity — in a word, gave it value.

His special interest was the fauna of Australasia, and his knowledge of the matter was as exhaustive as it was accurate. I already knew a good deal about the rabbits in Australasia and their marvelous fecundity, but in my talks with him I found that my estimate of the great hindrance and obstruction inflicted by the rabbit pest upon traffic and travel was far short of the facts. He told me that the first pair of rabbits imported into Australasia bred so wonderfully that within six months rabbits were so thick in the land that people had to dig trenches through them to get from town to town.

He told me a great deal about worms, and the kangaroo, and other coleoptera, and said he knew the history and ways of all such pachydermata. He said the kangaroo had pockets, and carried its young in them when it couldn't get apples. And he said that the emu was as big as an ostrich, and looked like one, and had an amorphous appetite and would eat bricks. Also, that the dingo was not a dingo at all, but just a wild dog; and that the only difference between a dingo and a dodo was that neither of them barked; otherwise they were just the same.

He said that the only game-bird in Australia was the wombat, and the only song-bird the larrikin, and that both were protected by government. The most beautiful of the native birds was the bird of Paradise. Next came the two kinds of lyres; not spelt the same. He said the one kind was dying out, the other thickening up. He explained that the "Sundowner" was not a bird, it was a man; sundowner was merely the Australian equivalent of our word, tramp. He is a loafer, a hard drinker, and a sponge. He tramps across the country in the sheep-shearing season, pretending to look for work; but he always times himself to arrive at a sheep-run just at sundown, when the day's labor ends; all he wants is whisky and supper and bed and breakfast; he gets them and then disappears. The naturalist spoke of the bell bird, the creature that at short intervals all day rings out its mellow and exquisite peal from the deeps of the forest. It is the favorite and best friend of the weary and thirsty sundowner ; for he knows that wherever the bell bird is, there is water ; and he goes somewhere else. The naturalist said that the oddest bird in Australasia was the Laughing Jackass, and the biggest the now extinct Great Moa.

The Moa stood thirteen feet high, and could step over an ordinary man's head or kick his hat off ; and his head, too, for that matter. He said it was wingless, but a swift runner. The natives used to ride it. It could make forty miles an hour, and keep it up for four hundred miles and come out reasonably fresh. It was still in existence when the railway was introduced into New Zealand ; still in existence, and carrying the mails. The railroad began with the same schedule it has now : two expresses a week — time, twenty miles an hour. The company exterminated the moa to get the mails.

Speaking of the indigenous coneys and bactrian camels, the naturalist said that the coniferous and bacteriological output of Australasia was remarkable for its many and curious departures from the accepted laws governing these species of tubercles, but that in his opinion Nature's fondness for dabbling in the erratic was most notably exhibited in that curious combination of bird, fish, amphibian, burrower, crawler, quadruped, and Christian called the Ornithorhyncus — grotesquest of animals, king of the animalculæ of the world for versatility of character and make-up. Said he —

"You can call it anything you want to, and be right. It is a fish, for it lives in the river half the time ; it is a land animal, for it resides on the land half the time ; it is an amphibian, since it likes both and does not know which it prefers ; it is a hybernian, for when times are dull and nothing much going on it buries itself under the mud at the bottom of a puddle and hybernates there a couple of weeks at a time ; it is a kind of duck, for it has a duck-bill and four webbed paddles ; it is a fish and quadruped together, for in the water it swims with the paddles and on shore it paws itself across country with them ; it is a kind of seal, for it has a seal's fur ; it is carnivorous, herbivorous, insectivorous, and vermifuginous, for it eats fish and grass and butterflies, and in the season digs worms out of the mud and devours them ; it is clearly

OFF GOES HIS HEAD.

a bird, for it lays eggs,
and hatches them ; it is
clearly a mammal, for it
nurses its young ; and it
is manifestly a kind of
Christian, for it keeps the
Sabbath when there is
anybody around, and
when there isn't, doesn't.
It has all the tastes there
are except refined ones,
it has all the habits there
are except good ones.

"It is a survival — a
survival of the fittest.
Mr. Darwin invented the
theory that goes by that
name, but the Ornitho-
rhyncus was the first to
put it to actual experi-
ment and prove that it
could be done. Hence it
should have as much of
the credit as Mr. Darwin
It was never in the Ark ;
you will find no mention
of it there ; it nobly
stayed out and worked the
theory. Of all creatures
in the world it was the
only one properly equip-
ped for the test. The
Ark was thirteen months
afloat, and all the globe
submerged ; no land visi-
ble above the flood, no
vegetation, no food for a
mammal to eat, nor water
for a mammal to drink ;
for all mammal food was
destroyed, and when the
pure floods from heaven
and the salt oceans of the
earth mingled their waters
and rose above the moun-
tain tops, the result was
a drink which no bird or

WAS NEVER IN THE ARK.

beast of ordinary construction could use and live.   But this combination was nuts for the Ornithorhyncus, if I may use a term like that without offense.   Its river home had always been salted by the flood tides of the sea.   On the face of the Noachian deluge innumerable forest trees were floating.   Upon these the Ornithorhyncus voyaged in peace ; voyaged fro.n clime to clime, from hemisphere to hemisphere, in contentment and comfort, in virile interest in the constant change of scene, in humble thankfulness for its privileges, in ever-increasing enthusiam in the development of the great theory upon whose validity it had staked its life, its fortunes, and its sacred honor, if I may use such expressions without impropriety in connection with an episode of this nature.

"It lived the tranquil and luxurious life of a creature of independent means.   Of things actually necessary to its existence and its happiness not a detail was wanting.   When it wished to walk, it scrambled along the tree-trunk ; it mused in the shade of the leaves by day, it slept in their shelter by night ; when it wanted the refreshment of a swim, it had it ; it ate leaves when it wanted a vegetable diet, it dug under the bark for worms and grubs; when it wanted fish it caught them, when it wanted eggs it laid them.   If the grubs gave out in one tree it swam to another ; and as for fish, the very opulence of the supply was an embarrassment.   And finally, when it was thirsty it smacked its chops in gratitude over a blend that would have slain a crocodile.

"When at last, after thirteen months of travel and research in all the Zones it went aground on a mountain-summit, it strode ashore, saying in its heart, 'Let them that come after me invent theories and dream dreams about the Survival of the Fittest if they like, but I am the first that has *done* it !

"This wonderful creature dates back like the kangaroo and many other Australian hydrocephalous invertebrates, to an age long anterior to the advent of man upon the earth ; they date back, indeed, to a time when a causeway hundreds of miles wide, and thousands of miles long, joined Australia to Africa, and the animals of the two countries were alike, and all belonged to that remote geological epoch known to science as the Old Red Grindstone Post-Pleosaurian.   Later the causeway sank under the sea ; subterranean convulsions lifted the African continent a thousand feet higher than it was before, but Australia kept her old level.   In Africa's new climate the animals necessarily began to develop and shade off into new forms and families and species, but the animals of Australia as necessarily remained stationary, and have so remained until this day.   In the course of some millions of years the African Ornithorhyncus developed and developed and developed, and sluffed off detail after detail of its make-up until at last the creature became wholly disintegrated and scattered.   Whenever you see a bird or a beast or a seal or an otter in Africa you know that he is merely a sorry surviving fragment of that sublime original of whom I have been speaking — that creature which was everything in general and nothing in particular — the opulently endowed *e pluribus unum* of the animal world.

"Such is the history of the most hoary, the most ancient, the most venerable creature that exists in the earth to-day — *Ornithorhyncus Platypus Extraordinariensis* — whom God preserve ! "

When he was strongly moved he could rise and soar like that with ease. And not only in the prose form, but in the poetical as well. He had written many pieces of poetry in his time, and these manuscripts he lent around among the passengers, and was willing to let them be copied. It seemed to me that the least technical one in the series, and the one which reached the loftiest note, perhaps, was his

INVOCATION.

"Come forth from thy oozy couch,
   O Ornithorhyncus dear !
And greet with a cordial claw
   The stranger that longs to hear

"From thy own own lips the tale
   Of thy origin all unknown:
Thy misplaced bone where flesh should be
   And flesh where should be bone ;

"And fishy fin where should be paw,
   And beaver-trowel tail,
And snout of beast equip'd with teeth
   Where gills ought *to* prevail.

"Come, Kangaroo, the good and true !
   Foreshortened as to legs,
And body tapered like a churn,
   And sack marsupial, i' fegs,

"And tells us why you linger here,
   Thou relic of a vanished time,
When all your friends as fossils sleep,
   Immortalized in lime ! "

Perhaps no poet is a conscious plagiarist; but there seems to be warrant for suspecting that there is no poet who is not at one time or another an unconscious one. The above verses are indeed beautiful, and, in a way, touching; but there is a haunting something about them which unavoidably suggests the Sweet Singer of Michigan. It can hardly be doubted that the author had read the works of that poet and been impressed by them. It is not apparent that he has borrowed from them any word or yet any phrase, but the style and swing

and mastery and melody of the Sweet Singer all are there.
Compare this Invocation with "Frank Dutton"—particularly
stanzas first and seventeenth — and I think the reader will feel
convinced that he who wrote the one had read the other : *

I.

"Frank Dutton was as fine a lad
    As ever you wish to see,
And he was drowned in Pine Island Lake
    On earth no more will he be,
His age was near fifteen years,
    And he was a motherless boy,
He was living with his grandmother
    When he was drowned, poor boy.

XVII.

" He was drowned on Tuesday afternoon,
    On Sunday he was found,
And the tidings of that drowned boy
    Was heard for miles around.
His form was laid by his mother's side,
    Beneath the cold, cold ground,
His friends for him will drop a tear
    When they view his little mound."

---

* The Sentimental Song Book.   By Mrs. Julia Moore, p. 36.

THE NATURALIST.

CHAPTER IX.

It is your human environment that makes climate.

—Pudd'nhead Wilson's New Calendar.

SEPT. 15 — *Night.* Close to Australia now. Sydney 50 miles distant.

That note recalls an experience. The passengers were sent for, to come up in the bow and see a fine sight. It was very dark. One could not follow with the eye the surface of the sea more than fifty yards in any direction — it dimmed away and became lost to sight at about that distance from us. But if you patiently gazed into the darkness a little while, there was a sure reward for you. Presently, a quarter of a mile away you would see a blinding splash or explosion of light on the water — a flash so sudden and so astonishingly brilliant that it would make you catch your breath; then that blotch of light would instantly extend itself and take the corkscrew shape and imposing length of the fabled sea-serpent, with every curve of its body and the " break " spreading away from its head, and the wake following behind its tail clothed in a fierce splendor of living fire. And my, but it was coming at a lightning gait! Almost before you could think, this monster of light, fifty feet long, would go flaming and storming by, and suddenly disappear. And out in the distance whence he came you would see another flash; and another and another and another, and see them turn into sea-serpents on the instant; and once sixteen flashed up at the same time and came tearing towards us, a swarm of wiggling curves, a moving conflagration, a vision of bewildering beauty, a spectacle of fire and energy whose equal the most of those people will not see again until after they are dead.

It was porpoises — porpoises aglow with phosphorescent light. They presently collected in a wild and magnificent jumble under the bows, and there they played for an hour, leaping and frollicking and carrying on, turning summersaults in front of the stem or across it and never getting hit, never making a miscalculation, though the stem missed them only about an inch, as a rule. They were porpoises of the ordinary length — eight or ten feet — but every twist of their bodies sent a long procession of united and glowing curves astern. That fiery jumble was an enchanting thing to look at, and we stayed out the performance; one cannot have such a show as that twice in a lifetime. The porpoise is the kitten of the sea; he never has a serious thought, he cares for nothing but fun and play. But I think I never saw him at his winsomest until that night. It was near a center of civilization, and he could have been drinking.

By and by, when we had approached to somewhere within thirty miles of Sydney Heads the great electric light that is posted on one of those lofty ramparts began to show, and in time the little spark grew to a great sun and pierced the firmament of darkness with a far-reaching sword of light.

Sydney Harbor is shut in behind a precipice that extends some miles like a wall, and exhibits no break to the ignorant stranger. It has a break in the middle, but it makes so little show that even Captain Cook sailed by it without seeing it. Near by that break is a false break which resembles it, and which used to make trouble for the mariner at night, in the early days before the place was lighted. It caused the memorable disaster to the *Duncan Dunbar*, one of the most pathetic tragedies in the history of that pitiless ruffian, the sea. The ship was a sailing vessel; a fine and favorite passenger packet, commanded by a popular captain of high reputation. She was due from England, and Sydney was waiting, and counting the hours; counting the hours, and making ready to give her a heart-stirring welcome; for she was bringing back a great company of mothers and daughters, the long-missed light and bloom of life of Sydney homes; daughters that had been years absent at school, and mothers that had been with them all that time watching over them. Of all the world only India and Australasia have by custom freighted ships and fleets with their hearts, and know the tremendous meaning of that phrase; only they know what the waiting is like when this freightage is entrusted to the fickle winds, not steam, and what the joy is like when the ship that is returning this treasure comes safe to port and the long dread is over.

On board the *Duncan Dunbar*, flying toward Sydney Heads in the waning afternoon, the happy home-comers made busy preparation, for it was not doubted that they would be in the arms of their friends before the day was done; they put away their sea-going clothes and put on clothes meeter for the meeting, their richest and their loveliest, these poor brides of the grave. But the wind lost force, or there was a miscalculation, and before the Heads were sighted the darkness came on. It was said that ordinarily the captain would have made a safe offing and waited for the morning; but this was no ordinary occasion; all about him were appealing faces, faces pathetic with disappointment. So his sympathy moved him to try the dangerous passage in the dark. He had entered the Heads seventeen times, and believed he knew the ground. So he steered straight for the false opening, mistaking it for the true one. He did not find out that he was wrong until it was too late. There was no saving the ship. The great seas swept her in and crushed her to splinters and rubbish upon the rock tushes at the base of the precipice. Not one of all that fair and gracious company was ever seen again alive. The tale is told to every stranger that passes the spot, and it will continue to be told to all that come, for generations; but it will never grow old, custom cannot stale it, the heart-break that is in it can never perish out of it.

There were two hundred persons in the ship, and but one survived the disaster. He was a sailor. A huge sea flung him up the face of the precipice and stretched him on a narrow shelf of rock midway between the top and the bottom, and there he lay all night. At any other time he would have lain there for the rest of his life, without chance of discovery; but the next morning the ghastly news swept through Sydney that the *Duncan Dunbar* had gone down in sight of home, and straightway the walls of the Heads were black with mourners; and one of these, stretching himself out over the precipice to spy out what might be seen below, discovered this miraculously preserved relic of the wreck. Ropes were brought and the nearly impossible feat of rescuing the man was accomplished. He was a person with a practical turn of mind, and he hired a hall in Sydney and exhibited himself at sixpence a head till he exhausted the output of the gold fields for that year.

We entered and cast anchor, and in the morning went oh-ing and ah-ing in admiration up through the crooks and turns of the spacious and beautiful harbor — a harbor which is the darling of Sydney and the wonder of the world. It is not surprising that the people are proud of it, nor that they put their enthusiasm into eloquent words. A returning citizen asked me what I thought of it, and I testified with a cordiality which I judged would be up to the market rate. I said it was beautiful — superbly beautiful. Then by a natural impulse I gave God the praise. The citizen did not seem altogether satisfied. He said:

"It *is* beautiful, of course it's beautiful — the Harbor;

but that isn't all of it, it's only half of it; Sydney's the
other half, and it takes both of them together to ring the
supremacy-bell. God made the Harbor, and that's all right;
but Satan made Sydney."

Of course I made an apology; and asked him to convey
it to his friend. He was right about Sydney being half
of it. It would be beautiful without Sydney, but not above
half as beautiful as it is now, with Sydney added. It is shaped
somewhat like an oak-leaf—a roomy sheet of lovely blue
water, with narrow off-shoots of water running up into the
country on both sides between long fingers of land, high
wooden ridges with sides sloped like graves. Handsome
villas are perched here and there on these ridges, snuggling
amongst the foliage, and one catches alluring glimpses of them
as the ship swims by toward the city. The city clothes a cluster
of hills and a ruffle of neighboring ridges with its undulating
masses of masonry, and out of these masses spring towers
and spires and other architectural dignities and grandeurs that break the flowing lines and give picturesqueness to the general effect.

The narrow inlets which I have men-

VIEW IN SYDNEY HARBOR.

·tioned go wandering out into the land everywhere and hiding
themselves in it, and pleasure-launches are always exploring

them with picnic parties on board. It is said by trustworthy people that if you explore them all you will find that you have covered 700 miles of water passage. But there are liars everywhere this year, and they will double that when their works are in good going order.

October was close at hand, spring was come. It was really spring—everybody said so; but you could have sold it for summer in Canada, and nobody would have suspected. It was the very weather that makes our home summers the perfection of climatic luxury; I mean, when you are out in the wood or by the sea. But these people said it was cool, now — a person ought to see Sydney in the summer time if he wanted to know what warm weather is; and he ought to go north ten or fifteen hundred miles if he wanted to know what hot weather is. They said that away up there toward the equator the hens laid fried eggs. Sydney is the place to go to get information about other people's climates. It seems to me that the occupation of Unbiased Traveler Seeking Information is the pleasantest and most irresponsible trade there is. The traveler can always find out anything he wants to, merely by asking. He can get at all the facts, and more. Everybody helps him, nobody hinders him. Anybody who has an old fact in stock that is no longer negotiable in the domestic market will let him have it at his own price. An accumulation of such goods is easily and quickly made. They cost almost nothing and they bring par in the foreign market. Travelers who come to America always freight up with the same old nursery tales that their predecessors selected, and they carry them back and always work them off without any trouble in the home market.

If the climates of the world were determined by parallels of latitude, then we could know a place's climate by its position on the map; and so we should know that the climate of Sydney was the counterpart of the climate of Columbia, S. C., and of Little Rock, Arkansas, since Sydney is about the same distance south of the equator that those other towns are north of it—thirty-four degrees. But no, climate disregards the parallels of latitude. In Arkansas they have a winter; in Sydney they have the name of it, but not the thing itself. I have seen the ice in the Mississippi floating past the mouth of the Arkansas river; and at Memphis, but a little way above, the Mississippi has been frozen over, from bank to bank. But they have never had a cold spell in Sydney which brought the mercury down to freezing point. Once in a mid-winter day there, in the month of July, the mercury went down to 36°, and that remains the memorable "cold day" in the history of the town. No doubt Little Rock has seen it below zero. Once, in Sydney, in mid-summer, about New Year's Day, the mercury went up to 106° in the shade, and that is Sydney's memorable hot day. That would about tally with Little Rock's hottest day also, I imagine. My Sydney figures are taken from a government report, and are trustworthy. In the matter of summer weather Arkansas has no advantage over Sydney, perhaps, but when it comes to winter weather, that is another affair. You could cut up an Arkansas winter into a hundred Sydney winters and have enough left for Arkansas and the poor.

The whole narrow, hilly belt of the Pacific side of New South Wales has the climate of its capital — a mean winter temperature of 54° and a mean summer one of 71°. It is a climate which cannot be improved upon for healthfulness. But the experts say that 90° in New South Wales is harder to bear than 112° in the neighboring colony of Victoria, because the atmosphere of the former is humid, and of the latter dry.

The mean temperature of the southernmost point of New South Wales is the same as that of Nice — 60°— yet Nice is further from the equator by 460 miles than is the former.

But Nature is always stingy of perfect climates; stingier in the case of Australia than usual. Apparently this vast continent has a really good climate nowhere but around the edges.

If we look at a map of the world we are surprised to see how big Australia is. It is about two-thirds as large as the United States was before we added Alaska.

But where as one finds a sufficiently good climate and fertile land almost everywhere in the United States, it seems settled that inside of the Australian border-belt one finds many deserts and in spots a climate which nothing can stand except a few of the hardier kinds of rocks. In effect, Australia is as yet unoccupied. If you take a map of the United States and

leave the Atlantic sea-board States in their places; also the fringe of Southern States from Florida west to the Mouth of the Mississippi; also a narrow, inhabited streak up the Mississippi half-way to its head waters; also a narrow, inhabited border along the Pacific coast: then take a brushful of paint and obliterate the whole remaining mighty stretch of country that lies between the Atlantic States and the Pacific-coast strip, your map will look like the latest map of Australia.

This stupendous blank is hot, not to say torrid; a part of it is fertile, the rest is desert; it is not liberally watered; it has no towns. One has only to cross the mountains of New South Wales and descend into the westward-lying regions to find that he has left the choice climate behind him, and found a new one of a quite different character. In fact, he would not know by the thermometer that he was not in the blistering Plains of India. Captain Sturt, the great explorer, gives us a sample of the heat.

"The wind, which had been blowing all the morning from the N.E., increased to a heavy gale, and I shall never forget its withering effect. I sought shelter behind a large gum-tree, but the blasts of heat were so terrific that I wondered *the very grass did not take fire.* This really was nothing ideal: everything both animate and inanimate gave way before it ; the horses stood with their backs to the wind and their noses to the ground, without the muscular strength to raise their heads ; the birds were mute, and the leaves of the trees under which we were sitting *fell like a snow shower around us.* At noon I took a thermometer graded to 127°, out of my box, and observed that the mercury was up to 125°. Thinking that it had been unduly influenced, I put it in the fork of a tree close to me, sheltered alike from the wind and the sun. I went to examine it about an hour afterwards, when I found the mercury had risen to the top of the instrument and had *burst the bulb,* a circumstance that I believe no traveler has ever before had to record. I cannot find language to convey to the reader's mind an idea of the intense and oppressive nature of the heat that prevailed."

That hot wind sweeps over Sydney sometimes, and brings with it what is called a "dust-storm." It is said that most Australian towns are acquainted with the dust-storm. I think I know what it is like, for the following description by Mr. Gane tallies very well with the alkali dust-storm of Nevada, if you leave out the "shovel" part. Still the shovel part is a pretty important part, and seems to indicate that my Nevada storm is but a poor thing, after all.

"As we proceeded the altitude became less, and the heat proportionately greater until we reached Dubbo, which is only 600 feet above sea-level. It is a pretty town, built on an extensive plain. . . . After the effects of a shower of rain have passed away the surface of the ground crumbles into a thick layer of dust, and occasionally, when the wind is in a particular quarter, *it is lifted bodily from the ground in one long opaque cloud.* In the midst of such a storm nothing can be seen a few yards ahead, and the unlucky person who happens to be out at the time is compelled to seek the nearest retreat at hand. When the thrifty housewife sees in the distance the dark column advancing in a steady whirl towards her house, she closes the doors and windows with all expedition. A drawing-room, the window of which has been carelessly left open during a dust-storm, is indeed an extraordinary sight. A lady who has resided in Dubbo for some years says that the dust lies so thick on the carpet that it is necessary to use a shovel to remove it."

And probably a wagon.   I was mistaken; I have not seen a proper dust-storm.   To my mind the exterior aspects and character of Australia are fascinating things to look at and think about, they are so strange, so weird, so new, so uncommonplace, such a startling and interesting contrast to the other sections of the planet, the sections that are known to us all, familiar to us all.   In the matter of particulars — a detail here, a detail there — we have had the choice climate of New South Wales' sea-coast; we have had the Australian heat as furnished by Captain Sturt; we have had the wonderful dust-storm; and we have considered the phenomenon of an almost empty hot wilderness half as big as the United States, with a narrow belt of civilization, population, and good climate around it.

A DUST STORM.

CHAPTER X.

CAPTAIN Cook found Australia in 1770, and eighteen years later the British Government began to transport convicts to it. Altogether, New South Wales received 83,000 in 53 years. The convicts wore heavy chains; they were ill-fed and badly treated by the officers set over them; they were heavily punished for even slight infractions of the rules; "the cruelest discipline ever known" is one historian's description of their life.*

English law was hard-hearted in those days. For trifling offenses which in our day would be punished by a small fine or a few days' confinement, men, women, and boys were sent to this other end of the earth to serve terms of seven and fourteen years; and for serious crimes they were transported for life. Children were sent to the penal colonies for seven years for stealing a rabbit!

When I was in London twenty-three years ago there was a new penalty in force for diminishing garroting and wife-beating — 25 lashes on the bare back with the cat-o'-nine-tails. It was said that this terrible punishment was able to bring the stubbornest ruffians to terms; and that no man had been found with grit enough to keep his emotions to himself beyond the ninth blow; as a rule the man shrieked earlier. That penalty had a great and wholesome effect upon the garroters and wife-beaters; but humane modern London could not endure it; it

---

* The Story of Australasia. J. S. Laurie.

got its law rescinded. Many a bruised and battered English wife has since had occasion to deplore that cruel achievement of sentimental " humanity."

Twenty-five lashes! In Australia and Tasmania they gave a convict fifty for almost any little offense; and sometimes a brutal officer would add fifty, and then another fifty, and so on, as long as the sufferer could endure the torture and live. In Tasmania I read the entry, in an old manuscript official record, of a case where a convict was given *three hundred* lashes—for stealing some silver spoons. And men got more than that, sometimes. Who handled the cat? Often it was another convict; sometimes it was the culprit's dearest comrade; and he had to lay on with all his might; otherwise he would get a flogging himself for his mercy — for he was under watch — and yet not do his friend any good: the friend would be attended to by another hand and suffer no lack in the matter of full punishment.

The convict life in Tasmania was so unendurable, and suicide so difficult to accomplish that once or twice despairing men got together and drew straws to determine which of them should kill another of the group — this murder to secure death to the perpetrator and to the witnesses of it by the hand of the hangman!

The incidents quoted above are mere hints, mere suggestions of what convict life was like — they are but a couple of details tossed into view out of a shoreless sea of such; or, to change the figure, they are but a pair of flaming steeples photographed from a point which hides from sight the burning city which stretches away from their bases on every hand.

Some of the convicts — indeed, a good many of them — were very bad people, even for that day; but the most of them were probably not noticeably worse than the average of the people they left behind them at home. We must believe this;

we cannot avoid it. We are obliged to believe that a nation that could look on, unmoved, and see starving or freezing women hanged for stealing twenty-six cents' worth of bacon or rags, and boys snatched from their mothers, and men from their families, and sent to the other side of the world for long terms of years for similar trifling offenses, was a nation to whom the term " civilized " could not in any large way be applied. And we must also believe that a nation that knew, during more than forty years, what was happening to those exiles and was still content with it, was not advancing in any showy way toward a higher grade of civilization.

If we look into the characters and conduct of the officers and gentlemen who had charge of the convicts and attended to their backs and stomachs, we must grant again that as between the convict and his masters, and between both and the nation at home, there was a quite noticeable monotony of sameness.

Four years had gone by, and many convicts had come. Respectable settlers were beginning to arrive. These two classes of colonists had to be protected, in case of trouble among themselves or with the natives. It is proper to mention the natives, though they could hardly count they were so scarce. At a time when they had not as yet begun to be much disturbed — not as yet being in the way — it was estimated that in New South Wales there was but one native to 45,000 acres of territory.

People had to be protected. Officers of the regular army did not want this service — away off there where neither honor nor distinction was to be gained. So England recruited and officered a kind of militia force of 1,000 uniformed civilians called the " New South Wales Corps " and shipped it.

This was the worst blow of all. The colony fairly staggered under it. The Corps was an object-lesson of the moral condition of England outside of the jails. The colonists trembled. It was feared that next there would be an importation of the nobility.

In those early days the colony was non-supporting. All the necessaries of life — food, clothing, and all — were sent out from England, and kept in great government store-houses,-and given to the convicts and sold to the settlers — sold at a trifling advance upon cost. The Corps saw its opportunity. Its officers went into commerce, and in a most lawless way. They went to importing rum, and also to manufacturing it in private stills, in defiance of the government's commands and protests. They leagued themselves together and ruled the market; they boycotted the government and the other dealers; they established a close monopoly and kept it strictly in their own hands. When a vessel arrived with spirits, they allowed nobody to buy but themselves, and they forced the owner to sell to them at a price named by themselves — and it was always low enough. They bought rum at an average of two dollars a gallon and sold it at an average of ten. They *made rum the currency of the country* — for there was little or no money — and they maintained their devastating hold and kept the colony under their heel for eighteen or twenty years before they were finally conquered and routed by the government.

Meantime, they had spread intemperance everywhere. And they had squeezed farm after farm out of the settlers' hands for rum, and thus had bountifully enriched themselves. When a farmer was caught in the last agonies of thirst they took advantage of him and sweated him for a drink.

In one instance they sold a man a gallon of rum worth two dollars for a piece of property which was sold some years later for $100,000.

When the colony was about eighteen or twenty years old it was discovered that the land was specially fitted for the wool culture. Prosperity followed, commerce with the world began, by and by rich mines of the noble metals were opened, immigrants flowed in, capital likewise. The result is the great and wealthy and enlightened commonwealth of New South Wales.

It is a country that is rich in mines, wool ranches, trams, railways, steamship lines, schools, newspapers, botanical gardens, art galleries, libraries, museums, hospitals, learned societies; it is the hospitable home of every species of culture and of every species of material enterprise, and there is a church at every man's door, and a race-track over the way.

NEW SOUTH WALES CORPS.

CHAPTER XI.

We should be careful to get out of an experience only the wisdom that is in it — and stop there; lest we be like the cat that sits down on a hot stove-lid. She will never sit down on a hot stove-lid again — and that is well; but also she will never sit down on a cold one any more.-- *Pudd'nhead Wilson's New Calendar*.

ALL English-speaking colonies are made up of lavishly hospitable people, and New South Wales and its capital are like the rest in this. The English-speaking colony of the United States of America is always called lavishly hospitable by the English traveler. As to the other English-speaking colonies throughout the world from Canada all around, I know by experience that the description fits them. I will not go more particularly into this matter, for I find that when writers try to distribute their gratitude here and there and yonder by detail they run across difficulties and do some ungraceful stumbling.

Mr. Gane (" New South Wales and Victoria in 1885 "), tried to distribute his gratitude, and was not lucky :

"The inhabitants of Sydney are renowned for their hospitality. The treatment which we experienced at the hands of this generous-hearted people will help more than anything else to make us recollect with pleasure our stay amongst them. In the character of hosts and hostesses they excel. The ' new chum ' needs only the acquaintanceship of one of their number, and he becomes at once the happy recipient of numerous complimentary invitations and thoughtful kindnesses. Of the towns it has been our good fortune to visit, none have portrayed home so faithfully as Sydney."

Nobody could say it finer than that. If he had put in his cork then, and stayed away from Dubbo — but no; heedless man, he pulled it again. Pulled it when he was away along in his book, and his memory of what he had said about Sydney had grown dim :

"We cannot quit the promising town of Dubbo without testifying, in warm praise, to the kind-hearted and hospitable usages of its inhabitants. Sydney, though well deserving the character it bears of its kindly treatment of strangers, possesses a little formality and reserve. In Dubbo, on the contrary, though the same congenial manners prevail, there is a pleasing degree of respectful familiarity which gives the town a homely comfort not often met with elsewhere. In laying on one side our pen we feel contented in having been able, though so late in this work, to bestow a panegyric, however unpretentious, on a town which, though possessing no picturesque natural surroundings, nor interesting architectural productions, has yet a body of citizens whose hearts cannot but obtain for their town a reputation for benevolence and kind-heartedness."

I wonder what soured him on Sydney. It seems strange that a pleasing degree of three or four fingers of respectful familiarity should fill a man up and give him the panegyrics so bad. For he *has* them, the worst way — any one can see that. A man who is perfectly at himself does not throw cold detraction at people's architectural productions and picturesque surroundings, and let on that what he prefers is a Dubbonese dust-storm and a pleasing degree of respectful familiarity No, these are old, old symptoms; and when they appear we know that the man has got the panegyrics.

HEEDLESS MAN.

Sydney has a population of 400,000. When a stranger from America steps ashore there, the first thing that strikes him is that the place is eight or nine times as large as he was expecting it to be; and the next thing that strikes him is that it is an English city with American trimmings. Later on, in Melbourne, he will find the American trimmings still more in evidence; there, even the architecture will often suggest America; a photograph of its stateliest business street might be passed upon him for a picture of the finest street in a large American city. I was told that the most of the fine residences were the city residences of squatters. The name seemed out of focus somehow. When the explanation came, it offered a new instance of the curious changes which words, as well as animals, undergo through change of habitat and climate. With us, when you speak of a squatter you are always supposed to be speaking of a poor man, but in Australia when you speak of a squatter you are supposed to be speaking of a millionaire; in America the word indicates the possessor of a few acres and a doubtful title, in Australia it indicates a man whose landfront is as long as a railroad, and whose title has been perfected in one way or another; in America the word indicates a man who owns a dozen head of live stock, in Australia a man who owns anywhere from fifty thousand up to half a million head; in America the word indicates a man who is obscure and not important, in Australia a man who is prominent and of the first importance; in America you take off your hat to no squatter, in Australia you do; in America if your uncle is a squatter you keep it dark, in Australia you advertise it; in America if your friend is a squatter nothing comes of it, but with a squatter for your friend in Australia you may sup with kings if there are any around.

In Australia it takes about two acres and a half of pastureland (some people say twice as many), to support a sheep; and when the squatter has half a million sheep his private domain is about as large as Rhode Island, to speak in general terms. His annual wool crop may be worth a quarter or a half million dollars.

He will live in a palace in Melbourne or Sydney or some other of the large cities, and make occasional trips to his sheep-kingdom several hundred miles away in the great plains to look after his battalions of riders and shepherds and other hands. He has a commodious dwelling out there, and if he approve of you he will invite you to spend a week in it, and

SQUATTER LIFE.

will make you at home and comfortable, and let you see the great industry in all its details, and feed you and slake you and smoke you with the best that money can buy.

On at least one of these vast estates there is a considerable town, with all the various businesses and occupations that go to make an important town; and the town and the land it stands upon are the property of the squatters. I have seen that town, and it is not unlikely that there are other squatter-owned towns in Australia.

Australia supplies the world not only with fine wool, but with mutton also. The modern invention of cold storage and its application in ships has created this great trade. In Sydney I visited a huge establishment where they kill and clean and solidly freeze a thousand sheep a day, for shipment to England.

The Australians did not seem to me to differ noticeably from Americans, either in dress, carriage, ways, pronunciation, inflections, or general appearance. There were fleeting and subtle suggestions of their English origin, but these were not pronounced enough, as a rule, to catch one's attention. The people have easy and cordial manners from the beginning — from the moment that the introduction is completed. This is American. To put it in another way, it is English friendliness with the English shyness and self-consciousness left out.

Now and then — but this is rare — one hears such words as *piper* for paper, *lydy* for lady, and *tyble* for table fall from lips whence one would not expect such pronunciations to come. There is a superstition prevalent in Sydney that this pronunciation is an Australianism, but people who have been "home" — as the native reverently and lovingly calls England — know better. It is "costermonger." All over Australasia this pronunciation is nearly as common among servants as it is in London among the uneducated and the partially educated of all

9

sorts and conditions of people.   That mislaid y is rather striking when a person gets enough of it into a short sentence to
enable it to show up.   In the hotel in Sydney the chambermaid said, one morning —

"The tyble is set, and here is the piper; and if the lydy is
ready I'll tell the wyter to bring up the breakfast."

I have made passing mention, a moment ago, of the native
Australasian's custom of speaking of England as "home."   It
was always pretty to hear it, and often it was said in an un-
consciously caressing way that made it touching; in a way
which transmuted a sentiment into an embodiment, and made
one seem to see Australasia as a young girl stroking mother
England's old gray head.

In the Australasian home the table-talk is vivacious and
unembarrassed; it is without stiffness or restraint.   This does
not remind one of England so much as it does of America.
But Australasia is strictly democratic, and reserves and restraints are things that are bred by differences of rank.

English and colonial audiences are phenomenally alert and
responsive.   Where masses of people are gathered together in
England, caste is submerged, and with it the English reserve;
equality exists for the moment, and every individual is free;
so free from any consciousness of fetters, indeed, that the Englishman's habit of watching himself and guarding himself
against any injudicious exposure of his feelings is forgotten,
and falls into abeyance — and to such a degree indeed, that he
will bravely applaud all by himself if he wants to — an exhibition of daring which is unusual elsewhere in the world.

But it is hard to move a new English acquaintance when he
is by himself, or when the company present is small, and new
to him.   He is on his guard then, and his natural reserve is to
the fore.   This has given him the false reputation of being
without humor and without the appreciation of humor.

Americans are not Englishmen, and American humor is not English humor; but both the American and his humor had their origin in England, and have merely undergone changes brought about by changed conditions and a new environment. About the best humorous speeches I have yet heard were a couple that were made in Australia at club suppers — one of them by an Englishman, the other by an Australian.

A DIFFERENCE.

CHAPTER XII.

There are those who scoff at the schoolboy, calling him frivolous and shallow
Yet it was the schoolboy who said " Faith is believing what you know ain't so."
—*Pudd'nhead Wilson's New Calendar.*

IN Sydney I had a large dream, and in the course of talk I told it to a missionary from India who was on his way to visit some relatives in New Zealand. I dreamed that the visible universe is the physical person of God; that the vast worlds that we see twinkling millions of miles apart in the fields of space are the blood corpuscles in His veins; and that we and the other creatures are the microbes that charge with multitudinous life the corpuscles.

Mr. X., the missionary, considered the dream awhile, then said :

"It is not surpassable for magnitude, since its metes and bounds are the metes and bounds of the universe itself ; and it seems to me that it almost accounts for a thing which is otherwise nearly unaccountable — the origin of the sacred legends of the Hindoos. Perhaps they dream them, and then honestly believe them to be divine revelations of fact. It looks like that, for the legends are built on so vast a scale that it does not seem reasonable that plodding priests would happen upon such colossal fancies when awake."

He told some of the legends, and said that they were implicitly believed by all classes of Hindoos, including those of high social position and intelligence; and he said that this universal credulity was a great hindrance to the missionary in his work. Then he said something like this :

"At home, people wonder why Christianity does not make faster progress in India. They hear that the Indians believe easily, and that they have a natural trust in miracles and give them a hospitable reception. Then they argue like this : since the Indian believes easily, place Christianity before them and they must believe ; confirm its truths by the biblical miracles, and they will no longer doubt. The natural deduction is, that as Christianity makes

(132)

WHY CHRISTIANITY MAKES SLOW PROGRESS.   133

but indifferent progress in India, the fault is with us : we are not fortunate in presenting the doctrines and the miracles.

"But the truth is, we are not by any means so well equipped as they think. We have *not* the easy task that they imagine.   To use a military figure, we are sent against the enemy with good powder in our guns, but only wads for bullets ; that is to say, our miracles are not effective ; the Hindoos do not care for them ; they have more extraordinary ones of their own.   All the details of their own religion are proven and established by miracles ; the details of ours must be proven in the same way.   When I first began my work in India I greatly underestimated the difficulties thus put upon my task.   A correction was not long in coming.   I thought as our friends think at home — that to prepare my childlike wonder-lovers to listen with favor to my grave message I only needed to charm the way to it with wonders, marvels, miracles.   With full confidence I told the wonders performed by Samson, the strongest man that had ever lived — for so I called him.

" At first I saw lively anticipation and strong interest in the faces of my people, but as I moved along from incident to incident of the great story, I was distressed to see that I was steadily losing the sympathy of my audience. I could not understand it.   It was a surprise to me, and a disappointment. Before I was through, the fading sympathy had paled to indifference. Thence to the end the indifference remained ; I was not able to make any impression upon it.

" A good old Hindoo gentleman told me where my trouble lay.   He said   'We Hindoos recognize a god by the work of his hands — we accept no other testimony.   Apparently, this is also the rule with you Christians.   And we know when a man has his power from a god by the fact that he does things which he could not do, as a man, with the mere powers of a man. Plainly, this is the Christian's way also, of knowing when a man is working by a god's power and not by his own.   You saw that there was a supernatural property in the hair of Samson ; for you perceived that when his hair was gone he was as other men.   It is our way, as I have said.   There are many nations in the world, and each group of nations has its own gods, and will pay no worship to the gods of the others.   Each group believes its own gods to be strongest, and it will not exchange them except for gods that shall be proven to be their superiors in power.   Man is but a weak creature, and needs the help of gods — he cannot do without it.   Shall he place his fate in the hands of weak gods when there may be stronger ones to be found ?   That would be foolish.   No, if he hear of gods that are stronger than his own, he should not turn a deaf ear, for it is not a light matter that is at stake.   How then shall he determine which gods are the stronger, his own or those that preside over the concerns of other nations ?   By comparing the known works of his own gods with the works of those others ; there is no other way.   Now, when we make this comparison, we are not drawn towards the gods of any other nation. Our gods are shown by their works to be the strongest, the most powerful. The Christians have but few gods, and they are new — new, and not strong, as it seems to us.   They will increase in number, it is true, for this has happened with all gods, but that time is far away, many ages and decades of ages away, for gods multiply slowly, as is meet for beings to whom a thousand years is but a single moment. Our own gods have been born millions of years apart. The process is slow, the gathering of strength and power is similarly slow. In the slow lapse of the ages the steadily accumulating power of our gods has at last become prodigious. We have a thousand proofs of this in the colossal character of their personal acts and the acts of ordinary men to whom they have given supernatural qualities. To your Samson was given supernatural power, and when he broke the withes, and slew the thousands with the jawbone of an ass, and carried away the gates of the city upon his shoulders, you were amazed — and also awed, for you recognized the divine source of his strength. But it could not profit to place these things before your Hindoo congregation and invite their wonder ; for they would compare them with the deed done by Hanuman, when our gods infused their divine strength into his muscles ; and they would be indifferent to them — as you saw. In the old, old times, ages and ages gone by, when our god Rama was warring with the demon god of Ceylon, Rama bethought him to bridge the sea and connect Ceylon with India, so that his armies might pass easily over ; and he sent his general, Hanuman, inspired like your own Samson with divine strength, to bring the materials for the bridge. In two days Hanuman strode fifteen hundred miles, to the Himalayas, and took upon his shoulder a range of those lofty mountains two hundred miles long, and started with it toward Ceylon. It was in the night ; and, as he passed along the plain, the people of Govardhun heard the thunder of his tread and felt the earth rocking under it, and they ran out, and there, with their snowy summits piled to heaven, they saw the Himalayas passing by. And as this huge continent swept along overshadowing the earth, upon its slopes they discerned the twinkling lights of a thousand sleeping villages, and it was as if the constellations were filing in procession through the sky. While they were looking, Hanuman stumbled, and a small ridge of red sandstone twenty miles long was jolted loose and fell. Half of its length has wasted away in the course of the ages, but the other ten miles of it remain in the plain by Govardhun to this day as proof of the might of the inspiration of our gods. You must know, yourself, that Hanuman could not have carried those mountains to Ceylon except by the strength of the gods. You know that it was not done by his own strength, therefore, you know that it *was* done by the strength of the gods, just as you know that Samson carried the gates by the divine strength and not by his own. I think you must concede two things : First, That in carrying the gates of the city upon his shoulders, Samson did not establish the superiority of his gods over ours ; secondly, That his feat is not supported by any but verbal evidence, while Hanuman's is not only supported by verbal evidence, but this evidence is confirmed, established, proven, by visible, tangible evidence, which is the strongest of all testimony. We have the sandstone ridge, and while it remains we cannot doubt, and shall not. Have you the gates ?' "

HANUMAN MOVING THE MOUNTAINS.

CHAPTER XIII.

ONE is sure to be struck by the liberal way in which Australasia spends money upon public works — such as legislative buildings, town halls, hospitals, asylums, parks, and botanical gardens. I should say that where minor towns in America spend a hundred dollars on the town hall and on public parks and gardens, the like towns in Australasia spend a thousand. And I think that this ratio will hold good in the matter of hospitals, also. I have seen a costly and well-equipped, and architecturally handsome hospital in an Australian village of fifteen hundred inhabitants. It was built by private funds furnished by the villagers and the neighboring planters, and its running expenses were drawn from the same sources. I suppose it would be hard to match this in any country. This village was about to close a contract for lighting its streets with the electric light, when I was there. That is ahead of London. London is still obscured by gas — gas pretty widely scattered, too, in some of the districts; so widely indeed, that except on moonlight nights it is difficult to find the gas lamps.

The botanical garden of Sydney covers thirty-eight acres, beautifully laid out and rich with the spoil of all the lands and all the climes of the world. The garden is on high ground in the middle of the town, overlooking the great harbor, and it adjoins the spacious grounds of Government House — fifty-six acres; and at hand also, is a recreation ground containing eighty-two acres. In addition, there are the zoölogical gardens, the race-course, and the great cricket-grounds where the international matches are played. Therefore there is plenty of room for reposeful lazying and lounging, and for exercise too, for such as like that kind of work.

There are four specialties attainable in the way of social pleasure. If you enter your name on the Visitor's Book at Government House you will receive an invitation to the next ball that takes place there, if nothing can be proven against you. And it will be very pleasant; for you will see everybody except the Governor, and add a number of acquaintances and several friends to your list. The Governor will be in England. He always is. The continent has four or five governors, and I do not know how many it takes to govern the outlying archipelago; but anyway you will not see them. When they are appointed they come out from England and get inaugurated, and give a ball, and help pray for rain, and get aboard ship and go back home. And so the Lieutenant-Governor has to do all the work. I was in Australasia three months and a half, and saw only one Governor. The others were at home.

The Australasian Governor would not be so restless, perhaps, if he had a war, or a veto, or something like that to call for his reserve-energies, but he hasn't. There isn't any war, and there isn't any veto in his hands. And so there is really little or nothing doing in his line. The country governs itself, and prefers to do it; and is so strenuous about it and so jealous of its independence that it grows restive if even the Imperial Government at home proposes to help; and so the Imperial veto, while a fact, is yet mainly a name.

Thus the Governor's functions are much more limited than are a Governor's functions with us. And therefore more fatiguing. He is the apparent head of the State, he is the real head of Society. He represents culture, refinement, elevated

SYDNEY'S FOUR ENTERTAINMENTS.

sentiment, polite life, religion; and by his example he propagates these, and they spread and flourish and bear good fruit. He creates the fashion, and leads it. His ball is the ball of balls, and his countenance makes the horse-race thrive.

He is usually a lord, and this is well; for his position compels him to lead an expensive life, and an English lord is generally well equipped for that.

Another of Sydney's social pleasures is the visit to the Admiralty House; which is nobly situated on high ground overlooking the water. The trim boats of the service convey the guests thither; and there, or on board the flag-ship, they have the duplicate of the hospitalities of Government House. The Admiral commanding a station in British waters is a magnate of the first degree, and he is sumptuously housed, as becomes the dignity of his office.

Third in the list of special pleasures is the tour of the harbor in a fine steam pleasure-launch. Your richer friends own boats of this kind, and they will invite you, and the joys of the trip will make a long day seem short.

And finally comes the shark-fishing. Sydney Harbor is populous with the finest breeds of man-eating sharks in the world. Some people make their living catching them; for the Government pays a cash bounty on them. The larger the shark the larger the bounty, and some of the sharks are twenty feet long. You not only get the bounty, but everything that is in the shark belongs to you. Sometimes the contents are quite valuable.

The shark is the swiftest fish that swims. The speed of the fastest steamer afloat is poor compared to his. And he is a great gad-about, and roams far and wide in the oceans, and visits the shores of all of them, ultimately, in the course of his restless excursions. I have a tale to tell now, which has not as yet been in print. In 1870 a young stranger arrived in

Sydney, and set about finding something to do; but he knew no one, and brought no recommendations, and the result was that he got no employment. He had aimed high, at first, but as time and his money wasted away he grew less and less exacting, until at last he was willing to serve in the humblest capacities if so he might get bread and shelter. But luck was still against him; he could find no opening of any sort. Finally his money was all gone. He walked the streets all day, thinking; he walked them all night, thinking, thinking, and growing hungrier and hungrier. At dawn he found himself well away from the town and drifting aimlessly along the harbor shore. As he was passing by a nodding shark-fisher the man looked up and said —

"Say, young fellow, take my line a spell, and change my luck for me."

"How do you know I won't make it worse?"

"Because you can't. It has been at its worst all night. If you can't change it, no harm's done; if you do change it, it's for the better, of course. Come."

"All right, what will you give?"

"I'll give you the shark, if you catch one."

"And I will eat it, bones and all. Give me the line."

"Here you are. I will get away, now, for awhile, so that my luck won't spoil yours; for many and many a time I've noticed that if — there, pull in, pull in, man, you've got a bite! *I* knew how it would be. Why, I knew you for a born son of luck the minute I saw you. All right — he's landed."

It was an unusually large shark — "a full nineteen-footer," the fisherman said, as he laid the creature open with his knife.

"Now you rob him, young man, while I step to my hamper for a fresh bait. There's generally something in them worth going for. You've changed my luck, you see. But my goodness, I hope you haven't changed your own."

"Oh, it wouldn't matter; don't worry about that. Get your bait. I'll rob him."

When the fisherman got back the young man had just finished washing his hands in the bay, and was starting away.

"What, you are not going?"

"Yes. Good-bye."

"But what about your shark?"

"The shark? Why, what use is he to me?"

"What *use* is he? I like that. Don't you know that we can go and report him to Government, and you'll get a clean solid eighty shillings bounty? Hard cash, you know. What do you think about it *now*?"

"Oh, well, you can collect it."

"And *keep* it? Is that what you mean?"

"Yes."

"Well, this is odd. You're one of those sort they call eccentrics, I judge. The saying is, you mustn't judge a man by his clothes, and I'm believing it now. Why yours are looking just ratty, don't you know; and yet you must be rich."

"I am."

The young man walked slowly back to the town, deeply musing as he went. He halted a moment in front of the best restaurant, then glanced at his clothes and passed on, and got his breakfast at a "stand-up." There was a good deal of it, and it cost five shillings. He tendered a sovereign, got his change, glanced at his silver, muttered to himself, "There isn't enough to buy clothes with," and went his way.

At half-past nine the richest wool-broker in Sydney was sitting in his morning-room at home, settling his breakfast with the morning paper. A servant put his head in and said:

"There's a sundowner at the door wants to see you, sir."

"What do you bring that kind of a message here for? Send him about his business."

"He won't go, sir. I've tried."

"He won't go? That's — why, that's unusual. He's one of two things, then: he's a remarkable person, or he's crazy. Is he crazy?"

"No, sir. He don't look it."

"Then he's remarkable. What does he say he wants?"

"He won't tell, sir; only says it's very important."

"And won't go. Does he *say* he won't go?"

"Says he'll stand there till he sees you, sir, if it's all day."

"And yet isn't crazy. Show him up."

The sundowner was shown in. The broker said to himself, "No, he's not crazy; that is easy to see; so he must be the other thing."

Then aloud, "Well, my good fellow, be quick about it; don't waste any words; what is it you want?"

"I want to borrow a hundred thousand pounds."

"Scott! (It's a mistake; he *is* crazy. . . . No — he *can't* be — not with that eye.) Why, you take my breath away. Come, who *are* you?"

"Nobody that you know."

"What is your name?"

"Cecil Rhodes."

"No, I don't remember hearing the name before. Now then — just for curiosity's sake — what has sent you to me on this extraordinary errand?"

"The intention to make a hundred thousand pounds for you and as much for myself within the next sixty days."

"Well, well, well. It is the most extraordinary idea that I — sit *down* — you interest me. And somehow you — well, you fascinate me; I think that that is about the word. And it isn't your proposition — no, that doesn't fascinate me; it's something else, I don't quite know what; something that's born in you and oozes out of you, I suppose. Now then —

just for curiosity's sake again, nothing more: as I understand
it, it is your desire to bor —"

"I said *intention.*"

"Pardon, so you did.   I thought it was an unheedful use of
the word — an unheedful valuing of its strength, you know."

"I knew its strength."

"Well, I must say — but look here, let me walk the floor a
little, my mind is getting into a sort of whirl, though *you*
don't seem disturbed any. (Plainly this young fellow isn't
crazy ; but as to his being remarkable — well, really he amounts
to that, and something over.)   Now then, I believe I am beyond the reach of further astonishment.   Strike, and spare
not.   What is your scheme?"

"To buy the wool crop — deliverable in sixty days."

"What, the *whole* of it?"

"The whole of it."

"No, I was not quite out of the reach of surprises, after
all.   Why, how you talk!   Do you know what our crop is
going to foot up?"

"Two and a half million sterling — maybe a little more."

"Well, you've got your statistics right, any way.   Now,
then, do you know what the margins would foot up, to buy it
at sixty days?"

"The hundred thousand pounds I came here to get."

"Right, once more.   Well, dear me, just to see what would
happen, I wish you had the money.   And if you had it, what
would you do with it?"

"I shall make two hundred thousand pounds out of it in
sixty days."

"You mean, of course, that you *might* make it if —"

"I said 'shall'."

"Yes, by George, you *did* say 'shall'!   You are the most
definite devil I ever saw, in the matter of language.   Dear,

10

dear, dear, look here! Definite speech means clarity of mind. Upon my word I believe you've got what you believe to be a rational *reason* for venturing into this house, an entire stranger, on this wild scheme of buying the wool crop of an entire colony on speculation. Bring it out — I am prepared — acclimatized, if I may use the word. *Why* would you buy the crop, and *why* would you make that sum out of it? That is to say, what makes you think you — "

"I don't think — I know."

"Definite again. *How* do you know?"

"Because France has declared war against Germany, and wool has gone up fourteen per cent. in London and is still rising."

"Oh, in-deed? *Now* then, I've *got* you! Such a thunderbolt as you have just let fly ought to have made me jump out of my chair, but it didn't stir me the least little bit, you see. And for a very simple reason: I have read the morning paper. You can look at it if you want to. The fastest ship in the service arrived at eleven o'clock last night, fifty days out from London. All her news is printed here. There are no war-clouds anywhere; and as for wool, why, it is the low-spiritedest commodity in the English market. It is your turn to jump, now. . . . Well, why don't you jump? Why do you sit there in that placid fashion, when — "

"Because I have later news."

"Later news? Oh, come — later news than fifty days, brought steaming hot from London by the — "

"My news is only ten days old."

"Oh, Mun-*chausen*, hear the maniac talk! Where did you get it?"

"Got it out of a shark."

"Oh, oh, oh, this is *too* much! Front! call the police — bring the gun — raise the town! All the asylums in Christendom have broken loose in the single person of — "

"GOT IT OUT OF A SHA┄

"Sit down! And collect yourself. Where is the use in getting excited? Am I excited? There is nothing to get excited *about*. When I make a statement which I cannot prove, it will be time enough for you to begin to offer hospitality to damaging fancies about me and my sanity."

"Oh, a thousand, thousand pardons! I ought to be ashamed of myself, and I *am* ashamed of myself for thinking that a little bit of a circumstance like sending a shark to England to fetch back a market report — "

"What does your middle initial stand for, sir?"

"Andrew. What are you writing?"

"Wait a moment. Proof about the shark — and another matter. Only ten lines. There — now it is done. Sign it."

"Many thanks — many. Let me see; it says — it says — oh, come, this is *interesting!* Why — why — look here! prove what you say here, and I'll put up the money, and double as much, if necessary, and divide the winnings with you, half and half. There, now — I've signed; make your promise good if you can. Show me a copy of the London *Times* only ten days old."

"Here it is — and with it these buttons and a memorandum book that belonged to the man the shark swallowed. Swallowed him in the Thames, without a doubt; for you will notice that the last entry in the book is dated ' London,' and is of the same date as the *Times*, and says, 'Per confequenz der Kriegeserflärung, reife ich heute nach Deutchland ab, auf daß ich mein Leben auf dem Altar meines Landes legen mag'— as clean native German as anybody can put upon paper, and means that in consequence of the declaration of war, this loyal soul is leaving for home *to-day*, to fight. And he did leave, too, but the shark had him before the day was done, poor fellow."

"And a pity, too. But there are times for mourning, and we will attend to this case further on; other matters are pressing, now. I will go down and set the machinery in motion in a quiet way and buy the crop. It will cheer the drooping spirits of the boys, in a transitory way. Everything is transitory in this world. Sixty days hence, when they are called to deliver the goods, they will think they've been struck by lightning. But there is a time for mourning, and we will attend to that case along with the other one. Come along, I'll take you to my tailor. What did you say your name is?"

"Cecil Rhodes."

"It is hard to remember. However, I think you will make it easier by and by, if you live. There are three kinds of people — Commonplace Men, Remarkable Men, and Lunatics. I'll classify you with the Remarkables, and take the chances."

The deal went through, and secured to the young stranger the first fortune he ever pocketed.

The people of Sydney ought to be afraid of the sharks, but for some reason they do not seem to be. On Saturdays the young men go out in their boats, and sometimes the water is fairly covered with the little sails. A boat upsets now and then, by accident, a result of tumultuous skylarking; sometimes the boys upset their boat for fun — such as it is — with sharks visibly waiting around for just such an occurrence. The young fellows scramble aboard whole — sometimes — not always. Tragedies have happened more than once. While I was in Sydney it was reported that a boy fell out of a boat in the mouth of the Paramatta river and screamed for help and a boy jumped overboard from another boat to save him from the assembling sharks; but the sharks made swift work with the lives of both.

The government pays a bounty for the shark; to get the bounty the fishermen bait the hook or the seine with agreeable mutton; the news spreads and the sharks come from all over the Pacific Ocean to get the free board. In time the shark culture will be one of the most successful things in the colony.

CHAPTER XIV.

We can secure other people's approval, if we do right and try hard; but our own is worth a hundred of it, and no way has been found out of securing that.
—*Pudd'nhead Wilson's New Calendar.*

M Y health had broken down in New York in May; it had remained in a doubtful but fairish condition during a succeeding period of 82 days; it broke again on the Pacific. It broke again in Sydney, but not until after I had had a good outing, and had also filled my lecture engagements. This latest break lost me the chance of seeing Queensland. In the circumstances, to go north toward hotter weather was not advisable.

So we moved south with a westward slant, 17 hours by rail to the capital of the colony of Victoria, Melbourne — that juvenile city of sixty years, and half a million inhabitants. On the map the distance looked small; but that is a trouble with all divisions of distance in such a vast country as Australia. The colony of Victoria itself looks small on the map — looks like a county, in fact — yet it is about as large as England, Scotland, and Wales combined. Or, to get another focus upon it, it is just 80 times as large as the state of Rhode Island, and one-third as large as the State of Texas.

Outside of Melbourne, Victoria seems to be owned by a handful of squatters, each with a Rhode Island for a sheep farm. That is the impression which one gathers from common talk, yet the wool industry of Victoria is by no means so great as that of New South Wales. The climate of Victoria is favorable to other great industries — among others, wheat-growing and the making of wine.

We took the train at Sydney at about four in the afternoon.
It was American in one way, for we had a most rational sleeping car; also the car was clean and fine and new — nothing
about it to suggest the rolling stock of the continent of Europe.
But our baggage was weighed, and extra weight charged for.
That was continental. Continental and troublesome. Any
detail of railroading that is not troublesome cannot honorably
be described as continental.

The tickets were round-trip ones — to Melbourne, and clear
to Adelaide in South Australia, and then all the way back
to Sydney. Twelve hundred more miles than we really expected to make; but then as the round trip wouldn't cost much
more than the single trip, it seemed well enough to buy as
many miles as one could afford, even if one was not likely to
need them. A human being has a natural desire to have more
of a good thing than he needs.

THE ODDEST THING IN AUSTRALASIA.

Now comes a singular thing: the oddest thing, the strangest
thing, the most baffling and unaccountable marvel that Australasia can show. At the frontier between New South Wales
and Victoria our multitude of passengers were routed out of
their snug beds by lantern-light in the morning in the biting

cold of a high altitude to change cars on a road that has no break in it from Sydney to Melbourne! Think of the paralysis of intellect that gave that idea birth; imagine the boulder it emerged from on some petrified legislator's shoulders.

It is a narrow-gauge road to the frontier, and a broader gauge thence to Melbourne. The two governments were the builders of the road and are the owners of it. One or two reasons are given for this curious state of things. One is, that it represents the jealousy existing between the colonies — the two most important colonies of Australasia. What the other one is, I have forgotten. But it is of no consequence. It could be but another effort to explain the inexplicable.

All passengers fret at the double-gauge; all shippers of freight must of course fret at it; unnecessary expense, delay, and annoyance are imposed upon everybody concerned, and no one is benefited.

Each Australian colony fences itself off from its neighbor with a custom-house. Personally, I have no objection, but it must be a good deal of inconvenience to the people. We have something resembling it here and there in America, but it goes by another name. The large empire of the Pacific coast requires a world of iron machinery, and could manufacture it economically on the spot if the imposts on foreign iron were removed. But they are not. Protection to Pennsylvania and Alabama forbids it. The result to the Pacific coast is the same as if there were several rows of custom-fences between the coast and the East. Iron carted across the American continent at luxurious railway rates would be valuable enough to be coined when it arrived.

We changed cars. This was at Albury. And it was there, I think, that the growing day and the early sun exposed the distant range called the Blue Mountains. Accurately named. "My word!" as the Australians say, but it was a stunning color, that blue.   Deep, strong, rich, exquisite; towering and majestic masses of blue — a softly luminous blue, a smouldering blue, as if vaguely lit by fires within.   It extinguished the blue of the sky — made it pallid and unwholesome, whitey and washed-out.   A wonderful color — just divine.

A resident told me that those were not mountains; he said they were rabbit-piles.   And explained that long exposure and the over-ripe condition of the rabbits was what made them look so blue.   This man may have been right, but much reading of books of travel has made me distrustful of gratis information furnished by unofficial residents of a country.   The facts which such people give to travelers are usually erroneous, and often intemperately so.   The rabbit-plague has indeed been very bad in Australia, and it could account for one mountain, but not for a mountain range, it seems to me.   It is too large an order.

We breakfasted at the station.   A good breakfast, except the coffee ; and cheap.   The Government establishes the prices and placards them.   The waiters were men, I think; but that is not usual in Australasia.   The usual thing is to have girls.   No, not girls, young ladies — generally duchesses.   Dress? They would attract attention at any royal levée in Europe. Even empresses and queens do not dress as they do.   Not that they could not afford it, perhaps, but they would not know how.

All the pleasant morning we slid smoothly along over the plains, through thin — not thick — forests of great melancholy gum trees, with trunks rugged with curled sheets of flaking bark — erysipelas convalescents, so to speak, shedding their dead skins.   And all along were tiny cabins, built sometimes of wood, sometimes of gray-blue corrugated iron ; and the door-steps and fences were clogged with children — rugged little simply-clad chaps that looked as if they had been imported from the banks of the Mississippi without breaking bulk.

And there were little villages, with neat stations well pla-
carded with showy advertisements — mainly of almost *too*
self-righteous brands of "sheep-dip." If that is the name —
and I think it is. It is a stuff like tar, and is dabbed on to
places where the shearer clips a piece out of the sheep. It
bars out the flies, and has healing properties, and a nip to it
which makes the sheep skip like the cattle on a thousand hills.
It is not good to eat. That is, it is not good to eat except
when mixed with railroad coffee. It improves railroad coffee.
Without it railroad coffee is too vague. But with it, it is quite
assertive and enthusiastic. By itself, railroad coffee is too pas-
sive; but sheep-dip makes it wake up and get down to business.
I wonder where they get railroad coffee?

THINGS NOT SEEN.

We saw birds, but not a kangaroo, not an emu, not an
ornithorhyncus, not a lecturer, not a native. Indeed, the land
seemed quite destitute of game. But I have misused the word
native. In Australia it is applied to Australian-born whites
only. I should have said that we saw no Aboriginals — no
"blackfellows." And to this day I have never seen one. In
the great museums you will find all the other curiosities, but in
the curio of chiefest interest to the stranger all of them are
lacking. We have at home an abundance of museums, and
not an American Indian in them. It is clearly an absurdity,
but it never struck me before.

CHAPTER XV.

Truth is stranger than fiction — to some people, but I am measurably familiar with it. — *Pudd'nhead Wilson's New Calendar.*

Truth *is* stranger than fiction, but it is because Fiction is obliged to stick to possibilities ; Truth isn't. — *Pudd'nhead Wilson's New Calendar.*

THE air was balmy and delicious, the sunshine radiant; it was a charming excursion. In the course of it we came to a town whose odd name was famous all over the world a quarter of a century ago — Wagga-Wagga. This was because the Tichborne Claimant had kept a butcher-shop there. It was out of the midst of his humble collection of sausages and tripe that he soared up into the zenith of notoriety and hung there in the wastes of space a time, with the telescopes of all nations leveled at him in unappeasable curiosity — curiosity as to which of the two long-missing persons he was: Arthur Orton, the mislaid roustabout of Wapping, or Sir Roger Tichborne, the lost heir of a name and estates as old as English history. We all know now, but not a dozen people knew then; and the dozen kept the mystery to themselves and allowed the most intricate and fascinating and marvelous real-life romance that has ever been played upon the world's stage to unfold itself serenely, act by act, in a British court by the long and laborious processes of judicial development.

When we recall the details of that great romance we marvel to see what daring chances truth may freely take in constructing a tale, as compared with the poor little conservative risks permitted to fiction. The fiction-artist could achieve no success with the materials of this splendid Tichborne romance.

(156)

He would have to drop out the chief characters; the public would say such people are impossible. He would have to drop out a number of the most picturesque incidents; the public would say such things could never happen. And yet the chief characters did exist, and the incidents did happen.

It cost the Tichborne estates $400,000 to unmask the Claimant and drive him out; and even after the exposure multitudes of Englishmen still believed in him. It cost the British Government another $400,000 to convict him of perjury; and after the conviction the same old multitudes still believed in him; and among these believers were many educated and intelligent men; and some of them had personally known the real Sir Roger. The Claimant was sentenced to 14 years' imprisonment. When he got out of prison he went to New York and kept a whisky saloon in the Bowery for a time, then disappeared from view.

He always claimed to be Sir Roger Tichborne until death called for him. This was but a few months ago — not very much short of a generation since he left Wagga-Wagga to go and possess himself of his estates. On his death-bed he yielded up his secret, and confessed in writing that he was only Arthur Orton of Wapping, able seaman and butcher — that and nothing more. But it is scarcely to be doubted that there are people whom even his dying confession will not convince. The old habit of assimilating incredibilities must have made strong food a necessity in their case; a weaker article would probably disagree with them.

I was in London when the Claimant stood his trial for perjury. I attended one of his showy evenings in the sumptuous quarters provided for him from the purses of his adherents and well-wishers. He was in evening dress, and I thought him a rather fine and stately creature. There were about twenty-five gentlemen present; educated men, men moving in good society, none of them commonplace; some of them were men of distinction, none of them were obscurities. They were his cordial friends and admirers. It was "S'r Roger," always "S'r Roger," on all hands; no one withheld the title, all turned it from the tongue with unction, and as if it tasted good.

For many years I had had a mystery in stock. Melbourne, and only Melbourne, could unriddle it for me. In 1873 I arrived in London with my wife and young child, and presently received a note from Naples signed by a name not familiar to me. It was not Bascom, and it was not Henry; but I will call it Henry Bascom for convenience's sake. This note, of about six lines, was written on a strip of white paper whose end-edges were ragged. I came to be familiar with those strips in later years. Their size and pattern were always the same. Their contents were usually to the same effect: would I and mine come to the writer's country-place in England on such and such a date, by such and such a train, and stay twelve days and depart by such and such a train at the end of the specified time? A carriage would meet us at the station.

These invitations were always for a long time ahead; if we were in Europe, three months ahead; if we were in America, six to twelve months ahead. They always named the exact date and train for the beginning and also for the end of the visit.

This first note invited us for a date three months in the future. It asked us to arrive by the 4.10 p. m. train from London, August 6th. The carriage would be waiting. The carriage would take us away seven days later — train specified. And there were these words: "Speak to Tom Hughes."

I showed the note to the author of "Tom Brown at Rugby," and he said: —

"Accept, and be thankful."

He described Mr. Bascom as being a man of genius, a man of fine attainments, a choice man in every way, a rare and beautiful character. He said that Bascom Hall was a particularly fine example of the stately manorial mansion of Elizabeth's days, and that it was a house worth going a long way to see — like Knowle; that Mr. B. was of a social disposition, liked the company of agreeable people, and always had samples of the sort coming and going.

We paid the visit. We paid others, in later years — the last one in 1879. Soon after that Mr. Bascom started on a voyage around the world in a steam yacht — a long and leisurely trip, for he was making collections, in all lands, of birds, butterflies, and such things.

The day that President Garfield was shot by the assassin Guiteau, we were at a little watering place on Long Island Sound; and in the mail matter of that day came a letter with the Melbourne post-mark on it. It was for my wife, but I recognized Mr. Bascom's handwriting on the envelope, and opened it. It was the usual note — as to paucity of lines — and was written on the customary strip of paper; but there was nothing usual about the contents. The note informed my wife that if it would be any assuagement of her grief to know that her husband's lecture-tour in Australia was a satisfactory venture from the beginning to the end, he, the writer, could testify that such was the case; also, that her husband's untimely death had been mourned by all classes, as she would already know by the press telegrams, long before the reception of this note; that the funeral was attended by the officials of the colonial and city governments; and that while he, the writer, her friend and mine, had not reached Melbourne in time to see the body, he had at least had the sad privilege of acting as one of the pall-bearers. Signed, "Henry Bascom."

My first thought was, why didn't he have the coffin opened? He would have seen that the corpse was an imposter, and he could have gone right ahead and dried up the most of those tears, and comforted those sorrowing governments, and sold the remains and sent me the money.

I did nothing about the matter. I had set the law after living lecture-doubles of mine a couple of times in America, and the law had not been able to catch them; others in my trade had tried to catch *their* impostor-doubles and had failed. Then where was the use in harrying a ghost? None — and so I did not disturb it. I had a curiosity to know about that man's lecture-tour and last moments, but that could wait. When I should see Mr. Bascom he would tell me all about it. But he passed from life, and I never saw him again. My curiosity faded away.

However, when I found that I was going to Australia it revived. And naturally : for if the people should say that I was a dull, poor thing compared to what I was before I died, it would have a bad effect on business. Well, to my surprise the Sydney journalists had *never heard of that impostor!* I pressed them, but they were firm — they had never heard of him, and didn't believe in him.

I could not understand it; still, I thought it would all come right in Melbourne. The government would remember ; and the other mourners. At the supper of the Institute of Journalists I should find out all about the matter. But no — it turned out that *they* had never heard of it.

So my mystery was a mystery still. It was a great disappointment. I believed it would never be cleared up — in this life — so I dropped it out of my mind.

But at last ! just when I was least expecting it —

However, this is not the place for the rest of it; I shall come to the matter again, in a far-distant chapter.

CHAPTER XVI.

There is a Moral Sense, and there is an Immoral Sense. History shows us that the Moral Sense enables us to perceive morality and how to avoid it, and that the Immoral Sense enables us to perceive immorality and how to enjoy it.
— Pudd'nhead Wilson's New Calendar.

MELBOURNE spreads around over an immense area of ground. It is a stately city architecturally as well as in magnitude. It has an elaborate system of cable-car service; it has museums, and colleges, and schools, and public gardens, and electricity, and gas, and libraries, and theaters, and mining centers, and wool centers, and centers of the arts and sciences, and boards of trade, and ships, and railroads, and a harbor, and social clubs, and journalistic clubs, and racing clubs, and a squatter club sumptuously housed and appointed, and as many churches and banks as can make a living. In a word, it is equipped with everything that goes to make the modern great city. It is the largest city of Australasia, and fills the post with honor and credit. It has one specialty; this must not be jumbled in with those other things. It is the mitred Metropolitan of the Horse-Racing Cult. Its raceground is the Mecca of Australasia. On the great annual day of sacrifice — the 5th of November, Guy Fawkes's Day — business is suspended over a stretch of land and sea as wide as from New York to San Francisco, and deeper than from the northern lakes to the Gulf of Mexico; and every man and woman, of high degree or low, who can afford the expense, put away their other duties and come. They begin to swarm in by ship and rail a fortnight before the day, and they swarm thicker and thicker day after day, until all the vehicles of

11 (161)

transportation are taxed to their uttermost to meet the demands of the occasion, and all hotels and lodgings are bulging outward because of the pressure from within. They come a hundred thousand strong, as all the best authorities say, and they pack the spacious grounds and grand-stands and make a spectacle such as is never to be seen in Australasia elsewhere.

It is the "Melbourne Cup" that brings this multitude together. Their clothes have been ordered long ago, at unlimited cost, and without bounds as to beauty and magnificence, and have been kept in concealment until now, for unto this day are they consecrate. I am speaking of the *ladies'* clothes; but one might know that.

And so the grand-stands make a brilliant and wonderful spectacle, a delirium of color, a vision of beauty. The champagne flows, everybody is vivacious, excited, happy; everybody bets, and gloves and fortunes change hands right along, all the time. Day after day the races go on, and the fun and the excitement are kept at white heat; and when each day is done, the people dance all night so as to be fresh for the race in the morning. And at the end of the great week the swarms secure lodgings and transportation for next year, then flock away to their remote homes and count their gains and losses, and order next year's Cup-clothes, and then lie down and sleep two weeks, and get up sorry to reflect that a whole year must be put in somehow or other before they can be wholly happy again.

The Melbourne Cup is the Australasian National Day. It would be difficult to overstate its importance. It overshadows all other holidays and specialized days of whatever sort in that congeries of colonies. Overshadows them? I might almost say it blots them out. Each of them gets attention, but not everybody's; each of them evokes interest, but not everybody's; each of them rouses enthusiasm, but not everybody's;

in each case a part of the attention, interest, and enthusiasm is a matter of habit and custom, and another part of it is official and perfunctory. Cup Day, and Cup Day only, commands an attention, an interest, and an enthusiasm which are universal — and spontaneous, not perfunctory. Cup Day is supreme — it has no rival. I can call to mind no specialized annual day, in any country, which can be named by that large name — Supreme. I can call to mind no specialized annual day, in any country, whose approach fires the whole land with a conflagration of conversation and preparation and anticipation and jubilation. No day save this one; but this one does it.

In America we have no annual supreme day; no day whose approach makes the whole nation glad. We have the Fourth of July, and Christmas, and Thanksgiving. Neither of them can claim the primacy; neither of them can arouse an enthusiasm which comes near to being universal. Eight grown Americans out of ten dread the coming of the Fourth, with its pandemonium and its perils, and they rejoice when it is gone — if still alive. The approach of Christmas brings harrassment and dread to many excellent people. They have to buy a cart-load of presents, and they never know what to buy to hit the various tastes; they put in three weeks of hard and anxious work, and when Christmas morning comes they are so dissatisfied with the result, and so disappointed that they want to sit down and cry. Then they give thanks that Christmas comes but once a year. The observance of Thanksgiving Day — as a function — has become general of late years. The Thankfulness is not so general. This is natural. Two-thirds of the nation have always had hard luck and a hard time during the year, and this has a calming effect upon their enthusiasm.

We *have* a supreme day — a sweeping and tremendous and tumultuous day, a day which commands an absolute universality of interest and excitement; but it is not annual. It comes but once in four years; therefore it cannot count as a rival of the Melbourne Cup.

In Great Britain and Ireland they have two great days—— Christmas and the Queen's birthday. But they are equally popular; there is no supremacy.

I think it must be conceded that the position of the Australasian Day is unique, solitary, unfellowed; and likely to hold that high place a long time.

The next things which interest us when we travel are, first, the people; next, the novelties; and finally the history of the places and countries visited. Novelties are rare in cities which represent the most advanced civilization of the modern day. When one is familiar with such cities in the other parts of the world he is in effect familiar with the cities of Australasia. The outside aspects will furnish little that is new. There will be new names, but the things which they represent will sometimes be found to be less new than their names. There may be shades of difference, but these can easily be too fine for detection by the incompetent eye of the passing stranger. In the larrikin he will not be able to discover a new species, but only an old one met elsewhere, and variously called loafer, rough, tough, bummer, or blatherskite, according to his geographical distribution. The larrikin differs by a shade from those others, in that he is more sociable toward the stranger than they, more kindly disposed, more hospitable, more hearty, more friendly. At least it seemed so to me, and I had opportunity to observe. In Sydney, at least. In Melbourne I had to drive to and from the lecture-theater, but in Sydney I was able to walk both ways, and did it. Every night, on my way home at ten, or a quarter past, I found the larrikin grouped in considerable force at several of the street corners, and he always gave me this pleasant salutation:

"HELLO, MARK!"

" Hello, Mark ! "

" Here's to you, old chap ! "

" Say — Mark ! — is he dead ? — a reference to a passage in some book of mine, though I did not detect, at that time, that that was its source. And I didn't detect it afterward in Melbourne, when I came on the stage for the first time, and the same question was dropped down upon me from the dizzy height of the gallery. It is always difficult to answer a sudden inquiry like that, when you have come unprepared and don't know what it means. I will remark here — if it is not an indecorum — that the welcome which an American lecturer gets from a British colonial audience is a thing which will move him to his deepest deeps, and veil his sight and break his voice. And from Winnipeg to Africa, experience will teach him nothing ; he will never learn to expect it, it will catch him as a surprise each time. The war-cloud hanging black over England and America made no trouble for me. I was a prospective prisoner of war, but at dinners, suppers, on the platform, and elsewhere, there was never anything to remind me of it. This was hospitality of the right metal, and would have been prominently lacking in some countries, in the circumstances.

And speaking of the war flurry, it seemed to me to bring to light the unexpected, in a detail or two. It seemed to relegate the war-talk to the politicians on both sides of the water ; whereas whenever a prospective war between two nations had been in the air theretofore, the public had done most of the talking and the bitterest. The attitude of the newspapers was new also. I speak of those of Australasia and India, for I had access to those only. They treated the subject argumentatively and with dignity, not with spite and anger. That was a new spirit, too, and not learned of the French and German press, either before Sedan or since. I heard many public speeches, and they reflected the moderation of the journals. The outlook is that the English-speaking race will dominate the earth a hundred years from now, if its sections do not get to fighting each other. It would be a pity to spoil that prospect by baffling and retarding wars when arbitration would settle their differences so much better and also so much more definitely.

No, as I have suggested, novelties are rare in the great capitals of modern times. Even the wool exchange in Melbourne could not be told from the familiar stock exchange of other countries. Wool brokers are just like stockbrokers; they all bounce from their seats and put up their hands and yell in unison — no stranger can tell what — and the president calmly says — " Sold to Smith & Co., threppence farthing — next !" — when probably nothing of the kind happened; for how should he know ?

In the museums you will find acres of the most strange and fascinating things; but all museums are fascinating, and they do so tire your eyes, and break your back, and burn out your vitalities with their consuming interest. You always say you will never go again, but you do go. The palaces of the rich, in Melbourne, are much like the palaces of the rich in America, and the life in them is the same; but there the resemblance ends. The grounds surrounding the American palace are not often large, and not often beautiful, but in the Melbourne case the grounds are often ducally spacious, and the climate and the gardeners together make them as beautiful as a dream. It is said that some of the country seats have grounds — domains — about them which rival in charm and magnitude those which surround the country mansion of an English lord; but I was not out in the country; I had my hands full in town.

And what was the origin of this majestic city and its efflorescence of palatial town houses and country seats ? Its first brick was laid and its first house built by a passing convict. Australian history is almost always picturesque; indeed, it is so curious and strange, that it is itself the chiefest novelty the country has to offer, and so it pushes the other novelties into second and third place. It does not read like history, but like the most beautiful lies. And all of a fresh new sort, no mouldy old stale ones. It is full of surprises, and adventures, and incongruities, and contradictions, and incredibilities; but they are all true, they all happened.

SUNRISE BLUE MOUNTAINS.

CHAPTER XVII.

WHEN we consider the immensity of the British Empire in territory, population, and trade, it requires a stern exercise of faith to believe in the figures which represent Australasia's contribution to the Empire's commercial grandeur. As compared with the landed estate of the British Empire, the landed estate dominated by any other Power except one — Russia — is not very impressive for size. My authorities make the British Empire not much short of a fourth larger than the Russian Empire. Roughly proportioned, if you will allow your entire hand to represent the British Empire, you may then cut off the fingers a trifle above the middle joint of the middle finger, and what is left of the hand will represent Russia. The populations ruled by Great Britain and China are about the same — 400,000,000 each. No other Power approaches these figures. Even Russia is left far behind.

The population of Australasia — 4,000,000 — sinks into nothingness, and is lost from sight in that British ocean of 400,000,000. Yet the statistics indicate that it rises again and shows up very conspicuously when its share of the Empire's commerce is the matter under consideration. The value of England's annual exports and imports is stated at three billions of dollars,* and it is claimed that more than one-tenth of this great aggregate is represented by Australasia's exports

---

* New South Wales Blue Book.

(170)

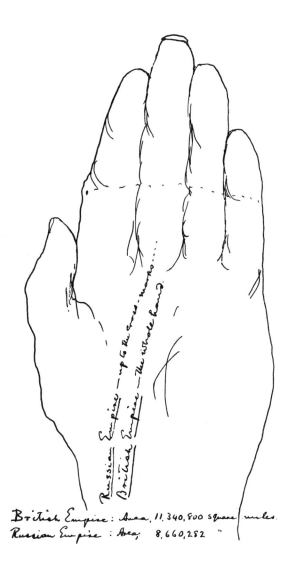

Russian Empire — up to the cross-marks ...
British Empire — The whole hand.

British Empire: Area, 11,340,800 square miles.
Russian Empire: Area, 8,660,282 "

to England and imports from England.* In addition to this, Australasia does a trade with countries other than England, amounting to a hundred million dollars a year, and a domestic intercolonial trade amounting to a hundred and fifty millions.*

In round numbers the 4,000,000 buy and sell about $600,-000,000 worth of goods a year. It is claimed that about half of this represents commodities of Australasian production. The products exported annually by India are worth a trifle over $500,000,000.† Now, here are some faith-straining figures:

Indian production (300,000,000 population), $500,000,000.

Australasian production (4,000,000 population), $300,000,000.

That is to say, the product of the individual Indian, annually (for export some whither), is worth $1.75; that of the individual Australasian (for export some whither), $75! Or, to put it in another way, the Indian family of man and wife and three children sends away an annual result worth $8.75, while the Australasian family sends away $375 worth.

There are trustworthy statistics furnished by Sir Richard Temple and others, which show that the individual Indian's whole annual product, both for export and home use, is worth in gold only $7.50; or, $37.50 for the family-aggregate. Ciphered out on a like ratio of multiplication, the Australasian family's aggregate production would be nearly $1,600. Truly, nothing is so astonishing as figures, if they once get started.

We left Melbourne by rail for Adelaide, the capital of the vast Province of South Australia — a seventeen-hour excursion. On the train we found several Sydney friends; among them a Judge who was going out on circuit, and was going to hold court at Broken Hill, where the celebrated silver mine is. It seemed a curious road to take to get to that region. Broken Hill is close to the western border of New South Wales, and Sydney is on the eastern border. A fairly straight line, 700

---

* D. M. Luckie.   † New South Wales Blue Book.

miles long, drawn westward from Sydney, would strike Broken Hill, just as a somewhat shorter one drawn west from Boston would strike Buffalo. The way the Judge was traveling would carry him over 2,000 miles by rail, he said; southwest from Sydney down to Melbourne, then northward up to Adelaide, then a cant back northeastward and over the border into New South Wales once more — to Broken Hill. It was like going from Boston southwest to Richmond, Virginia, then northwest up to Erie, Pennsylvania, then a cant back northeast and over the border — to Buffalo, New York.

But the explanation was simple. Years ago the fabulously rich silver discovery at Broken Hill burst suddenly upon an unexpectant world. Its stocks started at shillings, and went by leaps and bounds to the most fanciful figures. It was one of those cases where the cook puts a month's wages into shares, and comes next month and buys your house at your own price, and moves into it herself; where the coachman takes a few shares, and next month sets up a bank; and where the common sailor invests the price of a spree, and next month buys out the steamship company and goes into business on his own hook. In a word, it was one of those excitements which bring multitudes of people to a common center with a rush, and whose needs must be supplied, and at once. Adelaide was close by, Sydney was far away. Adelaide threw a short railway across the border before Sydney had time to arrange for a long one; it was not worth while for Sydney to arrange at all. The whole vast trade-profit of Broken Hill fell into Adelaide's hands, irrevocably. New South Wales furnishes law for Broken Hill and sends her Judges 2,000 miles — mainly through alien countries — to administer it, but Adelaide takes the dividends and makes no moan.

We started at 4.20 in the afternoon, and moved across level plains until night. In the morning we had a stretch of "scrub" country — the kind of thing which is so useful to the Australian novelist. In the scrub the hostile aboriginal lurks, and flits mysteriously about, slipping out from time to time to surprise and slaughter the settler; then slipping back again, and leaving no track that the white man can follow. In the scrub the novelist's heroine gets lost, search fails of result; she wanders here and there, and finally sinks down exhausted and unconscious, and the searchers pass within a yard or two of her, not suspecting that she is near, and by and by some rambler finds her bones and the pathetic diary which she had scribbled with her failing hand and left behind. Nobody can find a lost heroine in the scrub but the aboriginal "tracker," and he will not lend himself to the scheme if it will interfere with the novelist's plot. The scrub stretches miles and miles in all directions, and looks like a level roof of bush-tops without a break or a crack in it — as seamless as a blanket, to all appearance. One might as well walk under water and hope to guess out a route and stick to it, I should think. Yet it is claimed that the aboriginal "tracker" was able to hunt out people lost in the scrub. Also in the "bush"; also in the desert; and even follow them over patches of bare rocks and over alluvial ground which had to all appearance been washed clear of footprints.

From reading Australian books and talking with the people, I became convinced that the aboriginal tracker's performances evince a craft, a penetration, a luminous sagacity, and a minuteness and accuracy of observation in the matter of detective-work not found in nearly so remarkable a degree in any other people, white or colored. In an official account of the blacks of Australia published by the government of Victoria, one reads that the aboriginal not only notices the faint marks left on the bark of a tree by the claws of a climbing opossum, but knows in some way or other whether the marks were made to-day or yesterday.

And there is the case, on record, where A., a settler, makes a bet with B., that B. may lose a cow as effectually as he can, and A. will produce an aboriginal who will find her. B.

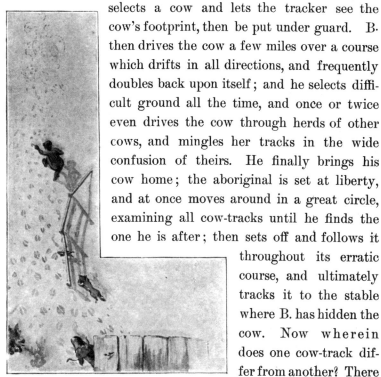

selects a cow and lets the tracker see the cow's footprint, then be put under guard. B. then drives the cow a few miles over a course which drifts in all directions, and frequently doubles back upon itself; and he selects diffi- cult ground all the time, and once or twice even drives the cow through herds of other cows, and mingles her tracks in the wide confusion of theirs. He finally brings his cow home; the aboriginal is set at liberty, and at once moves around in a great circle, examining all cow-tracks until he finds the one he is after; then sets off and follows it throughout its erratic course, and ultimately tracks it to the stable where B. has hidden the cow. Now wherein does one cow-track dif- fer from another? There must be a difference, or the tracker could not have performed the feat; a differ ence minute, shadowy, and not detectible by you or me, or by the late Sherlock Holmes, and yet discernible by a mem_ ber of a race charged by some people with occupying the bottom place in the gradations of human intelligence.

A TEST CASE.

CHAPTER XVIII.

It is easier to stay out than get out.— *Pudd'nhead Wilson's New Calendar.*

THE train was now exploring a beautiful hill country, and went twisting in and out through lovely little green valleys. There were several varieties of gum-trees; among them many giants. Some of them were bodied and barked like the sycamore; some were of fantastic aspect, and reminded one of the quaint apple trees in Japanese pictures. And there was one peculiarly beautiful tree whose name and breed I did not know. The foliage seemed to consist of big bunches of pine-spines, the lower half of each bunch a rich brown or old-gold color, the upper half a most vivid and strenuous and shouting green. The effect was altogether bewitching. The tree was apparently rare. I should say that the first and last samples of it seen by us were not more than half an hour apart. There was another tree of striking aspect, a kind of pine, we were told. Its foliage was as fine as hair, apparently, and its mass sphered itself above the naked straight stem like an explosion of misty smoke. It was not a sociable sort; it did not gather in groups or couples, but each individual stood far away from its nearest neighbor. It scattered itself in this spacious and exclusive fashion about the slopes of swelling grassy great knolls, and stood in the full flood of the wonderful sunshine; and as far as you could see the tree itself you could also see the ink-black blot of its shadow on the shining green carpet at its feet.

On some part of this railway journey we saw gorse and

broom — importations from England — and a gentleman who came into our compartment on a visit tried to tell me which was which; but as he didn't know, he had difficulty. He said he was ashamed of his ignorance, but that he had never been confronted with the question before during the fifty years and more that he had spent in Australia, and so he had never happened to get interested in the matter. But there was no need to be ashamed. The most of us have his defect. We take a natural interest in novelties, but it is against nature to take an interest in familiar things. The gorse and the broom were a fine accent in the landscape. Here and there they burst out in sudden conflagrations of vivid yellow against a background of sober or sombre color, with a so startling effect as to make a body catch his breath with the happy surprise of it. And then there was the wattle, a native bush or tree, an inspiring cloud of sumptuous yellow bloom. It is a favorite with the Australians, and has a fine fragrance, a quality usually wanting in Australian blossoms.

The gentleman who enriched me with the poverty of his information about the gorse and the broom told me that he came out from England a youth of twenty and entered the Province of South Australia with thirty-six shillings in his pocket — an adventurer without trade, profession, or friends, but with a clearly-defined purpose in his head: he would stay until he was worth £200, then go back home. He would allow himself five years for the accumulation of this fortune.

"That was more than fifty years ago," said he. "And here I am, yet."

As he went out at the door he met a friend, and turned and introduced him to me, and the friend and I had a talk and a smoke. I spoke of the previous conversation and said there was something very pathetic about this half century of exile, and that I wished the £200 scheme had succeeded.

12

"With *him?* Oh, it did. It's not so sad a case. He is modest, and he left out some of the particulars. The lad reached South Australia just in time to help discover the Burra-Burra copper mines. They turned out £700,000 in the first three years. Up to now they have yielded £20,000,000. He has had his share. Before that boy had been in the country two years he could have gone home and bought a village; he could go now and buy a city, I think. No, there is nothing very pathetic about his case. He and his copper arrived at just a handy time to save South Australia. It had got mashed pretty flat under the collapse of a land boom a while before."

"HERE I AM YET."

There it is again; picturesque history — Australia's specialty. In 1829 South Australia hadn't a white man in it. In 1836 the British Parliament erected it — still a solitude — into a Province, and gave it a governor and other governmental machinery. Speculators took hold, now, and inaugurated a vast land scheme, and invited immigration, encouraging it with lurid promises of sudden wealth. It was well worked in London; and bishops, statesmen, and all sorts of people made a rush for the land company's shares. Immigrants soon began to pour into the region of Adelaide and select town lots and farms in the sand and the mangrove swamps by the sea. The crowds continued to come, prices of land rose high, then higher and still higher, everybody was prosperous and happy, the boom swelled into gigantic proportions. A village of sheet-iron huts and clapboard sheds sprang up in the sand, and in these wigwams fashion made display; richly-dressed ladies played on costly pianos, London swells in evening dress and patent-leather boots were abundant, and this fine society drank champagne, and in other ways conducted itself in this capital of humble sheds as it had been accustomed to do in the aristocratic quarters of the metropolis of the world. The provincial government put up expensive buildings for its own use, and a palace with gardens for the use of its governor. The governor had a guard, and maintained a court. Roads, wharves, and hospitals were built. All this on credit, on paper, on wind, on inflated and fictitious values — on the boom's moonshine, in fact.

This went on handsomely during four or five years. Then all of a sudden came a smash. Bills for a huge amount drawn by the governor upon the Treasury were dishonored, the land company's credit went up in smoke, a panic followed, values fell with a rush, the frightened immigrants seized their gripsacks and fled to other lands, leaving behind them a good imitation of a solitude, where lately had been a buzzing and populous hive of men.

Adelaide was indeed almost empty; its population had fallen to 3,000. During two years or more the death-trance continued. Prospect of revival there was none; hope of it ceased. Then, as suddenly as the paralysis had come, came the resurrection from it. Those astonishingly rich copper mines were discovered, and the corpse got up and danced.

The wool production began to grow; grain-raising followed — followed so vigorously, too, that four or five years after the copper discovery, this little colony, which had had to import its breadstuffs formerly, and pay hard prices for them — once $50 a barrel for flour — had become an exporter of grain. The prosperities continued. After many years Providence, desiring to show especial regard for New South Wales and exhibit a loving interest in its welfare which should certify to all

nations the recognition of that colony's conspicuous righteous-
ness and distinguished well-deserving, conferred upon it that
treasury of inconceivable riches, Broken Hill; and South
Australia went over the border and took it, giving thanks.

Among our passengers was an American with a unique
vocation. Unique is a strong word, but I use it justifiably if I
did not misconceive what the American told me; for I understood him to say that in the world there was not another man
engaged in the business which he was following. He was buying the kangaroo-skin crop; buying all of it, both the Australian crop and the Tasmanian; and buying it for an American
house in New York. The prices were not high, as there was
no competition, but the year's aggregate of skins would cost
him £30,000. I had had the idea that the kangaroo was about
extinct in Tasmania and well thinned out on the continent.
In America the skins are tanned and made into shoes. After
the tanning, the leather takes a new name — which I have forgotten — I only remember that the new name does not indicate
that the kangaroo furnishes the leather. There was a German
competition for a while, some years ago, but that has ceased.
The Germans failed to arrive at the secret of tanning the skins
successfully, and they withdrew from the business. Now then,
I suppose that I have seen a man whose occupation is really entitled to bear that high epithet — unique. And I suppose that
there is not another occupation in the world that is restricted
to the hands of a sole person. I can think of no instance of it.
There is more than one Pope, there is more than one Emperor,
there is even more than one living god, walking upon the earth
and worshiped in all sincerity by large populations of men.
I have seen and talked with two of these Beings myself in
India, and I have the autograph of one of them. It can come
good, by and by, I reckon, if I attach it to a "permit."

Approaching Adelaide we dismounted from the train, as

the French say, and were driven in an open carriage over the hills and along their slopes to the city. It was an excursion of an hour or two, and the charm of it could not be overstated, I think. The road wound around gaps and gorges, and offered all varieties of scenery and prospect — mountains, crags, country homes, gardens, forests — color, color, color every-where, and the air fine and fresh, the skies blue, and not a shred of cloud to mar the downpour of the brilliant sunshine. And finally the mountain gateway opened, and the immense plain lay spread out below and stretching away into dim dis-tances on every hand, soft and delicate and dainty and beauti-ful. On its near edge reposed the city.

We descended and entered. There was nothing to remind one of the humble capital of huts and sheds of the long-van-ished day of the land-boom. No, this was a modern city, with wide streets, compactly built; with fine homes everywhere, embowered in foliage and flowers, and with imposing masses of public buildings nobly grouped and architecturally beautiful.

There was prosperity in the air; for another boom was on. Providence, desiring to show especial regard for the neighbor-ing colony on the west — called Western Australia — and ex-hibit a loving interest in its welfare which should certify to all nations the recognition of that colony's conspicuous righteous-ness and distinguished well-deserving, had recently conferred upon it that majestic treasury of golden riches, Coolgardie; and now South Australia had gone around the corner and taken it, giving thanks. Everything comes to him who is patient and good, and waits.

But South Australia deserves much, for apparently she is a hospitable home for every alien who chooses to come ; and for his religion, too. She has a population, as per the latest census, of only 320,000-odd, and yet her varieties of religion indicate the presence within her borders of samples of people from pretty nearly every part of the globe you can think of. Tabulated, these varieties of religion make a remarkable show. One would have to go far to find its match. I copy here this cosmopolitan curiosity, and it comes from the published census:

Church of England,	. .	89,271	Society of Friends,	. . 100
Roman Catholic,	. .	47,179	Salvation Army,	. . 4,356
Wesleyan,	. . .	49,159	New Jerusalem Church,	. 168
Lutheran,	. . .	23,328	Jews, ...	. 840
Presbyterian,	. .	18,206	Protestants (undefined),	. 5,532
Congregationalist,	. .	11,882	Mohammedans,	. . 299
Bible Christian,	. .	15,762	Confucians, etc.,	. . 3,884
Primitive Methodist,	. .	11,654	Other religions,	. . 1,719
Baptist,	. . .	17,547	Object, ...	. 6,940
Christian Brethren,	. .	465	Not stated, . . .	. 8,046
Methodist New Connexion,	.	39		
Unitarian,	. . .	688	Total, . . .	. 320,431
Church of Christ,	. .	3,367		

The item in the above list "Other religions" includes the following as returned:

Agnostics, . . .	50	Memnonists, ...	1
Atheists, . . .	22	Moravians, ...	139
Believers in Christ, . .	4	Mormons, ...	4
Buddhists, . . .	52	Naturalists, . . .	2
Calvinists, . . .	46	Orthodox, ...	4
Christadelphians, . .	134	Others (indefinite), . .	17
Christians, . . .	308	Pagans, ...	20
Christ's Chapel, . .	9	Pantheists, ...	3
Christian Israelites, . .	2	Plymouth Brethren, . .	111
Christian Socialists, . .	6	Rationalists, ...	4
Church of God, . .	6	Reformers, ...	7
Cosmopolitans, . . .	3	Secularists, ...	12
Deists, ...	14	Seventh-day Adventists, .	203
Evangelists, . . .	60	Shaker, ...	1
Exclusive Brethren, . .	8	Shintoists, ...	24
Free Church, . . .	21	Spiritualists, . . .	37
Free Methodists, . . .	5	Theosophists, . . .	9
Freethinkers, . . .	258	Town (City) Mission, . .	16
Followers of Christ, . .	8	Welsh Church, . . .	27
Gospel Meetings, . . .	11	Huguenot, ...	2
Greek Church, . . .	44	Hussite, ...	1
Infidels, ...	9	Zoroastrians, . . .	2
Maronites, . . .	2	Zwinglian, ...	1

About 64 roads to the other world.   You see how healthy the religious atmosphere is.   Anything can live in it.   Agnostics, Atheists, Freethinkers, Infidels, Mormons, Pagans, Indefinites: they are all there.   And all the big sects of the world can do more than merely live in it : they can spread, flourish, prosper. All except the Spiritualists and the Theosophists.   That is the most curious feature of this curious table.   What is the matter with the specter?   Why do they puff him away?   He is a welcome toy everywhere else in the world.

"NOT WANTED HERE."

CHAPTER XIX.

Pity is for the living, envy is for the dead.— *Pudd'nhead Wilson's New Calendar.*

THE successor of the sheet-iron hamlet of the mangrove marshes has that other Australian specialty, the Botanical Gardens. We cannot have these paradises. The best we could do would be to cover a vast acreage under glass and apply steam heat. But it would be inadequate, the lacks would still be so great: the confined sense, the sense of suffocation, the atmospheric dimness, the sweaty heat — these would all be there, in place of the Australian openness to the sky, the sunshine and the breeze. Whatever will grow under glass with us will flourish rampantly out of doors in Australia.* When the white man came the continent was nearly as poor, in variety of vegetation, as the desert of Sahara; now it has everything that grows on the earth. In fact, not Australia only, but all Australasia has levied tribute upon the flora of the rest of the world; and wherever one goes the results appear, in gardens private and public, in the woodsy walls of the highways, and in even the forests. If you see a curious or beautiful tree or bush or flower, and ask about it, the people, answering, usually name a foreign country as the place of its origin — India, Africa, Japan, China, England, America, Java, Sumatra, New Guinea, Polynesia, and so on.

In the Zoölogical Gardens of Adelaide I saw the only laughing jackass that ever showed any disposition to be courteous

---

* The greatest heat in Victoria, that there is an authoritative record of, was at Sandhurst, in January, 1862. The thermometer then registered 117 degrees in the shade. In January, 1880, the heat at Adelaide, South Australia, was 172 degrees in the sun.

to me. This one opened his head wide and laughed like a demon; or like a maniac who was consumed with humorous scorn over a cheap and degraded pun. It was a very human laugh. If he had been out of sight I could have believed that the laughter came from a man. It is an odd-looking bird, with a head and beak that are much too large for its body. In time man will extermi-nate the rest of the wild crea-tures of Aus-tralia, but this one will prob-ably survive, for man is his friend and lets him alone. Man always has a good reason for his charities to-wards wild things, human or animal — when he has any. In this case the bird is

LAUGHING JACKASS.

spared because he kills snakes. If L. J. will take my advice he will not kill all of them.

In that garden I also saw the wild Australian dog — the dingo. He was a beautiful creature — shapely, graceful, a little wolfish in some of his aspects, but with a most friendly eye and sociable disposition. The dingo is not an importation; he was present in great force when the whites first came to the continent. It may be that he is the oldest dog in the universe; his origin, his descent, the place where his ancestors first appeared, are as unknown and as untraceable as are the camel's. He is the most precious dog in the world, for he does not bark. But in an evil hour he got to raiding the sheep-runs to appease his hunger, and that sealed his doom. He is hunted, now, just as if he were a wolf. He has been sentenced to extermination, and the sentence will be carried out. This is all right, and not objectionable. The world was made for man — the white man.

South Australia is confusingly named. All of the colonies have a southern exposure except one — Queensland. Properly speaking, South Australia is *middle* Australia. It extends straight up through the center of the continent like the middle board in a center-table. It is 2,000 miles high, from south to north, and about a third as wide. A wee little spot down in its southeastern corner contains eight or nine-tenths of its population; the other one or two-tenths are elsewhere — as elsewhere as they could be in the United States with all the country between Denver and Chicago, and Canada and the Gulf of Mexico to scatter over. There is plenty of room.

A telegraph line stretches straight up north through that 2,000 miles of wilderness and desert from Adelaide to Port Darwin on the edge of the upper ocean. South Australia built the line; and did it in 1871–2 when her population numbered only 185,000. It was a great work; for there were no roads, no paths; 1,300 miles of the route had been traversed but once before by white men; provisions, wire, and poles had to be carried over immense stretches of desert; wells had to be dug along the route to supply the men and cattle with water.

A cable had been previously laid from Port Darwin to Java and thence to India, and there was telegraphic communication with England from India. And so, if Adelaide could make connection with Port Darwin it meant connection with the whole world. The enterprise succeeded. One could watch

THE WHITE MAN'S WORLD.

the London markets daily, now; the profit to the wool-growers of Australia was instant and enormous.

A telegram from Melbourne to San Francisco covers approximately 20,000 miles — the equivalent of five-sixths of the way around the globe. It has to halt along the way a good many times and be repeated; still, but little time is lost. These halts, and the distances between them, are here tabulated.*

	Miles.				Miles.
Melbourne — Mount Gambier,	300	Madras — Bombay,	.	.	650
Mount Gambier — Adelaide,	270	Bombay — Aden, .	.	.	1,662
Adelaide — Port Augusta, .	200	Aden — Suez, .	.	.	1,346
Port Augusta — Alice Springs,	1,036	Suez — Alexandria,	.	.	224
Alice Springs — Port Darwin,	898	Alexandria — Malta,	.	.	828
Port Darwin — Banjoewangie,	1,150	Malta — Gibraltar,	.	.	1,008
Banjoewangie — Batavia, .	480	Gibraltar — Falmouth, .	.	1,061
Batavia — Singapore, .	.	553	Falmouth — London,	.	.	350
Singapore — Penang, .	.	399	London — New York, .	.	2,500
Penang — Madras,	.	.	1,280	New York — San Francisco,	3,500

I was in Adelaide again, some months later, and saw the multitudes gather in the neighboring city of Glenelg to commemorate the Reading of the Proclamation — in 1836 — which founded the Province. If I have at any time called it a Colony, I withdraw the discourtesy. It is not a Colony, it is a Province; and officially so. Moreover, it is the only one so named in Australasia. There was great enthusiasm; it was the Province's national holiday, its Fourth of July, so to speak. It is the pre-eminent holiday; and that is saying much, in a country where they seem to have a most un-English mania for holidays. Mainly they are workingmen's holidays; for in South Australia the workingman is sovereign; his vote is the desire of the politician — indeed, it is the very breath of the politician's being; the parliament exists to deliver the will of the workingman, and the government exists to execute it. The workingman is a great power everywhere in Australia, but South Australia is his paradise. He has had a hard time in this world, and has earned a paradise. I am glad he has found it. The holidays there are frequent enough to be bewil-

---

*From " Round the Empire." (George R. Parkin), all but the last two.

dering to the stranger. I tried to get the hang of the system, but was not able to do it.

You have seen that the Province is tolerant, religious-wise. It is so politically, also. One of the speakers at the Commemoration banquet — the Minister of Public Works — was an American, born and reared in New England. There is nothing narrow about the Province, politically, or in any other way that I know of. Sixty-four religions and a Yankee cabinet minister. No amount of horse-racing can damn this community.

The mean temperature of the Province is 62°. The death-rate is 13 in the 1,000 — about half what it is in the city of New York, I should think, and New York is a healthy city. Thirteen is the death-rate for the average citizen of the Province, but there seems to be no death-rate for the old people. There were people at the Commemoration banquet who could remember Cromwell. There were six of them. These Old Settlers had all been present at the original Reading of the Proclamation, in 1836. They showed signs of the blightings and blastings of time, in their outward aspect, but they were young within; young and cheerful, and ready to talk; ready to talk, and talk all you wanted; in their turn, and out of it. They were down for six speeches, and they made 42. The governor and the cabinet and the mayor were down for 42 speeches, and they made 6. They have splendid grit, the Old Settlers, splendid staying power. But they do not hear well, and when they see the mayor going through motions which they recognize as the introducing of a speaker, they think they are the one, and they all get up together, and begin to respond, in the most animated way; and the more the mayor gesticulates, and shouts " Sit down! Sit down! " the more they take it for applause, and the more excited and reminiscent and enthusiastic they get; and next, when they see the whole house laughing and crying, three of them think it is about the bitter old-time hardships they are describing, and the other three

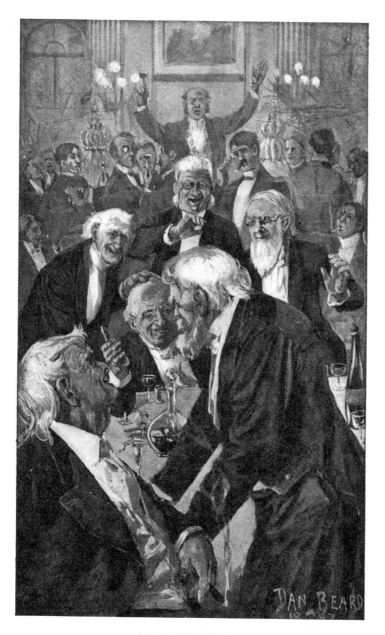

THE OLD SETTLERS.

think the laughter is caused by the jokes they have been un-corking — jokes of the vintage of 1836 — and then the way they *do* go on! And finally when ushers come and plead, and beg, and gently and reverently crowd them down into their seats, they say, "Oh, I'm not tired — I could bang along a week!" and they sit there looking simple and childlike, and gentle, and proud of their oratory, and wholly unconscious of what is going on at the other end of the room. And so one of the great dignitaries gets a chance, and begins his carefully-prepared speech, impressively and with solemnity —

"When we, now great and prosperous and powerful, bow our heads in reverent wonder in the contemplation of those sublimities of energy, of wisdom, of forethought, of — "

Up come the immortal six again, in a body, with a joyous "Hey, I've thought of another one!" and at it they go, with might and main, hearing not a whisper of the pandemonium that salutes them, but taking all the visible violences for applause, as before, and hammering joyously away till the imploring ushers pray them into their seats again. And a pity, too; for those lovely old boys did so enjoy living their heroic youth over, in these days of their honored antiquity; and certainly the things they had to tell were usually worth the telling and the hearing.

It was a stirring spectacle; stirring in more ways than one, for it was amazingly funny, and at the same time deeply pathetic; for they had seen so much, these time-worn veterans, and had suffered so much; and had built so strongly and well, and laid the foundations of their commonwealth so deep, in liberty and tolerance; and had lived to see the structure rise to such state and dignity and hear themselves so praised for their honorable work.

One of these old gentlemen told me some things of interest afterward; things about the aboriginals, mainly. He thought them intelligent — remarkably so in some directions — and he said that along with their unpleasant qualities they had some exceedingly good ones; and he considered it a great pity that the race had died out. He instanced their invention of the boomerang and the "weet-weet" as evidences of their brightness; and as another evidence of it he said he had never seen a white man who had cleverness enough to learn to do the miracles with those two toys that the aboriginals achieved. He said that even the smartest whites had been obliged to confess that they could not learn the trick of the boomerang in perfection; that it had possibilities which they could not master. The white man could not control its motions, could not make it obey him; but the aboriginal could. He told me some wonderful things — some almost incredible things — which he had seen the blacks do with the boomerang and the weet-weet. They have been confirmed to me since by other early settlers and by trustworthy books.

It is contended — and may be said to be conceded — that the boomerang was known to certain savage tribes in Europe in Roman times. In support of this, Virgil and two other Roman poets are quoted. It is also contended that it was known to the ancient Egyptians.

One of two things either some one with  is then apparent: a boomerang arrived in Australia in the days of antiquity before European knowledge of the thing had been lost, or the Australian aboriginal reinvented it. It will

ADAM AT PRACTICE.

take some time to find out which of these two propositions is the fact. But there is no hurry.

CHAPTER XX.

It is by the goodness of God that in our country we have those three unspeakably precious things: freedom of speech, freedom of conscience, and the prudence never to practice either of them. — *Pudd'nhead Wilson's New Calendar.*

FROM diary:

Mr. G. called. I had not seen him since Nauheim, Germany — several years ago; the time that the cholera broke out at Hamburg. We talked of the people we had known there, or had casually met; and G. said:

" Do you remember my introducing you to an earl — the Earl of C.?"

" Yes. That was the last time I saw you. You and he were in a carriage, just starting — belated — for the train. I remember it."

" I remember it too, because of a thing which happened then which I was not looking for. He had told me a while before, about a remarkable and interesting Californian whom he had met and who was a friend of yours, and said that if he should ever meet you he would ask you for some particulars about that Californian. The subject was not mentioned that day at Nauheim, for we were hurrying away, and there was no time; but the thing that surprised me was this: when I introduced you, you said, 'I am glad to meet your lordship — again.' The 'again' was the surprise. He is a little hard of hearing, and didn't catch that word, and I thought you hadn't intended that he should. As we drove off I had only time to say, 'Why, what do you know about him?' and I understood you to say, 'Oh, nothing, except that he is the quickest judge of —' Then we were gone, and I didn't get the rest. I wondered what it was that he was such a quick judge of. I have thought of it many times since, and still wondered what it could be. He and I talked it over, but could not guess it out. He thought it must be fox-hounds or horses, for he is a good judge of those — no one is a better. But *you* couldn't know that, because you didn't know *him ;* you had mistaken him for some one else; it must be that, he said, because he knew you had never met him before. And of course you hadn't — had you ? "

" Yes, I had."

" Is that so ?   Where ? "

" At a fox-hunt, in England."

" How curious that is.   Why, he hadn't the least recollection of it.   Had you any conversation with him ? "

" Some — yes."

" Well, it left not the least impression upon him.   What did you talk about ? "

" About the fox.   I think that was all."

" Why, *that* would interest him ; that ought to have left an impression.   What did *he* talk about ? "

" The fox."

" It's very curious.   I don't understand it.   Did what he said leave an impression upon you ? "

" Yes.   It showed me that he was a quick judge of — however, I will tell you all about it, then you will understand.   It was a quarter of a century ago — 1873 or '74.   I had an American friend in London named F., who was fond of hunting, and his friends the Blanks invited him and me to come out to a hunt and be their guests at their country place.   In the morning the mounts were provided, but when I saw the horses I changed my mind and asked permission to walk.   I had never seen an English hunter before, and it seemed to me that I

could hunt a fox safer on the ground. I had always been diffident about horses, anyway, even those of the common altitudes, and I did not feel competent to hunt on a horse that
went on stilts. So then Mrs. Blank came to my help and said
I could go with her in the dog-cart and we would drive to a
place she knew of, and there we should have a good glimpse of
the hunt as it went by.

"When we got to that place I got out and went and leaned
my elbows on a low stone wall which enclosed a turfy and
beautiful great field with heavy wood on all its sides except
ours. Mrs. Blank sat in the dog-cart fifty yards away, which
was as near as she could get with the vehicle. I was full of
interest, for I had never seen a fox-hunt. I waited, dreaming
and imagining, in the deep stillness and impressive tranquility
which reigned in that retired spot. Presently, from away off
in the forest on the left, a mellow bugle-note came floating;
then all of a sudden a multitude of dogs burst out of that forest
and went tearing by and disappeared in the forest on the right;
there was a pause, and then a cloud of horsemen in black caps
and crimson coats plunged out of the left-hand forest and went
flaming across the field like a prairie-fire, a stirring sight to
see. There was one man ahead of the rest, and he came spur-
ring straight at me. He was fiercely excited. It was fine to
see him ride; he was a master horseman. He came like a
storm till he was within seven feet of me, where I was leaning
on the wall, then he stood his horse straight up in the air on
his hind toe-nails, and shouted like a demon :

"'Which way'd the fox go?'

"I didn't much like the tone, but I did not let on; for he
was excited, you know. But I was calm; so I said softly, and
without acrimony :

"'*Which* fox?'

"It seemed to anger him. I don't know why; and he thundered out :

" '*Which* fox ? Why, *the* fox ? Which way did the *fox* go ?'

" I said, with great gentleness — even argumentatively :

" 'If you could be a little more definite — a little less vague — because I am a stranger, and there are many foxes, as you will know even better than I, and unless I know which one it is that you desire to identify, and — '

" 'You're certainly the damdest idiot that has escaped in a thousand years !' and he snatched his great horse around as easily as I would snatch a cat, and was away like a hurricane. A very excitable man.

" I went back to Mrs. Blank, and *she* was excited, too — oh, all alive. She said :

" 'He *spoke* to you ! — *didn't* he ?'

" 'Yes, it is what happened.'

" 'I *knew* it ! I couldn't hear what he said, but I *knew* he spoke to you ! Do you know who it was ? It was Lord C., — and he is Master of the Buckhounds ! Tell me — what do you think of him ?'

" 'Him ? Well, for sizing-up a stranger, he's got the most sudden and accurate judgment of any man I ever saw.'

"It pleased her. I thought it would."

G. got away from Nauheim just in time to escape being shut in by the quarantine-bars on the frontiers; and so did we, for we left the next day. But G. had a great deal of trouble in getting by the Italian custom-house, and we should have fared likewise but for the thoughtfulness of our consul-general in Frankfort. He introduced me to the Italian consul-general, and I brought away from that consulate a letter which made our way smooth. It was a dozen lines merely commending me in a general way to the courtesies of servants in his Italian Majesty's service, but it was more powerful than it looked. In addition to a raft of ordinary baggage, we had six or eight trunks which were filled exclusively with dutiable stuff — household goods purchased in Frankfort for use in Florence, where we had taken a house. I was going to ship these through by express; but at the last moment an order went throughout Germany forbidding the moving of any parcels by train unless the owner went with them. This was a bad outlook. We must take these things along, and the delay sure to be caused by the examination of them in the custom-house might lose us our train. I imagined all sorts of terrors, and enlarged them steadily as we approached the Italian frontier. We were six in number, clogged with all that baggage, and I was courier for the party — the most incapable one they ever employed.

We arrived, and pressed with the crowd into the immense custom-house, and the usual worries began; everybody crowding to the counter and begging to have his baggage examined first, and all hands clattering and chattering at once. It seemed to me that I could do nothing; it would be better to give it all up and go away and leave the baggage. I couldn't speak the language; I should never accomplish anything. Just then a tall handsome man in a fine uniform was passing by and I knew he must be the station-master — and that reminded me of my letter. I ran to him and put it into his hands. He took it out of the envelope, and the moment his eye caught the royal coat of arms printed at its top, he took off his cap and made a beautiful bow to me, and said in English —

"Which is your baggage? Please show it to me."

I showed him the mountain. Nobody was disturbing it; nobody was interested in it; all the family's attempts to get attention to it had failed — except in the case of one of the trunks containing the dutiable goods. It was just being opened. My officer said —

"There, let that alone! Lock it. Now chalk it. Chalk

all of the lot. Now please come and show me the hand-baggage."

He plowed through the waiting crowd, I following, to the counter, and he gave orders again, in his emphatic military way —

"Chalk these. Chalk *all* of them."

Then he took off his cap and made that beautiful bow again, and went his way. By this time these attentions had attracted the wonder of that acre of passengers, and the whisper had gone around that the royal family were present getting their baggage chalked; and as we passed down in review on our way to the door, I was conscious of a pervading atmosphere of envy which gave me deep satisfaction.

But soon there was an accident. My overcoat pockets were stuffed with German cigars and linen packages of American smoking tobacco, and a porter was following us around with this overcoat on his arm, and gradually getting it upside down. Just as I, in the rear of my family, moved by the sentinels at the door, about three hatfuls of the tobacco tumbled out on the floor. One of the soldiers pounced upon it, gathered it up in his arms, pointed back whence I had come, and marched me ahead of him past that long wall of passengers again — he chattering and exulting like a devil, they smiling in peaceful joy, and I trying to look as if my pride was not hurt, and as if I did not mind being brought to shame before these pleased people who had so lately envied me. But at heart I was cruelly humbled.

When I had been marched two-thirds of the long distance and the misery of it was at the worst, the stately station-master stepped out from somewhere, and the soldier left me and darted after him and overtook him; and I could see by the soldier's excited gestures that he was betraying to him the whole shabby business. The station-master was plainly very

"WE MARCHED THROUGH THE CROWD"

angry. He came striding down toward me, and when he was come near he began to pour out a stream of indignant Italian; then suddenly took off his hat and made that beautiful bow and said —

"Oh, it is *you!* I beg a thousands pardons! This idiot here — " He turned to the exulting soldier and burst out with a flood of white-hot Italian lava, and the next moment he was bowing, and the soldier and I were moving in procession again — *he* in the lead and ashamed, this time, I with my chin up. And so we marched by the crowd of fascinated passengers, and I went forth to the train with the honors of war. Tobacco and all.

THE ROYAL LETTER.

CHAPTER XXI.

Man will do many things to get himself loved, he will do all things to get himself envied. — *Puddn'head Wilson's New Calendar.*

BEFORE I saw Australia I had never heard of the "weet-weet" at all. I met but few men who had seen it thrown — at least I met but few who mentioned having seen it thrown. Roughly described, it is a fat wooden cigar with its butt-end fastened to a flexible twig. The whole thing is only a couple of feet long, and weighs less than two ounces. This feather — so to call it — is not thrown through the air, but is flung with an underhanded throw and made to strike the ground a little way in front of the thrower; then it glances and makes a long skip; glances again, skips again, and again and again, like the flat stone which a boy sends skating over the water. The water is smooth, and the stone has a good chance; so a strong man may make it travel fifty or seventy-five yards; but the weet-weet has no such good chance, for it strikes sand, grass, and earth in its course. Yet an expert aboriginal has sent it a measured distance of *two hundred and twenty yards.* It would have gone even further but it encountered rank ferns and underwood on its passage and they damaged its speed. Two hundred and twenty yards; and so weightless a toy — a mouse on the end of a bit of wire, in effect; and not sailing through the accomodating air, but encountering grass and sand and stuff at every jump. It looks wholly impossible; but Mr. Brough Smyth saw the feat and did the measuring, and set down the facts in his book about aboriginal life, which he wrote by command of the Victorian Government.

What is the secret of the feat? No one explains. It cannot be physical strength, for that could not drive such a feather-weight any distance. It must be art. But no one explains what the art of it is; nor how it gets around that law of nature which says you shall not throw any two-ounce thing 220 yards, either through the air or bumping along the ground. Rev. J. G. Woods says:

"The distance to which the weet-weet or kangaroo-rat can be thrown is truly astonishing. I have seen an Australian stand at one side of Kennington Oval and throw the kangaroo-rat completely across it." (Width of Kennington Oval not stated.) "It darts through the air with the sharp and menacing hiss of a rifle-ball, its greatest height from the ground being some seven or eight feet. . . . . . When properly thrown it looks just like a living animal leaping along. . . . . . Its movements have a wonderful resemblance to the long leaps of a kangaroo-rat fleeing in alarm, with its long tail trailing behind it."

The Old Settler said that he had seen distances made by the weet-weet, in the early days, which almost convinced him that it was as extraordinary an instrument as the boomerang.

There must have been a large distribution of acuteness among those naked skinny aboriginals, or they couldn't have been such unapproachable trackers and boomerangers and weet-weeters. It must have been race-aversion that put upon them a good deal of the low-rate intellectual reputation which they bear and have borne this long time in the world's estimate of them.

They were lazy—always lazy. Perhaps that was their trouble. It is a killing defect. Surely they could have invented and built a competent house, but they didn't. And they could have invented and developed the agricultural arts, but they didn't. They went naked and houseless, and lived on fish and grubs and worms and wild fruits, and were just plain savages, for all their smartness.

With a country as big as the United States to live and multiply in, and with no epidemic diseases among them till the white man came with those and his other appliances of civilization, it is quite probable that there was never a day in his history when he could muster 100,000 of his race in all Australia. He diligently and deliberately kept population down by infanticide — largely; but mainly by certain other methods. He did not need to practise these artificialities any more after the white man came. The white man knew ways of keeping down population which were worth several of his. The white man knew ways of reducing a native population 80 per cent in 20 years. The native had never seen anything as fine as that before.

THE WHITE MAN'S APPLIANCES.

For example, there is the case of the country now called Victoria — a country eighty times as large as Rhode Island, as I have already said. By the best official guess there were 4,500 aboriginals in it when the whites came along in the middle of the 'Thirties. Of these, 1,000 lived in Gippsland, a patch of territory the size of fifteen or sixteen Rhode Islands: they did not diminish as fast as some of the other communities; indeed, at the end of forty years there were still 200 of them left. The Geelong tribe diminished more satisfactorily: from 173 persons it faded to 34 in twenty years; at the end of another twenty the tribe numbered one person altogether. The two Melbourne tribes could muster almost 300 when the white man came; they could muster but twenty, thirty-seven years later, in 1875. In that year there were still odds and ends of tribes scattered about the colony of Victoria, but I was told that natives of full blood are very scarce now. It is said that the aboriginals continue in some force in the huge territory called Queensland.

The early whites were not used to savages. They could not understand the primary law of savage life: that if a man do you a wrong, his whole tribe is responsible — each individual of it — and you may take your change out of any individual of it, without bothering to seek out the guilty one. When a white killed an aboriginal, the tribe applied the ancient law, and killed the first white they came across. To the whites this was a monstrous thing. Extermination seemed to be the proper medicine for such creatures as this. They did not kill all the blacks, but they promptly killed enough of them to make their own persons safe. From the dawn of civilization down to this day the white man has always used that very precaution. Mrs. Campbell Praed lived in Queensland, as a child, in the early days, and in her "Sketches of Australian life," we get informing pictures of the early struggles of the white and the black to reform each other.

Speaking of pioneer days in the mighty wilderness of Queensland, Mrs. Praed says:

"At first the natives retreated before the whites; and, except that they every now and then speared a beast in one of the herds, gave little cause for uneasiness. But, as the number of squatters increased, each one taking up miles of country and bringing two or three men in his train, so that shepherds' huts and stockmen's camps lay far apart, and defenseless in the midst of hostile tribes, the Blacks' depredations became more frequent and murder was no unusual event.

"The loneliness of the Australian bush can hardly be painted in words. Here extends mile after mile of primeval forest where perhaps foot of white man has never trod — interminable vistas where the eucalyptus trees rear their lofty trunks and spread forth their lanky limbs, from which the red gum oozes and hangs in fantastic pendants like crimson stalactites; ravines along the sides of which the long-bladed grass grows rankly; level untimbered plains alternating with undulating tracts of pasture, here and there broken by a stony ridge, steep gully, or dried-up creek. All wild, vast and desolate; all the same monotonous gray coloring, except where the wattle, when in

14

blossom, shows patches of feathery gold, or a belt of scrub lies green, glossy, and impenetrable as Indian jungle.

"The solitude seems intensified by the strange sounds of reptiles, birds, and insects, and by the absence of larger creatures; of which in the day-time, the only audible signs are the stampede of a herd of kangaroo, or the rustle of a wallabi, or a dingo stirring the grass as it creeps to its lair. But there are the whirring of locusts, the demoniac chuckle of the laughing jack-ass, the screeching of cockatoos and parrots, the hissing of the frilled lizard, and the buzzing of innumerable insects hidden under the dense undergrowth. And then at night, the melancholy wailing of the curlews, the dismal howling of dingoes, the discordant croaking of tree-frogs, might well shake the nerves of the solitary watcher."

That is the theater for the drama. When you comprehend one or two other details, you will perceive how well suited for trouble it was, and how loudly it invited it. The cattlemen's stations were scattered over that profound wilderness miles and miles apart — at each station half a dozen persons. There was a plenty of cattle, the black natives were always ill-nourished and hungry. The land belonged to *them*. The whites had not bought it, and couldn't buy it; for the tribes had no chiefs, nobody in authority, nobody competent to sell and convey; and the tribes themselves had no comprehension of the idea of transferable ownership of land. The ousted owners were despised by the white interlopers, and this opinion was not hidden under a bushel. More promising materials for a tragedy could not have been collated. Let Mrs. Praed speak:

"At Nie Nie station, one dark night, the unsuspecting hut-keeper, having, as he believed, secured himself against assault, was lying wrapped in his blankets sleeping profoundly. The Blacks crept stealthily down the chimney and battered in his skull while he slept."

One could guess the whole drama from that little text. The curtain was up. It would not fall until the mastership of one party or the other was determined — and permanently:

"There was treachery on both sides. The Blacks killed the Whites when they found them defenseless, and the Whites slew the Blacks in a wholesale and promiscuous fashion which offended against my childish sense of justice.
. . . They were regarded as little above the level of brutes, and in some cases *were destroyed like vermin.*

"Here is an instance. A squatter, whose station was surrounded by Blacks, whom he suspected to be hostile and from whom he feared an attack, parleyed with them from his house-door. He told them it was Christmas-time — a time at which all men, black or white, feasted ; that there were flour, sugar-plums, good things in plenty in the store, and that he would make for them such a pudding as they had never dreamed of — a great pudding of which all might eat and be filled. The Blacks listened and were lost. The pudding was made and distributed. Next morning there was howling in the camp, for it had been sweetened with sugar and arsenic ! "

The white man's spirit was right, but his method was wrong. His spirit was the spirit which the civilized white has always exhibited toward the savage, but the use of poison was a departure from custom. True, it was merely a technical departure, not a real one; still, it was a departure, and therefore a mistake, in my opinion. It was better, kinder, swifter,

THE USUAL SPIRIT.

and much more humane than a number of the methods which have been sanctified by custom, but that does not justify its employment. That is, it does not wholly justify it. Its unusual nature makes it stand out and attract an amount of attention which it is not entitled to. It takes hold upon morbid imaginations and they work it up into a sort of exhibition of cruelty, and this smirches the good name of our civilization, whereas one of the old harsher methods would have had no such effect because usage has made those methods familiar to us and innocent. In many countries we have chained the savage and starved him to death; and this we do not care for, because custom has inured us to it; yet a quick death by poison is lovingkindness to it. In many countries we have burned the savage at the stake; and this we do not care for, because custom has inured us to it; yet a quick death is lovingkindness to it. In more than one country we have hunted the savage and his little children and their mother with dogs and guns through the woods and swamps for an afternoon's sport, and filled the region with happy laughter over their sprawling and stumbling flight, and their wild supplications for mercy; but this method we do not mind, because custom has inured us to it; yet a quick death by poison is lovingkindness to it. In many countries we have taken the savage's land from him, and made him our slave, and lashed him every day, and broken his pride, and made death his only friend, and overworked him till he dropped in his tracks; and this we do not care for, because custom has inured us to it; yet a quick death by poison is lovingkindness to it. In the Matabeleland to-day — why, there we are confining ourselves to sanctified custom, we Rhodes-Beit millionaires in South Africa and Dukes in London; and nobody cares, because we are used to the old holy customs, and all we ask is that no notice-inviting new ones shall be intruded upon the attention of our comfortable consciences. Mrs. Praed says of the poisoner, "That squatter deserves to have his name handed down to the contempt of posterity."

I am sorry to hear her say that. I myself blame him for one thing, and severely, but I stop there. I blame him for the indiscretion of introducing a novelty which was calculated to attract attention to our civilization. There was no occasion to do that. It was his duty, and it is every loyal man's duty to protect that heritage in every way he can; and the best way to do that is to attract attention elsewhere. The squatter's judgment was bad — that is plain; but his heart was right. He is almost the only pioneering representative of civilization in history who has risen above the prejudices of his caste and his heredity and tried to introduce the element of mercy into the superior race's dealings with the savage. His name is lost, and it is a pity; for it deserves to be handed down to posterity with homage and reverence.

This paragraph is from a London journal:

"To learn what France is doing to spread the blessings of civilization in her distant dependencies we may turn with advantage to New Caledonia. With a view to attracting free settlers to that penal colony, M. Feillet, the Governor, forcibly expropriated the Kanaka cultivators from the best of their plantations, with a derisory compensation, in spite of the protests of the Council General of the island. Such immigrants as could be induced to cross the seas thus found themselves in possession of thousands of coffee, cocoa, banana, and bread-fruit trees, the raising of which had cost the wretched natives years of toil, whilst the latter had a few five-franc pieces to spend in the liquor stores of Noumea."

You observe the combination? It is robbery, humiliation, and slow, slow murder, through poverty and the white man's whisky. The savage's gentle friend, the savage's noble friend, the only magnanimous and unselfish friend the savage has ever had, was not there with the merciful swift release of his poisoned pudding.

There are many humorous things in the world; among them the white man's notion that he is less savage than the other savages.*

---

*See Chapter on Tasmania, *post*.

CHAPTER XXII.

Nothing is so ignorant as a man's left hand, except a lady's watch.
— *Pudd'nhead Wilson's New Calendar.*

YOU notice that Mrs. Praed knows her art. She can place a thing before you so that you can see it. She is not alone in that. Australia is fertile in writers whose books are faithful mirrors of the life of the country and of its history. The materials were surprisingly rich, both in quality and in mass, and Marcus Clarke, Ralph Boldrewood, Gordon, Kendall, and the others, have built out of them a brilliant and vigorous literature, and one which must endure. Materials — there is no end to them! Why, a literature might be made out of the aboriginal all by himself, his character and ways are so freckled with varieties — varieties not staled by familiarity, but new to us. You do not need to invent any picturesquenesses; whatever you want in that line he can furnish you; and they will not be fancies and doubtful, but realities and authentic. In his history, as preserved by the white man's official records, he is everything — everything that a human creature can be. He covers the entire ground. He is a coward — there are a thousand fact to prove it. He is brave — there are a thousand facts to prove it. He is treacherous — oh, beyond imagination! he is faithful, loyal, true — the white man's records supply you with a harvest of instances of it that are noble, worshipful, and pathetically beautiful. He kills the starving stranger who comes begging for food and shelter — there is proof of it. He succors, and feeds, and guides to safety, to-day, the lost stranger who fired on him only yester-

day — there is proof of it. He takes his reluctant bride by
force, he courts her with a club, then loves her faithfully
through a long life — it is of record. He gathers to himself
another wife by the same processes, beats and bangs her
as a daily diversion, and by and by lays down his life in
defending her from some outside harm — it is of record.
He will face a hundred hostiles to rescue one of his children,
and will kill another of his children because the family is large
enough without it. His delicate stomach turns, at certain details of the white man's food; but he likes over-ripe fish, and
brazed dog, and cat, and rat, and will eat his own uncle with
relish. He is a sociable animal, yet he turns aside and hides
behind his shield when his mother-in-law goes by. He is
childishly afraid of ghosts and other trivialities that menace
his soul, but dread of physical pain is a weakness which he is
not acquainted with. He knows all the great and many of the
little constellations, and has names for them; he has a symbol-
writing by means of which he can convey messages far and
wide among the tribes; he has a correct eye for form and expression, and draws a good picture; he can track a fugitive by
delicate traces which the white man's eye cannot discern, and
by methods which the finest white intelligence cannot master; he makes a missile which science itself cannot duplicate
without the model — if with it; a missile whose secret baffled
and defeated the searchings and theorizings of the white
mathematicians for seventy years; and by an art all his own
he performs miracles with it which the white man cannot approach untaught, nor parallel after teaching. Within certain
limits this savage's intellect is the alertest and the brightest
known to history or tradition; and yet the poor creature was
never able to invent a counting system that would reach above
five, nor a vessel that he could boil water in. He is the prize-
curiosity of all the races. To all intents and purposes he is

dead — in the body; but he has features that will live in literature.

Mr. Philip Chauncy, an officer of the Victorian Government, contributed to its archives a report of his personal observations of the aboriginals which has in it some things which I wish to condense slightly and insert here. He speaks of the quickness of their eyes and the accuracy of their judgment of the direction of approaching missiles as being quite extraordinary, and of the answering suppleness and accuracy of limb and muscle in avoiding the missile as being extraordinary also. He has seen an aboriginal stand as a target for cricket-balls thrown with great force ten or fifteen yards, by professional bowlers, and successfully dodge them or parry them with his shield during about half an hour. One of those balls, properly placed, could have killed him; "Yet he depended, with the utmost self-possession, on the quickness of his eye and his agility."

The shield was the customary war-shield of his race, and would not be a protection to you or to me. It is no broader than a stovepipe, and is about as long as a man's arm. The opposing surface is not flat, but slopes away from the center-line like a boat's bow. The difficulty about a cricket-ball that has been thrown with a scientific "twist" is, that it suddenly changes it course when it is close to its target and comes straight for the mark when apparently it was going overhead or to one side. I should not be able to protect myself from such balls for half-an-hour, or less.

Mr. Chauncy once saw "a little native man" throw a cricket-ball 119 yards. This is said to beat the English professional record by thirteen yards.

We have all seen the circus-man bound into the air from a spring-board and make a somersault over eight horses standing side by side. Mr. Chauncy saw an aboriginal do it over eleven;

and was assured that he had sometimes done it over fourteen. But what is that to this:

"I saw the same man leap from the *ground*, and in going over he dipped his head, unaided by his hands, into a hat placed in an inverted position on the top of the head of another man sitting upright on horseback — both man and horse being of the average size. The native landed on the other side of the horse with the hat fairly on his head. The prodigious height of the leap, and the precision with which it was taken so as to enable him to dip his head into the hat, exceeded any feat of the kind I have ever beheld."

I should think so! On board a ship lately I saw a young Oxford athlete *run four steps* and spring into the air and squirm his hips by a side-twist over a bar that was five and one-half feet high; but he could not have stood still and cleared a bar that was *four* feet high. I know this, because I tried it myself.

One can see now where the kangaroo learned its art.

Sir George Grey and Mr. Eyre testify that the natives dug wells fourteen or fifteen feet deep and two feet in diameter at the bore — dug them in the *sand* — wells that were "quite circular, carried straight down, and the work beautifully executed."

Their tools were their hands and feet. How did they throw sand out from such a depth? How could they stoop down and get it, with only two feet of space to stoop in? How did they keep that sand-pipe from caving in on them? I do not know. Still, they did manage those seeming impossibilities. Swallowed the sand, may be.

Mr. Chauncy speaks highly of the patience and skill and alert intelligence of the native huntsman when he is stalking the emu, the kangaroo, and other game:

"As he walks through the bush his step is light, elastic, and noiseless; every track on the earth catches his keen eye; a leaf, or fragment of a stick turned, or a blade of grass recently bent by the tread of one of the lower animals, instantly arrests his attention; in fact, nothing escapes his quick and powerful sight on the ground, in the trees, or in the distance, which may supply him with a meal or warn him of danger. A little examination of the trunk of a tree which may be nearly covered with the scratches of opossums ascending and descending is sufficient to inform him whether one *went up the night before without coming down again* or not."

Fennimore Cooper lost his chance. He would have known how to value these people. He wouldn't have traded the dullest of them for the brightest Mohawk he ever invented.

All savages draw outline pictures upon bark; but the resemblances are not close, and expression is usually lacking. But the Australian aboriginal's pictures of animals were nicely accurate in form, attitude, carriage; and he put spirit into them, and expression. And his pictures of white people and natives were pretty nearly as good as his pictures of the other animals. He dressed his whites in the fashion of their day, both the ladies and the gen-

HIS PLACE IN ART.

tlemen. As an untaught wielder of the pencil it is not likely that he has had his equal among savage people.

His place in art — as to drawing, not color-work — is well up, all things considered. His art is not to be classified with savage art at all, but on a plane two degrees above it and one degree above the lowest plane of civilized art. To be exact, his place in art is between Botticelli and De Maurier. That is to say, he could not draw as well as De Maurier but better than Boticelli. In feeling, he resembles both; also in grouping and in his preferences in the matter of subjects. His "corrobboree" of the Australian wilds reappears in De Maurier's Belgravian ballrooms, with clothes and the smirk of civilization added; Botticelli's "Spring" is the corrobboree further idealized, but with fewer clothes and more smirk. And well enough as to intention, *but* — my word!

The aboriginal can make a fire by friction. I have tried that.

All savages are able to stand a good deal of physical pain. The Australian aboriginal has this quality in a well-developed degree. Do not read the following instances if horrors are not pleasant to you. They were recorded by the Rev. Henry N. Wolloston, of Melbourne, who had been a surgeon before he became a clergyman:

1. "In the summer of 1852 I started on horseback from Albany, King George's Sound, to visit at Cape Riche, accompanied by a native on foot. We traveled about forty miles the first day, then camped by a water-hole for the night. After cooking and eating our supper, I observed the native, who had said nothing to me on the subject, collect the hot embers of the fire together, and deliberately place his right foot in the glowing mass for a moment, then suddenly withdraw it, stamping on the ground and uttering a long-drawn guttural sound of mingled pain and satisfaction. This operation he repeated several times. On my inquiring the meaning of his strange conduct, he only said, 'Me carpenter-make 'em' ('I am mending my foot'), and then showed me his charred great toe, the nail of which had been torn off by a tea-tree stump, in which it had been caught during the journey, and the pain of which he had borne with stoical composure until the evening, when he had an opportunity of cauterizing the wound in the primitive manner above described."

And he proceeded on the journey the next day, "as if nothing had happened" — and walked thirty miles. It was a strange idea, to keep a surgeon and then do his own surgery.

2. "A native about twenty-five years of age once applied to me, as a doctor, to extract the wooden barb of a spear, which, during a fight in the bush some four months previously, had entered his chest, just missing the heart, and penetrated the viscera to a considerable depth. The spear had been cut off, leaving the barb behind, which continued to force its way by muscular action gradually toward the back; and when I examined him I could feel a hard substance between the ribs below the left blade-bone. I made a deep incision, and with a pair of forceps extracted the barb, which was made, as usual, of hard wood about four inches long and from half an inch to an inch thick. It was very smooth, and partly digested, so to speak, by

the maceration to which it had been exposed during its four months' journey through the body. The wound made by the spear had long since healed, leaving only a small cicatrix ; and after the operation, which the native bore without flinching, he appeared to suffer no pain. Indeed, judging from his good state of health, the presence of the foreign matter did not materially annoy him. He was perfectly well in a few days."

But No. 3 is my favorite. Whenever I read it I seem to enjoy all that the patient enjoyed — whatever it was :

3. "Once at King George's Sound a native presented himself to me with one leg only, and requested me to supply him with a wooden leg. He had traveled in this maimed state about ninety-six miles, for this purpose. I examined the limb, which had been severed just below the knee, and found that it had been charred by fire, while about two inches of the partially calcined

"NO FEELING IN IT."

bone protruded through the flesh. I at once removed this with the saw ; and having made as presentable a stump of it as I could, covered the amputated end of the bone with a surrounding of muscle, and kept the patient a few days under my care to allow the wound to heal. On inquiring, the native told me that in a fight with other blackfellows a spear had struck his leg and penetrated the bone below the knee. Finding it was serious, he had recourse to the following crude and barbarous operation, which it appears is not uncommon among these people in their native state. He made a fire, and dug a hole in the earth only sufficiently large to admit his leg, and deep enough to allow the wounded part to be on a level with the surface of the ground. He then *surrounded the limb with the live coals* or charcoal, which was replenished until the leg was literally burnt off. The cauterization thus applied completely checked the hemorrhage, and he was able in a day or two to hobble down to the Sound, with the aid of a long stout stick, although he was more than a week on the road."

But he was a fastidious native. He soon discarded the wooden leg made for him by the doctor, because "it had no feeling in it." It must have had as much as the one he burnt off, I should think.

So much for the Aboriginals. It is difficult for me to let them alone. They are marvelously interesting creatures. For a quarter of a century, now, the several colonial governments have housed their remnants in comfortable stations, and fed them well and taken good care of them in every way. If I had found this out while I was in Australia I could have seen some of those people — but I didn't. I would walk thirty miles to see a stuffed one.

Australia has a slang of its own. This is a matter of course. The vast cattle and sheep industries, the strange aspects of the country, and the strange native animals, brute and human, are matters which would naturally breed a local slang. I have notes of this slang somewhere, but at the moment I can call to mind only a few of the words and phrases. They are expressive ones. The wide, sterile, unpeopled deserts have created eloquent phrases like "No Man's Land" and the "Never-never Country." Also this felicitous form : "She lives in the Never-never Country" — that is, she is an old maid. And this one is not without merit: "heifer-paddock" — young ladies' seminary. "Bail up" and "stick up" — equivalent of our highwayman-term to "hold up" a stage-coach or a train. "New-chum" is the equivalent of our "tenderfoot" — new arrival.

And then there is the immortal "My word!" We must import it. "M-y *word!*" In cold print it is the equivalent of our "Ger-*reat Cæsar!*" but spoken with the proper Australian unction and fervency, it is worth six of it for grace and charm and expressiveness. Our form is rude and explosive; it is not suited to the drawing-room or the heifer-paddock; but "M-y *word!* " is, and is music to the ear, too, when the utterer knows how to say it. I saw it in print several times on the Pacific Ocean, but it struck me coldly, it aroused no sympathy. That was because it was the dead corpse of the thing, the soul was not there — the tones were lacking — the informing spirit — the deep feeling — the eloquence. But the first time I heard an Australian say it, it was positively thrilling.

CHAPTER XXIII.

WE left Adelaide in due course, and went to Horsham, in the colony of Victoria; a good deal of a journey, if I remember rightly, but pleasant. Horsham sits in a plain which is as level as a floor — one of those famous dead levels which Australian books describe so often; gray, bare, sombre, melancholy, baked, cracked, in the tedious long drouths, but a horizonless ocean of vivid green grass the day after a rain. A country town, peaceful, reposeful, inviting, full of snug homes, with garden plots, and plenty of shrubbery and flowers.

"*Horsham, October 17.* At the hotel. The weather divine. Across the way, in front of the London Bank of Australia, is a very handsome cottonwood. It is in opulent leaf, and every leaf perfect. The full power of the on-rushing spring is upon it, and I imagine I can see it grow. Alongside the bank and a little way back in the garden there is a row of soaring fountain-sprays of delicate feathery foliage quivering in the breeze, and mottled with flashes of light that shift and play through the mass like flash-lights through an opal — a most beautiful tree, and a striking contrast to the cottonwood. Every leaf of the cottonwood is distinctly defined — it is a kodak for faithful, hard, unsentimental detail; the other an impressionist picture, delicious to look upon, full of a subtle and exquisite charm, but all details fused in a swoon of vague and soft loveliness."

It turned out, upon inquiry, to be a pepper tree — an im-

portation from China. It has a silky sheen, soft and rich.
I saw some that had long red bunches of currant-like berries
ambushed among the foliage.   At a distance, in certain lights,
they give the tree a pinkish tint and a new charm.

There is an agricultural college eight miles from Horsham.
We were driven out to it by its chief.   The conveyance was
an open wagon; the time, noonday; no wind; the sky without a cloud, the sunshine brilliant — and the mercury at 92° in
the shade.   In some countries an indolent unsheltered drive of
an hour and a half under such conditions would have been a

sweltering and prostrating experience; but there was nothing of that
in this case.   It is a climate that
is perfect.   There was no sense of
heat; indeed, there was no heat;
the air was fine and pure and exhilarating; if the drive had lasted half
a day I think we should not have
felt any discomfort, or grown silent
or droopy or tired.   Of course, the
secret of it was the exceeding dry-
ness of the atmosphere.   In that
plain 112° in the shade is without
doubt no harder upon a man than is
88° or 90° in New York.

A WIDE SPACE.

The road lay through the middle
of an empty space which seemed to me to be a hundred yards
wide between the fences.   I was not given the width in yards,
but only in chains and perches — and furlongs, I think.   I
would have given a good deal to know what the width was,
but I did not pursue the matter.   I think it is best to put
up with information the way you get it; and seem satisfied
with it, and surprised at it, and grateful for it, and say, " My

word!" and never let on. It was a wide space; I could tell you how wide, in chains and perches and furlongs and things, but that would not help you any. Those things sound well, but they are shadowy and indefinite, like troy weight and avoirdupois; nobody knows what they mean. When you buy a pound of a drug and the man asks you which you want, troy or avoirdupois, it is best to say "Yes," and shift the subject.

They said that the wide space dates from the earliest sheep and cattle-raising days. People had to drive their stock long distances — immense journeys — from worn-out places to new ones where were water and fresh pasturage; and this wide space had to be left in grass and unfenced, or the stock would have starved to death in the transit.

On the way we saw the usual birds — the beautiful little green parrots, the magpie, and some others; and also the slender native bird of modest plumage and the eternally-forgetable name — the bird that is the smartest among birds, and can give a parrot 30 to 1 in the game and then talk him to death. I cannot recall that bird's name. I think it begins with M. I wish it began with G. or something that a person can remember.

The magpie was out in great force, in the fields and on the fences. He is a handsome large creature, with snowy white decorations, and is a singer; he has a murmurous rich note that is lovely. He was once modest, even diffident; but he lost all that when he found out that he was Australia's sole musical bird. He has talent, and cuteness, and impudence; and in his tame state he is a most satisfactory pet — never coming when he is called, always coming when he isn't, and studying disobedience as an accomplishment. He is not confined, but loafs all over the house and grounds, like the laughing jackass. I

15

think he learns to talk, I know he learns to sing tunes, and his friends say that he knows how to steal without learning. I was acquainted with a tame magpie in Melbourne. He had lived in a lady's house several years, and believed he owned it. The lady had tamed him, and in return he had tamed the lady. He was always on deck when not wanted, always having his own way, always tyrannizing over the dog, and always making the cat's life a slow sorrow and a martyrdom. He knew a number of tunes and could sing them in perfect time and tune; and would do it, too, at any time that silence was wanted; and then encore himself and do it again; but if he was asked to sing he would go out and take a walk.

It was long believed that fruit trees would not grow in that baked and waterless plain around Horsham, but the agricultural college has dissipated that idea. Its ample nurseries were producing oranges, apricots, lemons, almonds, peaches, cherries, 48 varieties of apples — in fact, all manner of fruits, and in abundance. The trees did not seem to miss the water; they were in vigorous and flourishing condition.

Experiments are made with different soils, to see what things thrive best in them and what climates are best for them. A man who is ignorantly trying to produce upon his farm things not suited to its soil and its other conditions can make a journey to the college from anywhere in Australia, and go back with a change of scheme which will make his farm productive and profitable.

There were forty pupils there — a few of them farmers, re-learning their trade, the rest young men mainly from the cities — novices. It seemed a strange thing that an agricultural college should have an attraction for city-bred youths, but such is the fact. They are good stuff, too; they are above the agricultural average of intelligence, and they come without any inherited prejudices in favor of hoary ignorances made sacred by long descent.

The students work all day in the fields, the nurseries, and the shearing-sheds, learning and doing all the practical work of the business — three days in a week. On the other three they study and hear lectures. They are taught the beginnings of such sciences as bear upon agriculture — like chemistry, for instance. We saw the sophomore class in sheep-shearing shear a dozen sheep. They did it by hand, not with the machine. The sheep was seized and flung down on his side and held there; and the students took off his coat with great celerity and adroitness. Sometimes they clipped off a sample of the sheep, but that is customary with shearers, and they don't mind it; they don't even mind it as much as the sheep. They dab a splotch of sheep-dip on the place and go right ahead.

The coat of wool was unbelievably thick. Before the shearing the sheep looked like the fat woman in the circus; after it he looked like a bench. He was clipped to the skin; and smoothly and uniformly. The fleece comes from him all in one piece and has the spread of a blanket.

The college was flying the Australian flag — the gridiron of England smuggled up in the northwest corner of a big red field that had the random stars of the Southern Cross wandering around over it.

From Horsham we went to Stawell. By rail. Still in the colony of Victoria. Stawell is in the gold-mining country. In the bank-safe was half a peck of surface-gold — gold dust, grain gold; rich; pure in fact, and pleasant to sift through one's fingers; and would be pleasanter if it would stick. And there were a couple of gold bricks, very heavy to handle, and worth $7,500 a piece. They were from a very valuable quartz mine; a lady owns two-thirds of it; she has an income of $75,000 a month from it, and is able to keep house.

The Stawell region is not productive of gold only; it has great vineyards, and produces exceptionally fine wines. One of these vineyards — the Great Western, owned by Mr. Irving — is regarded as a model. Its product has reputation abroad. It yields a choice champagne and a fine claret, and its hock took a prize in France two or three years ago. The champagne is kept in a maze of passages under ground, cut in the rock, to secure it an even temperature during the three-year term required to perfect it. In those vaults I saw 120,000 bottles of champagne. The colony of Victoria has a population of 1,000,-000, and those people are said to drink 25,000,000 bottles of

THE THREE SISTERS.

champagne per year. The dryest community on the earth. The government has lately reduced the duty upon foreign wines. That is one of the unkindnesses of Protection. A man invests years of work and a vast sum of money in a worthy enterprise, upon the faith of existing laws; then the law is changed, and the man is robbed by his own government.

On the way back to Stawell we had a chance to see a group of boulders called the Three Sisters — a curiosity oddly located; for it was upon high ground, with the land sloping away from it, and no height above it from whence the boulders could have rolled down. Relics of an early ice-drift, perhaps. They are noble boulders. One of them has the size and smoothness and plump sphericity of a balloon of the biggest pattern.

The road led through a forest of great gum-trees, lean and scraggy and sorrowful. The road was cream-white — a clayey kind of earth, apparently. Along it toiled occasional freight wagons, drawn by long double files of oxen. Those wagons were going a journey of two hundred miles, I was told, and were running a successful opposition to the railway! The railways are owned and run by the government.

Those sad gums stood up out of the dry white clay, pictures of patience and resignation. It is a tree that can get along without water; still it is fond of it — ravenously so. It is a very intelligent tree and will detect the presence of hidden water at a distance of fifty feet, and send out slender long root-fibres to prospect it. They will find it; and will also get at it — even through a cement wall six inches thick. Once a cement water-pipe under ground at Stawell began to gradually reduce its output, and finally ceased altogether to deliver water. Upon examining into the matter it was found stopped up, wadded compactly with a mass of root-fibres, delicate and hair-like. How this stuff had gotten into the pipe was a puzzle for some little time; finally it was found that it had crept in through a crack that was almost invisible to the eye. A gum tree forty feet away had tapped the pipe and was drinking the water.

CHAPTER XXIV.

There is no such thing as "the Queen's English." The property has gone into the hands of a joint stock company and we own the bulk of the shares!
— *Pudd'nhead Wilson's New Calendar.*

FREQUENTLY, in Australia, one has cloud-effects of an unfamiliar sort. We had this kind of scenery, finely staged, all the way to Ballarat. Consequently we saw more sky than country on that journey. At one time a great stretch of the vault was densely flecked with wee ragged-edged flakes of painfully white cloud-stuff, all of one shape and size, and equidistant apart, with narrow cracks of adorable blue showing between. The whole was suggestive of a hurricane of snow-flakes drifting across the skies. By and by these flakes fused themselves together in interminable lines, with shady faint hollows between the lines, the long satin-surfaced rollers following each other in simulated movement, and enchantingly counterfeiting the majestic march of a flowing sea. Later, the sea solidified itself; then gradually broke up its mass into innumerable lofty white pillars of about one size, and ranged these across the firmament, in receding and fading perspective, in the similitude of a stupendous colonnade — a mirage without a doubt flung from the far Gates of the Hereafter.

The approaches to Ballarat were beautiful. The features, great green expanses of rolling pasture-land, bisected by eye-contenting hedges of commingled new-gold and old-gold gorse — and a lovely lake. One must put in the pause, there, to fetch the reader up with a slight jolt, and keep him from gliding by without noticing the lake. One *must* notice it; for a lovely lake is not as common a thing along the railways of

Australia as are the dry places. Ninety-two in the shade again, but balmy and comfortable, fresh and bracing. A perfect climate.

Forty-five years ago the site now occupied by the City of Ballarat was a sylvan solitude as quiet as Eden and as lovely. Nobody had ever heard of it. On the 25th of August, 1851, the first *great* gold-strike made in Australia was made here. The wandering prospectors who made it scraped up two pounds and a half of gold the first day — worth $600. A few days later the place was a hive — a town. The news of the strike spread everywhere in a sort of instantaneous way — spread like a flash to the very ends of the earth. A celebrity so prompt and so universal has hardly been paralleled in history, perhaps. It was as if the name BALLARAT had suddenly been written on the sky, where all the world could read it at once.

The smaller discoveries made in the colony of New South Wales three months before had already started emigrants toward Australia; they had been coming as a stream, but they came as a flood, now. A hundred thousand people poured into Melbourne from England and other countries in a single month, and flocked away to the mines. The crews of the ships that brought them flocked with them; the clerks in the government offices followed; so did the cooks, the maids, the coachmen, the butlers, and the other domestic servants; so did the carpenters, the smiths, the plumbers, the painters, the reporters, the editors, the lawyers, the clients, the barkeepers, the bummers, the blacklegs, the thieves, the loose women, the grocers, the butchers, the bakers, the doctors, the druggists, the nurses; so did the police; even officials of high and hitherto envied place threw up their positions and joined the procession. This roaring avalanche swept out of Melbourne and left it desolate, Sunday-like, paralyzed, everything at a stand-still, the ships lying idle at anchor, all signs of life departed, all sounds stilled save the rasping of the cloud-shadows as they scraped across the vacant streets.

That grassy and leafy paradise at Ballarat was soon ripped open, and lacerated and scarified and gutted, in the feverish search for its hidden riches. There is nothing like surface-mining to snatch the graces and beauties and benignities out of a paradise, and make an odious and repulsive spectacle of it.

What fortunes were made! Immigrants got rich while the ship unloaded and reloaded — and went back home for good in the same cabin they had come out in! Not all of them. Only some. I saw the others in Ballarat myself, forty-five years later — what were left of them by time and death and the disposition to rove. They were young and gay, then; they are patriarchal and grave, now; and they do not get excited any more. They talk of the Past. They live in it. Their life is a dream, a retrospection.

Ballarat was a great region for " nuggets." No such nuggets were found in California as Ballarat produced. In fact, the Ballarat region has yielded the largest ones known to history. Two of them weighed about 180 pounds each, and together were worth $90,000. They were offered to any poor person who would shoulder them and carry them away. Gold was so plentiful that it made people liberal like that.

Ballarat was a swarming city of tents in the early days. Everybody was happy, for a time, and apparently prosperous. Then came trouble. The government swooped down with a mining tax. And in its worst form, too; for it was not a tax upon what the miner had taken out, but upon what he was *going* to take out — if he could find it. It was a license-tax — license to work his claim — and it had to be paid before he could begin digging.

Consider the situation. No business is so uncertain as

surface-mining. Your claim may be good, and it may be worthless. It may make you well off in a month; and then again you may have to dig and slave for half a year, at heavy expense, only to find out at last that the gold is not there in cost-paying quantity, and that your time and your hard work have been thrown away. It might be wise policy to advance the miner a monthly sum to encourage him to develop the country's riches; but to tax him monthly in advance instead — why, such a thing was never dreamed of in America. There, neither the claim itself nor its products, howsoever rich or poor, were taxed.

The Ballarat miners protested, petitioned, complained — it was of no use; the government held its ground, and went on collecting the tax. And not by pleasant methods, but by ways which must have been very galling to free people. The rumblings of a coming storm began to be audible.

By and by there was a result; and I think it may be called the finest thing in Australasian history. It was a revolution — small in size, but great politically; it was a strike for liberty, a struggle for a principle, a stand against injustice and oppression. It was the Barons and John, over again; it was Hampden and Ship-Money; it was Concord and Lexington; small beginnings, all of them, but all of them great in political results, all of them epoch-making. It is another instance of a victory won by a lost battle. It adds an honorable page to history; the people know it and are proud of it. They keep green the memory of the men who fell at the Eureka Stockade, and Peter Lalor has his monument.

The surface-soil of Ballarat was full of gold. This soil the miners ripped and tore and trenched and harried and disembowled, and made it yield up its immense treasure. Then they went down into the earth with deep shafts, seeking the gravelly beds of ancient rivers and brooks — and found them. They followed the courses of these streams, and gutted them, sending the gravel up in buckets to the upper world, and washing out of it its enormous deposits of gold. The next biggest of the two monster nuggets mentioned above came from an old river-channel 180 feet under ground.

Finally the quartz lodes were attacked. That is not poorman's mining. Quartz-mining and milling require capital, and staying-power, and patience. Big companies were formed, and for several decades, now, the lodes have been successfully worked, and have yielded great wealth. Since the gold discovery in 1853 the Ballarat mines — taking the three kinds of mining together — have contributed to the world's pocket something over *three hundred millions of dollars*, which is to say that this nearly invisible little spot on the earth's surface has yielded about one-fourth as much gold in forty-four years as all California has yielded in forty-seven. The Californian aggregate, from 1848 to 1895, inclusive, as reported by the Statistician of the United States Mint, is $1,265,217,217.

A citizen told me a curious thing about those mines. With all my experience of mining I had never heard of anything of the sort before. The main gold reef runs about north and south — of course — for that is the custom of a rich gold reef. At Ballarat its course is between walls of slate. Now the citizen told me that throughout a stretch of twelve miles along the reef, the reef is crossed at intervals by a straight black streak of a carbonaceous nature — a streak in the slate; a streak no thicker than a pencil — and that wherever it crosses the reef you will certainly find gold at the junction. It is called the Indicator. Thirty feet on each side of the Indicator (and down in the slate, of course) is a still finer streak — a streak as fine as a pencil mark; and indeed, that is its name — Pencil Mark. Whenever you find the Pencil Mark you know that thirty feet from it is the Indicator; you measure the dis-

BALLARAT STATUARY.

tance, excavate, find the Indicator, trace it straight to the reef, and sink your shaft; your fortune is made, for certain. If that is true, it is curious. And it is curious any way.

Ballarat is a town of only 40,000 population; and yet, since it is in Australia, it has every essential of an advanced and enlightened big city. This is pure matter of course. I must stop dwelling upon these things. It is hard to keep from dwelling upon them, though; for it is difficult to get away from the surprise of it. I will let the other details go, this time, but I must allow myself to mention that this little town has a park of 326 acres; a flower garden of 83 acres, with an elaborate and expensive fernery in it and some costly and unusually fine statuary; and an artificial lake covering 600 acres, equipped with a fleet of 200 shells, small sail boats, and little steam yachts.

At this point I strike out some other praiseful things which I was tempted to add. I do not strike them out because they were not true or not well said, but because I find them better said by another man — and a man more competent to testify, too, because he belongs on the ground, and knows. I clip them from a chatty speech delivered some years ago by Mr. William Little, who was at that time mayor of Ballarat:

" The language of our citizens, in this as in other parts of Australasia, is mostly healthy Anglo-Saxon, free from Americanisms, vulgarisms, and the conflicting dialects of our Fatherland, and is pure enough to suit a Trench or a Latham. Our youth, aided by climatic influence, are in point of physique and comeliness unsurpassed in the Sunny South. Our young men are well ordered ; and our maidens, ' not stepping over the bounds of modesty,' are as fair as Psyches, dispensing smiles as charming as November flowers."

The closing clause has the seeming of a rather frosty compliment, but that is apparent only, not real. November is summer-time there.

His compliment to the local purity of the language is warranted. It is quite free from impurities; this is acknowledged far and wide. As in the German Empire all cultivated people claim to speak Hanovarian German, so in Australasia all culti- vated people claim to speak Ballarat English. Even in Eng- land this cult has made considerable progress, and now that it is favored by the two great Universities, the time is not far away when Ballarat English will come into general use among the educated classes of Great Britain at large. Its great merit is, that it is shorter than ordinary English — that is, it is more compressed. At first you have some difficulty in understand- ing it when it is spoken as rapidly as the orator whom I have quoted speaks it. An illustration will show what I mean. When he called and I handed him a chair, he bowed and said :

"Q."

Presently, when we were lighting our cigars, he held a match to mine and I said :

"Thank you," and he said :

"Km."

Then I saw. Q is the end of the phrase "I thank you" Km is the end of the phrase "You are welcome." Mr. Little puts no emphasis upon either of them, but delivers them so reduced that they hardly have a sound. All Ballarat English is like that, and the effect is very soft and pleasant ; it takes all the hardness and harshness out of our tongue and gives to it a delicate whispery and vanishing cadence which charms the ear like the faint rustling of the forest leaves.

"DO YOU REMEMBER THAT TRIP?"

CHAPTER XXV.

" Classic." A book which people praise and don't read.
—*Pudd'nhead Wilson's New Calendar.*

ON the rail again — bound for Bendigo. From diary :
October 23. Got up at 6, left at 7.30 ; soon reached Castlemaine, one of the rich gold-fields of the early days ; waited several hours for a train ; left at 3.40 and reached Bendigo in an hour. For comrade, a Catholic priest who was better than I was, but didn't seem to know it — a man full of graces of the heart, the mind, and the spirit ; a lovable man. He will rise. He will be a bishop some day. Later an Archbishop. Later a Cardinal. Finally an Archangel, I hope. And then he will recall me when I say, " Do you remember that trip we made from Ballarat to Bendigo, when you were nothing but Father C., and I was nothing to what I am now ? " It has actually taken nine hours to come from Ballarat to Bendigo. We could have saved seven by walking. However, there was no hurry.

Bendigo was another of the rich strikes of the early days. It does a great quartz-mining business, now — that business which, more than any other that I know of, teaches patience, and requires grit and a steady nerve. The town is full of towering chimney-stacks, and hoisting-works, and looks like a petroleum-city. Speaking of patience ; for example, one of the local companies went steadily on with its deep borings and searchings without show of gold or a penny of reward for *eleven years* — then struck it, and became suddenly rich. The eleven years' work had cost $55,000, and the first gold found

16   (241)

was a grain the size of a pin's head. It is kept under locks and bars, as a precious thing, and is reverently shown to the visitor, ",hats off."   When I saw it I had not heard its history.

" It is gold. Examine it — take the glass. Now how much should you say it is worth ? "

I said —

"I should say about two cents; or in your English dialect, four farthings."

" Well, it cost £11,000."

" Oh, come ! "

" Yes, it did. Ballarat and Bendigo have produced the three monumental nuggets of the world, and this one is the monumentalest one of the three.   The other two represent £9,000 a piece; this one a couple of thousand more. It is small, and not much to look at, but it is entitled to it name — Adam.   It is the Adam-nugget of this mine, and its children run up into the millions."

Speaking of patience again, another of the mines was worked, under heavy expenses, during 17 years before pay was struck, and still another one compelled a wait of 21 years before pay was struck ; then, in both instances, the outlay was all back in a year or two, with compound interest.

Bendigo has turned out even more gold than Ballarat. The two together have produced $650,000,000 worth — which is half as much as California has produced.

It was through Mr. Blank — not to go into particulars about his name — it was mainly through Mr. Blank that my stay in Bendigo was made memorably pleasant and interesting. He explained this to me himself.   He told me that it was through his influence that the city government invited me to the town-hall to hear complimentary speeches and respond to them ; that it was through his influence that I had been taken on a long pleasure-drive through the city and shown its notable features; that it was through his influence that I was invited to visit the great mines; that it was through his influence that I was taken to the hospital and allowed to see the

convalescent Chinaman who had been attacked at midnight in his lonely hut eight weeks before by robbers, and stabbed forty-six times and scalped besides; that it was through his influence that when I arrived this awful spectacle of piecings and patchings and bandagings was sitting up in his cot letting on to read one of my books; that it was through his influence that efforts had been made to get the Catholic Archbishop of Bendigo to

ALL THROUGH HIS INFLUENCE.

invite me to dinner; that it was through his influence that efforts had been made to get the Anglican Bishop of Bendigo to ask me to supper; that it was through his influence that the dean of the editorial fraternity had driven me through the woodsy outlying country and shown me, from the summit of Lone Tree Hill, the mightiest and loveliest expanse of forest-clad mountain and valley that I had seen in all Australia. And when he asked me what had most impressed me in Bendigo and I answered and said it was the taste and the public spirit which had adorned the streets with 105 miles of shade trees, he said that it was through his influence that it had been done.

But I am not representing him quite correctly. He did not *say* it was through his influence that all these things had happened — for that would have been coarse; he merely *conveyed* that idea; conveyed it so subtly that I only caught it fleetingly, as one catches vagrant faint breaths of perfume when one traverses the meadows in summer; conveyed it without offense and without any suggestion of egoism or ostentation — but *conveyed* it, nevertheless.

He was an Irishman; an educated gentleman; grave, and kindly, and courteous; a bachelor, and about forty-five or possibly fifty years old, apparently. He called upon me at the hotel, and it was there that we had this talk. He made me like him, and did it without trouble. This was partly through his winning and gentle ways, but mainly through the amazing familiarity with my books which his conversation showed. He was down to date with them, too; and if he had made them the study of his life he could hardly have been better posted as to their contents than he was. He made me better satisfied with myself than I had ever been before. It was plain that he had a deep fondness for humor, yet he never laughed; he never even chuckled; in fact, humor could not win to outward expression on his face at all. No, he was always grave — tenderly, pensively grave; but he made *me* laugh, all along; and this was very trying — and very pleasant at the same time — for it was at quotations from my own books.

When he was going, he turned and said —

"You don't remember me?"

"I? Why, no. Have we met before?"

"No, it was a matter of correspondence."

"Correspondence?"

"Yes, many years ago. Twelve or fifteen. Oh, longer than that. But of course you —" A musing pause. Then he said —

"Do you remember Corrigan Castle?"

"N – no, I believe I don't. I don't seem to recall the name."

He waited a moment, pondering, with the door-knob in his hand, then started out; but turned back and said that I had once been interested in Corrigan Castle, and asked me if I would go with him to his quarters in the evening and take a hot Scotch and talk it over. I was a teetotaler and liked relaxation, so I said I would.

We drove from the lecture-hall together about half-past ten. He had a most comfortably and tastefully furnished parlor, with good pictures on the walls, Indian and Japanese ornaments on the mantel, and here and there, and books everywhere — largely mine; which made me proud. The light was brilliant, the easy chairs were deep-cushioned, the arrangements for brewing and smoking were all there. We brewed and lit up; then he passed a sheet of note-paper to me and said —

"Do you remember that?"

"Oh, yes, indeed!"

The paper was of a sumptuous quality. At the top was a twisted and interlaced monogram printed from steel dies in gold and blue and red, in the ornate English fashion of long years ago; and under it, in neat gothic capitals was this — printed in blue:

THE MARK TWAIN CLUB
CORRIGAN CASTLE

............187..

"My!" said I, "how did you come by this?"

"I was President of it."

"No! — you don't mean it."

"It is true. I was its first President. I was re-elected annually as long as its meetings were held in my castle — Corrigan — which was five years."

Then he showed me an album with twenty-three photographs of me in it. Five of them were of old dates, the others of various later crops; the list closed with a picture taken by Falk in Sydney a month before.

"You sent us the first five; the rest were bought."

This was paradise! We ran late, and talked, talked, talked — subject, the Mark Twain Club of Corrigan Castle, Ireland.

My first knowledge of that Club dates away back; all of twenty years, I should say. It came to me in the form of a courteous letter, written on the note-paper which I have described, and signed "By order of the President; C. PEMBROKE, Secretary." It conveyed the fact that the Club had been created in my honor, and added the hope that this token of appreciation of my work would meet with my approval.

I answered, with thanks; and did what I could to keep my gratification from over-exposure.

It was then that the long correspondence began. A letter came back, by order of the President, furnishing me the names of the members — thirty-two in number. With it came a copy of the Constitution and By-Laws, in pamphlet form, and artistically printed. The initiation fee and dues were in their proper place; also, schedule of meetings — monthly — for essays upon works of mine, followed by discussions; quarterly for business and a supper, without essays, but with after-supper speeches also, there was a list of the officers: President, Vice-President, Secretary, Treasurer, etc. The letter was brief, but it was pleasant reading, for it told me about the strong interest which the membership took in their new venture, etc., etc. It also asked me for a photograph — a special one. I went down and sat for it and sent it — with a letter, of course.

Presently came the badge of the Club, and very dainty and and pretty it was; and very artistic. It was a frog peeping out from a graceful tangle of grass-sprays and rushes, and was done in enamels on a gold basis, and had a gold pin back of it. After I had petted it, and played with it, and caressed it, and enjoyed it a couple of hours, the light happened to fall upon it at a new angle, and revealed to me a cunning new detail; with the light just right, certain delicate shadings of the grass-blades and rush-stems wove themselves into a monogram — mine! You can see that that

THE CLUB BADGE.

jewel was a work of art. And when you come to consider the intrinsic value of it, you must concede that it is not every literary club that could afford a badge like that. It was easily worth $75, in the opinion of Messrs. Marcus and Ward of New York. They said they could not duplicate it for that and make a profit.

By this time the Club was well under way; and from that time forth its secretary kept my off-hours well supplied with business. He reported the Club's discussions of my books with laborious fullness, and did his work with great spirit and ability. As a rule, he synopsized; but when a speech was especially brilliant, he short-handed it and gave me the best passages from it, written out. There were five speakers whom he particularly favored in that way: Palmer, Forbes, Naylor, Norris, and Calder. Palmer and Forbes could never get through a speech without attacking each other, and each in his own way was formidably effective — Palmer in virile and eloquent abuse, Forbes in courtly and elegant but scalding satire. I could always tell which of them was talking without looking for his name. Naylor had a polished style and a happy knack at felicitous metaphor; Norris's style was wholly without ornament, but enviably compact, lucid, and strong. But after all, Calder was the gem. He never spoke when sober, he spoke continuously when he wasn't. And certainly they were the drunkest speeches that a man ever uttered. They were full of good things, but so incredibly mixed up and wandering that it made one's head swim to follow him. They were not intended to be funny, but they were, — funny for the very gravity which the speaker put into his flowing miracles of incongruity. In the course of five years I came to know the styles of the five orators as well as I knew the style of any speaker in my own club at home.

These reports came every month. They were written on foolscap, 600 words to the page, and usually about twenty-five pages in a report — a good 15,000 words, I should say, — a solid week's work. The reports were absorbingly entertaining, long as they were; but, unfortunately for me, they did not come alone. They were always accompanied by a lot of questions about passages and purposes in my books, which the Club wanted answered; and additionally accompanied every quarter by the Treasurer's report, and the Auditor's report, and the Committee's report, and the President's review, and my opinion of these was always desired; also suggestions for the good of the Club, if any occurred to me.

By and by I came to dread those things; and this dread grew and grew and grew; grew until I got to anticipating them with a cold horror. For I was an indolent man, and not fond of letter-writing, and whenever these things came I had to put everything by and sit down — for my own peace of mind — and dig and dig until I got something out of my head which would answer for a reply. I got along fairly well the first year; but for the succeeding four years the Mark Twain Club of Corrigan Castle was my curse, my nightmare, the grief and misery of my life. And I got so, *so* sick of sitting for photographs. I sat every year for five years, trying to satisfy that insatiable organization. Then at last I rose in revolt. I could endure my oppressions no longer. I pulled my fortitude together and tore off my chains, and was a free man again, and happy. From that day I burned the secretary's fat envelopes the moment they arrived, and by and by they ceased to come.

Well, in the sociable frankness of that night in Bendigo I brought this all out in full confession. Then Mr. Blank came out in the same frank way, and with a preliminary word of gentle apology said that *he* was the Mark Twain Club, and the only member it had ever had!

Why, it was matter for anger, but I didn't feel any. He said he never had to work for a living, and that by the time he was thirty life had become a bore and a weariness to him. He had no interests left; they had paled and perished, one by one, and left him desolate. He had begun to think of suicide. Then all of a sudden he thought of that happy idea of starting an imaginary club, and went straightway to work at it, with enthusiasm and love. He was charmed with it; it gave him something to do. It elaborated itself on his hands; it became twenty times more complex and formidable than was his first rude draft of it. Every new addition to his original plan which cropped up in his mind gave him a fresh interest and a new pleasure. He designed the Club badge himself, and worked over it, altering and improving it, a number of days and nights; then sent to London and had it made. It was the

only one that was made. It was made for me; the "rest of
the Club" went without.

He invented the thirty-two members and their names. He
invented the five favorite speakers and their five separate styles.
He invented their speeches, and reported them himself. He
would have kept that Club going until now, if I hadn't deserted, he said. He said he worked like a slave over those reports;
each of them cost him from a week to a fortnight's work, and
the work gave him pleasure and kept him alive and willing to
be alive. It was a bitter blow to him when the Club died.

Finally, there wasn't any Corrigan Castle. He had invented that, too.

It was wonderful — the whole thing; and altogether the
most ingenious and laborious and cheerful and painstaking
practical joke I have ever heard of. And I liked it; liked to
hear him tell about it; yet I have been a hater of practical
jokes from as long back as I can remember. Finally he said —

"Do you remember a note from Melbourne fourteen or fifteen years ago, telling about your lecture tour in Australia,
and your death and burial in Melbourne? — a note from Henry
Bascomb, of Bascomb Hall, Upper Holywell Hants."

"Yes."

"I wrote it."

"M-y — word!"

"Yes, I did it. I don't know why. I just took the notion,
and carried it out without stopping to think. It was wrong.
It could have done harm. I was always sorry about it afterward. You must forgive me. I was Mr. Bascom's guest
on his yacht, on his voyage around the world. He often spoke
of you, and of the pleasant times you had had together in his
home; and the notion took me, there in Melbourne, and I
imitated his hand, and wrote the letter."

So the mystery was cleared up, after so many, many years.

CHAPTER XXVI.

There are people who can do all fine and heroic things but one ! keep from telling their happinesses to the unhappy.—*Pudd'nhead Wilson's New Calendar.*

AFTER visits to Maryborough and some other Australian towns, we presently took passage for New Zealand. If it would not look too much like showing off, I would tell the reader where New Zealand is; for he is as I was; he thinks he knows. And he thinks he knows where Hertzegovina is; and how to pronounce *pariah ;* and how to use the word *unique* without exposing himself to the derision of the dictionary. But in truth, he knows none of these things. There are but four or five people in the world who possess this knowledge, and these make their living out of it. They travel from place to place, visiting literary assemblages, geographical societies, and seats of learning, and springing sudden bets that these people do not know these things. Since all people think they know them, they are an easy prey to these adventurers. Or rather they were an easy prey until the law interfered three months ago, and a New York court decided that this kind of gambling is illegal, " because it traverses Article IV, Section 9, of the Constitution of the United States, which forbids betting on a sure thing." This decision was rendered by the full Bench of the New York Supreme Court, after a test sprung upon the court by counsel for the prosecution, which showed that none of the nine Judges was able to answer any of the four questions.

All people think that New Zealand is close to Australia or

Asia, or somewhere, and that you cross to it on a bridge. But
that is not so. It is not close to anything, but lies by itself,
out in the water. It is nearest to Australia, but still not near.
The gap between is very wide. It will be a surprise to the
reader, as it was to me, to learn that the distance from Australia to New Zealand is really twelve or thirteen hundred
miles, and that there is no bridge. I learned this from Professor X., of Yale University, whom I met in the steamer on the
great lakes when I was crossing the continent to sail across the
Pacific. I asked him about New Zealand, in order to make
conversation. I supposed he would generalize a little without
compromising himself, and then turn the subject to something
he was acquainted with, and my object would then be attained;
the ice would be broken, and we could go smoothly on, and get
acquainted, and have a pleasant time. But, to my surprise, he
was not only not embarrassed by my question, but seemed to
welcome it, and to take a distinct interest in it. He began to
talk — fluently, confidently, comfortably ; and as he talked, my
admiration grew and grew ; for as the subject developed under
his hands, I saw that he not only knew where New Zealand
was, but that he was minutely familiar with every detail of its
history, politics, religions, and commerce, its fauna, flora, geol-
ogy, products, and climatic peculiarities. When he was done,
I was lost in wonder and admiration, and said to myself, he
knows everything ; in the domain of human knowledge he is king.

I wanted to see him do more miracles ; and so, just for the
pleasure of hearing him answer, I asked him about Hertzegovina, and pariah, and unique. But he began to generalize
then, and show distress. I saw that with New Zealand gone,
he was a Samson shorn of his locks ; he was as other men.
This was a curious and interesting mystery, and I was frank
with him, and asked him to explain it.

He tried to avoid it at first ; but then laughed and said

that after all, the matter was not worth concealment, so he would let me into the secret. In substance, this is his story:

"Last autumn I was at work one morning at home, when a card came up — the card of a stranger. Under the name was printed a line which showed that this visitor was Professor of Theological Engineering in Wellington University, New Zealand. I was troubled — troubled, I mean, by the shortness of the notice. College etiquette required that he be at once invited to dinner by some member of the Faculty — invited to dine on *that* day — not put off till a subsequent day. I did not quite know what to do. College etiquette requires, in the case of a foreign guest, that the dinner-talk shall begin with complimentary references to his country, its great men, its services to civilization, its seats of learning, and things like that; and of course the host is responsible, and must either begin this talk himself or see that it is done by some one else. I was in great difficulty; and the more I searched my memory, the more my trouble grew. I found that I knew nothing about New Zealand. I thought I knew where it was, and that was all. I had an impression that it was close to Australia, or Asia, or somewhere, and that one went over to it on a bridge. This might turn out to be incorrect; and even if correct, it would not furnish matter enough

THE SCHEME WORKED.

for the purpose at the dinner, and I should expose my College to shame before my guest; he would see that I, a member of the Faculty of the first University in America, was wholly ignorant of his country, and he would go away and tell this, and laugh at it. The thought of it made my face burn.

"I sent for my wife and told her how I was situated, and asked for her help, and she thought of a thing which I might have thought of myself, if I had not been excited and worried. She said she would go and tell the visitor that I was out but would be in in a few minutes; and she would talk, and keep him busy while I got out the back way and hurried over and make Professor Lawson give the dinner. For Lawson knew everything, and could meet the guest in a creditable way and save the reputation of the

University. I ran to Lawson, but was disappointed. He did not know anything about New Zealand. He said that, as far as his recollection went
it was close to Australia, or Asia, or somewhere, and you
go over to it on a bridge; but that was all he knew. It
was too bad. Lawson was a perfect encyclopedia of
abstruse learning; but now in this hour of our need,
it turned out that he did not know any useful thing.

"WHAT DO YOU?"

"We consulted. He saw that the reputation of the
University was in very real peril, and he walked the floor
in anxiety, talking, and trying to think out some way
to meet the difficulty. Presently he decided that we
must try the rest of the Faculty — some of them might
know about New Zealand. So we went to the telephone
and called up the professor of astronomy and asked him,
and he said that all he knew was, that it was close to
Australia, or Asia, or somewhere, and you went over to
it on —

"We shut him off and called up the professor of
biology, and he said that all he knew was that it was
close to Aus —.

"We shut him off, and sat down, worried and dis-
heartened, to see if we could think up some other scheme. We shortly hit
upon one which promised well, and this one we adopted, and set its machinery going at once. It was this. Lawson must give the dinner. The
Faculty must be notified by telephone to prepare. We must all get to work
diligently, and at the end of eight hours and a half we must come to dinner
acquainted with New Zealand; at least well enough in-
formed to appear without discredit before this native. To
seem properly intelligent we should have to know about
New Zealand's population, and politics, and form of government, and commerce, and taxes, and products, and ancient
history, and modern history, and varieties of religion, and
nature of the laws, and their codification, and amount of
revenue, and whence drawn, and methods of collection, and
percentage of loss, and character of climate, and — well, a
lot of things like that; we must suck the maps and cyclo-
pedias dry. And while we posted up in this way, the
Faculty's wives must flock over, one after the other, in a
studiedly casual way, and help my wife keep the New
Zealander quiet, and not let him get out and come interfering with our studies. The scheme worked admirably; but
it stopped business, stopped it entirely.

" CLOSE TO AUS-."

"It is in the official log-book of Yale, to be read and
wondered at by future generations — the account of the
Great Blank Day — the memorable Blank Day — the day
wherein the wheels of culture were stopped, a Sunday silence prevailed
all about, and the whole University stood still while the Faculty read-up

and qualified itself to sit at meat, without shame, in the presence of the Professor of Theological Engineering from New Zealand.

" When we assembled at the dinner we were miserably tired and worn — but we were posted. Yes, it is fair to claim that. In fact, erudition is a pale name for it. New Zealand was the only subject; and it was just beautiful to hear us ripple it out. And with such an air of unembarrassed ease, and unostentatious familiarity with detail, and trained and seasoned mastery of the subject — and oh, the grace and fluency of it!

" Well, finally somebody happened to notice that the guest was looking dazed, and wasn't saying anything. So they stirred him up, of course. Then that man came out with a good, honest, eloquent compliment that made the Faculty blush. He said he was not worthy to sit in the company of men like these; that he had been silent from admiration; that he had been silent from another cause also — silent from shame — silent from *ignorance!* 'For,' said he, ' I, who have lived eighteen years in New Zealand and have served five in a professorship, and ought to know much about that country, perceive. now, that I know almost nothing about it. I say it with shame, that I have learned fifty times, yes, a hundred times more about New Zealand in these two hours at this table than I ever knew before in all the eighteen years put together. I was silent because I could not help myself. What I knew about taxes, and policies, and laws, and revenue,

HE LOOKED DAZED.

and products, and history, and all that multitude of things, was but general, and ordinary, and vague — unscientific, in a word — and it would have been insanity to expose it here to the searching glare of your amazingly accurate and all-comprehensive knowledge of those matters, gentlemen. I beg you to let me sit silent — as becomes me. But do not change the subject; I can at least follow you, in this one; whereas if you change to one which shall call out the full strength of your mighty erudition, I shall be as one lost. If you know all this about a remote little inconsequent patch like New Zealand, ah, what *wouldn't* you know about any other subject!' '

CHAPTER XXVII.

Man is the Only Animal that Blushes. Or needs to.
— *Pudd'nhead Wilson's New Calendar.*

The universal brotherhood of man is our most precious possession, what there is of it. — *Pudd'nhead Wilson's New Calendar.*

FROM DIARY:

NOVEMBER 1 — *noon*. A fine day, a brilliant sun. Warm in the sun, cold in the shade — an icy breeze blowing out of the south. A solemn long swell rolling up northward. It comes from the South Pole, with nothing in the way to obstruct its march and tone its energy down. I have read somewhere that an acute observer among the early explorers — Cook? or Tasman? — accepted this majestic swell as trustworthy circumstantial evidence that no important land lay to the southward, and so did not waste time on a useless quest in that direction, but changed his course and went searching elsewhere.

Afternoon. Passing between Tasmania (formerly Van Diemen's Land) and neighboring islands — islands ·whence the poor exiled Tasmanian savages used to gaze at their lost home-land and cry; and die of broken hearts. How glad I am that all these native races are dead and gone, or nearly so. The work was mercifully swift and horrible in some portions of Australia. As far as Tasmania is concerned, the extermination was complete: not a native is left. It was a strife of years, and decades of years. The Whites and the Blacks hunted each other, ambushed each other, butchered each other. The Blacks were not numerous. But they were wary, alert, cunning, and

(256)

"Why—Massa Gubernor"—said Black Jack—"You Proflamation all gammon, how blackfellow read him ?—eh! He no read him book." "Read that then," said the Governor, pointing to a picture.

GOVERNOR'S PROCLAMATION.

they knew their country well. They lasted a long time, few as they were, and inflicted much slaughter upon the Whites.

The Government wanted to save the Blacks from ultimate extermination, if possible. One of its schemes was to capture them and coop them up, on a neighboring island, under guard. Bodies of Whites volunteered for the hunt, for the pay was good — £5 for each Black captured and delivered, but the success achieved was not very satisfactory. The Black was naked, and his body was greased. It was hard to get a grip on him that would hold. The Whites moved about in armed bodies, and surprised little families of natives, and did make captures; but it was suspected that in these surprises half a dozen natives were killed to one caught — and that was not what the Government desired.

Another scheme was to drive the natives into a corner of the island and fence them in by a cordon of men placed in line across the country; but the natives managed to slip through, constantly, and continue their murders and arsons.

The governor warned these unlettered savages *by printed proclamation* that they must stay in the desolate region officially appointed for them! The proclamation was a dead letter; the savages could not read it. Afterward a *picture-*proclamation was issued. It was painted up on boards, and these were nailed to trees in the forest. Herewith is a photographic reproduction of this fashion-plate. Substantially it means:

1. The Governor wishes the Whites and the Blacks to love each other;
2. He loves his black subjects;
3. Blacks who kill Whites will be hanged;
4. Whites who kill Blacks will be hanged.

Upon its several schemes the Government spent £30,000 and employed the labors and ingenuities of several thousand Whites for a long time — with failure as a result. Then, at last, a quarter of a century after the beginning of the troubles between the two races, the right man was found. No, he found himself. This was George Augustus Robinson, called in history "The Conciliator." He was not educated, and not conspicuous in any way. He was a working bricklayer, in Hobart Town. But he must have been an amazing personality; a man worth traveling far to see. It may be his counterpart appears in history, but I do not know where to look for it.

He set himself this incredible task: to go out into the wilderness, the jungle, and the mountain-retreats where the hunted and implacable savages were hidden, and appear among them unarmed, speak the language of love and of kindness to them, and persuade them to forsake their homes and the wild free life that was so dear to them, and go with him and surrender to the hated Whites and live under their watch and ward, and upon their charity the rest of their lives! On its face it was the dream of a madman.

In the beginning, his moral-suasion project was sarcastically dubbed the *sugar-plum speculation*. If the scheme was striking, and new to the world's experience, the situation was not less so. It was this. The White population numbered 40,000 in 1831; the Black population numbered *three hundred*. Not 300 warriors, but 300 men, women, and children. The Whites were armed with guns, the Blacks with clubs and spears. The Whites had fought the Blacks for a quarter of a century, and had tried every thinkable way to capture, kill, or subdue them; and could not do it. If white men of any race *could* have done it, these would have accomplished it. But every scheme had failed, the splendid 300, the matchless 300 were unconquered, and manifestly unconquerable. They would not yield, they would listen to no terms, they would fight to the bitter end. Yet they had no poet to keep up their heart, and sing the marvel of their magnificent patriotism.

At the end of five-and-twenty years of hard fighting, the surviving 300 naked patriots were still defiant, still persistent, still efficacious with their rude weapons, and the Governor and the 40,000 knew not which way to turn, nor what to do.

Then the Bricklayer — that wonderful man — proposed to go out into the wilderness, with no weapon but his tongue, and no protection but his honest eye and his humane heart; and track those embittered savages to their lairs in the gloomy forests and among the mountain snows. Naturally, he was considered a crank. But he was not quite that. In fact, he was a good way short of that. He was building upon his long and intimate knowledge of the native character. The deriders of his project were right — from their standpoint — for they believed the natives to be mere wild beasts; and Robinson was right, from his standpoint — for he believed the natives to be human beings. The truth did really lie between the two. The event proved that Robinson's judgment was soundest; but about once a month for four years the event came near to giving the verdict to the deriders, for about that frequently Robinson barely escaped falling under the native spears.

But history shows that he had a thinking head, and was not a mere wild sentimentalist. For instance, he wanted the war parties call in before he started unarmed upon his mission of peace. He wanted the best chance of success — not a half-chance. And he was very willing to have help; and so, high rewards were advertised, for any who would go unarmed with him. This opportunity was declined. Robinson persuaded some tamed natives of both sexes to go with him — a strong evidence of his persuasive powers, for those natives well knew that their destruction would be almost certain. As it turned out, they had to face death over and over again.

Robinson and his little party had a difficult undertaking upon their hands. They could not ride off, horseback, comfortably into the woods and call Leonidas and his 300 together for a talk and a treaty the following day; for the wild men were not in a body; they were scattered, immense distances apart, over regions so desolate that even the birds could not make a living with the chances offered — scattered in groups of twenty, a dozen, half a dozen, even in groups of three. And the mission must go on foot. Mr. Bonwick furnishes a description of those horrible regions, whereby it will be seen that even fugitive gangs of the hardiest and choicest human devils the world has seen — the convicts set apart to people the "Hell of Macquarrie Harbor Station"— were never able, but once, to survive the horrors of a march through them, but starving and struggling, and fainting and failing, ate each other, and died:

"Onward, still onward, was the order of the indomitable Robinson. No one ignorant of the western country of Tasmania can form a correct idea of the traveling difficulties. While I was resident in Hobart Town, the Governor, Sir John Franklin, and his lady, undertook the western journey to Macquarrie Harbor, and suffered terribly. One man who assisted to carry her ladyship through the swamps, gave me his bitter experience of its miseries. Several were disabled for life. No wonder that but one party, escaping from Macquarrie Harbor convict settlement, arrived at the civilized region in safety. Men perished in the scrub, were lost in snow, or were devoured by their companions. This was the territory traversed by Mr. Robinson and his Black guides. All honor to his intrepidity, and their wonderful fidelity! When they had, in the depth of winter, to cross deep and rapid rivers, pass among mountains six thousand feet high, pierce dangerous thickets, and find food in a country forsaken even by birds, we can realize their hardships.

"After a frightful journey by Cradle Mountain, and over the lofty plateau of Middlesex Plains, the travelers experienced unwonted misery, and the circumstances called forth the best qualities of the noble little band. Mr. Robinson wrote afterwards to Mr. Secretary Burnett some details of this passage of horrors. In that letter, of Oct 2, 1834, he states that his Natives were very reluctant to go over the dreadful mountain passes; that 'for seven successive days we continued traveling over one solid body of snow;' that 'the snows were of incredible depth;' that 'the Natives were frequently up to their middle in snow.' But still the ill-clad, ill-fed, diseased, and wayworn men and women were sustained by the cheerful voice of their unconquerable friend, and responded most nobly to his call."

Mr. Bonwick says that Robinson's friendly capture of the Big River tribe — remember, it was a whole tribe —"was by far the grandest feature of the war, and the crowning glory of his efforts." The word "war" was not well chosen, and is misleading. There *was* war still, but only the Blacks were conducting it — the Whites were holding off until Robinson could give his scheme a fair trial. I think that we are to understand that the friendly capture of that tribe was by far the most important thing, the highest in value, that happened during the whole thirty years of truceless hostilities; that it was a decisive thing, a peaceful Waterloo, the surrender of the native Napoleon and his dreaded forces, the happy ending of the long strife. For "that tribe was the terror of the colony," its chief "the Black Douglas of Bush households."

Robinson knew that these formidable people were lurking somewhere, in some remote corner of the hideous regions just described, and he and his unarmed little party started on a tedious and perilous hunt for them. At last, "there, under the shadows of the Frenchman's Cap, whose grim cone rose five thousand feet in the uninhabited westward interior," they were found. It was a serious moment. Robinson himself believed, for once, that his mission, successful until now, was to end here in failure, and that his own death-hour had struck.

The redoubtable chief stood in menacing attitude, with his eighteen-foot spear poised; his warriors stood massed at his back, armed for battle, their faces eloquent with their long-cherished loathing for white men. "They rattled their spears and shouted their war-cry." Their women were back of them, laden with supplies of weapons, and keeping their 150 eager dogs quiet until the chief should give the signal to fall on.

"I think we shall soon be in the resurrection," whispered a member of Robinson's little party.

"I think we shall," answered Robinson; then plucked up heart and began his persuasions — in the tribe's own dialect, which surprised and pleased the chief. Presently there was an interruption by the chief:

"Who are you?"

"We are gentlemen."

"Where are your guns?"

"We have none."

The warrior was astonished.

"Where your little guns?" (pistols).

"We have none."

A few minutes passed — in by-play — suspense — discussion among the tribesmen — Robinson's tamed squaws ventured to cross the line and begin persuasions upon the wild squaws. Then the chief stepped back "to confer with the old women — the real arbiters of savage war." Mr. Bonwick continues:

"As the fallen gladiator in the arena looks for the signal of life or death from the president of the amphitheatre, so waited our friends in anxious suspense while the conference continued. In a few minutes before a word was uttered, the women of the tribe threw up their arms three times. This was the inviolable sign of peace! Down fell the spears. Forward, with a heavy sigh of relief, and upward glance of gratitude, came the friends of peace. The impulsive natives rushed forth with tears and cries, as each saw in the other's rank a loved one of the past. . . .

"It was a jubilee of joy. A festival followed. And, while tears flowed at the recital of woe, a corrobory of pleasant laughter closed the eventful day."

In four years, without the spilling of a drop of blood, Robinson brought them all in, willing captives, and delivered them to the white governor, and ended the war which powder and bullets, and thousands of men to use them, had prosecuted without result since 1804.

Marsyas charming the wild beasts with his music — that is fable; but the miracle wrought by Robinson is fact. It is history — and authentic; and surely, there is nothing greater, nothing more reverence-compelling in the history of any country, ancient or modern.

And in memory of the greatest man Australasia ever developed or ever will develop, there is a stately monument to George Augustus Robinson, the Conciliator in — no, it is to another man, I forget his name.

However, Robertson's own generation honored him, and in manifesting it honored themselves. The Government gave him a money-reward and a thousand acres of land; and the people held mass-meetings and praised him and emphasized their praise with a large subscription of money.

A good dramatic situation; but the curtain fell on another:

"When this desperate tribe was thus captured, there was much surprise to find that the £30,000 of a little earlier day had been spent, and the whole population of the colony placed under arms, in contention with an opposing force of *sixteen men with wooden spears!* Yet such was the fact. The celebrated Big River tribe, that had been raised by European fears to a host, consisted of *sixteen men, nine women and one child.* With a knowledge of the mischief done by these few, their wonderful marches and their widespread aggressions, their enemies cannot deny to them the attributes of courage and military tact. A Wallace might harass a large army with a small and determined band; but the contending parties were at least equal in arms and civilization The Zulus who fought us in Africa, the Maories in New Zealand, the Arabs in the Soudan, were far better provided with weapons, more advanced in the science of war, and considerably more numerous, than the naked Tasmanians. Governor Arthur rightly termed them a *noble race.*"

These were indeed wonderful people, the natives. They ought not to have been wasted. They should have been crossed with the Whites. It would have improved the Whites and done the Natives no harm.

But the Natives *were* wasted, poor heroic wild creatures. They were gathered together in little settlements on neighboring islands, and paternally cared for by the Government, and instructed in religion, and deprived of tobacco, because the superintendent of the Sunday-school was not a smoker, and so considered smoking immoral.

The Natives were not used to clothes, and houses, and regular hours, and church, and school, and Sunday-school, and work, and the other misplaced persecutions of civilization, and they pined for their lost home and their wild free life. Too late they repented that they had traded that heaven for this hell. They sat homesick on their alien crags, and day by day gazed out through their tears over the sea with unappeasable longing toward the hazy bulk which was the specter of what had been their paradise; one by one their hearts broke and they died.

In a very few years nothing but a scant remnant remained

THE LAST OF HER RACE.

alive. A handful lingered along into age. In 1864 the last man died, in 1876 the last woman died, and the Spartans of Australasia were extinct.

The Whites always mean well when they take human fish out of the ocean and try to make them dry and warm and happy and comfortable in a chicken coop; but the kindest-hearted white man can always be depended on to prove himself inadequate when he deals with savages. He cannot turn the situation around and imagine how he would like it to have a well-meaning savage transfer him from his house and his church and his clothes and his books and his choice food to a hideous wilderness of sand and rocks and snow, and ice and sleet and storm and blistering sun, with no shelter, no bed, no covering for his and his family's naked bodies, and nothing to eat but snakes and grubs and offal. This would be a hell to him; and if he had any wisdom he would know that his own civilization is a hell to the savage — but he hasn't any, and has never had any; and for lack of it he shut up those poor natives in the unimaginable perdition of his civilization, committing his crime with the very best intentions, and saw those poor creatures waste away under his tortures; and gazed at it, vaguely troubled and sorrowful, and wondered what could be the matter with them. One is almost betrayed into respecting those criminals, they were so sincerely kind, and tender, and humane, and well-meaning.

They didn't know why those exiled savages faded away, and they did their honest best to reason it out. And one man, in a like case in New South Wales, *did* reason it out and arrive at a solution:

"*It is from the wrath of God, which is revealed from heaven against all ungodliness and unrighteousness of men.*"

That settles it.

CHAPTER XXVIII.

Let us be thankful for the fools. But for them the rest of us could not succeed.
— *Pudd'nhead Wilson's New Calendar.*

THE aphorism does really seem true: "Given the Circumstances, the Man will appear." But the man musn't appear ahead of time, or it will spoil everything. In Robinson's case the Moment had been approaching for a quarter of a century — and meantime the future Conciliator was tranquilly laying bricks in Hobart. When all other means had failed, the Moment had arrived, and the Bricklayer put down his trowel and came forward. Earlier he would have been jeered back to his trowel again. It reminds me of a tale that was told me by a Kentuckian on the train when we were crossing Montana. He said the tale was current in Louisville years ago. He thought it had been in print, but could not remember. At any rate, in substance it was this, as nearly as I can call it back to mind.

A few years before the outbreak of the Civil War it began to appear that Memphis, Tennessee, was going to be a great tobacco *entrepot* — the wise could see the signs of it. At that time Memphis had a wharfboat, of course. There was a paved sloping wharf, for the accommodation of freight, but the steamers landed on the outside of the wharfboat, and all loading and unloading was done across it, between steamer and shore. A number of wharfboat clerks were needed, and part of the time, every day, they were very busy, and part of the time tediously idle. They were boiling over with youth and spirits, and they had to make the intervals of idleness endurable in some way; and as a rule, they did it by contriving practical jokes and playing them upon each other.

The favorite butt for the jokes was Ed Jackson, because he played none himself, and was easy game for other people's — for he always believed whatever was told him.

One day he told the others his scheme for his holiday. He was not going fishing or hunting this time — no, he had thought out a better plan. Out of his $40 a month he had saved enough for his purpose, in an economical way, and he was going to have a look at New York.

It was a great and surprising idea. It meant travel — immense travel — in those days it meant seeing the world; it was the equivalent of a voyage around it in ours. At first the other youths thought his mind was affected, but when they found that he was in earnest, the next thing to be thought of was, what sort of opportunity this venture might afford for a practical joke.

The young men studied over the matter, then held a secret consultation and made a plan. The idea was, that one of the conspirators should offer Ed a letter of introduction to Commodore Vanderbilt, and trick him into delivering it. It would be easy to do this. But what would Ed do when he got back to Memphis? That was a serious matter. He was good-hearted, and had always taken the jokes patiently; but they had been jokes which did not humiliate him, did not bring him to shame; whereas, this would be a cruel one in that way, and to play it was to meddle with fire; for with all his good nature, Ed was a Southerner — and the English of that was, that when he came back he would kill as many of the conspirators as he could before falling himself. However, the chances must be taken — it wouldn't do to waste such a joke as that.

So the letter was prepared with great care and elaboration. It was signed Alfred Fairchild, and was written in an easy and friendly spirit. It stated that the bearer was the bosom friend of the writer's son, and was of good parts and sterling character, and it begged the Commodore to be kind to the young stranger for the writer's sake. It went on to say, "You may have forgotten me, in this long stretch of time, but you will easily call me back out of your boyhood memories when I remind you of how we robbed old Stevenson's orchard that night; and how, while he was chasing down the road after us, we cut across the field and doubled back and sold his own apples to his own cook for a hatfull of doughnuts; and the time that we —" and so forth and so on, bringing in names of imaginary comrades, and detailing all sorts of wild and absurd and, of course, wholly imaginary schoolboy pranks and adventures, but putting them into lively and telling shape.

With all gravity Ed was asked if he would like to have a letter to Commodore Vanderbilt, the great millionaire. It was expected that the question would astonish Ed, and it did.

"What? Do *you* know that extraordinary man?"

"No; but my father does. They were schoolboys together. And if you like, I'll write and ask father. I know he'll be glad to give it to you for my sake."

Ed could not find words capable of expressing his gratitude and delight. The three days passed, and the letter was put into his hands. He started on his trip, still pouring out his thanks while he shook good-bye all around. And when he was out of sight his comrades let fly their laughter in a storm of happy satisfaction — and then quieted down, and were less happy, less satisfied. For the old doubts as to the wisdom of this deception began to intrude again.

Arrived in New York, Ed found his way to Commodore Vanderbilt's business quarters, and was ushered into a large anteroom, where a score of people were patiently awaiting their turn for a two-minute interview with the millionaire in his private office. A servant asked for Ed's card, and got the letter instead. Ed was sent for a moment later, and found Mr. Vanderbilt alone, with the letter — open — in his hand.

"Pray sit down, Mr. — er — "

"Jackson."

"Ah — sit down, Mr. Jackson. By the opening sentences it seems to be a letter from an old friend. Allow me — I will run my eye through it. He says — he says — why, who *is* it ?" He turned the sheet and found the signature. "Alfred Fairchild — hm — Fairchild — I don't recall the name. But that is nothing — a thousand names have gone from me. He says — he says — hm — hm — oh, dear, but it's good ! Oh, it's rare ! I don't *quite* remember it, but I *seem* to — it'll all come back to me presently. He says — he says — hm — hm — oh, but that was a game ! Oh, spl–endid ! How it carries me back ! It's all dim, of course — it's a long time ago — and the names — *some* of the names are wavery and indistinct — but sho', I know it happened — I can *feel* it ! and lord, how it warms my heart, and brings back my lost youth ! Well, well, well, I've got to come back into this work-a-day world now — business presses and people are waiting — I'll keep the rest for bed to-night, and live my youth over again. And you'll thank Fairchild for me when you see him — I used to call him Alf, I think — and you'll give him my gratitude for what this letter has done for the tired spirit of a hard-worked man ; and tell him there isn't anything that I can do for him or any friend of his that I won't do. And as for you, my lad, you are my guest ; you can't stop at any hotel in New York. Sit where you are a little while, till I get through with these people, then we'll go home. I'll take care of *you*, my boy — make yourself easy as to that."

Ed stayed a week, and had an immense time — and never suspected that the Commodore's shrewd eye was on him, and that he was daily being weighed and measured and analyzed and tried and tested.

Yes, he had an immense time; and never wrote home, but saved it all up to tell when he should get back. Twice, with proper modesty and decency, he proposed to end his visit, but the Commodore said, "No — wait; leave it to me; I'll tell you when to go."

In those days the Commodore was making some of those vast combinations of his — consolidations of warring odds and ends of railroads into harmonious systems, and concentrations of floating and rudderless commerce in effective centers — and among other things his far-seeing eye had detected the convergence of that huge tobacco-commerce, already' spoken of, toward Memphis, and he had resolved to set his grasp upon it and make it his own.

The week came to an end. Then the Commodore said:

"Now you can start home. But first we will have some more talk about that tobacco matter. I know you now. I know your abilities as well as you know them yourself — perhaps better. You understand that tobacco matter; you understand that I am going to take possession of it, and you also understand the plans which I have matured for doing it. What I want is a man who knows my mind, and is qualified to represent me in Memphis, and be in supreme command of that important business — and I appoint you."

" Me!"

" Yes. Your salary will be high — of course — for you are representing me. Later you will earn increases of it, and will get them. You will need a small army of assistants; choose them yourself — and carefully. Take no man for friendship's sake; but, all things being equal, take the man you know, take your friend, in preference to the stranger." After some further talk under this head, the Commodore said:

"Good-bye, my boy, and thank Alf for me, for sending you to me."

When Ed reached Memphis he rushed down to the wharf in a fever to tell his great news and thank the boys over and over again for thinking to give him the letter to Mr. Vanderbilt. It happened to be one of those idle times. Blazing hot noonday, and no sign of life on the wharf. But as Ed threaded his way among the freight piles, he saw a white linen figure stretched in slumber upon a pile of grain-sacks under an awning, and said to himself, "That's one of them," and hastened his step; next, he said, "It's Charley — it's Fairchild — good"; and the next moment laid an affectionate hand on the sleeper's shoulder. The eyes opened lazily, took one glance, the face blanched, the form whirled itself from the sack-pile, and in an instant Ed was alone and Fairchild was flying for the wharfboat like the wind!

Ed was dazed, stupefied. Was Fairchild crazy? What could be the meaning of this? He started slow and dreamily down toward the wharfboat; turned the corner of a freight-pile and came suddenly upon two of the boys. They were lightly laughing over some pleasant matter; they heard his step, and glanced up just as he discovered them; the laugh died abruptly; and before Ed could speak they were off, and sailing over barrels and bales like hunted deer. Again Ed was paralyzed. Had the boys all gone mad? What *could* be the explanation of this extraordinary conduct? And so, dreaming along, he reached the wharfboat, and stepped aboard — nothing but silence there, and vacancy. He crossed the deck, turned the corner to go down the outer guard, heard a fervent —

"O lord!" and saw a white linen form plunge overboard.

The youth came up coughing and strangling, and cried out —

"Go 'way from here! You let me alone. *I* didn't do it, I swear I didn't!"

18

" Didn't do *what?* "

" Give you the — "

" Never mind what you didn't do — come out of that! What makes you all act so? What have *I* done?"

" You? Why *you* haven't done anything. But — "

" Well, then, what have you got against me? What do you all treat me so for?"

" I — er — but haven't you got anything against *us?*"

" Of course not. What

" GO 'WAY FROM HERE ! "

put such a thing into your head?"

" Honor bright — you haven't?"

" Honor bright."

"Swear it!"

"I don't know what in the *world* you mean, but I swear it, anyway."

"And you'll shake hands with me?"

"Goodness knows I'll be *glad* to! Why, I'm just starving to shake hands with *somebody!*"

The swimmer muttered, "Hang him, he smelt a rat and never delivered the letter! — but it's all right, I'm not going to fetch up the subject." And he crawled out and came dripping and draining to shake hands. First one and then another of the conspirators showed up cautiously — armed to the teeth — took in the amicable situation, then ventured warily forward and joined the love-feast.

And to Ed's eager inquiry as to what made them act as they had been acting, they answered evasively, and pretended that they had put it up as a joke, to see what he would do. It was the best explanation they could invent at such short notice. And each said to himself, "He never delivered that letter, and the joke is on *us*, if he only knew it or we were dull enough to come out and tell."

Then, of course, they wanted to know all about the trip; and he said —

"Come right up on the boiler deck and order the drinks — it's my treat. I'm going to tell you all about it. And to-night it's my treat again — and we'll have oysters and a time!"

When the drinks were brought and cigars lighted, Ed said —

"Well, when I delivered the letter to Mr. Vanderbilt —"

"Great Scott!"

"Gracious, how you scared me. What's the matter?"

"Oh — er — nothing. Nothing — it was a tack in the chair-seat," said one.

"But you *all* said it. However, no matter. When I delivered the letter —"

" *Did* you deliver it ? "   And they looked at each other as people might who thought that maybe they were dreaming.

Then they settled to listening ; and as the story deepened and its marvels grew, the amazement of it made them dumb, and the interest of it took their breath.   They hardly uttered a whisper during two hours, but sat like petrifactions and drank in the immortal romance.   At last the tale was ended, and Ed said —

" And it's all owing to *you*, boys, and you'll never find *me* ungrateful — bless your hearts, the best friends a fellow ever had !   You'll all have places ; I want every one of you.   I *know* you — I know you ' by the *back*,' as the gamblers say. You're jokers, and all that, but you're *sterling*, with the hall-mark *on*.   And Charley Fairchild, you shall be my first assistant and right hand, because of your first-class ability, and because you got me the letter, and for your father's sake who wrote it for me, and to please Mr. Vanderbilt, who *said* it would !   And here's to that great man — drink hearty ! "

Yes, when the Moment comes, the Man appears — even if he is a thousand miles away, and has to be discovered by a practical joke.

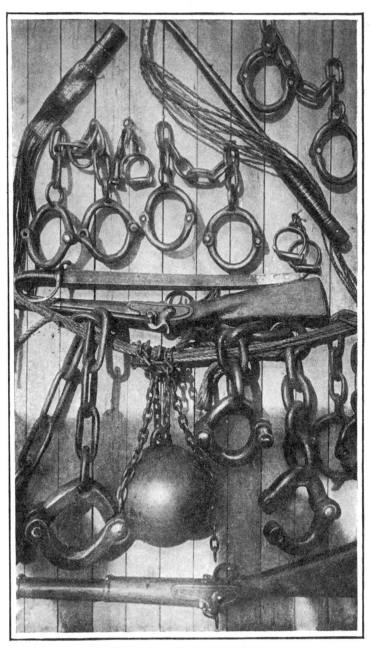

RELICS OF CONVICT DISCIPLINE.

CHAPTER XXIX.

When people do not respect us we are sharply offended ; yet deep down in his private heart no man much respects himself. —*Pudd'nhead Wilson's New Calendar.*

ECESSARILY, the human interest is the first interest in the log-book of any country. The annals of Tasmania, in whose shadow we were sailing, are lurid with that feature. Tasmania was a convict-dump, in old times; this has been indicated in the account of the Conciliator, where reference is made to vain attempts of desperate convicts to win to permanent freedom, after escaping from Macquarrie Harbor and the "Gates of Hell." In the early days Tasmania had a great population of convicts, of both sexes and all ages, and a bitter hard life they had. In one spot there was a settlement of juvenile convicts — children — who had been sent thither from their home and their friends on the other side of the globe to expiate their "crimes."

In due course our ship entered the estuary called the Derwent, at whose head stands Hobart, the capital of Tasmania. The Derwent's shores furnish scenery of an interesting sort. The historian Laurie, whose book, "The Story of Australasia," is just out, invoices its features with considerable truth and intemperance: "The marvelous picturesqueness of every point of view, combined with the clear balmy atmosphere and the transparency of the ocean depths, must have delighted and deeply impressed" the early explorers. "If the rock-bound coasts, sullen, defiant, and lowering, seemed uninviting, these were occasionally broken into charmingly alluring coves floored with golden sand, clad with evergreen shrubbery, and

(279)

adorned with every variety of indigenous wattle, she-oak, wild flower, and fern, from the delicately graceful 'maiden-hair' to the palm-like 'old man'; while the majestic gum-tree, clean and smooth as the mast of 'some tall ammiral' pierces the clear air to the height of 230 feet or more."

It looked so to me. "Coasting along Tasman's Peninsula, what a shock of pleasant wonder must have struck the early mariner on suddenly sighting Cape Pillar, with its cluster of black-ribbed basaltic columns rising to a height of 900 feet, the hydra head wreathed in a turban of fleecy cloud, the base lashed by jealous waves spouting angry fountains of foam."

That is well enough, but I did not suppose those snags were 900 feet high. Still they were a very fine show. They stood boldly out by themselves, and made a fascinatingly odd spectacle. But there was nothing about their appearance to suggest the heads of a hydra. They looked like a row of lofty slabs with their upper ends tapered to the shape of a carving-knife point; in fact, the early voyager, ignorant of their great height, might have mistaken them for a rusty old rank of piles that had sagged this way and that out of the perpendicular.

The Peninsula is lofty, rocky, and densely clothed with scrub, or brush, or both. It is joined to the main by a low neck. At this junction was formerly a convict station called Port Arthur — a place hard to escape from. Behind it was the wilderness of scrub, in which a fugitive would soon starve; in front was the narrow neck, with a cordon of chained dogs across it, and a line of lanterns, and a fence of living guards, armed. We saw the place as we swept by — that is, we had a glimpse of what we were told was the entrance to Port Arthur. The glimpse was worth something, as a remembrancer, but that was all.

"The voyage thence up the Derwent Frith displays a grand succession of fairy visions, in its entire length elsewhere unequaled. In gliding over the deep blue sea studded with lovely islets luxuriant to the water's edge, one is at a loss which scene to choose for contemplation and to admire most. When the Huon and Bruni have been passed, there seems no possible chance of a rival; but suddenly Mount Wellington, massive and noble like his brother Etna, literally heaves in sight, sternly guarded on either hand by Mounts Nelson and Rumney; presently we arrive at Sullivan's Cove — Hobart!"

It is an attractive town. It sits on low hills that slope to the harbor — a harbor that looks like a river, and is as smooth as one. Its still surface is pictured with dainty reflections of boats and grassy banks and luxuriant foliage. Back of the town rise highlands that are clothed in woodland loveliness, and over the way is that noble mountain, Wellington, a stately bulk, a most majestic pile. How beautiful is the whole region, for form, and grouping, and opulence, and freshness of foliage, and variety of color, and grace and shapeliness of the hills, the capes, the promontories; and then, the splendor of the sunlight, the dim rich distances, the charm of the water-glimpses! And it was in this paradise that the yellow-liveried convicts were landed, and the Corps-bandits quartered, and the wanton slaughter of the kangaroo-chasing black innocents con-summated on that autumn day in May, in the brutish old time. It was all out of keeping with the place, a sort of bring-ing of heaven and hell together.

The remembrance of this paradise reminds me that it was at Hobart that we struck the head of the procession of Junior Englands. We were to encounter other sections of it in New Zealand, presently, and others later in Natal. Wherever the exiled Englishman can find in his new home resemblances to his old one, he is touched to the marrow of his being; the love that is in his heart inspires his imagination, and these allied forces transfigure those resemblances into authentic duplicates of the revered originals. It is beautiful, the feeling which works this enchantment, and it compels one's homage; compels it, and also compels one's assent — compels it always — even when, as happens sometimes, one does not see the resemblances as clearly as does the exile who is pointing them out.

The resemblances do exist, it is quite true; and often they cunningly approximate the originals — but after all, in the matter of certain physical patent rights there is only one England. Now that I have sampled the globe, I am not in doubt. There is a beauty of Switzerland, and it is repeated in the glaciers and snowy ranges of many parts of the earth; there is a beauty of the fiord, and it is repeated in New Zealand and Alaska; there is a beauty of Hawaii, and it is repeated in ten thousand islands of the Southern seas; there is a beauty of the prairie and the plain, and it is repeated here and there in the earth; each of these is worshipful, each is perfect in its way, yet holds no monopoly of its beauty; but that beauty which is England is alone — it has no duplicate.

It is made up of very simple details — just grass, and trees, and shrubs, and roads, and hedges, and gardens, and houses, and vines, and churches, and castles, and here and there a ruin — and over it all a mellow dream-haze of history. But its beauty is incomparable, and all its own.

Hobart has a peculiarity — it is the neatest town that the sun shines on; and I incline to believe that it is also the cleanest. However that may be, its supremacy in neatness is not to be questioned. There cannot be another town in the world that has no shabby exteriors; no rickety gates and fences, no neglected houses crumbling to ruin, no crazy and unsightly sheds, no weed-grown front-yards of the poor, no back-yards littered with tin cans and old boots and empty bottles, no rubbish in the gutters, no clutter on the sidewalks, no outer-borders fraying out into dirty lanes and tin-patched huts. No, in Hobart all the aspects are tidy, and all a comfort to the eye; the modestest cottage looks combed and brushed, and has its vines, its flowers, its neat fence, its neat gate, its comely cat asleep on the window ledge.

We had a glimpse of the museum, by courtesy of the American gentleman who is curator of it. It has samples of half-a-dozen different kinds of marsupials* — one, the " Tasmanian devil;" that is, I *think* he was one of them. And there was a fish with lungs. When the water dries up it can live in the mud. Most curious of all was a parrot that kills sheep. On one great sheep-run this bird killed a thousand sheep in a whole year. He doesn't want the whole sheep, but only the kidney-fat. This restricted taste makes him an expensive bird to support. To get the fat he drives his beak in and rips it out; the wound is mortal. This parrot furnishes a notable example of evolution brought about by changed conditions. When the sheep culture was introduced, it presently brought famine to the parrot by exterminating a kind of grub which had always thitherto been the parrot's diet. The miseries of hunger made the bird willing to eat raw flesh, since it could get no other food, and it began to pick remnants of meat from sheep skins hung out on the fences to dry. It soon came to prefer sheep meat to any other food, and by and by it came to prefer the kidney-fat to any other detail of the sheep. The parrot's bill was not well shaped for digging out the fat, but Nature fixed that matter ; she altered the bill's shape, and now the parrot can dig out kidney-fat better than the Chief Justice of the Supreme Court, or anybody else, for that matter — even an Admiral.

And there was another curiosity — quite a stunning one, I

---

* A marsupial is a plantigrade vertebrate whose specialty is its pocket. In some countries it is extinct, in the others it is rare. The first American marsupials were Stephen Girard, Mr. Astor, and the opossum ; the principal marsupials of the Southern Hemisphere are Mr. Rhodes, and the kangaroo. I, myself, am the latest marsupial. Also, I might boast that I have the largest pocket of them all. But there is nothing in that.

thought: Arrow-heads and knives just like those which Primeval Man made out of flint, and thought he had done such a wonderful thing — yes, and has been humored and coddled in that superstition by this age of admiring scientists until there is probably no living with him in the other world by now. Yet here is his finest and nicest work exactly duplicated in our day; and by people who have never heard of him or his works: by aborigines who lived in the islands of these seas, within our time. And they not only duplicated those works of art but did it in the brittlest and most treacherous of substances — *glass*: made them out of old brandy bottles flung out of the British camps; millions of tons of them. It is time for Primeval Man to make a little less noise, now. He has had his day. He is not what he used to be.

We had a drive through a bloomy and odorous fairy-land, to the Refuge for the Indigent — a spacious and comfortable home, with hospitals, etc., for both sexes. There was a crowd there, of the oldest people I have ever seen. It was like being suddenly set down in a new world — a weird world where Youth has never been, a world sacred to Age, and bowed forms, and wrinkles. Out of the 359 persons present, 223 were ex-convicts, and could have told stirring tales, no doubt, if they had been minded to talk; 42 of the 359 were past 80, and several were close upon 90; the average age at death there is 76 years. As for me, I have no use for that place; it is too healthy. Seventy is old enough — after that, there is too much risk. Youth and gaiety might vanish, any day — and then, what is left? Death in life; death without its privileges, death without its benefits. There were 185 women in that Refuge, and 81 of them were ex-convicts.

The steamer disappointed us. Instead of making a long visit at Hobart, as usual, she made a short one. So we got but a glimpse of Tasmania, and then moved on.

CHAPTER XXX.

WE spent part of an afternoon and a night at sea, and reached Bluff, in New Zealand, early in the morning. Bluff is at the bottom of the middle island, and is away down south, nearly forty-seven degrees below the equator. It lies as far south of the line as Quebec lies north of it, and the climates of the two should be alike; but for some reason or other it has not been so arranged. Quebec is hot in the summer and cold in the winter, but Bluff's climate is less intense; the cold weather is not very cold, the hot weather is not very hot; and the difference between the hottest month and the coldest is but 17 degrees Fahrenheit.

In New Zealand the rabbit plague began at Bluff. The man who introduced the rabbit there was banqueted and lauded; but they would hang him, now, if they could get him. In England the natural enemy of the rabbit is detested and persecuted; in the Bluff region the natural enemy of the rabbit is honored, and his person is sacred. The rabbit's natural enemy in England is the poacher, in Bluff its natural enemy is the stoat, the weasel, the ferret, the cat, and the mongoose. In England any person below the Heir who is caught with a rabbit in his possession must satisfactorily explain how it got there, or he will suffer fine and imprisonment, together with extinction of his peerage; in Bluff, the cat found with a rabbit in its possession does not have to explain — everybody looks the other way; the person caught noticing

(285)

would suffer fine and imprisonment, with extinction of peerage. This is a sure way to undermine the moral fabric of a cat. Thirty years from now there will not be a moral cat in New Zealand. Some think there is none there now. In England the poacher is watched, tracked, hunted — he dare not show his face; in Bluff the cat, the weasel, the stoat, and the mongoose go up and down, whither they will, unmolested. By a law of the legislature, posted where all may read, it is decreed that any person found in possession of one of these creatures (dead) must satisfactorily explain the circumstances or pay a fine of not less than £5, nor more than £20. The revenue from this source is not large. Persons who want to pay a hundred dollars for a dead cat are getting rarer and rarer every day. This is bad, for the revenue was·to go to the endowment of a University. All governments are more or less short-sighted: in England they fine the poacher, whereas he ought to be banished to New Zealand. New Zealand would pay his way, and give him wages.

LAKE MANAPOURI.

It was from Bluff that we ought to have cut across to the west coast and visited the New Zealand Switzerland, a land of superb scenery, made up of snowy grandeurs, and mighty glaciers, and beautiful lakes; and over there, also, are the wonderful rivals of the Norwegian and Alaskan fiords; and for neighbor, a waterfall of 1,900 feet; but we were obliged to postpone the trip to some later and indefinite time.

November 6. A lovely summer morning; brilliant blue

sky. A few miles out from Invercargill, passed through vast level green expanses snowed over with sheep. Fine to see. The green, deep and very vivid sometimes; at other times less so, but delicate and lovely. A passenger reminds me that I am in "the England of the Far South."

Dunedin, same date. The town justifies Michael Davitt's praises. The people are Scotch. They stopped here on their way from home to heaven — thinking they had arrived. The population is stated at 40,000, by Malcolm Ross, journalist; stated by an M. P. at 60,000. A journalist cannot lie.

To the residence of Dr. Hockin. He has a fine collection of books relating to New Zealand; and his house is a museum of Maori art and antiquities. He has pictures and prints in color of many native chiefs of the past — some of them of note in history. There is nothing of the savage in the faces; nothing could be finer than these men's features, nothing more intellectual than these faces, nothing more masculine, nothing nobler than their aspect. The aborigi- nals of Australia and Tasmania looked the savage, but these chiefs looked like Roman patricians.

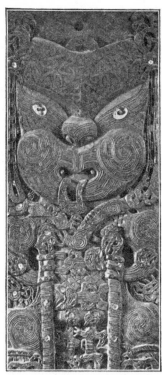

Monument to a Maori robber who went about on stilts ; thus avoiding de- tection for a long time. So great was the public admiration of his shrewdness that this monument was erected. It is supposed to be a portrait of the man and the stilts.

The tattooing in these portraits ought to suggest the savage, of course, but it does not. The designs are so flowing and graceful and beautiful that they are a most satisfactory decoration. It takes but fifteen minutes to get reconciled to the tattooing, and but fifteen more to

perceive that it is just the thing.   After that, the undecorated
European face is unpleasant and ignoble.

Dr. Hockin gave us a ghastly curiosity — a lignified caterpillar with a plant growing out of the back of its neck — a
plant with a slender stem 4 inches high.   It happened not by
accident, but by design — Nature's design.   This caterpillar
was in the act of loyally carrying out a law inflicted upon
him by Nature — a law purposely inflicted upon him to get
him into trouble — a law which was a trap; in pursuance of
this law he made the proper preparations for turning himself
into a night-moth; that is to say, he dug a little trench, a little
grave, and then stretched himself out in it on his stomach and
partially buried himself — then Nature was ready for him.
She blew the spores of a peculiar fungus through the air —
with a purpose.   Some of them fell into a crease in the back
of the caterpillar's neck, and began to sprout and grow — for
there was soil there — he had not washed his neck.   The
roots forced themselves down into the worm's person, and
rearward along through its body, sucking up the creature's
juices for sap; the worm slowly died, and turned to wood.
And here he was now, a wooden caterpillar, with every detail
of his former physique delicately and exactly preserved and
perpetuated, and with that stem standing up out of him for
his monument — monument commemorative of his own loyalty
and of Nature's unfair return for it.

Nature is always acting like that.   Mrs. X. said (of
course) that the caterpillar was not conscious and didn't
suffer.   She should have known better.   No caterpillar can
deceive Nature.   If this one couldn't suffer, Nature would
have known it and would have hunted up another caterpillar.
Not that she would have let this one go, merely because it
was defective.   No.   She would have waited and let him turn
into a night-moth; and then fried him in the candle.

Nature cakes a fish's eyes over with parasites, so that it

shan't be able to avoid its enemies or find its food. She sends parasites into a star-fish's system, which clog up its prongs and swell them and make them so uncomfortable that the poor creature delivers itself from the prong to ease its misery; and presently it has to part with another prong for the sake of comfort, and finally with a third. If it re-grows the prongs, the parasite returns and the same thing is repeated. And finally, when the ability to reproduce prongs is lost through age, that poor old star-fish can't get around any more, and so it dies of starvation.

In Australia is prevalent a horrible disease due to an "unperfected tape-worm." Unperfected — that is what they call it, I do not know why, for it transacts business just as well as if it were finished and frescoed and gilded, and all that.

November 9. To the museum and public picture gallery with the president of the Society of Artists. Some fine pictures there, lent by the S. of A. — several of them they bought, the others came to them by gift. Next, to the gallery of the S. of A. — annual exhibition — just opened. Fine. Think of a town like this having two such collections as this, and a Society of Artists. It is so all over Australasia. If it were a monarchy one might understand it. I mean an absolute monarchy, where it isn't necessary to vote money, but take it. Then art flourishes. But these colonies are republics — republics with a wide suffrage; voters of both sexes, this one of New Zealand. In republics, neither the government nor the rich private citizen is much given to propagating art. All over Australasia pictures by famous European artists are bought for the public galleries by the State and by societies of citizens. Living citizens — not dead ones. They rob *themselves* to give, not their heirs. This S. of A. here owns its buildings — built it by subscription.

19

CHAPTER XXXI.

The spirit of wrath — not the words — is the sin; and the spirit of wrath is cursing. We begin to swear before we can talk.

— *Pudd'nhead Wilson's New Calendar.*

NOVEMBER 11. *On the road.* This train — express — goes twenty and one-half miles an hour, schedule time; but it is fast enough, the outlook upon sea and land is so interesting, and the cars so comfortable. They are not English, and not American; they are the Swiss combination of the two. A narrow and railed porch along the side, where a person can walk up and down. A lavatory in each car. This is progress; this is nineteenth-century spirit. In New Zealand, these fast expresses run twice a week. It is well to know this if you want to be a bird and fly through the country at a 20-mile gait; otherwise you may start on one of the five wrong days, and then you will get a train that can't overtake its own shadow.

By contrast, these pleasant cars call to mind the branch-road cars at Maryborough, Australia, and the passengers' talk about the branch-road and the hotel.

Somewhere on the road to Maryborough I changed for a while to a smoking-carriage. There were two gentlemen there; both riding backward, one at each end of the compartment. They were acquaintances of each other. I sat down facing the one that sat at the starboard window. He had a good face, and a friendly look, and I judged from his dress that he was a dissenting minister. He was along toward fifty. Of his own motion he struck a match, and shaded it with his hand for me to light my cigar. I take the rest from my diary:

In order to start conversation I asked him something about Maryborough.   He said, in a most pleasant — even musical — voice, but with quiet and cultured decision :

" It's a charming town, with a hell of a hotel."

I was astonished.   It seemed so odd to hear a minister swear out loud.   He went placidly on:

"It's the worst hotel in Australia.   Well, one may go further, and say in Australasia."

" Bad beds ? "

" No — none at all.   Just sand-bags."

" The pillows, too ? "

" Yes, the pillows, too.   Just sand.   And not a good quality of sand.   It packs too hard, and has never been screened. There is too much gravel in it.   It is like sleeping on nuts."

" Isn't there any good sand ? "

" Plenty of it.   There is as good bed-sand in this region as the world can furnish.   Aerated sand — and loose; but they won't buy it.   They want something that will pack solid, and petrify."

" How are the rooms ? "

" Eight feet square; and a sheet of iced oil-cloth to step on in the morning when you get out of the sand-quarry."

" As to lights ? "

" Coal-oil lamp."

" A good one ? "

" No.   It's the kind that sheds a gloom."

" I like a lamp that burns all night."

" This one won't.   You must blow it out early."

" That is bad.   One might want it again in the night. Can't find it in the dark."

" There's no trouble; you can find it by the stench."

" Wardrobe ? "

"Two nails on the door to hang seven suits of clothes on — if you've got them."

"Bells?"

"There aren't any."

"What do you do when you want service?"

"Shout. But it won't fetch anybody."

"Suppose you want the chambermaid to empty the slop-jar?"

"There isn't any slop-jar. The hotels don't keep them. That is, outside of Sydney and Melbourne."

"Yes, I knew that. I was only talking. It's the oddest thing in Australia. Another thing: I've got to get up in the dark, in the morning, to take the 5 o'clock train. Now if the boots — "

"There isn't any."

"Well, the porter."

"There isn't any."

"But who will call me?"

"Nobody. You'll call yourself. And you'll light yourself, too. There'll not be a light burning in the halls or anywhere. And if you don't carry a light, you'll break your neck."

"But who will help me down with my baggage?"

A THEOLOGICAL STUDENT.

"Nobody. However, I will tell you what to do. In Maryborough there's an American who has lived there half a life-time; a fine man, and prosperous and popular. He will be on the lookout for you; you won't have any trouble. Sleep in peace; he will rout you out, and you will make your train. Where is your manager?"

"I left him at Ballarat, studying the language. And besides, he had to go to Melbourne and get us ready for New Zealand. I've not tried to pilot myself before, and it doesn't look easy."

"Easy! You've selected the very most difficult piece of railroad in Australia for your experiment. There are twelve miles of this road which no man without good executive ability can ever hope — tell me, have you good executive ability? — first-rate executive ability?"

"I — well, I think so, but — "

"That settles it. The tone of — oh, *you* wouldn't ever make it in the world. However, that American will point you right, and you'll go. You've got tickets?"

"Yes — round trip; all the way to Sydney."

"Ah, there it is, you see! You are going in the 5 o'clock by Castlemaine — twelve miles — instead of the 7.15 by Ballarat — in order to save two hours of fooling along the road. Now then, don't interrupt — let me have the floor. You're going to save the government a deal of hauling, but that's nothing; your ticket is by Ballarat, and it isn't good over that twelve miles, and so — "

"But why should the government care which way I go?"

"Goodness knows! Ask of the winds that far away with fragments strewed the sea, as the boy that stood on the burning deck used to say. The government chooses to do its railway business in its own way, and it doesn't know as much about it as the French. In the beginning they tried idiots; then they imported the French — which was going backwards, you see; now it runs the roads itself — which is going backwards again, you see. Why, do you know, in order to curry favor with the voters, the government puts down a road wherever anybody wants it — anybody that owns two sheep and a dog; and by consequence we've got, in the colony of Victoria, 800 railway stations, and the business done at eighty of them doesn't foot up twenty shillings a week."

"Five dollars? Oh, come!"

"It's true. It's the absolute truth."

"Why, there are three or four men on wages at every station."

"I know it. And the station-business doesn't pay for the sheep-dip to sanctify their coffee with. It's just as I say. And accommodating? Why, if you shake a rag the train will stop in the midst of the wilderness to pick you up. All that kind of politics costs, you see. And then, besides, any town that has a good many votes and wants a fine station, gets it. Don't you overlook that Maryborough station, if you take an interest in governmental curiosities. Why, you can put the whole population of Maryborough into it, and give them a sofa apiece, and have room for more. You haven't fifteen stations in America that are as big, and you probably haven't five that are half as fine. Why, it's per-fectly elegant. And the clock! Everybody will show you the clock. There isn't a station in Europe that's got such a clock. It doesn't strike — and that's one mercy. It hasn't any bell; and as you'll have cause to remember, if you keep your reason, all Australia is simply be-damned with bells. On every quarter-hour, night and day, they jingle a tiresome chime of half a dozen notes — all the clocks in town at once, all the clocks in Australasia at once, and all the *very same* notes; first, downward scale: *mi, re, do, sol* — then upward scale: *sol, si, re, do* — down again: *mi, re, do, sol* — up again: *sol, si, re, do* — then the clock — say at

midnight:
clang—clang—
clang — clang!
hello, what's
Oh, I see—
ed by the
wouldn't
could scare
of course,
and run
at a loss,
stations
like Mary-
at another
the govern-
has got to econ-
somewhere,
it? Very well—
the rolling stock!
where they save the
Why, that train from
ough will consist of
freight-cars and
senger-kennels;
cheap, poor,
shabby, slovenly;
no drinking water,
no sanitary arrange-
ments, every imaginable
inconvenience; and slow?—
oh, the gait of cold molasses;
no air-brake, no springs, and they'll
jolt your head off every time they start or

clang — clang — clang — clang —
clang — clang — clang — clang —
—and, by that time you're —
all this excitement about?
a runaway — scar-
train; why, you
think *this* train
anything. Well,
when they build
eighty stations
and a lot of palace-
and clocks
borough's
loss,
ment
omize
hasn't
look at
That's
money.
Marybor-
eighteen
two pas-

stop. That's where they make their little economies, you see.
They spend tons of money to house you palatially while you
wait fifteen minutes for a train, then degrade you to six hours'
convict-transportation to get the foolish outlay back. What
a rational man really needs is discomfort while he's waiting,
then his journey in a nice train would be a grateful change.
But no, that would be common sense — and out of place in a
government. And then, besides, they save in that other little
detail, you know — repudiate their own tickets, and collect a
poor little illegitimate extra shilling out of you for that
twelve miles, and — "

"Well, in any case — "

"Wait — there's more. Leave that American out of the
account and see what would happen. There's nobody on hand
to examine your ticket when you arrive. But the conductor
will come and examine it when the train is ready to start. It
is too late to buy your extra ticket now; the train can't wait,
and won't. You must climb out."

"But can't I pay the conductor?"

"No, he is not authorized to receive the money, and he
won't. You must climb out. There's no other way. I tell
you, the railway management is about the only thoroughly
European thing here — continentally European I mean, not
English. It's the continental business in perfection; down *fine*.
Oh, yes, even to the peanut-commerce of weighing baggage."

The train slowed up at his place. As he stepped out he said:

"Yes, you'll like Maryborough. Plenty of intelligence
there. It's a charming place — with a hell of a hotel."

Then he was gone. I turned to the other gentleman:

"Is your friend in the ministry?"

"No — studying for it."

CHAPTER XXXII.

The man with a new idea is a Crank until the idea succeeds.
— *Pudd'nhead Wilson's New Calendar.*

IT was Junior England all the way to Christchurch — in fact, just a garden. And Christchurch is an English town, with an English-park annex, and a winding English brook just like the Avon — and named the Avon; but from a man, not from Shakespeare's river. Its grassy banks are bordered by the stateliest and most impressive weeping willows to be found in the world, I suppose. They continue the line of a great ancestor; they were grown from sprouts of the willow that sheltered Napoleon's grave in St. Helena. It is a settled old community, with all the serenities, the graces, the conveniences, and the comforts of the ideal home-life. If it had an established Church and social inequality it would be England over again with hardly a lack.

In the museum we saw many curious and interesting things; among others a fine native house of the olden time, with all the details true to the facts, and the showy colors right and in their proper places. All the details: the fine mats and rugs and things; the elaborate and wonderful wood carvings — wonderful, surely, considering who did them — wonderful in design and particularly in execution, for they were done with admirable sharpness and exactness, and yet with no better tools than flint and jade and shell could furnish; and the totem-posts were there, ancestor above ancestor, with tongues protruded and hands clasped comfortably over bellies containing other people's ancestors — grotesque and ugly devils, every one, but lovingly carved, and ably; and the stuffed natives were present, in their proper places, and looking as natural as life; and the housekeeping utensils were there, too, and close at hand the carved and finely ornamented war canoe.

And we saw little jade gods, to hang around the neck — not everybody's, but sacred to the necks of natives of rank. Also jade weapons, and many kinds of jade trinkets — all made out of that excessively hard stone without the help of any tool of iron. And some of these things had small round holes bored through them — nobody knows how it was done; a mystery, a lost art. I think it was said that if you want such a hole bored in a piece of jade now, you must send it to London or Amsterdam where the lapidaries are.

CARVED CANOE PROW.

Also we saw a complete skeleton of the giant Moa. It stood ten feet high, and must have been a sight to look at when it was a living bird. It was a kicker, like the ostrich; in fight it did not use its beak, but its foot. It must have been a convincing kind of kick. If a person had his back to the bird and did not see who it was that did it, he would think he had been kicked by a wind-mill.

There must have been a sufficiency of moas in the old forgotten days when his breed walked the earth. His bones are found in vast masses, all crammed together in huge graves. They are not in caves, but in the ground. Nobody knows how they happened to get concentrated there. Mind, they are bones, not fossils. This means that the moa has not been extinct very long. Still, this is the only New Zealand creature

which has no mention in that otherwise comprehensive literature, the native legends. This is a significant detail, and is
good circumstantial evidence that the moa has been extinct
500 years, since the Maori has himself — by tradition — been
in New Zealand since the end of the fifteenth century. He
came from an unknown land — the first Maori did — then
sailed back in his canoe and brought his tribe, and they removed the aboriginal peoples into the sea and into the ground
and took the land. That is the tradition. That that first
Maori could come, is understandable, for anybody can come to
a place when he isn't trying to; but how that discoverer found
his way back home again without a compass is his secret, and
he died with it in him. His language indicates that he came
from Polynesia. He *told* where he came from, but he couldn't
spell well, so one can't find the place on the map, because
people who could spell better than he could, spelt the resemblance all out of it when they made the map. However, it is
better to have a map that is spelt right than one that has
information in it.

In New Zealand women have the right to vote for members of the legislature, but they cannot be members themselves. The law extending the suffrage to them went into
effect in 1893. The population of Christchurch (census of
1891) was 31,454. The first election under the law was held
in November of that year. Number of men who voted, 6,313;
number of women who voted, 5,989. These figures ought to
convince us that women are not as indifferent about politics as
some people would have us believe. In New Zealand as a
whole, the estimated adult female population was 139,915; of
these 109,461 qualified and registered their names on the rolls
— 78.23 per cent. of the whole. Of these, 90,290 went to the
polls and voted — 85.18 per cent. Do men ever turn out
better than that — in America or elsewhere? Here is a re-

mark to the other sex's credit, too — I take it from the official
report:

"A feature of the election was the orderliness and sobriety
of the people. Women were in no way molested."

At home, a standing argument against woman suffrage has
always been that women could not go to the polls without being
insulted. The arguments against woman suffrage have always
taken the easy form of prophecy. The prophets have been
prophesying ever since the woman's rights movement began in
1848 — and in forty-seven years they have never scored a hit.

Men ought to begin to feel a sort of respect for their
mothers and wives and sisters by this time. The women
deserve a change of attitude like that, for they have wrought
well. In forty-seven years they have swept an imposingly
large number of unfair laws from the statute books of America.
In that brief time these serfs have set themselves free —
essentially. Men could not have done so much for themselves
in that time without bloodshed — at least they never have;
and that is argument that they didn't know how. The women
have accomplished a peaceful revolution, and a very beneficent
one; and yet that has not convinced the average man that
they are intelligent, and have courage and energy and perse-
verance and fortitude. It takes much to convince the average
man of anything; and perhaps nothing can ever make him
realize that he is the average woman's inferior — yet in several
important details the evidences seems to show that that is
what he is. Man has ruled the human race from the beginning — but he should remember that up to the middle of the
present century it was a dull world, and ignorant and stupid;
but it is not such a dull world now, and is growing less and
less dull all the time. This is woman's opportunity — she has
had none before. I wonder where man will be in another
forty-seven years?

In the New Zealand law occurs this: "The word *person* wherever it occurs throughout the Act includes *woman*."

That is promotion, you see. By that enlargement of the word, the matron with the garnered wisdom and experience of fifty years becomes at one jump the political equal of her callow kid of twenty-one. The white population of the colony is 626,000, the Maori population is 42,000. The whites elect seventy members of the House of Representatives, the Maoris four. The Maori women vote for their four members.

November 16. After four pleasant days in Christchurch, we are to leave at midnight to-night. Mr. Kinsey gave me an ornithorhyncus, and I am taming it.

Sunday, 17th. Sailed last night in the *Flora*, from Lyttelton.

So we did. I remember it yet. The people who sailed in the *Flora* that night may forget some other things if they live a good while, but they will not live long enough to forget that. The *Flora* is about the equivalent of a cattle-scow; but when the Union Company find it inconvenient to keep a contract and lucrative to break it, they smuggle her into passenger service, and "keep the change."

They give no notice of their projected depredation; you innocently buy tickets for the advertised passenger boat, and when you get down to Lyttelton at midnight, you find that they have substituted the scow. They have plenty of good boats, but no competition — and that is the trouble. It is too late now to make other arrangements if you have engagements ahead.

It is a powerful company, it has a monopoly, and everybody is afraid of it — including the government's representative, who stands at the end of the stage-plank to tally the passengers and see that no boat receives a greater number than the law allows her to carry. This conveniently-blind representative saw the scow receive a number which was far in excess of its privilege, and winked a politic wink and said nothing. The passengers bore with meekness the cheat which had been put upon them, and made no complaint.

It was like being at home in America, where abused passengers act in just the same way. A few days before, the Union Company had discharged a captain for getting a boat into danger, and had advertised this act as evidence of its vigilance in looking after the safety of the passengers — for thugging a captain costs the company nothing, but when opportunity offered to send this dangerously overcrowded tub to sea and save a little trouble and a tidy penny by it, it forgot to worry about the passenger's safety.

The first officer told me that the *Flora* was privileged to carry 125 passengers. She must have had all of 200 on board. All the cabins were full, all the cattle-stalls in the main stable were full, the spaces at the heads of companionways were full, every inch of floor and table in the swill-room was packed with sleeping men and remained so until the place was required for breakfast, all the chairs and benches on the hurricane deck were occupied, and *still* there were people who had to walk about all night !

If the *Flora* had gone down that night, half of the people on board would have been wholly without means of escape.

The owners of that boat were not technically guilty of conspiracy to commit murder, but they were morally guilty of it.

I had a cattle-stall in the main stable — a cavern fitted up with a long double file of two-storied bunks, the files separated by a calico partition — twenty men and boys on one side of it, twenty women and girls on the other. The place was as dark as the soul of the Union Company, and smelt like a kennel. When the vessel got out into the heavy seas and began to pitch and wallow, the cavern prisoners became immediately sea-sick, and then the peculiar results that ensued laid all my previous experiences of the kind well away in the shade. And the wails, the groans, the cries, the shrieks, the strange ejaculations — it was wonderful.

The women and children and some of the men and boys s p e n t t h e night in that place, for they were too ill to leave it; but the rest of us got up, by and by, and finished the night on the hurricane-deck.

That boat was the foulest I was ever in; and the smell of the breakfast saloon when we threaded our way among the layers of steaming passengers

CATTLE STALLS ON THE FLORA.

stretched upon its floor and its tables was incomparable for efficiency.

A good many of us got ashore at the first way-port to seek another ship. After a wait of three hours we got good rooms in the *Mahinapua*, a wee little bridal-parlor of a boat — only 205 tons burthen; clean and comfortable; good service; good beds; good table, and no crowding. The seas danced her about like a duck, but she was safe and capable.

Next morning early she went through the French Pass — a narrow gateway of rock, between bold headlands — so narrow, in fact, that it seemed no wider than a street. The current tore through there like a mill-race, and the boat darted through like a telegram. The passage was made in half a minute; then we were in a wide place where noble vast eddies swept grandly round and round in shoal water, and I wondered what they would do with the little boat. They did as they pleased with her. They picked her up and flung her around like nothing and landed her gently on the solid, smooth bottom of sand—so gently, indeed, that we barely felt her touch it, barely felt her quiver when she came to a standstill. The water was as clear as glass, the sand on the bottom was vividly distinct, and the fishes seemed to be swimming about in nothing. Fishing lines were brought out, but before we could bait the hooks the boat was off and away again.

CHAPTER XXXIII.

Let us be grateful to Adam our benefactor. He cut us out of the "blessing" of idleness and won for us the "curse" of labor.

— *Pudd'nhead Wilson's New Calendar.*

WE soon reached the town of Nelson, and spent the most of the day there, visiting acquaintances and driving with them about the garden — the whole region is a garden, excepting the scene of the "Maungatapu Murders," of thirty years ago. That is a wild place — wild and lonely; an ideal place for a murder. It is at the base of a vast, rugged, densely timbered mountain. In the deep twilight of that forest solitude four desperate rascals — Burgess, Sullivan, Levy, and Kelley — ambushed themselves beside the mountain trail to murder and rob four travelers — Kempthorne, Mathieu, Dudley, and De Pontius, the latter a New Yorker. A harmless old laboring man came wandering along, and as his presence was an embarrassment, they choked him, hid him, and then resumed their watch for the four. They had to wait a while, but eventually everything turned out as they desired.

That dark episode is the one large event in the history of Nelson. The fame of it traveled far. Burgess made a confession. It is a remarkable paper. For brevity, succinctness, and concentration, it is perhaps without its peer in the literature of murder. There are no waste words in it; there is no obtrusion of matter not pertinent to the occasion, nor any departure from the dispassionate tone proper to a formal business statement — for that is what it is: a business statement of a murder, by the chief engineer of it, or superintendent, or foreman, or whatever one may prefer to call him.

20   (305)

"We were getting impatient, when we saw four men and a pack-horse coming. I left my cover and had a look at the men, for Levy had told me that Mathieu was a small man and wore a large beard, and that it was a chestnut horse. I said, 'Here they come.' They were then a good distance away; I took the caps off my gun, and put fresh ones on. I said, 'You keep where you are, I'll put them up, and you give me your gun while you tie them.' It was arranged as I have described. The men came'; they arrived within about fifteen yards when I stepped up and said, 'Stand! bail up!' That means all of them to get together I made them fall back on the upper side of the road with their faces up the range, and Sullivan brought me his gun, and then tied their hands behind them. The horse was very quiet all the time, he did not move. When they were all tied. Sullivan took the horse up the hill, and put him in the bush; he cut the rope and let the swags* fall on the ground, and then came to me. We then marched the men down the incline to the creek; the water at this time barely running. Up this creek we took the men; we went, I daresay, five or six hundred yards up it, which took us nearly half-an-hour to accomplish. Then we turned to the right up the range; we went. I daresay, one hundred and fifty yards from the creek, and there we sat down with the men. I said to Sullivan, 'Put down your gun and search these men.' which he did. I asked them their several names; they told me. I asked them if they were expected at Nelson. They said, 'No.' If such their lives would have been spared. In money we took £60 odd. I said, 'Is this all you have? You had better tell me.' Sullivan said, 'Here is a bag of gold.' I said, 'What's on that pack-horse? Is there any gold?' when Kempthorne said, 'Yes, my gold is in the portmanteau, and I trust you will not take it all.' 'Well,' I said, 'we must take you away one at a time, because the range is steep just here, and then we will let you go.' They said, 'All right,' most cheerfully. We tied their feet, and took Dudley with us; we went about sixty yards with him. This was through a scrub. It was arranged the night previously that it would be best to choke them, in case the report of the arms might be heard from the road, and if they were missed they never would be found. So we tied a handkerchief over his eyes, when Sullivan took the sash off his waist, put it round his neck, and so strangled him. Sullivan, after I had killed the old laboring man, found fault with the way he was choked. He said, 'The *next* we do I'll show you *my* way.' I said, 'I have never done such a thing before. I have shot a man. but never choked one.' We returned to the others, when Kempthorne said, 'What noise was that?' I said it was caused by breaking through the scrub. This was taking too much time, so it was agreed to shoot them. With that I said, 'We'll take you no further, but separate you, and then loose one of you, and he can relieve the others.' So with that, Sullivan took De Pontius to the left of where Kempthorne was sitting. I took Mathieu to the right I tied a strap round his legs, and shot him with a revolver. He yelled, I ran from him with my gun in my hand, I sighted Kempthorne, who had risen to his feet. I presented the gun, and shot him

---

* A "swag" is a kit, a pack, small baggage.

behind the right ear; his life's blood welled from him, and he died instantaneously. Sullivan had shot De Pontius in the meantime, and then came to me. I said, 'Look to Mathieu,' indicating the spot where he lay. He shortly returned and said, 'I had to "chiv" that fellow, he was not dead,' a cant word, meaning that he had to stab him. Returning to the road we passed where De Pontius lay and was dead Sullivan said, 'This is the digger, the others were all storekeepers; this is the digger, let's cover him up, for should the others be found, they'll think he done it and sloped,' meaning he had gone. So with that we threw all the stones on him, and then left him. This bloody work took nearly an hour and a half from the time we stopped the men."

Anyone who reads that confession will think that the man who wrote it was destitute of emotions, destitute of feeling. That is partly true. As regarded others he was plainly without feeling — utterly cold and pitiless; but as regarded himself the case was different. While he cared nothing for the future of the murdered men, he cared a great deal for his own. It makes one's flesh creep to read the introduction to his confession. The judge on the bench characterized it as " scandalously blasphemous," and it certainly reads so, but Burgess meant no blasphemy. He was merely a brute, and whatever he said or wrote was sure to expose the fact. His redemption was a very real thing to him, and he was as jubilantly happy on the gallows as ever was Christian martyr at the stake. We dwellers in this world are strangely made, and mysteriously circumstanced. We have to suppose that the murdered men are lost, and that Burgess is saved; but we cannot suppress our natural regrets:

" Written in my dungeon drear this 7th of August, in the year of Grace, 1866. To God be ascribed all power and glory in subduing the rebellious spirit of a most guilty wretch, who has been brought, through the instrumentality of a faithful follower of Christ, to see his wretched and guilty state, inasmuch as hitherto he has led an awful and wretched life, and through the assurance of this faithful soldier of Christ, he has been led and also believes that Christ will yet receive and cleanse him from all his deep-dyed and bloody sins. I lie under the imputation which says, 'Come now and let us reason together, saith the Lord: though your sins be as scarlet, they shall be as white as snow; though they be red like crimson, they shall be as wool.' On this promise I rely."

We sailed in the afternoon late, spent a few hours at New

Plymouth, then sailed again and reached Auckland the next day, November 20th, and remained in that fine city several days.   Its situation is commanding, and the sea-view is superb. There are charming drives all about, and by courtesy of friends we had opportunity to enjoy them.   From the grassy crater-summit of Mount Eden one's eye ranges over a grand sweep and variety of scenery — forests clothed in luxuriant foliage, rolling green fields, conflagrations of flowers, receding and dimming stretches of green plain, broken by lofty and symmetrical old craters — then the blue bays twinkling and sparkling away into the dreamy distances where the mountains loom spiritual in their veils of haze.

It is from Auckland that one goes to Rotorua, the region of the renowned hot lakes and geysers — one of the chief wonders of New Zealand; but I was not well enough to make the trip.   The government has a sanitorium there, and everything is comfortable for the tourist and the invalid.   The government's official physician is almost over-cautious in his estimates of the efficacy of the baths, when he is talking about rheumatism, gout, paralysis, and such things; but when he is talking about the effectiveness of the waters in eradicating the whisky-habit, he seems to have no reserves.   The baths will cure the drinking-habit no matter how chronic it is — and cure it so effectually that even the *desire* to drink intoxicants will come no more.   There should be a rush from Europe and America to that place; and when the victims of alcoholism find out what they can get by going there, the rush will begin.

The Thermal-springs District of New Zealand comprises an area of upwards of 600,000 acres, or close on 1,000 square miles.   Rotorua is the favorite place.   It is the center of a rich field of lake and mountain scenery; from Rotorua as a base the pleasure-seeker makes excursions.   The crowd of sick people is great, and growing.   Rotorua is the Carlsbad of Australasia.

THE HOT SPRINGS REGION

NATIVES BATHING AND COOKING IN THE SPRINGS.

HOT SPRINGS, AND GEYSERS.

It is from Auckland that the Kauri gum is shipped. For a long time now about 8,000 tons of it have been brought into the town per year. It is worth about $300 per ton, unassorted; assorted, the finest grades are worth about $1,000. It goes to America, chiefly. It is in lumps, and is hard and smooth, and looks like amber — the light-colored like new amber, and the dark brown like rich old amber. And it has the pleasant feel of amber, too. Some of the light-colored samples were a tolerably fair counterfeit of uncut South African diamonds, they were so perfectly smooth and polished and transparent. It is manufactured into varnish; a varnish which answers for copal varnish and is cheaper.

The gum is dug up out of the ground; it has been there for ages. It is the sap of the Kauri tree. Dr. Campbell of Auckland told me he sent a cargo of it to England fifty years ago, but nothing came of the venture. Nobody knew what to do with it; so it was sold at £5 a ton, to light fires with.

November 26 — 3 P. M., sailed. Vast and beautiful harbor. Land all about for hours. Tangariwa, the mountain that "has the same shape from *every* point of view." That is the common belief in Auckland. And so it has — from every point of view except thirteen. . . . Perfect summer weather. Large school of whales in the distance. Nothing could be daintier than the puffs of vapor they spout up, when seen against the pink glory of the sinking sun, or against the dark mass of an island reposing in the deep blue shadow of a storm-cloud. . . . Great Barrier rock standing up out of the sea away to the left. Sometime ago a ship hit it full speed in a fog — 20 miles out of her course — 140 lives lost; the captain committed suicide without waiting a moment. He knew that, whether he was to blame or not, the company owning the vessel would discharge him and make a devotion-to-passengers'-safety advertisement out of it, and his chance to make a livelihood would be permanently gone.

CHAPTER XXXIV.

Let us not be too particular. It is better to have old second-hand diamonds than none at all. —*Pudd'nhead Wilson's New Calendar.*

NOVEMBER 27. To-day we reached Gisborne, and anchored in a big bay; there was a heavy sea on, so we remained on board.

We were a mile from shore; a little steam-tug put out from the land; she was an object of thrilling interest; she would climb to the summit of a billow, reel drunkenly there a moment, dim and gray in the driving storm of spindrift, then make a plunge like a diver and remain out of sight until one had given her up, then up she would dart again, on a steep slant toward the sky, shedding Niagaras of water from her forecastle — and this she kept up, all the way out to us. She brought twenty-five passengers in her stomach — men and women — mainly a traveling dramatic company. In sight on deck were the crew, in sou'westers, yellow waterproof canvas suits, and boots to the thigh. The deck was never quiet for a moment, and seldom nearer level than a ladder, and noble were the seas which leapt aboard and went flooding aft. We rove a long line to the yard-arm, hung a most primitive basket-chair to it and swung it out into the spacious air of heaven, and there it swayed, pendulum-fashion, waiting for its chance — then down it shot, skillfully aimed, and was grabbed by the two men on the forecastle. A young fellow belonging to our crew was in the chair, to be a protection to the lady-comers. At once a couple of ladies appeared from below, took seats in his lap, we hoisted them into the sky, waited a moment till the

PROTECTING THE LADIES.

roll of the ship brought them in, overhead, then we lowered suddenly away, and seized the chair as it struck the deck. We took the twenty-five aboard, and delivered twenty-five into the tug — among them several aged ladies, and one blind one — and all without accident. It was a fine piece of work.

Ours is a nice ship, roomy, comfortable, well-ordered, and satisfactory. Now and then we step on a rat in a hotel, but we have had no rats on shipboard lately; unless, perhaps in the *Flora;* we had more serious things to think of there, and did not notice. I have noticed that it is only in ships and hotels which still employ the odious Chinese gong, that you find rats. The reason would seem to be, that as a rat cannot tell the time of day by a clock, he won't stay where he cannot find out when dinner is ready.

November 29. The doctor tells me of several old drunkards, one spiritless loafer, and several far-gone moral wrecks who have been reclaimed by the Salvation Army and have remained staunch people and hard workers these two years. Wherever one goes, these testimonials to the Army's efficiency are forthcoming. . . . This morning we had one of those whizzing green Ballarat flies in the room, with his stunning buzz-saw noise — the swiftest creature in the world except the lightning-flash. It is a stupendous force that is stored up in that little body. If we had it in a ship in the same proportion, we could spin from Liverpool to New York in the space of an hour — the time it takes to eat luncheon. The New Zealand express train is called the Ballarat Fly. . . . Bad teeth in the colonies. A citizen told me they don't have teeth filled, but pull them out and put in false ones, and that now and then one sees a young lady with a full set. She is fortunate. I wish I had been born with false teeth and a false liver and false carbuncles. I should get along better.

December 2. Monday. Left Napier in the Ballarat Fly —

the one that goes twice a week. From Napier to Hastings, twelve miles; time, fifty-five minutes — not so far short of thirteen miles an hour. . . . A perfect summer day; cool breeze, brilliant sky, rich vegetation. Two or three times during the afternoon we saw wonderfully dense and beautiful forests, tumultuously piled skyward on the broken highlands — not the customary roof-like slant of a hillside, where the trees are all the same height. The noblest of these trees were of the Kauri breed, we were told — the timber that is now furnishing the wood-paving for Europe, and is the best of all wood for that purpose. Sometimes these towering upheavals of forestry were festooned and garlanded with vine-cables, and sometimes the masses of undergrowth were cocooned in another sort of vine of a delicate cobwebby texture — they call it the "supple-jack," I think. Tree ferns everywhere — a stem fifteen feet high, with a graceful chalice of fern-fronds sprouting from its top — a lovely forest ornament. And there was a ten-foot reed with a flowing suit of what looked like yellow hair hanging from its upper end. I do not know its name, but if there is such a thing as a scalp-plant, this is it. A romantic gorge, with a brook flowing in its bottom, approaching Palmerston North.

Waitukurau. Twenty minutes for luncheon. With me sat my wife and daughter, and my manager, Mr. Carlyle Smythe. I sat at the head of the table, and could see the right-hand wall; the others had their backs to it. On that wall, at a good distance away, were a couple of framed pictures. I could not see them clearly, but from the groupings of the figures I fancied that they represented the killing of Napoleon III's son by the Zulus in South Africa. I broke into the conversation, which was about poetry and cabbage and art, and said to my wife —

"Do you remember when the news came to Paris —"

" Of the killing of the Prince ? "

(Those were the very words I had in my mind.)   " Yes,
but *what* Prince ? "

" Napoleon.   Lulu."

" What made you think of that ? "

" I don't know."

There was no collusion.   She had not seen the pictures, and
they had not been mentioned.   She ought to have thought
of some *recent* news that came to Paris, for we were but seven
months from there and had been living there a couple of years
when we started on this trip; but instead of that she thought
of an incident of our brief sojourn in Paris of sixteen years
before.

Here was a clear case of mental telegraphy; of mind-trans-
ference; of my mind telegraphing a thought into hers.   How
do I know ?   Because I telegraphed an *error*.   For it turned
out that the pictures did not represent the killing of Lulu at all,
nor anything connected with Lulu.   She had to get the error
from my head — it existed nowhere else.

CHAPTER XXXV.

The Autocrat of Russia possesses more power than any other man in the earth ;
but he cannot stop a sneeze. — *Pudd'nhead Wilson's New Calendar.*

WAUGANUI, *December 3.* A pleasant trip, yesterday, per Ballarat Fly. Four hours. I do not know the distance, but it must have been well along toward fifty miles. The Fly could have spun it out to eight hours and not discommoded me ; for where there is comfort, and no need for hurry, speed is of no value — at least to me ; and nothing that goes on wheels can be more comfortable, more satisfactory, than the New Zealand trains. Outside of America there are no cars that are so rationally devised. When you add the constant presence of charming scenery and the nearly constant absence of dust — well, if one is not content then, he ought to get out and walk. That would change his spirit, perhaps? I think so. At the end of an hour you would find him waiting humbly beside the track, and glad to be taken aboard again.

Much horseback riding, in and around this town ; many comely girls in cool and pretty summer gowns ; much Salvation Army ; lots of Maoris ; the faces and bodies of some of the old ones very tastefully frescoed. Maori Council House over the river — large, strong, carpeted from end to end with matting, and decorated with elaborate wood carvings, artistically executed. The Maoris were very polite.

I was assured by a member of the House of Representatives that the native race is not decreasing, but actually increasing slightly. It is another evidence that they are a superior breed of savages. I do not call to mind any savage race that built such good houses, or such strong and ingenious and scientific fortresses, or gave so much attention to agriculture, or had military arts and devices which so nearly approached the white man's. These, taken together with their high abilities in boat-building, and their tastes and capacities in the ornamental arts, modify their savagery to a semi-civilization — or at least to a quarter-civilization.

It is a compliment to them that the British did not exterminate them, as they did the Australians and the Tasmanians, but were content with subduing them, and showed no desire to go further. And it is another compliment to them that the British did not take the whole of their choicest lands, but left them a considerable part, and then went further and protected them from the rapacities of land-sharks — a protection

MAORI WOMEN WITH FEATHER ROBES.

which the New Zealand Government still extends to them. And it is still another compliment to the Maoris that the Government allows native representation in both the legislature and the cabinet, and gives both sexes the vote. And in doing these things the Government also compliments itself; it has not been the custom of the world for conquerors to act in this large spirit toward the conquered.

The highest class white men who lived among the Maoris in the earliest time had a high opinion of them and a strong affection for them. Among the whites of this sort was the author of "Old New Zealand;" and Dr. Campbell of Auckland was another. Dr. Campbell was a close friend of several chiefs, and has many pleasant things to say of their fidelity, their magnanimity, and their generosity. Also of their quaint notions about the white man's queer civilization, and their equally quaint comments upon it. One of them thought the missionary had got everything wrong end first and upside down. "Why, he wants us to stop worshiping and supplica-

NOSE RUBBING, FORM OF SALUTATION.

ting the evil gods, and go to worshiping and supplicating the Good One! There is no sense in that. A *good* god is not going to do us any harm."

The Maoris had the *tabu;* and had it on a Polynesian scale of comprehensiveness and elaboration. Some of its features could have been importations from India and Judea. Neither the Maori nor the Hindoo of common degree could cook by a fire that a person of higher caste had used, nor could the high Maori or high Hindoo employ fire that had served a man of low grade; if a low-grade Maori or Hindoo drank from a vessel belonging to a high-grade man, the vessel was defiled, and had to be destroyed. There were other resemblances between Maori *tabu* and Hindoo caste-custom.

Yesterday a lunatic burst into my quarters and warned me that the Jesuits were going to " cook " (poison) me in my food, or kill me on the stage at night. He said a mysterious sign ◊ was visible upon my posters and meant my death. He said he saved Rev. Mr. Haweis's life by warning him that there were three men on his platform who would kill him if he took his eyes off them for a moment during his lecture. The same men were in my audience last night, but they saw that *he* was there. "Will they be there again to-night?" He hesitated; then said no, *he thought they would rather take a rest* and chance the poison. This lunatic has no delicacy. But he was not uninteresting. He told me a lot of things. He said he had "saved so many *lecturers* in twenty years, that *they put him in the asylum.*" I think he has less refinement than any lunatic I have met.

December 8. A couple of curious war-monuments here at Wanganui. One is in honor of white men "who fell in defence of law and order against fanaticism and barbarism." Fanaticism. We Americans are English in blood, English in speech, English in religion, English in the essentials of our governmental system, English in the essentials of our civilization; and so, let us hope, for the honor of the blend, for the honor of the blood, for the honor of the race, that that word got there through lack of heedfulness, and will not be suffered to remain. If you carve it at Thermopylae, or where Winkelried died, or upon Bunker Hill monument, and read it again — "who fell in defence of law and order against fanaticism"— you will perceive what the word means, and how mischosen it is. Patriotism is Patriotism. Calling it Fanaticism cannot degrade it; nothing can degrade it. Even though it be a political mistake, and a thousand times a political mistake, that does not affect it; it is honorable — always honorable, always noble — and privileged to hold its head up and look the nations in the face. It is right to praise these brave white men who fell in the Maori war — they deserve it; but the presence of that word detracts from the dignity of their cause and their deeds, and makes them appear to have spilt their blood in a conflict with ignoble men, men not worthy

of that costly sacrifice.   But the men *were* worthy.   It was
no shame to fight them.   They fought for their homes, they
fought for their country; they bravely fought and bravely
fell; and it would take nothing from the honor of the brave
Englishmen who lie under the monument, but *add* to it, to say
that they died in defense of English laws and English homes
against men worthy of the sacrifice — the Maori patriots.

The other monument cannot be rectified.   Except with dynamite.   It is a mistake all through, and a strangely thoughtless one.   It is a monument erected by white men to Maoris
who fell fighting with the whites and *against their own people*,
in the Maori war.   " Sacred to the memory of the brave men
who fell on the 14th of May, 1864," etc.   On one side are the
names of about twenty Maoris.   It is not a fancy of mine; the
monument exists.   I saw it.   It is an object-lesson to the rising
generation.   It invites to treachery, disloyalty, unpatriotism.
Its lesson, in frank terms is, " Desert your flag, slay your
people, burn their homes, shame your nationality — we honor
such."

December 9.   Wellington.   Ten hours from Wanganui by
the Fly.

December 12.   It is a fine city and nobly situated.   A busy
place, and full of life and movement.   Have spent the three
days partly in walking about, partly in enjoying social privileges, and largely in idling around the magnificent garden at
Hutt, a little distance away, around the shore.   I suppose we
shall not see such another one soon.

We are packing to-night for the return-voyage to Australia.
Our stay in New Zealand has been too brief; still, we are not
unthankful for the glimpse which we have had of it.

The sturdy Maoris made the settlement of the country by
the whites rather difficult.   Not at first — but later.   At first
they welcomed the whites, and were eager to trade with them

— particularly for muskets; for their pastime was internecine war, and they greatly preferred the white man's weapons to their own. War *was* their pastime — I use the word advisedly. They often met and slaughtered each other just for a lark, and when there was no quarrel. The author of "Old New Zealand" mentions a case where a victorious army could have followed up its advantage and exterminated the opposing army, but declined to do it; explaining naïvely that "if we did that, there couldn't be any more fighting." In another battle one army sent word that it was out of ammunition, and would be obliged to stop unless the opposing army would send some. It was sent, and the fight went on.

In the early days things went well enough. The natives sold land without clearly understanding the terms of exchange, and the whites bought it without being much disturbed about the native's confusion of mind. But by and by the Maori began to comprehend that he was being wronged; then there was trouble, for he was not the man to swallow a wrong and go aside and cry about it. He had the Tasmanian's spirit and endurance, and a notable share of military science besides; and so he rose against the oppressor, did this gallant "fanatic," and started a war that was not brought to a definite end until more than a generation had sped.

CHAPTER XXXVI.

There are several good protections against temptations, but the surest is cowardice.—*Pudd'nhead Wilson's New Calendar.*

Names are not always what they seem. The common Welsh name Bzjxxllwcp is pronounced Jackson.—*Pudd'nhead Wilson's New Calendar.*

FRIDAY, *December 13.* Sailed, at 3 p. m., in the *Mararoa.* Summer seas and a good ship — life has nothing better.

Monday. Three days of paradise. Warm and sunny and smooth; the sea a luminous Mediterranean blue. . . . One lolls in a long chair all day under deck-awnings, and reads and smokes, in measureless content. One does not read prose at such a time, but poetry. I have been reading the poems of Mrs. Julia A. Moore, again, and I find in them the same grace and melody that attracted me when they were first published, twenty years ago, and have held me in happy bonds ever since. "The Sentimental Song Book" has long been out of print, and has been forgotten by the world in general, but not by me. I carry it with me always — it and Goldsmith's deathless story. . . . Indeed, it has the same deep charm for me that the Vicar of Wakefield has, and I find in it the same subtle touch — the touch that makes an intentionally humorous episode pathetic and an intentionally pathetic one funny. In her time Mrs. Moore was called "the Sweet Singer of Michigan," and was best known by that name. I have read her book through twice to-day, with the purpose of determining which of her pieces has most merit, and I am persuaded that for wide grasp and sustained power, "William Upson" may claim first place:

WILLIAM UPSON.

AIR — "*The Major's Only Son.*"

Come all good people far and near,
Oh, come and see what you can hear,
It's of a young man true and brave,
That is now sleeping in his grave.

Now, William Upson was his name —
If it's not that, it's all the same —
He did enlist in a cruel strife,
And it caused him to lose his life.

He was Perry Upson's eldest son,
His father loved his noble son,
This son was nineteen years of age
When first in the rebellion he engaged.

His father said that he might go,
But his dear mother she said no,
"Oh! stay at home, dear Billy," she said.
But she could not turn his head.

He went to Nashville, in Tennessee,
There his kind friends he could not see;
He died among strangers, so far away,
They did not know where his body lay.

He was taken sick and lived four weeks,
And Oh! how his parents weep,
But now they must in sorrow mourn,
For Billy has gone to his heavenly home.

Oh! if his mother could have seen her son,
For she loved him, her darling son;
If she could heard his dying prayer,
It would ease her heart till she met him there

How it would relieve his mother's heart
To see her son from this world depart,
And hear his noble words of love,
As he left this world for that above.

Now it will relieve his mother's heart,
For her son is laid in our graveyard;
For now she knows that his grave is near,
She will not shed so many tears.

Although she knows not that it was her son,
For his coffin could not be opened —
It might be someone in his place,
For she could not see his noble face.

December 17. Reached Sydney.

December 19. In the train. Fellow of 30 with four valises; a slim creature, with teeth which made his mouth look like a neglected churchyard. He had solidified hair — solidified with pomatum; it was all one shell. He smoked the most extraordinary cigarettes — made of some kind of manure, apparently. These and his hair made him smell like the very nation. He had a low-cut vest on, which exposed a deal of frayed and broken and unclean shirt-front. Showy studs, of imitation gold — they had made black disks on the linen. Oversized sleeve buttons of imitation gold, the copper base showing through. Ponderous watch-chain of imitation gold. I judge that he couldn't tell the time by it, for he asked Smythe what time it was, once. He wore a coat which had been gay when it was young; 5-o'clock-tea-trousers of a light tint, and marvelously soiled; yellow mustache with a dashing upward whirl at the ends; foxy shoes, imitation patent leather. He was a novelty — an imitation dude. He would have been a real one if he could have afforded it. But he was satisfied with himself. You could see it in his expression, and in all his attitudes and movements. He was living in a dude dreamland where all his squalid shams were genuine, and himself a sincerity. It disarmed criticism, it mollified spite, to see him so enjoy his imitation languors, and arts, and airs, and his studied daintinesses of gesture and misbegotten refinements. It was plain to me that he was imagining himself the Prince of Wales, and was doing everything the way he thought the Prince would do it. For bringing his four valises aboard and stowing them in the nettings, he gave his porter four cents, and lightly apologized for the smallness of the gratuity — just with the condescendingest little royal air in the world. He stretched himself out on the front seat and rested his pomatum-cake on the middle arm, and stuck his feet out of the window, and began to pose as the Prince and work his dreams and languors for exhibition; and he would indolently watch the blue films curling up from his cigarette, and inhale the stench, and look so grateful; and would flip the ash away with the daintiest gesture, unintentionally displaying his brass ring in the most intentional way; why, it was as good as being in Marlborough House itself to see him do it so like.

There was other scenery in the trip. That of the Hawksbury river, in the National Park region, fine — extraordinarily fine, with spacious views of stream and lake imposingly framed in woody hills; and every now and then the noblest groupings of mountains, and the most enchanting re-arrangements of the

SO LIKE THE PRINCE.

water effects. Further along, green flats, thinly covered with gum forests, with here and there the huts and cabins of small farmers engaged in raising children. Still further along, arid stretches, lifeless and melancholy. Then Newcastle, a rushing town, capital of the rich coal regions. Approaching Scone, wide farming and grazing levels, with pretty frequent glimpses of a troublesome plant — a particularly devilish little prickly pear, daily damned in the orisons of the agriculturist; imported by a lady of sentiment, and contributed gratis to the colony.
. . . Blazing hot, all day.

December 20. Back to Sydney. Blazing hot again. From

the newspaper, and from the map, I have made a collection of curious names of Australasian towns, with the idea of making a poem out of them :

Tumut	Waitpinga	Wollongong
Takee	Goelwa	Woolloomooloo
Murriwillumba	Munno Para	Bombola
Bowral	Nangkita	Coolgardie
Ballarat	Myponga	Bendigo
Mullengudgery	Kapunda	Coonamble
Murrurundi	Kooringa	Cootamundra
Wagga-Wagga	Penola	Woolgoolga
Wyalong	Nangwarry	Mittagong
Murrumbidgee	Kongorong	Jamberoo
Goomeroo	Comaum	Kondoparinga
Wolloway	Koolywurtie	Kuitpo
Wangary	Killanoola	Tungkillo
Wanilla	Naracoorte	Oukaparinga
Worrow	Muloowurtie	Talunga
Koppio	Binnum	Yatala
Yankalilla	Wallaroo	Parawirra
Yaranyacka	Wirrega	Moorooroo
Yackamoorundie	Mundoora	Whangarei
Kaiwaka	Hauraki	Woolundunga
Coomooroo	Rangiriri	Booleroo
Tauranga	Teawamute	Pernatty
Geelong	Taranaki	Parramatta
Tongariro	Toowoomba	Taroom
Kaikoura	Goondiwindi	Narrandera
Wakatipu	Jerrilderie	Deniliquin
Oohipara	Whangaroa	Kawakawa.

It may be best to build the poem now, and make the weather help :

A SWELTERING DAY IN AUSTRALIA.

(To be read soft and low, with the lights turned down.)

The Bombola faints in the hot Bowral tree,
    Where fierce Mullengudgery's smothering fires
Far from the breezes of Coolgardie
    Burn ghastly and blue as the day expires ;

And Murriwillumba complaineth in song
    For the garlanded bowers of Woolloomooloo,
And the Ballarat Fly and the lone Wollongong
    They dream of the gardens of Jamberoo ;

The wallabi sighs for the Murrubid*gee*,
    For the velvety sod of the Munno Pa*rah*,
Where the waters of healing from Muloowur*tie*
    Flow dim in the gloaming by Yaranyac*kah ;*

The Koppio sorrows for lost Wolloway,
    And sigheth in secret for Murruru*ndi*,
The Whangeroo wombat lamenteth the day
    That made him an exile from Jerrilde*rie ;*

The Teawamute Tumut from Wirrega's glade,
    The Nangkita swallow, the Wallaroo swan,
They long for the peace of the Timaru shade
    And thy balmy soft airs, O sweet Mittagong !

The Kooringa buffalo pants in the sun,
    The Kondoparinga lies gaping for breath,
The Kongorong Camaum to the shadow has won,
    But the Goomeroo sinks in the slumber of death ;

In the weltering hell of the Moorooroo plain
    The Yatala Wangary withers and dies,
And the Worrow Wanilla, demented with pain,
    To the Woolgoolga woodlands despairingly flies ;

Sweet Nangwarry's desolate, Coonamble wails,
    And Tungkillo Kuito in sables is drest,
For the Whangerei winds fall asleep in the sails
    And the Booleroo life-breeze is dead in the west.

Mypongo, Kapunda, O slumber no more !
    Yankalilla, Parawirra, be warned !
There's death in the air ! Killanoola, wherefore
    Shall the prayer of Penola be scorned ?

Cootamundra, and Takee, and Wakatipu,
    Toowoomba, Kaikoura are lost !
From Onkaparinga to far Oamaru
    All burn in this hell's holocaust !

Paramatta and Binnum are gone to their rest
    In the vale of Tapanni Taroom,
Kawakawa, Deniliquin — all that was best
    In the earth are but graves and a tomb !

Narrandera mourns, Cameroo answers not
    When the roll of the scathless we cry :
Tongariro, Goondiwindi, Woolundunga, the spot
    Is mute and forlorn where ye lie.

Those are good words for poetry. Among the best I have
ever seen. There are 81 in the list. I did not need them all,

but I have knocked down 66 of them; which is a good bag, it seems to me, for a person not in the business. Perhaps a poet laureate could do better, but a poet laureate gets wages, and that is different. When I write poetry I do not get any wages; often I lose money by it. The best word in that list, and the most musical and gurgly, is Woolloomoolloo. It is a place near Sydney, and is a favorite pleasure-resort. It has eight O's in it.

CHAPTER XXXVII.

MONDAY, *December 23, 1895.* Sailed from Sydney for Ceylon in the P. & O. steamer *Oceana*. A Lascar crew mans this ship — the first I have seen. White cotton petticoat and pants; barefoot; red shawl for belt; straw cap, brimless, on head, with red scarf wound around it; complexion a rich dark brown; short straight black hair; whiskers fine and silky; lustrous and intensely black. Mild, good faces; willing and obedient people; capable, too; but are said to go into hopeless panics when there is danger. They are from Bombay and the coast thereabouts. . . . Left some of the trunks in Sydney, to be shipped to South Africa by a vessel advertised to sail three months hence. The proverb says: "Separate not yourself from your baggage." . . . This *Oceana* is a stately big ship, luxuriously appointed. She has spacious promenade decks. Large rooms; a surpassingly comfortable ship. The officers' library is well selected; a ship's library is not usually that. . . . For meals, the bugle call, man-of-war fashion; a pleasant change from the terrible gong. . . . Three big cats — very friendly loafers; they wander all over the ship; the white one follows the chief steward around like a dog. There is also a basket of kittens. One of these cats goes ashore, in port, in England, Australia, and India, to see how his various families are getting along, and is seen no more till the ship is ready to sail. No one knows how he finds out the sailing date, but no doubt he comes down to

the dock every day and takes a look, and when he sees baggage and passengers flocking in, recognizes that it is time to
get aboard.   This is what the sailors believe.   .   .   .   The
Chief Engineer has been in the China and India trade thirty-
three years, and has had but three Christmases at home in that
time.   .   .   .   Conversational items at dinner, " Mocha! sold
all over the world!   It is not true.   In fact, very few foreigners

WHAT THE SAILORS BELIEVE.

except the Emperor of Russia have ever seen a grain of it, or
ever will, while they live."   Another man said: " There is no
sale in Australia for Australian wine.   But it goes to France
and comes back with a French label on it, and then they buy
it."   I have heard that the most of the French-labeled claret
in New York is made in California.   And I remember what
Professor S. told me once about Veuve Cliquot — if that was
the wine, and I think it was.   He was the guest of a great wine
merchant whose town was quite near that vineyard, and this
merchant asked him if very much V. C. was drunk in
America.

"Oh, yes," said S., "a great abundance of it."

"Is it easy to be had?"

"Oh, yes — easy as water. All first and second-class hotels have it."

"What do you pay for it?"

"It depends on the style of the hotel — from fifteen to twenty-five francs a bottle."

"Oh, fortunate country! Why, it's worth 100 francs right here on the ground."

"No!"

"Yes!"

"Do you mean that we are drinking a bogus Veuve Cliquot over there?"

"Yes — and there was never a bottle of the genuine in America since Columbus's time. That wine all comes from a little bit of a patch of ground which isn't big enough to raise many bottles; and all of it that is produced goes every year to one person — the Emperor of Russia. He takes the whole crop in advance, be it big or little."

January 4, 1896. Christmas in Melbourne, New Year's Day in Adelaide, and saw most of the friends again in both places. . . . Lying here at anchor all day — Albany (King George's Sound), Western Australia. It is a perfectly land-locked harbor, or roadstead — spacious to look at, but not deep water. Desolate-looking rocks and scarred hills. Plenty of ships arriving now, rushing to the new gold-fields. The papers are full of wonderful tales of the sort always to be heard in connection with new gold diggings. A sample: a youth staked out a claim and tried to sell half for £5; no takers; he stuck to it fourteen days, starving, then struck it rich and sold out for £10,000. . . . About sunset, strong breeze blowing, got up the anchor. We were in a small deep puddle, with a narrow channel leading out of it, minutely buoyed, to the sea.

I stayed on deck to see how we were going to manage it with such a big ship and such a strong wind. On the bridge our giant captain, in uniform; at his side a little pilot in elaborately gold-laced uniform; on the forecastle a white mate and quartermaster or two, and a brilliant crowd of lascars standing by for business. Our stern was pointing straight at the head of the channel; so we must turn entirely around in the puddle — and the wind blowing as described. It was done, and beautifully. It was done by help of a jib. We stirred up much mud, but did not touch the bottom. We turned right around in our tracks — a seeming impossibility. We had several casts of quarter-less 5, and one cast of half 4 — 27 feet; we were drawing 26 astern. By the time we were entirely around and *pointed*, the first buoy was not more than a hundred yards in front of us. It was a fine piece of work, and I was the only passenger that saw it. However, the others got their dinner; the P. & O. Company got mine. . . . More cats developed. Smythe says it is a British law that they must be carried; and he instanced a case of a ship not allowed to sail till she sent for a couple. The bill came, too: " Debtor, to 2 cats, 20 shillings." . . . News comes that within this week Siam has acknowledged herself to be, in effect, a French province. It seems plain that all savage and semi-civilized countries are going to be grabbed. . . . A vulture on board; bald, red, queer-shaped head, featherless red places here and there on his body, intense great black eyes set in featherless rims of inflamed flesh; dissipated look; a business-like style, a selfish, conscienceless, murderous aspect — the very look of a professional assassin, and yet a bird which does no murder. What was the use of getting him up in that tragic style for so innocent a trade as his? For this one isn't the sort that wars upon the living, his diet is offal — and the more out of date it is the better he likes it. Nature should give him a suit of rusty black;

then he would be all right, for he would look like an undertaker and would harmonize with his business; whereas the way he is now he is horribly out of true.

January 5. At 9 this morning we passed Cape Leeuwin (lioness) and ceased from our long due-west course along the southern shore of Australia. Turning this extreme southwestern corner, we now take a long straight slant nearly N. W., without a break, for Ceylon. As we speed northward it will grow hotter very fast — but it isn't chilly, now. . . . The vulture is from the public menagerie at Adelaide — a great and interesting collection. It was there that we saw the baby tiger solemnly spreading its mouth and trying to roar like its majestic mother. It swaggered, scowling, back and forth on its short legs just as it had seen her do on her long ones, and now and then snarling viciously, exposing its teeth, with a threatening lift of its upper lip and bristling moustache; and when it thought it was impressing the visitors, it would spread its mouth wide and do that screechy cry which it meant for a roar, but which did not deceive. It took itself quite seriously, and was lovably comical. And there was a hyena — an ugly creature; as ugly as the tiger-kitty was pretty. It repeatedly arched its back and delivered itself of *such* a human cry; a startling resemblance; a cry which was just that of a grown person badly hurt. In the dark one would assuredly go to its assistance — and be disappointed. . . . Many friends of Australasian Federation on board. They feel sure that the good day is not far off, now. But there seems to be a party that would go further — have Australasia cut loose from the British Empire and set up housekeeping on her own hook. It seems an unwise idea. They point to the United States, but it seems to me that the cases lack a good deal of being alike. Australasia governs herself wholly — there is no interference; and her commerce and manufactures are not oppressed in any way.

If our case had been the same we should not have gone out when we did.

January 13. Unspeakably hot. The equator is arriving again. We are within eight degrees of it. Ceylon present. Dear me, it is beautiful! And most sumptuously tropical, as to character of foliage and opulence of it. "What though the spicy breezes blow soft o'er Ceylon's isle" — an eloquent line, an incomparable line; it says little, but conveys whole libraries of sentiment, and Oriental charm and mystery, and tropic deliciousness — a line that quivers and tingles with a thousand unexpressed and inexpressible things, things that haunt one and find no articulate voice. . . . Colombo, the capital. An Oriental town, most manifestly; and fascinating. . . . In this palatial ship the passengers dress for dinner. The ladies' toilettes make a fine display of color, and this is in keeping with the elegance of the vessel's furnishings and the flooding brilliancies of the electric light. On the stormy Atlantic one never sees a man in evening dress, except at the rarest intervals; and then there is only one, not two; and he shows up but once on the voyage — the night before the ship makes port — the night when they have the "concert" and do the amateur wailings and recitations. He is the tenor, as a rule. . . . There has been a deal of cricket-playing on board; it seems a queer game for a ship, but they enclose the promenade deck with nettings and keep the ball from flying overboard, and the sport goes very well, and is properly violent and exciting. . . . We must part from this vessel here.

January 14. Hotel Bristol. Servant Brompy. Alert, gentle, smiling, winning young brown creature as ever was. Beautiful shining black hair combed back like a woman's, and knotted at the back of his head — tortoise-shell comb in it, sign that he is a Singhalese; slender, shapely form; jacket; under it is a beltless and flowing white cotton gown — from

SERVANT BROMPY.

neck straight to heel; he and his outfit quite unmasculine. It was an embarassment to undress before him.

We drove to the market, using the Japanese jinriksha — our first acquaintanceship with it. It is a light cart, with a native to draw it. He makes good speed for half-an-hour, but it is hard work for him; he is too slight for it. After the half-hour there is no more pleasure for you; your attention is all on the man, just as it would be on a tired horse, and necessarily your sympathy is there too. There's a plenty of these 'rickshas, and the tariff is incredibly cheap.

I was in Cairo years ago. That was Oriental, but there was a lack. When you are in Florida or New Orleans you are in the South — that is granted; but you are not in *the* South; you are in a modified South, a tempered South. Cairo was a tempered Orient — an Orient with an indefinite something wanting. That feeling was not present in Ceylon. Ceylon was Oriental in the last measure of completeness — utterly Oriental; also utterly tropical; and indeed to one's unreasoning spiritual sense the two things belong together. All the requisites were present. The costumes were right; the black and brown exposures, unconscious of immodesty, were right; the juggler was there, with his basket, his snakes, his mongoose, and his arrangements for growing a tree from seed to foliage and ripe fruitage before one's eyes; in sight were plants and flowers familiar to one on books but in no other way — celebrated, desirable, strange, but in production restricted to the hot belt of the equator; and out a little way in the country were the proper deadly snakes, and fierce beasts of prey, and the wild elephant and the monkey. And there was that swoon in the air which one associates with the tropics, and that smother of heat, heavy with odors of unknown flowers, and that sudden invasion of purple gloom fissured with lightnings, — then the tumult of crashing thunder and the downpour —

and presently all sunny and smiling again; all these things were there; the conditions were complete, nothing was lacking. And away off in the deeps of the jungle and in the remotenesses of the mountains were the ruined cities and mouldering temples, mysterious relics of the pomps of a forgotten time and a vanished race — and this was as it should be, also, for nothing is quite satisfyingly Oriental that lacks the somber and impressive qualities of mystery and antiquity.

The drive through the town and out to the Gallè Face by the seashore, what a dream it was of tropical splendors of bloom and blossom, and Oriental conflagrations of costume! The walking groups of men, women, boys, girls, babies — each individual was a flame, each group a house afire for color. And such stunning colors, such intensely vivid colors, such rich and exquisite minglings and fusings of rainbows and lightnings! And all harmonious, all in perfect taste; never a discordant note; never a color on any person swearing at another color on him or failing to harmonize faultlessly with the colors of any group the wearer might join. The stuffs were silk — thin, soft, delicate, clinging; and, as a rule, each piece a solid color: a splendid green, a splendid blue, a splendid yellow, a splendid purple, a splendid ruby, deep, and rich with smouldering fires — they swept continuously by in crowds and legions and multitudes, glowing, flashing, burning, radiant; and every five seconds came a burst of blinding red that made a body catch his breath, and filled his heart with joy. And then, the unimaginable grace of those costumes! Sometimes a woman's whole dress was but a scarf wound about her person and her head, sometimes a man's was but a turban and a careless rag or two — in both cases generous areas of polished dark skin showing — but always the arrangement compelled the homage of the eye and made the heart sing for gladness.

I can see it to this day, that radiant panorama, that wilder-

THOMAS FOGARTY

ABUSED CREATURES.

ness of rich color, that incomparable dissolving-view of harmonious tints, and lithe half-covered forms, and beautiful brown faces, and gracious and graceful gestures and attitudes and movements, free, unstudied, barren of stiffness and restraint, and —

Just then, into this dream of fairyland and paradise a grating dissonance was injected. Out of a missionary school came marching, two and two, sixteen prim and pious little Christian black girls, Europeanly clothed — dressed, to the last detail, as they would have been dressed on a summer Sunday in an English or American village. Those clothes — oh, they were unspeakably ugly! Ugly, barbarous, destitute of taste, destitute of grace, repulsive as a shroud. I looked at my women-folk's clothes — just full-grown duplicates of the outrages disguising those poor little abused creatures — and was ashamed to be seen in the street with them. Then I looked at my own clothes, and was ashamed to be seen in the street with myself.

However, we must put up with our clothes as they are — they have their reason for existing. They are on us to expose us — to advertise what we wear them to conceal. They are a sign; a sign of insincerity; a sign of suppressed vanity; a pretense that we despise gorgeous colors and the graces of harmony and form; and we put them on to propagate that lie and back it up. But we do not deceive our neighbor; and when we step into Ceylon we realize that we have not even deceived ourselves. We do love brilliant colors and graceful costumes; and at home we will turn out in a storm to see them when the procession goes by — and envy the wearers. We go to the theater to look at them and grieve that we can't be clothed like that. We go to the King's ball, when we get a chance, and are glad of a sight of the splendid uniforms and the glittering orders. When we are granted permission to attend an imperial drawing-room we shut ourselves up in private and parade around in the theatrical court-dress by the hour, and admire ourselves in the glass, and are utterly happy; and every member of every governor's staff in democratic America does the same with his grand new uniform — and if he is not watched he will get himself photographed in it, too. When I see the Lord Mayor's footman I am dissatisfied with my lot. Yes, our clothes are a lie, and have been nothing short of that these hundred years. They are insincere, they are the ugly and appropriate outward exposure of an inward sham and a moral decay.

The last little brown boy I chanced to notice in the crowds and swarms of Colombo had nothing on but a twine string around his waist, but in my memory the frank honesty of his costume still stands out in pleasant contrast with the odious flummery in which the little Sunday-school dowdies were masquerading.

CHAPTER XXXVIII.

Prosperity is the best protector of principle.— *Pudd'nhead Wilson's New Calendar.*

EVENING — *14th*. Sailed in the *Rosetta*. This is a poor old ship, and ought to be insured and sunk. As in the *Oceana*, just so here: everybody dresses for dinner; they make it a sort of pious duty. These fine and formal costumes are a rather conspicuous contrast to the poverty and shabbiness of the surroundings. . . . If you want a slice of a lime at four o'clock tea, you must sign an order on the bar. Limes cost 14 cents a barrel.

January 18th. We have been running up the Arabian Sea, latterly. Closing up on Bombay now, and due to arrive this evening.

January 20th. *Bombay!* A bewitching place, a bewildering place, an enchanting place — the Arabian Nights come again! It is a vast city; contains about a million inhabitants. Natives, they are, with a slight sprinkling of white people — not enough to have the slightest modifying effect upon the massed dark complexion of the public. It is winter here, yet the weather is the divine weather of June, and the foliage is the fresh and heavenly foliage of June. There is a rank of noble great shade trees across the way from the hotel, and under them sit groups of picturesque natives of both sexes; and the juggler in his turban is there with his snakes and his magic; and all day long the cabs and the multitudinous varieties of costumes flock by. It does not seem as if one could ever get tired of watching this moving show, this shining and shifting spectacle. . . . In the great bazar the pack and jam of

natives was marvelous, the sea of rich-colored turbans and
draperies an inspiring sight, and the quaint and showy Indian
architecture was just the right setting for it. Toward sunset
another show; this is the drive around the sea-shore to Malabar Point, where Lord Sandhurst, the Governor of the Bombay Presidency, lives. Parsee palaces all along the first part
of the drive; and past them all the world is driving; the
private carriages of wealthy Englishmen and natives of rank
are manned by a driver and three footmen in stunning oriental
liveries — two of these turbaned statues standing up behind, as
fine as monuments. Sometimes even the public carriages
have this superabundant crew, slightly modified — one to
drive, one to sit by and see it done, and one to stand up
behind and yell — yell when there is anybody in the way, and
for practice when there isn't. It all helps to keep up the live-
liness and augment the general sense of swiftness and energy
and confusion and pow-wow.

In the region of Scandal Point — felicitous name — where
there are handy rocks to sit on and a noble view of the sea on
the one hand, and on the other the passing and repassing
whirl and tumult of gay carriages, are great groups of comfort-
ably-off Parsee women — perfect flower-beds of brilliant color,
a fascinating spectacle. Tramp, tramp, tramping along the
road, in singles, couples, groups, and gangs, you have the
working-man and the working-woman — but not clothed like
ours. Usually the man is a nobly-built great athlete, with not
a rag on but his loin-handkerchief; his color a deep dark
brown, his skin satin, his rounded muscles knobbing it as if it
had eggs under it. Usually the woman is a slender and
shapely creature, as erect as a lightning-rod, and she has but
one thing on — a bright-colored piece of stuff which is wound
about her head and her body down nearly half-way to her
knees, and which clings like her own skin. Her legs and feet

are bare, and so are her arms, except for her fanciful bunches of loose silver rings on her ankles and on her arms. She has jewelry bunched on the side of her nose also, and showy cluster-rings on her toes. When she undresses for bed she takes off her jewelry, I suppose. If she took off anything more she would catch cold. As a rule she has a large shiney brass water-jar of graceful shape on her head, and one of her naked arms curves up and the hand holds it there. She is so straight, so erect, and she steps with such style, and such easy grace and dignity; and her curved arm and her brazen jar are such a help to the picture — indeed, our working-women cannot begin with her as a road-decoration.

A ROAD-DECORATION.

It is all color, bewitching color, e n c h a n t i n g color — everywhere — all around — all the way around the curving great opaline bay clear to Government House, where the turbaned big native *chuprassies* stand grouped in state at the door in their robes of fiery red, and do most properly and stunningly finish up the splendid show and make it theatrically complete. I wish I were a chuprassy.

This is indeed India! the land of dreams and romance, of fabulous wealth and fabulous poverty, of splendor and rags, of palaces and hovels, of famine and pestilence, of genii and giants and Aladdin lamps, of tigers and elephants, the cobra and the jungle, the country of a hundred nations and a hundred tongues, of a thousand religions and two million gods, cradle of the human race, birthplace of human speech, mother of history, grandmother of legend, great-grandmother of tradition, whose yesterdays bear date with the mouldering antiquities of the rest of the nations — the one sole country under the sun that is endowed with an imperishable interest for alien prince and alien peasant, for lettered and ignorant, wise and fool, rich and poor, bond and free, the one land that *all* men desire to see, and having seen once, by even a glimpse, would not give that glimpse for the shows of all the rest of the globe combined.

Even now, after the lapse of a year, the delirium of those days in Bombay has not left me, and I hope never will. It was all new, no detail of it hackneyed. And India did not wait for morning, it began at the hotel — straight away. The lobbies and halls were full of turbaned, and fez'd and embroidered, cap'd, and barefooted, and cotton-clad dark natives, some of them rushing about, others at rest squatting, or sitting on the ground ; some of them chattering with energy, others still and dreamy ; in the dining-room every man's own private native servant standing behind his chair, and dressed for a part in the Arabian Nights.

Our rooms were high up, on the front. A white man — he was a burly German — went up with us, and brought three natives along to see to arranging things. About fourteen others followed in procession, with the hand-baggage ; each carried an article — and only one ; a bag, in some cases, in other cases less. One strong native carried my overcoat, another a parasol, another a box of cigars, another a novel, and the last man in the procession had no load but a fan. It was all done with earnestness and sincerity, there was not a smile in the pro-

FOURTEEN FOLLOWED.

cession from the head of it to the tail of it. Each man waited patiently, tranquilly, in no sort of hurry, till one of us found time to give him a copper, then he bent his head reverently, touched his forehead with his fingers, and went his way. They seemed a soft and gentle race, and there was something both winning and touching about their demeanor.

There was a vast glazed door which opened upon the balcony. It needed closing, or cleaning, or something, and a native got down on his knees and went to work at it. He seemed to be doing it well enough, but perhaps he wasn't, for the burly German put on a look that betrayed dissatisfaction, then without *explaining* what was wrong, gave the native a brisk cuff on the jaw and *then* told him where the defect was. It seemed such a shame to do that before us all. The native took it with meekness, saying nothing, and not showing in his face or manner any resentment. I had not seen the like of this for fifty years. It carried me back to my boyhood, and flashed upon me the forgotten fact that this was the *usual* way of explaining one's desires to a slave. I was able to re-member that the method seemed right and natural to me in those days, I being born to it and unaware that elsewhere there were other methods; but I was also able to remember that those unresented cuffings made me sorry for the victim and ashamed for the punisher. My father was a refined and kindly gentleman, very grave, rather austere, of rigid probity, a sternly just and upright man, albeit he attended no church and never spoke of religious matters, and had no part nor lot in the pious joys of his Presbyterian family, nor ever seemed to suffer from this deprivation. He laid his hand upon me in punishment only twice in his life, and then not heavily; once for telling him a lie — which surprised me, and showed me how unsuspicious he was, for that was not my maiden effort. He punished me those two times only, and never any other

member of the family at all; yet every now and then he cuffed
our harmless slave boy, Lewis, for trifling little blunders and
awkardnesses.   My father had passed his life among the slaves
from his cradle up, and his cuffings proceeded from the custom
of the time, not from his nature.   When I was ten years old
I saw a man fling a lump of iron-ore at a slave-man in anger,
for merely doing something awkwardly — as if that were a
crime.   It bounded from the man's skull, and the man fell and
never spoke again.   He was dead in an hour.   I knew the man
had a right to kill his slave if he wanted to, and yet it seemed
a pitiful thing and somehow wrong, though why wrong I was
not deep enough to explain if I had been asked to do it.
Nobody in the village approved of that murder, but of course
no one said much about it.

It is curious — the space-annihilating power of thought.
For just one second, all that goes to make the *me* in me was
in a Missourian village, on the other side of the globe, vividly
seeing again these forgotten pictures of fifty years ago, and
wholly unconscious of all things but just those; and in the next
second I was back in Bombay, and that kneeling native's
smitten cheek was not done tingling yet! Back to boyhood
— fifty years; back to age again, another fifty; and a flight
equal to the circumference of the globe — all in two seconds
by the watch!

Some natives — I don't remember how many — went into
my bedroom, now, and put things to rights and arranged the
mosquito-bar, and I went to bed to nurse my cough.   It was
about nine in the evening.   What a state of things! For three
hours the yelling and shouting of natives in the hall continued,
along with the velvety patter of their swift bare feet — what a
racket it was! They were yelling orders and messages down
three flights.   Why, in the matter of noise it amounted to a
riot, an insurrection, a revolution.   And then there were other

noises mixed up with these and at intervals tremendously accenting them — roofs falling in, I judged, windows smashing, persons being murdered, crows squawking, and deriding, and cursing, canaries screeching, monkeys jabbering, macaws blaspheming, and every now and then fiendish bursts of laughter and explosions of dynamite. By midnight I had suffered all the different kinds of shocks there are, and knew that I could never more be disturbed by them, either isolated or in combination. Then came peace — stillness deep and solemn — and lasted till five.

Then it all broke loose again. And who re-started it? The Bird of Birds — the Indian crow. I came to know him well, by and by, and be infatuated with him. I suppose he is the hardest lot that wears feathers. Yes, and the cheerfulest, and the best satisfied with himself. He never arrived at what he is by any careless process, or any sudden one; he is a work of art, and "art is long"; he is the product of immemorial ages, and of deep calculation; one can't make a bird like that in a day. He has been re-incarnated more times than Shiva; and he has kept a sample of each incarnation, and fused it into his constitution. In the course of his evolutionary promotions, his sublime march toward ultimate perfection, he has been a gambler, a low comedian, a dissolute priest, a fussy woman, a blackguard, a scoffer, a liar, a thief, a spy, an informer, a trading politician, a swindler, a professional hypocrite, a patriot for cash, a reformer, a lecturer, a lawyer, a conspirator, a rebel, a royalist, a democrat, a practicer and propagator of irreverence, a meddler, an intruder, a busybody, an infidel, and a wallower in sin for the mere love of it. The strange result, the incredible result, of this patient accumulation of all damnable traits is, that he does not know what care is, he does not know what sorrow is, he does not know what remorse is, his life is one long thundering ecstasy of happiness, and he will go to his

23

death untroubled, knowing that he will soon turn up again as an author or something, and be even more intolerably capable and comfortable than ever he was before.

In his straddling wide forward-step, and his springy side-wise series of hops, and his impudent air, and his cunning way of canting his head to one side upon occasion, he reminds one of the American blackbird. But the sharp resemblances stop there. He is much bigger than the blackbird; and he lacks the blackbird's trim and slender and beautiful build and shapely beak; and of course his sober garb of gray and rusty black is a poor and humble thing compared with the splendid lustre of the blackbird's metallic sables and shifting and flashing bronze glories. The blackbird is a perfect gentleman, in deportment and attire, and is not noisy, I believe, except when holding religious services and political conventions in a tree; but this Indian sham Quaker is just a rowdy, and is always noisy when awake — always chaffing, scolding, scoffing, laughing, ripping, and cursing, and carrying on about something or other. I never saw such a bird for delivering opinions. Nothing escapes him; he notices everything that happens, and brings out his opinion about it, particularly if it is a matter that is none of his business. And it is never a mild opinion, but always violent — violent and profane — the presence of ladies does not affect him. His opinions are not the outcome of reflection, for he never thinks about anything, but heaves out the opinion that is on top in his mind, and which is often an opinion about some quite different thing and does not fit the case. But that is his way; his main idea is to get out an opinion, and if he stopped to think he would lose chances.

I suppose he has no enemies among men. The whites and Mohammedans never seemed to molest him; and the Hindoos, because of their religion, never take the life of any creature, but spare even the snakes and tigers and fleas and rats. If I sat on one end of the balcony, the crows would gather on the railing at the other end and talk about me ; and edge closer, little by little, till I could almost reach them ; and they would sit there, in the most unabashed way, and talk about my clothes, and my hair, and my complexion, and probable character and vocation and politics, and how I came to be in India, and what I had been doing, and how m a n y

OPPRESSIVELY SOCIABLE.

days I had got for it, and how I had happened to go unhanged so long, and when would it probably come off, and might there be more of my sort where I came from, and when would *they* be hanged, — and so on, and so on, until I could not longer endure the embarrassment of it; then I would shoo them away, and they would circle around in the air a little while, laughing and deriding and mocking, and presently settle on the rail and do it all over again.

They were very sociable when there was anything to eat — oppressively so. With a little encouragement they would come in and light on the table and help me eat my breakfast; and once when I was in the other room and they found themselves alone, they carried off everything they could lift; and they were particular to choose things which they could make no use of after they got them.   In India their number is beyond estimate, and their noise is in proportion.   I suppose they cost the country more than the government does; yet that is not a light matter.   Still, they pay; their company pays; it would sadden the land to take their cheerful voice out of it.

CHAPTER XXXIX.

By trying we can easily learn to endure adversity. Another man's, I mean.
—*Pudd'nhead Wilson's New Calendar.*

YOU soon find your long-ago dreams of India rising in a sort of vague and luscious moonlight above the horizon-rim of your opaque consciousness, and softly lighting up a thousand forgotten details which were parts of a vision that had once been vivid to you when you were a boy, and steeped your spirit in tales of the East. The barbaric gorgeousnesses, for instance; and the princely titles, the sumptuous titles, the sounding titles,— how good they taste in the mouth! The Nizam of Hyderabad; the Maharajah of Travancore; the Nabob of Jubbelpore; the Begum of Bhopal; the Nawab of Mysore; the Ranee of Gulnare; the Ahkoond of Swat's; the Rao of Rohilkund; the Gaikwar of Baroda. Indeed, it is a country that runs richly to name. The great god Vishnu has 108 — 108 special ones — 108 peculiarly holy ones — names just for Sunday use only. I learned the whole of Vishnu's 108 by heart once, but they wouldn't stay; I don't remember any of them now but John W.

And the romances connected with those princely native houses — to this day they are always turning up, just as in the old, old times. They were sweating out a romance in an English court in Bombay a while before we were there. In this case a native prince, $16\frac{1}{2}$ years old, who has been enjoying his titles and dignities and estates unmolested for fourteen years, is suddenly haled into court on the charge that he is rightfully no prince at all, but a pauper peasant; that the real prince died when two and one-half years old; that the death was concealed, and a peasant child smuggled into the royal cradle, and that this present incumbent was that smuggled substitute. This is the very material that so many oriental tales have been made of.

The case of that great prince, the Gaikwar of Baroda, is a reversal of the theme. When that throne fell vacant, no heir could be found for some time, but at last one was found in the person of a peasant child who was making mud pies in a village street, and having an innocent good time. But his pedigree was straight; he was the true prince, and he has reigned ever since, with none to dispute his right.

Lately there was another hunt for an heir to another princely house, and one was found who was circumstanced about as the Gaikwar had been. His fathers were traced back, in humble life, along a branch of the ancestral tree to the point where it joined the stem fourteen generations ago, and his heirship was thereby squarely established. The tracing was done by means of the records of one of the great Hindoo shrines, where princes on pilgrimage record their names and the date of their visit. This is to keep the prince's religious account straight, and his spiritual person safe; but the record has the added value of keeping the pedigree authentic, too.

When I think of Bombay now, at this distance of time, I seem to have a kaleidoscope at my eye; and I hear the clash of the glass bits as the splendid figures change, and fall apart, and flash into new forms, figure after figure, and with the birth of each new form I feel my skin crinkle and my nerve-web tingle with a new thrill of wonder and delight. These remembered pictures float past me in a sequence of contracts; following the same order always, and always whirling by and disappearing with the swiftness of a dream, leaving me with the sense that the actuality was the experience of an hour, at most, whereas it really covered days, I think.

The series begins with the hiring of a "bearer"—native man-servant—a person who should be selected with some care, because as long as he is in your employ he will be about as near to you as your clothes.

In India your day may be said to begin with the "bearer's" knock on the bedroom door, accompanied by a formula of words—a formula which is intended to mean that the bath is ready. It doesn't really seem to mean anything at all. But that is because you are not used to "bearer" English. You will presently understand.

Where he gets his English is his own secret. There is nothing like it elsewhere in the earth; or even in paradise, perhaps, but the other place is probably full of it. You hire him as soon as you touch Indian soil; for no matter what your sex is, you cannot do without him. He is messenger, valet, chambermaid, table-waiter, lady's maid, courier—he is everything. He carries a coarse linen clothes-bag and a quilt; he sleeps on the stone floor outside your chamber door, and gets his meals you do not know where nor when; you only know that he is not fed on the premises, either when you are in a hotel or when you are a guest in a private house. His wages are large —from an Indian point of view—and he feeds and clothes himself out of them. We had three of him in two and a half months. The first one's rate was thirty rupees a month— that is to say, twenty-seven cents a day; the rate of the others, Rs. 40 (40 rupees) a month. A princely sum; for the native switchman on a railway and the native servant in a private family get only Rs. 7 per month, and the farm-hand only 4. The two former feed and clothe themselves and their families on their $1.90 per month; but I cannot believe that the farm-hand has to feed himself on his $1.08. I think the farm probably feeds him, and that the whole of his wages, except a trifle for the priest, go to the support of his family. That is, to the feeding of his family; for they live in a mud hut, hand-made, and, doubtless, rent-free, and they wear no clothes; at least, nothing more than a rag. And not much of a rag at that, in the case of the males. However, these are handsome times for the farm-hand; he was not always the child of luxury that he is now. The Chief Commissioner of the Central Provinces, in a recent official utterance wherein he was rebuking a native deputation for complaining of hard times, reminded them that they could easily remember when a farm-hand's wages were only half a rupee (former value) a month — that is to say, less than a cent a day; nearly $2.90 a year. If such a wage-earner had a good deal of a family — and they all have that, for God is very good to these poor natives in some ways — he would save a profit of fifteen cents, clean and clear, out of his year's toil; I mean a frugal, thrifty person would, not one given to display and ostentation. And if he owed $13.50 and took good care of his health, he could pay it off in ninety years. Then he could hold up his head, and look his creditors in the face again.

Think of these facts and what they mean. India does not consist of cities. There are no cities in India — to speak of. Its stupendous population consists of farm-laborers. India is one vast farm — one almost interminable stretch of fields with mud fences between. Think of the above facts; and consider what an incredible aggregate of poverty they place before you.

The first Bearer that applied, waited below and sent up his recommendations. That was the first morning in Bombay. We read them over; carefully, cautiously, thoughtfully. There was not a fault to find with them — except one; they were all from Americans. Is that a slur? If it is, it is a deserved one. In my experience, an American's recommendation of a servant is not usually valuable. We are too good-natured a race; we hate to say the unpleasant thing; we shrink from speaking the unkind truth about a poor fellow whose bread depends upon our verdict; so we speak of his good points only, thus not scrupling to tell a lie — a *silent* lie — for in not mentioning his bad ones we as good as say he hasn't any. The only difference that I know of between a silent lie and a spoken one is, that the silent lie is a less respectable one than the other. And it can deceive, whereas the other can't — as a rule. We not only tell the silent lie as to a servant's faults, but we sin in another way : we overpraise his merits; for when it comes to writing recommendations of servants we are a nation of gushers. And we have not the Frenchman's excuse. In France you *must* give the departing servant a good recommendation; and you *must* conceal his faults; you have no choice. If you mention his faults for the protection of the next candidate for his services, he can sue you for damages; and the court will award them, too; and, moreover, the judge will give you a sharp dressing-down from the bench for trying to destroy a poor man's character, and rob him of his bread. I do not state this on my own authority, I got it from a French physician of fame and repute — a man who was born in Paris, and had practiced there all his life. And he said that he spoke not merely from common knowledge, but from exasperating personal experience.

As I was saying, the Bearer's recommendations were all from American tourists; and St. Peter would have admitted him to the fields of the blest on them — I mean if he is as unfamiliar with our people and our ways as I suppose he is. According to these recommendations, Manuel X. was supreme in all the arts connected with his complex trade; and these manifold arts were mentioned — and praised — in detail. His English was spoken of in terms of warm admiration — admiration verging upon rapture. I took pleased note of that, and hoped that some of it might be true.

We had to have some one right away ; so the family went
down stairs and took him a week on trial; then sent him up
to me and departed on their affairs. I was shut up in my
quarters with a bronchial cough, and glad to have something
fresh to look at, something new to play with. Manuel filled
the bill; Manuel was very welcome. He was toward fifty
years old, tall, slender, with a slight stoop — an artificial stoop,

a deferential stoop, a stoop rigidi-
fied by long habit — with face of
European mould ; short hair intensely black; gentle black eyes,
timid black eyes, indeed; complexion very dark, nearly black
in fact; face smooth-shaven. He
was bareheaded and barefooted,
and was never otherwise while his
week with us lasted; his clothing
was European, cheap, flimsy, and
showed much wear.

He stood before me and inclined
his head (and body) in the pathetic
Indian way, touching his forehead
with the finger-ends of his right
hand, in salute. I said : —

MANUEL.

"Manuel, you are evidently Indian, but you seem to have a
Spanish name when you put it all together. How is that ? "

A perplexed look gathered in his face; it was plain that he
had not understood — but he didn't let on. He spoke back
placidly.

"Name, Manuel. Yes, master."

"I know ; but how did you *get* the name ? "

"Oh, yes, I suppose. Think happen so. Father same
name, not mother."

I saw that I must simplify my language and spread my words apart, if I would be understood by this English scholar.

"Well — then — how — did —your — father — get — *his* — name?"

"Oh, he,"— brightening a little — "he Christian — Portygee; live in Goa; I born Goa; mother not Portygee, mother native — high-caste Brahmin — Coolin Brahmin; highest caste; no other so high caste. I high-caste Brahmin, too. Christian, too, same like father; high-caste Christian Brahmin, master — Salvation Army."

All this haltingly, and with difficulty. Then he had an inspiration, and began to pour out a flood of words that I could make nothing of; so I said: —

"There — don't do that. I can't understand Hindostani."

"Not Hindostani, master — English. Always I speaking English sometimes when I talking every day all the time at you."

"Very well, stick to that; that is intelligible. It is not up to my hopes, it is not up to the promise of the recommendations, still it is English, and I understand it. Don't elaborate it; I don't like elaborations when they are crippled by uncertainty of touch."

"Master?"

"Oh, never mind; it was only a random thought; I didn't expect you to understand it. How did you get your English; is it an acquirement, or just a gift of God?"

After some hesitation — piously:

"Yes, he very good. Christian god very good, Hindoo god very good, too. Two million Hindoo god, one Christian god — make two million and one. All mine; two million and one god. I got a plenty. Sometime I pray all time at those, keep it up, go all time every day; give something at shrine, all good for me, make me better man; good for me, good for my family, dam good."

Then he had another inspiration, and went rambling off into fervent confusions and incoherencies, and I had to stop him again. I thought we had talked enough, so I told him to go to the bathroom and clean it up and remove the slops — this to get rid of him. He went away, seeming to understand, and got out some of my clothes and began to brush them. I repeated my desire several times, simplifying and re-simplifying it, and at last he got the idea. Then he went away and put a coolie at the work, and explained that he would lose caste if he did it himself; it would be pollution, by the law of his caste, and it would cost him a deal of fuss and trouble to purify himself and accomplish his rehabilitation. He said that that kind of work was strictly forbidden to persons of caste, and as strictly restricted to the very bottom layer of Hindoo society — the despised *Sudra* (the toiler, the laborer). He was right; and apparently the poor Sudra has been content with his strange lot, his insulting distinction, for ages and ages — clear back to the beginning of things, so to speak. Buckle says that his name — laborer — is a term of contempt; that it is ordained by the Institutes of Menu (900 B.C.) that *if a Sudra sit on a level with his superior he shall be exiled or branded* [*] . . . ; if he speak contemptuously of his superior or insult him *he shall suffer death; if he listen to the reading of the sacred books he shall have burning oil poured in his ears;* if he memorize passages from them *he shall be killed;* if he marry his daughter to a Brahmin *the husband shall go to hell for defiling himself by contact with a woman so infinitely his inferior;* and that it is *forbidden to a Sudra to acquire wealth.* "The bulk of the population of India," says Buckle[†] "is the Sudras — the *workers, the farmers, the creators of wealth.*"

Manuel was a failure, poor old fellow. His age was against

---

[*] Without going into particulars I will remark that as a rule they wear no clothing that would conceal the brand. — M. T.

[†] Population to-day, 300,000,000.

him. He was desperately slow and phenomenally forgetful. When he went three blocks on an errand he would be gone two hours, and then forget what it was he went for. When he packed a trunk it took him forever, and the trunk's contents were an unimaginable chaos when he got done. He couldn't wait satisfactorily at table — a prime defect, for if you haven't your own servant in an Indian hotel you are likely to have a slow time of it and go away hungry. We couldn't understand his English; he couldn't understand ours; and when we found that he couldn't understand his own, it seemed time for us to part. I had to discharge him; there was no help for it. But I did it as kindly as I could, and as gently. We must part, said I, but I hoped we should meet again in a better world. It was not true, but it was only a little thing to say, and saved his feelings and cost me nothing.

But now that he was gone, and was off my mind and heart, my spirits began to rise at once, and I was soon feeling brisk and ready to go out and have adventures. Then his newly-hired successor flitted in, touched his forehead, and began to fly around here, there, and everywhere, on his velvet feet, and in five minutes he had everything in the room "ship-shape and Bristol fashion," as the sailors say, and was standing at the salute, waiting for orders. Dear me, what a rustler he was after the slumbrous way of Manuel, poor old slug! All my heart, all my affection, all my admiration, went out spontaneously to this frisky little forked black thing, this compact and compressed incarnation of energy and force and promptness and celerity and confidence, this smart, smily, engaging, shiney-eyed little devil, feruled on his upper end by a gleaming fire-coal of a fez with a red-hot tassel dangling from it. I said, with deep satisfaction —

"You'll suit. What is your name?"

He reeled it mellowly off.

"Let me see if I can make a selection out of it — for business uses, I mean; we will keep the rest for Sundays. Give it to me in installments."

He did it. But there did not seem to be any short ones, except Mousa — which suggested mouse. It was out of character; it was too soft, too quiet, too conservative; it didn't fit his splendid style. I considered, and said:

"Mousa is short enough, but I don't quite like it. It seems colorless — inharmonious — inadequate; and I am sensitive to such things. How do you think Satan would do?"

"Yes, master. Satan do wair good."

It was his way of saying "very good."

There was a rap at the door. Satan covered the ground with a single skip; there was a word or two of Hindostani, then he disappeared. Three minutes later he was before me again, militarily erect, and waiting for me to speak first.

"What is it, Satan?"

"God want to see you."

"*Who?*"

"God. I show him up, master?"

"Why, this is so unusual, that — that — well, you see — indeed I am so unprepared — I don't quite know what I *do* mean. Dear me, can't you explain? Don't you see that this is a most ex—"

"Here his card, master."

Wasn't it curious — and amazing, and tremendous, and all that? Such a personage going around calling on such as I, and sending up his card, like a mortal — sending it up by Satan. It was a bewildering collision of the impossibles. But this was the land of the Arabian Nights, this was India! and what is it that cannot happen in India?

We had the interview. Satan was right — the Visitor was indeed a God in the conviction of his multitudinous followers, and was worshiped by them in sincerity and humble adoration. They are troubled by no doubts as to his divine origin and office. They believe in him, they pray to him, they make offerings to him, they beg of him remission of sins; to them his person, together with everything connected with it, is sacred; from his barber they buy the parings of his nails and set them in gold, and wear them as precious amulets.

I tried to seem tranquilly conversational and at rest, but I was not. Would you have been? I was in a suppressed frenzy of excitement and curiosity and glad wonder. I could not keep my eyes off him. I was looking upon a *god*, an actual god, a recognized and accepted god; and every detail of his person and his dress had a consuming interest for me. And the thought went floating through my head, " He is worshiped — think of it — he is not a recipient of the pale homage called compliment, wherewith the highest human clay must make shift to be satisfied, but of an infinitely richer spiritual food: adoration, worship! — men and women lay their cares and their griefs and their broken hearts at his feet; and he gives them his peace, and they go away healed."

And just then the Awful Visitor said, in the simplest way —

"There is a feature of the philosophy of Huck Finn which" — and went luminously on with the construction of a compact and nicely-discriminated literary verdict.

It *is* a land of surprises — India! I had had my ambitions — I had hoped, and almost expected, to be read by kings and presidents and emperors — but I had never looked so high as That. It would be false modesty to pretend that I was not inordinately pleased. I was. I was much more pleased than I should have been with a compliment from a man.

He remained half an hour, and I found him a most courteous and charming gentleman. The godship has been in his

family a good while, but I do not know how long. He is a
Mohammedan deity; by earthly rank he is a prince; not an
Indian but a Persian prince. He is a direct descendant of the
Prophet's line. He is comely; also young — for a god; not
forty, perhaps not above thirty-five years old. He wears his
immense honors with tranquil grace, and with a dignity proper
to his awful calling. He speaks English with the ease and
purity of a person born to it. I think I am not overstating
this. He was the only god I had ever seen, and I was very
favorably impressed. When he rose to say good-bye, the door
swung open and I caught the flash of a red fez, and heard
these words, reverently said —

"Satan see God out?"

"Yes." And these mis-mated Beings passed from view —
Satan in the lead and The Other following after.

CHAPTER XL.

Few of us can stand prosperity. Another man's, I mean.
— *Pudd'nhead Wilson's New Calendar.*

THE next picture in my mind is Government House, on Malabar Point, with the wide sea-view from the windows and broad balconies; abode of His Excellency the Governor of the Bombay Presidency — a residence which is European in everything but the native guards and servants, and is a home and a palace of state harmoniously combined.

That was England, the English power, the English civilization, the modern civilization — with the quiet elegancies and quiet colors and quiet tastes and quiet dignity that are the outcome of the modern cultivation. And following it came a picture of the ancient civilization of India — an hour in the mansion of a native prince: Kumar Schri Samatsinhji Bahadur of the Palitana State.

The young lad, his heir, was with the prince; also, the lad's sister, a wee brown sprite, very pretty, very serious, very winning, delicately moulded, costumed like the daintiest butterfly, a dear little fairyland princess, gravely willing to be friendly with the strangers, but in the beginning preferring to hold her father's hand until she could take stock of them and determine how far they were to be trusted. She must have been eight years old; so in the natural (Indian) order of things she would be a bride in three or four years from now, and then this free contact with the sun and the air and the other belongings of out-door nature and comradeship with visiting male folk would end, and she would shut herself up in the zenana for life, like

24 (369)

her mother, and by inherited habit of mind would be happy in that seclusion and not look upon it as an irksome restraint and a weary captivity.

The game which the prince amuses his leisure with — however, never mind it, I should never be able to describe it intelligibly. I tried to get an idea of it while my wife and daughter visited the princess in the zenana, a lady of charming graces and a fluent speaker of English, but I did not make it out. It is a complicated game, and I believe it is said that nobody can learn to play it well but an Indian. And I was not able to learn how to wind a turban. It seemed a simple art and easy; but that was a deception. It is a piece of thin, delicate stuff a foot wide or more, and forty or fifty feet long; and the exhibitor of the art takes one end of it in his two hands, and winds it in and out intricately about his head,

KUMAR SCHRI SAMATSINHJI
BAHADUR.

twisting it as he goes, and in a minute or two the thing is finished, and is neat and symmetrical and fits as snugly as a mould.

We were interested in the wardrobe and the jewels, and in the silverware, and its grace of shape and beauty and delicacy of ornamentation. The silverware is kept locked up, except at meal-times, and none but the chief butler and the prince have keys to the safe. I did not clearly understand why, but it was not for the protection of the silver. It was either to protect the prince from the contamination which his caste would suffer if the vessels were touched by low-caste hands, or it was to protect his highness from poison. Possibly it was both. I believe a salaried taster has to taste everything before the prince ventures it — an ancient and judicious custom in the East, and has thinned out the tasters a good deal, for of course it is the cook that puts the poison in. If I were an Indian prince I would not go to the expense of a taster, I would eat with the cook.

Ceremonials are always interesting; and I noted that the Indian good-morning is a ceremonial, whereas ours doesn't amount to that. In salutation the son reverently touches the father's forehead with a small silver implement tipped with vermillion paste which leaves a red spot there, and in return the son receives the father's blessing. Our good morning is well enough for the rowdy West, perhaps, but would be too brusque for the soft and ceremonious East.

After being properly necklaced, according to custom, with great garlands made of yellow flowers, and provided with betel-nut to chew, this pleasant visit closed, and we passed thence to a scene of a different sort: from this glow of color and this sunny life to those grim receptacles of the Parsee dead, the Towers of Silence. There is something stately about that name, and an impressiveness which sinks deep; the hush of death is in it. We have the Grave, the Tomb, the Mausoleum, God's Acre, the Cemetery; and association has made them elo-quent with solemn meaning; but we have no name that is so majestic as that one, or lingers upon the ear with such deep and haunting pathos.

On lofty ground, in the midst of a paradise of tropical foliage and flowers, remote from the world and its turmoil and noise, they stood — the Towers of Silence; and away below was spread the wide groves of cocoa palms, then the city, mile on mile, then the ocean with its fleets of creeping ships — all steeped in a stillness as deep as the hush that hallowed this high place of the dead. The vultures were there. They stood close together in a great circle all around the rim of a massive low tower — waiting; stood as motionless as sculptured ornaments, and indeed almost deceived one into the belief that that was what they were. Presently there was a slight stir among the score of persons present, and all moved reverently out of the path and ceased from talking. A funeral procession entered the great gate, marching two and two, and moved silently by, toward the Tower. The corpse lay in a shallow shell, and was under cover of a white cloth, but was otherwise naked. The bearers of the body were separated by an interval of thirty feet from the mourners. They, and also the mourners, were draped all in pure white, and each couple of mourners was figuratively bound together by a piece of white rope or a handkerchief — though they merely held the ends of it in their hands. Behind the procession followed a dog, which was led in a leash. When the mourners had reached the neighborhood of the Tower — neither they nor any other human being but the bearers of the dead must approach within thirty feet of it — they turned and went back to one of the prayerhouses within the gates, to pray for the spirit of their dead. The bearers unlocked the Tower's sole door and disappeared from view within. In a little while they came out bringing the bier and the white covering-cloth, and locked the door again. Then the ring of vultures rose, flapping their wings, and swooped down into the Tower to devour the body. Nothing was left of it but a clean-picked skeleton when they flocked out again a few minutes afterward.

The principle which underlies and orders everything connected with a Parsee funeral is Purity. By the tenets of the Zoroastrian religion, the elements, Earth, Fire, and Water, are sacred, and must not be contaminated by contact with a dead

ONE OF THE TOWERS OF SILENCE.

body. Hence corpses must not be burned, neither must they be buried. None may touch the dead or enter the Towers where they repose except certain men who are officially appointed for that purpose. They receive high pay, but theirs is a dismal life, for they must live apart from their species, because their commerce with the dead defiles them, and any who should associate with them would share their defilement. When they come out of the Tower the clothes they are wearing are exchanged for others, in a building within the grounds, and the ones which they have taken off are left behind, for they are contaminated, and must never be used again or suffered to go outside the grounds. These bearers come to every funeral in new garments. So far as is known, no human being, other than an official corpse-bearer — save one — has ever entered a Tower of Silence after its consecration. Just a hundred years ago a European rushed in behind the bearers and fed his brutal curiosity with a glimpse of the forbidden mysteries of the place. This shabby savage's name is not given; his quality is also concealed. These two details, taken in connection with the fact that for his extraordinary offense the only punishment he got from the East India Company's Government was a solemn official " reprimand "— suggest the suspicion that he was a European of consequence. The same public document which contained the reprimand gave warning that future offenders of his sort, if in the Company's service, would be dismissed; and if merchants, suffer revocation of license and exile to England.

The Towers are not tall, but are low in proportion to their circumference, like a gasometer. If you should fill a gasometer half way up with solid granite masonry, then drive a wide and deep well down through the center of this mass of masonry, you would have the idea of a Tower of Silence. On the masonry surrounding the well the bodies lie, in shallow trenches which radiate like wheel-spokes from the well. The trenches slant toward the well and carry into it the rainfall. Underground drains, with charcoal filters in them, carry off this water from the bottom of the well.

When a skeleton has lain in the Tower exposed to the rain and the flaming sun a month it is perfectly dry and clean. Then the same bearers that brought it there come gloved and take it up with tongs and throw it into the well. There it turns to dust. It is never seen again, never touched again, in the world. Other peoples separate their dead, and preserve and continue social distinctions in the grave — the skeletons of kings and statesmen and generals in temples and pantheons proper to skeletons of their degree, and the skeletons of the commonplace and the poor in places suited to their meaner estate; but the Parsees hold that all men rank alike in death — all are humble, all poor, all destitute. In sign of their poverty they are sent to their grave naked, in sign of their equality the bones of the rich, the poor, the illustrious and the obscure are flung into the common well together. At a Parsee funeral there are no vehicles; all concerned must walk, both rich and poor, howsoever great the distance to be traversed may be. In the wells of the Five Towers of Silence is mingled the dust of all the Parsee men and women and children who have died in Bombay and its vicinity during the two centuries which have elapsed since the Mohammedan conquerors drove the Parsees out of Persia, and into that region of India. The earliest of the five towers was built by the Modi family something more than 200 years ago, and it is now reserved to the heirs of that house; none but the dead of that blood are carried thither.

The origin of at least one of the details of a Parsee funeral is not now known— the presence of the dog. Before a corpse is borne from the house of mourning it must be uncovered and exposed to the gaze of a dog; a dog must also be led in the rear of the funeral. Mr. Nusserwanjee Byramjee, Secretary to the Parsee Punchayet, said that these formalities had once had a meaning and a reason for their institution, but that they were survivals whose origin none could now account for. Custom and tradition continue them in force, antiquity hallows them. It is thought that in ancient times in Persia the dog was a sacred animal and could guide souls to heaven; also that his eye had the power of purifying objects which had been contaminated by the touch of the dead; and that hence his presence with the funeral cortege provides an ever-applicable remedy in case of need.

The Parsees claim that their method of disposing of the dead is an effective protection of the living; that it disseminates no corruption, no impurities of any sort, no disease-germs; that no wrap, no garment which has touched the dead is allowed to touch the living afterward; that from the Towers of Silence nothing proceeds which can carry harm to the outside world. These are just claims, I think. As a sanitary measure, their system seems to be about the equivalent of cremation, and as sure. We are drifting slowly — but hopefully — toward cremation in these days. It could not be expected that this progress should be swift, but if it be steady and continuous, even if slow, that will suffice. When cremation becomes the rule we shall cease to shudder at it; we should shudder at burial if we allowed ourselves to think what goes on in the grave.

The dog was an impressive figure to me, representing as he did a mystery whose key is lost. He was humble, and apparently depressed; and he let his head droop pensively, and looked as if he might be trying to call back to his mind what it was that he had used to symbolize ages ago when he began his function. There was another impressive thing close at

hand, but I was not privileged to see it. That was the sacred
fire — a fire which is supposed to have been burning without
interruption for more than two centuries; and so, living by
the same heat that was imparted to it so long ago.

The Parsees are a remarkable community. There are only
about 60,000 in Bombay, and only about half as many as that
in the rest of India; but they make up in importance what
they lack in numbers. They are highly educated, energetic,

enterprising, progressive, rich, and
the Jew himself is not more lavish
or catholic in his charities and
benevolences. The Parsees build
and endow hospitals, for both men
and animals; and they and their
womenkind keep an open purse for
all great and good objects. They
are a political force, and a valued
support to the government. They
have a pure and lofty religion, and
they preserve it in its integrity and
order their lives by it.

A PARSEE.

We took a final sweep of the
wonderful view of plain and city
and ocean, and so ended our visit to the garden and the
Towers of Silence; and the last thing I noticed was another
symbol — a voluntary symbo this one; it was a vulture
standing on the sawed-off top of a tall and slender and branch-
less palm in an open space in the ground; he was perfectly
motionless, and looked like a piece of sculpture on a pillar.
And he had a mortuary look, too, which was in keeping with
the place.

CHAPTER XLI.

There is an old-time toast which is golden for its beauty. "When you ascend the hill of prosperity may you not meet a friend."— *Pudd'nhead Wilson's New Calendar.*

THE next picture that drifts across the field of my memory is one which is connected with religious things. We were taken by friends to see a Jain temple. It was small, and had many flags or streamers flying from poles standing above its roof; and its little battlements supported a great many small idols or images. Up stairs, inside, a solitary Jain was praying or reciting aloud in the middle of the room. Our presence did not interrupt him, nor even incommode him or modify his fervor. Ten or twelve feet in front of him was the idol, a small figure in a sitting posture. It had the pinkish look of a wax doll, but lacked the doll's roundness of limb and approximation to correctness of form and justness of proportion. Mr. Gandhi explained everything to us. He was delegate to the Chicago Fair Congress of Religions. It was lucidly done, in masterly English, but in time it faded from me, and now I have nothing left of that episode but an impression: a dim idea of a religious belief clothed in subtle intellectual forms, lofty and clean, barren of fleshly grossnesses; and with this another dim impression which connects that intellectual system somehow with that crude image, that inadequate idol — how, I do not know. Properly they do not seem to belong together. Apparently the idol symbolized a person who had become a saint or a god through accessions of steadily augmenting holiness acquired through a series of reincarnations and promotions extending over many ages; and was now at last a saint and qualified to vicariously receive worship and transmit it to heaven's chancellery. Was that it?

And thence we went to Mr. Premchand Roychand's bungalow, in Lovelane, Byculla, where an Indian prince was to receive a deputation of the Jain community who desired to congratulate him upon a high honor lately conferred upon him by his sovereign, Victoria, Empress of India. She had made him a knight of the order of the Star of India. It would seem that even the grandest Indian prince is glad to add the modest title "Sir" to his ancient native grandeurs, and is willing to do valuable service to win it. He will remit taxes liberally, and will spend money freely upon the betterment of the condition of his subjects, if there is a knighthood to be gotten by it. And he will also do good work and a deal of it to get a gun added to the salute allowed him by the British Government. Every year the Empress distributes knighthoods and adds guns for public services done by native princes. The salute of a small prince is three or four guns; princes of greater consequence have salutes that run higher and higher, gun by gun,—oh, clear away up to eleven; possibly more, but I did not hear of any above eleven-gun princes. I was told that when a four-gun prince gets a gun added, he is pretty troublesome for a while, till the novelty wears off, for he likes the music, and keeps hunting up pretexts to get himself saluted. It may be that supremely grand folk, like the Nyzam of Hyderabad and the Gaikwar of Baroda, have more than eleven guns, but I don't know.

When we arrived at the bungalow, the large hall on the ground floor was already about full, and carriages were still flowing into the grounds. The company present made a fine show, an exhibition of human fireworks, so to speak, in the matters of costume and comminglings of brilliant color. The variety of form noticeable in the display of turbans was remarkable. We were told that the explanation of this was, that this Jain delegation was drawn from many parts of India, and that each man wore the turban that was in vogue in his own region. This diversity of turbans made a beautiful effect.

I could have wished to start a rival exhibition there, of Christian hats and clothes. I would have cleared one side of the room of its Indian splendors and repacked the space with Christians drawn from America, England, and the Colonies, dressed in the hats and habits of now, and of twenty and forty and fifty years ago. It would have been a hideous exhibition, a thoroughly devilish spectacle. Then there would have been the added disadvantage of the white complexion. It is not an unbearably unpleasant complexion when it keeps to itself, but when it comes into competition with masses of brown and black the fact is betrayed that it is endurable only because we are used to it. Nearly all black and brown skins are beautiful, but a beautiful white skin is rare. How rare, one may learn by walking down a street in Paris, New York, or London on a week-day — particularly an unfashionable street — and keeping count of the satisfactory complexions encountered in the course of a mile. Where dark complexions are massed, they make the whites look bleached-out, unwholesome, and sometimes frankly ghastly. I could notice this as a boy, down South in the slavery days before the war. The splendid black satin skin of the South African Zulus of Durban seemed to me to come very close to perfection. I can see those Zulus yet — 'ricksha athletes waiting in front of the hotel for custom; handsome and intensely black creatures, moderately clothed in loose summer stuffs whose snowy whiteness made the black all the blacker by contrast. Keeping that group in my mind, I can compare those complexions with the white ones which are streaming past this London window now:

A lady. Complexion, new parchment.

Another lady. Complexion, old parchment.

Another. Pink and white, very fine.

Man. Grayish skin, with purple areas.

Man. Unwholesome fish-belly skin.

Girl. Sallow face, sprinkled with freckles.

Old woman. Face whitey-gray.

Young butcher. Face a general red flush.

Jaundiced man — mustard yellow.

Elderly lady. Colorless skin, with two conspicuous moles.

Elderly man — a drinker. Boiled-cauliflower nose in a flabby face veined with purple crinklings.

Healthy young gentleman. Fine fresh complexion.

Sick young man. His face a ghastly white.

No end of people whose skins are dull and characterless modifications of the tint which we miscall white. Some of these faces are pimply; some exhibit other signs of diseased blood; some show scars of a tint out of a harmony with the surrounding shades of color. The white man's complexion makes no concealments. It can't. It seemed to have been designed as a catch-all for everything that can damage it. Ladies have to paint it, and powder it, and cosmetic it, and diet it with arsenic, and enamel it, and be always enticing it, and persuading it, and pestering it, and fussing at it, to make it beautiful; and they do not succeed. But these efforts show what they think of the natural complexion, as distributed. As distributed it needs these helps. The complexion which they try to counterfeit is one which nature restricts to the few — to the very few. To ninety-nine persons she gives a bad complexion, to the hundredth a good one. The hundredth can keep it — how long? Ten years, perhaps.

The advantage is with the Zulu, I think. He starts with a beautiful complexion, and it will last him through. And as for the Indian brown — firm, smooth, blemishless, pleasant and restful to the eye, afraid of no color, harmonizing with all colors and adding a grace to them all — I think there is no sort of chance for the average white complexion against that rich and perfect tint.

To return to the bungalow. The most gorgeous costumes present were worn by some children. They seemed to blaze, so bright were the colors, and so brilliant the jewels strung over the rich materials. These children were professional nautch-dancers, and looked like girls, but they were boys. They got up by ones and twos and fours, and danced and sang to an accompaniment of weird music. Their posturings and gesturings were elaborate and graceful, but their voices were stringently raspy and unpleasant, and there was a good deal of monotony about the tune.

By and by there was a burst of shouts and cheers outside and the prince with his train entered in fine dramatic style. He was a stately man, he was ideally costumed, and fairly festooned with ropes of gems; some of the ropes were of pearls, some were of uncut great emeralds — emeralds renowned in Bombay for their quality and value. Their size was marvelous, and enticing to the eye, those rocks. A boy — a princeling — was with the prince, and he also was a radiant exhibition.

The ceremonies were not tedious. The prince strode to his throne with the port and majesty — and the sternness — of a Julius Cæsar coming to receive and receipt for a back-country kingdom and have it over and get out, and no fooling. There was a throne for the young prince, too, and the two sat there, side by side, with their officers grouped at either hand and most accurately and creditably reproducing the pictures which one sees in the books — pictures which people in the prince's line of business have been furnishing ever since Solomon received the Queen of Sheba and showed her his things. The chief of the Jain delegation read his paper of congratulations, then pushed it into a beautifully engraved silver cylinder, which was delivered with ceremony into the prince's hands and at once delivered by him without ceremony into the hands of an officer. I will copy the address here. It is interesting, as showing what an Indian prince's subject may have opportunity to thank him for in these days of modern English rule, as contrasted with what his ancestor would have given them opportunity to thank him for a century and a half ago — the days of freedom unhampered by English interference. A century and a half ago an address of thanks could have been put into small space. It would have thanked the prince —

1. For not slaughtering too many of his people upon mere caprice;
2. For not stripping them bare by sudden and arbitrary tax levies, and bringing famine upon them;
3. For not upon empty pretext destroying the rich and seizing their property;

4. For not killing, blinding, imprisoning, or banishing the relatives of the royal house to protect the throne from possible plots ;

5. For not betraying the subject secretly, for a bribe, into the hands of bands of professional Thugs, to be murdered and robbed in the prince's back lot.

Those were rather common princely industries in the old times, but they and some others of a harsh sort ceased long ago under English rule. Better industries have taken their place, as this Address from the Jain community will show :

"Your Highness,— We the undersigned members of the Jain community of Bombay have the pleasure to approach your Highness with the expression of our heartfelt congratulations on the recent conference on your Highness of the Knighthood of the Most Exalted Order of the Star of India Ten years ago we had the pleasure and privilege of welcoming your Highness to this city under circumstances which have made a memorable epoch in the history of your State, for had it not been for a generous and reasonable spirit that your Highness displayed in the negotiations between the Palitana Durbar and the Jain community, the conciliatory spirit that animated our people could not have borne fruit. That was the first step in your Highness's administration, and it fitly elicited the praise of the Jain community, and of the Bombay Government. A decade of your Highness's administration, combined with the abilities, training, and acquirements that your Highness brought to bear upon it, has justly earned for your Highness the unique and honourable distinction — the Knighthood of the Most Exalted Order of the Star of India, which we understand your Highness is the first to enjoy among Chiefs of your Highness's rank and standing. And we assure your Highness that for this mark of honour that has been conferred on you by Her Most Gracious Majesty, the Queen-Empress, we feel no less proud than your Highness. Establishment of commercial factories, schools, hospitals, etc., by your Highness in your State has marked your Highness's career during these ten years, and we trust that your Highness will be spared to rule over your people with wisdom and foresight and foster the many reforms that your Highness has been pleased to introduce in your State. We again offer your Highness our warmest felicitations for the honour that has been conferred on you. We beg to remain your Highness's obedient servants."

Factories, schools, hospitals, reforms. The prince propagates that kind of things in the modern times, and gets knighthood and guns for it.

After the address the prince responded with snap and brevity ; spoke a moment with half a dozen guests in English, and with an official or two in a native tongue ; then the garlands were distributed as usual, and the function ended.

25

CHAPTER XLII.

Each person is born to one possession which outvalues all his others — his last breath. — *Pudd'nhead Wilson's New Calendar.*

TOWARD midnight, that night, there was another function. This was a Hindoo wedding — no, I think it was a betrothal ceremony. Always before, we had driven through streets that were multitudinous and tumultuous with picturesque native life, but now there was nothing of that. We seemed to move through a city of the dead. There was hardly a suggestion of life in those still and vacant streets. Even the crows were silent. But everywhere on the ground lay sleeping natives — hundreds and hundreds. They lay stretched at full length and tightly wrapped in blankets, heads and all. Their attitude and their rigidity counterfeited death. The plague was not in Bombay then, but it is devasting the city now. The shops are deserted, now, half of the people have fled, and of the remainder the smitten perish by shoals every day. No doubt the city looks now in the daytime as it looked then at night. When we had pierced deep into the native quarter and were threading its narrow dim lanes, we had to go carefully, for men were stretched asleep all about and there was hardly room to drive between them. And every now and then a swarm of rats would scamper across past the horses' feet in the vague light — the forbears of the rats that are carrying the plague from house to house in Bombay now. The shops were but sheds, little booths open to the street; and the goods had been removed, and on the counters families were sleeping, usually with an oil lamp present. Recurrent dead-watches, it looked like. (386)

MIDNIGHT IN A BOMBAY STREET.

But at last we turned a corner and saw a great glare of light ahead. It was the home of the bride, wrapped in a perfect conflagration of illuminations, — mainly gas-work designs, gotten up specially for the occasion. Within was abundance of brilliancy — flames, costumes, colors, decorations, mirrors — it was another Aladdin show.

The bride was a trim and comely little thing of twelve years, dressed as we would dress a boy, though more expensively than we should do it, of course. She moved about very much at her ease, and stopped and talked with the guests and allowed her wedding jewelry to be examined. It was very fine. Particularly a rope of great diamonds, a lovely thing to look at and handle. It had a great emerald hanging to it.

The bridegroom was not present.   He was having betrothal festivities of his own at his father's house.   As I understood it, he and the bride were to entertain company every night and nearly all night for a week or more, then get married, if alive. Both of the children were a little elderly, as brides and grooms go, in India — twelve ; they ought to have been married a year or two sooner; still to a stranger twelve seems quite young enough.

A while after midnight a couple of celebrated and high-priced nautch-girls appeared in the gorgeous place, and danced and sang.   With them were men who played upon strange instruments which made uncanny noises of a sort to make one's flesh creep.   One of these instruments was a pipe, and to its music the girls went through a performance which represented snake-charming.   It seemed a doubtful sort of music to charm anything with, but a native gentleman assured me

A HIGH PRICED NAUTCH GIRL.

that snakes like it and will come out of their holes and listen to it with every evidence of refreshment and gratitude.   He said that at an entertainment in his grounds once, the pipe brought out half a dozen snakes, and the music had to be stopped before they would be persuaded to go.   Nobody wanted their company, for they were bold, familiar, and dangerous; but no one would kill them, of course, for it is sinful for a Hindoo to kill any kind of a creature.

We withdrew from the festivities at two in the morning. Another picture, then — but it has lodged itself in my memory rather as a stage-scene than as a reality. It is of a porch and short flight of steps crowded with dark faces and ghostly-white draperies flooded with the strong glare from the dazzling concentration of illuminations; and midway of the steps one conspicuous figure for accent — a turbaned giant, with a name according to his size: Rao Bahadur Baskirao Balinkanje Pitale, Vakeel to his Highness the Gaikwar of Baroda. Without him the picture would not have been complete; and if his name had been merely Smith, *he* wouldn't have answered. Close at hand on house-fronts on both sides of the narrow street were illuminations of a kind commonly employed by the natives — scores of glass tumblers (containing tapers) fastened a few inches apart all over great latticed frames, forming starry constellations which showed out vividly against their black backgrounds. As we

NAUTCH DANCING.

drew away into the distance down the dim lanes the illuminations gathered together into a single mass, and glowed out of the enveloping darkness like a sun.

Then again the deep silence, the skurrying rats, the dim forms stretched everywhere on the ground; and on either hand those open booths counterfeiting sepulchres, with counterfeit corpses sleeping motionless in the flicker of the counterfeit death lamps. And now, a year later, when I read the cablegrams I seem to be reading of what I myself partly saw — saw before it happened — in a prophetic dream, as it were. One cablegram says, " Business in the native town is about

suspended.   Except the wailing and the tramp of the funerals.
There is but little life or movement.   The closed shops exceed
in number those that remain open."   Another says that 325,-
000 of the people have fled the city and are carrying the plague
to the country.   Three days later comes the news, " The population is reduced by *half*."   The refugees have carried the
disease to Karachi; " 220 cases, 214 deaths."   A day or two
later, " 52 fresh cases, *all* of which proved fatal."

The plague carries with it a terror which no other disease
can excite; for of all diseases known to men it is the deadliest
— by far the deadliest.   " Fifty-two fresh cases — *all* fatal."
It is the Black Death alone that slays like that.   We can all
imagine, after a fashion, the desolation of a plague-stricken
city, and the stupor of stillness broken at intervals by distant
bursts of wailing, marking the passing of funerals, here and there
and yonder, but I suppose it is not possible for us to realize
to ourselves the nightmare of dread and fear that possesses
the living who are present in such a place and cannot get away.
That half million fled from Bombay in a wild panic suggests
to us something of what they were feeling, but perhaps not
even they could realize what the half million were feeling
whom they left stranded behind to face the stalking horror
without chance of escape.   Kinglake was in Cairo many years
ago during an epidemic of the Black Death, and he has
imagined the terrors that creep into a man's heart at such a
time and follow him until they themselves breed the fatal sign
in the armpit, and then the delirium with confused images, and
home-dreams, and reeling billiard-tables, and then the sudden
blank of death :

" To the contagionist, filled as he is with the dread of final causes, having
no faith in destiny, nor in the fixed will of God, and with none of the devil-
may-care indifference which might stand him instead of creeds — to such one,
every rag that shivers in the breeze of a plague-stricken city has this sort of
sublimity.   If by any terrible ordinance he be forced to venture forth, he sees
death dangling from every sleeve ; and, as he creeps forward, he poises his

shuddering limbs between the imminent jacket that is stabbing at his right elbow and the murderous pelisse that threatens to mow him clean down as it sweeps along on his left. But most of all he dreads that which most of all he should love — the touch of a woman's dress; for mothers and wives, hurrying forth on kindly errands from the bedsides of the dying, go slouching along through the streets more willfully and less courteously than the men. For a while it may be that the caution of the poor Levantine may enable him to avoid contact, but sooner or later, perhaps, the dreaded chance arrives; that bundle of linen, with the dark tearful eyes at the top of it, that labors along with the voluptuous clumsiness of Grisi — she has touched the poor Levantine with the hem of her sleeve ! From that dread moment his peace is gone ; his mind for ever hanging upon the fatal touch invites the blow which he fears ; he watches for the symptoms of plague so carefully, that sooner or later they come in truth. The parched mouth is a sign — his mouth *is* parched ; the throbbing brain — his brain *does* throb ; the rapid pulse — he touches his own wrist (for he dares not ask counsel of any man lest he be deserted), he touches his wrist, and feels how his frighted blood goes galloping out of his heart. There is nothing but the fatal swelling that is wanting to make his sad conviction complete ; immediately, he has an odd feel under the arm — no pain, but a little straining of the skin ; he would to God it were his fancy that were strong enough to give him that sensation ; this is the worst of all. It now seems to him that he could be happy and contented with his parched mouth, and his throbbing brain, and his rapid pulse, if only he could know that there were no swelling under the left arm ; but dares he try ? — in a moment of calmness and deliberation he dares not; but when for a while he has writhed under the torture of suspense, a sudden strength of will drives him to seek and know his fate ; he touches the gland, and finds the skin sane and sound but under the cuticle there lies a small lump like a pistol-bullet, that moves as he pushes it. Oh ! but is this for all certainty, is this the sentence of death ? Feel the gland of the other arm. There is not the same lump exactly, yet something a little like it. Have not some people glands naturally enlarged ? — would to heaven he were one ! So he does for himself the work of the plague, and when the Angel of Death thus courted does indeed and in truth come, he has only to finish that which has been so well begun ; he passes his fiery hand over the brain of the victim, and lets him rave for a season, but all chance-wise, of people and things once dear, or of people and things indifferent. Once more the poor fellow is back at his home in fair Provence, and sees the sun-dial that stood in his childhood's garden — sees his mother, and the long-since-forgotten face of that little dear sister — (he sees her, he says, on a Sunday morning, for all the church bells are ringing); he looks up and down through the universe, and owns it well piled with bales upon bales of cotton, and cotton eternal — so much so — that he feels — he knows — he swears he could make that winning hazard, if the billiard-table would not slant upwards, and if the cue were a cue worth playing with ; but it is not — it's a cue that won't move — his own arm won't move — in short, there's the devil to pay in the brain of the poor Levantine ; and perhaps, the next night but one he becomes the "life and the soul" of some squalling jackal family, who fish him out by the foot from his shallow and sandy grave."

CHAPTER XLIII.

Hunger is the handmaid of genius.— Pudd'nhead Wilson's New Calendar.

ONE day during our stay in Bombay there was a criminal trial of a most interesting sort, a terribly realistic chapter out of the "Arabian Nights," a strange mixture of simplicities and pieties and murderous practicalities, which brought back the forgotten days of Thuggee and made them live again; in fact, even made them believable. It was a case where a young girl had been assassinated for the sake of her trifling ornaments, things not worth a laborer's day's wages in America. This thing could have been done in many other countries, but hardly with the cold business-like depravity, absence of fear, absence of caution, destitution of the sense of horror, repentance, remorse, exhibited in this case. Elsewhere the murderer would have done his crime secretly, by night, and without witnesses; his fears would have allowed him no peace while the dead body was in his neighborhood; he would not have rested until he had gotten it safe out of the way and hidden as effectually as he could hide it. But this Indian murderer does his deed in the full light of day, cares nothing for the society of witnesses, is in no way incommoded by the presence of the corpse, takes his own time about disposing of it, and the whole party are so indifferent, so phlegmatic, that they take their regular sleep as if nothing was happening and no halters hanging over them; and these five bland people close the episode with a religious service. The thing reads like a Meadows-Taylor Thug-tale of half a century ago, as may be seen by the official report of the trial:

" At the Mazagon Police Court yesterday, Superintendent Nolan again charged Tookaram Suntoo Savat Baya, woman, her daughter Krishni, and Gopal Vithoo Bhanayker, before Mr. Phiroze Hoshang Dastur, Fourth Presidency Magistrate, under sections 302 and 109 of the Code, with having on the night of the 30th of December last murdered a Hindoo girl named Cassi, aged 12, by strangulation, in the room of a chawl at Jakaria Bunder, on the Sewri-road, and also with aiding and abetting each other in the commission of the offense.

" Mr. F. A. Little, Public Prosecutor, conducted the case on behalf of the Crown, the accused being undefended.

" Mr. Little applied under the provisions of the Criminal Procedure Code to tender pardon to one of the accused, Krishni, woman, aged 22, on her undertaking to make a true and full statement of facts under which the deceased girl Cassi was murdered.

" The Magistrate having granted the Public Prosecutor's application, the accused Krishni went into the witness-box, and, on being examined by Mr. Little, made the following confession : — I am a mill-hand employed at the Jubilee Mill. I recollect the day (Tuesday on which the body of the deceased Cassi was found. Previous to that I attended the mill for half a day, and then returned home at 3 in the afternoon, when I saw five persons in the house, viz. : the first accused Tookaram, who is my paramour, my mother, the second accused Baya, the accused Gopal, and two guests named Ramji Daji and Annaji Gungaram. Tookaram rented the room of the chawl situated at Jakaria Bunder-road from its owner, Girdharilal Radhakishan, and in that room I, my paramour, Tookaram, and his younger brother, Yesso Mahadhoo, live. Since his arrival in Bombay from his native country Yesso came and lived with us. When I returned from the mill on the afternoon of that day, I saw the two guests seated on a cot in the veranda, and a few minutes after the accused Gopal came and took his seat by their side, while I and my mother were seated inside the room. Tookaram, who had gone out to fetch some *pan* and betelnuts, on his return home had brought the two guests with him. After returning home he gave them *pan supari*. While they were eating it my mother came out of the room and inquired of one of the guests, Ramji, what had happened to his foot, when he replied that he had tried many remedies, but they had done him no good. My mother then took some rice in her hand and prophesied that the disease which Ramji was suffering from would not be cured until he returned to his native country. In the meantime the deceased Cassi came from the direction of an out-house, and stood in front on the threshhold of our room with a *lota* in her hand. Tookaram then told his two guests to leave the room, and they then went up the steps towards the quarry. After the guests had gone away, Tookaram seized the deceased, who had come into the room, and he afterwards put a waistband around her, and tied her to a post which supports a loft. After doing this, he pressed the girl's throat, and, having tied her mouth with the *dhotur* (now shown in Court), fastened it to the post. Having killed the girl, Tookaram removed her gold head ornament and a gold *putlee*, and also took charge of her *lota*. Besides these two ornaments Cassi had on her person ear-studs, a nose-ring, some silver toe-rings, two necklaces, a pair of silver anklets and bracelets.   Tookaram afterwards tried to remove the silver amulets, the ear-studs, and the nose-ring ; but he failed in his attempt.   While he was doing so, I, my mother, and Gopal were present   After removing the two gold ornaments, he handed them over to Gopal, who was at the time standing near me.   When he killed Cassi, Tookaram threatened to strangle me also if I informed any one of this.   Gopal and myself were then standing at the door of our room, and we both were threatened by Tookaram.   My mother, Baya, had seized the legs of the deceased at the time she was killed, and whilst she was being tied to the post.   Cassi then made a noise.   Tookaram and my mother took part in killing the girl.   After the murder her body was wrapped up in a mattress and kept on the loft over the door of our room. When Cassi was strangled, the door of the room was fastened from the inside by Tookaram.   This deed was committed shortly after my return home from work in the mill.   Tookaram put the body of the deceased in the mattress, and, after it was left on the loft, he went to have his head shaved by a barber named Sambhoo Raghoo, who lives only one door away from me.   My mother and myself then remained in the possession of the information.   I was clapped and threatened by my paramour, Tookaram, and that was the only reason why I did not inform any one at that time.   When I told Tookaram that I would give information of the occurrence, he slapped me.   The accused Gopal was asked by Tookaram to go back to his room, and he did so, taking away with him the two gold ornaments and the *lota*.   Yesso Mahadhoo, a brother-in-law of Tookaram, came to the house and asked Tookaram why he was washing, the water-pipe being just opposite.   Tookaram replied that he was washing his *dhotur*, as a fowl had polluted it.   About 6 o'clock of the evening of that day my mother gave me three pice and asked me to buy a cocoanut, and I gave the money to Yessoo, who went and fetched a cocoanut and some betel leaves.   When Yessoo and others were in the room I was bathing, and, after I finished my bath, my mother took the cocoanut and the betel leaves from Yessoo, and we five went to the sea.   The party consisted of Tookaram, my mother, Yessoo, Tookaram's younger brother, and myself.   On reaching the seashore, my mother made the offering to the sea, and prayed to be pardoned for what we had done.   Before we went to the sea, some one came to inquire after the girl Cassi.   The police and other people came to make these inquiries both before and after we left the house for the seashore.   The police questioned my mother about the girl, and she replied that Cassi had come to her door, but had left.   The next day the police questioned Tookaram, and he, too, gave a similar reply.   This was said the same night when the search was made for the girl.   After the offering was made to the sea, we partook of the cocoanut and returned home, when my mother gave me some food ; but Tookaram did not partake of any food that night.   After dinner I and my mother slept inside the room, and Tookaram slept on a cot near his brother-in-law, Yessoo Mahadhoo, just outside the door.   That was not the usual place where Tookaram slept.   He usually slept inside the room.   The body of the deceased remained on the loft when I went to sleep.   The room in which we slept was locked, and I heard that my paramour, Tookaram, was restless outside. About 3 o'clock the following morning Tookaram knocked at the door, when both myself and my mother opened it. He then told me to go to the steps leading to the quarry, and see if any one was about. Those steps lead to a stable, through which we go to the quarry at the back of the compound. When I got to the steps I saw no one there. Tookaram asked me if any one was there, and I replied that I could see no one about. He then took the body of the deceased from the loft, and, having wrapped it up in his *saree*, asked me to accompany him to the steps of the quarry, and I did so. The *saree* now produced here was the same. Besides the *saree*, there was also a *cholee* on the body. He then carried the body in his arms, and went up the steps, through the stable, and then to the right hand towards a *sahib's* bungalow, where Tookaram placed the body near a wall. All the time I and my mother were with him. When the body was taken down, Yessoo was lying on the cot. After depositing the body under the wall, we all returned home, and soon after 5 a. m. the police again came and took Tookaram away. About an hour after they returned and took me and my mother away. We were questioned about it, when I made a statement. Two hours later I was taken to the room, and I pointed out this waistband, the *dhotur*, the mattress, and the wooden post to Superintendent Nolan and Inspectors Roberts and Rashanali, in the presence of my mother and Tookaram. Tookaram killed the girl Cassi for her ornaments, which he wanted for the girl to whom he was shortly going to be married. The body was found in the same place where it was deposited by Tookaram."

The criminal side of the native has always been picturesque, always readable. The Thuggee and one or two other particularly outrageous features of it have been suppressed by the English, but there is enough of it left to keep it darkly interesting. One finds evidence of these survivals in the newspapers. Macaulay has a light-throwing passage upon this matter in his great historical sketch of Warren Hastings, where he is describing some effects which followed the temporary paralysis of Hastings' powerful government brought about by Sir Philip Francis and his party:

"The natives considered Hastings as a fallen man; and they acted after their kind. Some of our readers may have seen, in India, a cloud of crows pecking a sick vulture to death — no bad type of what happens in that country as often as fortune deserts one who has been great and dreaded. In an instant all the sycophants, who had lately been ready to lie for him, to forge for him, to pander for him, to poison for him, hasten to purchase the favor of his victorious enemies by accusing him. An Indian government has only to let it be understood that it wishes a particular man to be ruined, and in twenty-four hours it will be furnished with grave charges, supported by depositions so full and circumstantial that any person unaccustomed to Asiatic mendacity would regard them as decisive.   It is well if the signature of the destined victim is not counterfeited at the foot of some illegal compact, and if some treasonable paper is not slipped into a hiding-place in his house."

That was nearly a century and a quarter ago.   An article in one of the chief journals of India (the *Pioneer*) shows that in some respects the native of to-day is just what his ancestor was then.   Here are niceties of so subtle and delicate a sort that they lift their breed of rascality to a place among the fine arts, and almost entitle it to respect :

" The records of the Indian courts might certainly be relied upon to prove that swindlers as a class in the East come very close to, if they do not surpass, in brilliancy of execution and originality of design the most expert of their fraternity in Europe and America.   India in especial is the home of forgery.   There are some particular districts which are noted as marts for the finest specimens of the forger's handiwork.   The business is carried on by firms who possess *stores of stamped papers to suit every emergency*.   They habitually lay in a store of fresh stamped papers every year, and some of the older and more thriving houses can supply documents *for the past forty years, bearing the proper water-mark and possessing the genuine appearance of age*.   Other districts have earned notoriety for skilled perjury, a pre-eminence that excites a respectful admiration when one thinks of *the universal prevalence of the art*, and persons desirous of succeeding in false suits are ready to pay handsomely to avail themselves of the services of these local experts as witnesses."

Various instances illustrative of the methods of these swindlers are given.   They exhibit deep cunning and total depravity on the part of the swindler and his pals, and more obtuseness on the part of the victim than one would expect to find in a country where suspicion of your neighbor must surely be one of the earliest things learned.   The favorite subject is the young fool who has just come into a fortune and is trying to see how poor a use he can put it to.   I will quote one example:

"Sometimes another form of confidence trick is adopted, which is invariably successful.   The particular pigeon is spotted, and, his acquaintance having been made, he is encouraged in every form of vice.   When the friendship is thoroughly established, the swindler remarks to the young man that he has a brother who has asked him to lend him Rs.10,000.   The swindler says he has the money and would lend it ; but, as the borrower is his brother, he cannot charge interest.   So he proposes that he should hand the dupe the money, and the latter should lend it to the swindler's brother, exacting a heavy pre-payment of interest which, it is pointed out, they may equally enjoy in dissipation. The dupe sees no objection, and on the appointed day receives Rs.7,000 from the swindler, which he hands over to the confederate. The latter is profuse in his thanks, and executes a promissory note for Rs.10,000, payable to bearer. The swindler allows the scheme to remain quiescent for a time, and then suggests that, as the money has not been repaid and as it would be unpleasant to sue his brother, it would be better to sell the note in the bazaar. The dupe hands the note over, for the money he advanced was not his, and, on being informed that it would be necessary to have his signature on the back so as to render the security negotiable, he signs without any hesitation. The swindler passes it on to confederates, and the latter employ a respectable firm of solicitors to ask the dupe if his signature is genuine He admits it at once, and his fate is sealed. A suit is filed by a confederate against the dupe, two accomplices being made co-defendants. They admit their signatures as indorsers, and the one swears he bought the note for value from the dupe The latter has no defense, for no court would believe the apparently idle explanation of the manner in which he came to endorse the note."

There is only one India! It is the only country that has a monopoly of grand and imposing specialties. When another country has a remarkable thing, it cannot have it all to itself —some other country has a duplicate. But India—that is different. Its marvels are its own; the patents cannot be infringed; imitations are not possible. And think of the size of them, the majesty of them, the weird and outlandish character of the most of them!

There is the Plague, the Black Death: India invented it; India is the cradle of that mighty birth.

The Car of Juggernaut was India's invention.

So was the Suttee; and within the time of men still living eight hundred widows willingly, and, in fact, rejoicingly, burned themselves to death on the bodies of their dead husbands in a single year. Eight hundred would do it this year if the British government would let them.

Famine is India's specialty. Elsewhere famines are inconsequential incidents — in India they are devastating cataclysms; in one case they annihilate hundreds; in the other, millions.

India has 2,000,000 gods, and worships them all. In religion all other countries are paupers; India is the only millionaire.

With her everything is on a giant scale—even her poverty; no other country can show anything to compare with it. And she has been used to wealth on so vast a scale that she has to shorten to single words the expressions describing great sums. She describes 100,000 with one word—a *lakh;* she describes ten millions with one word—a *crore.*

In the bowels of the granite mountains she has patiently carved out dozens of vast temples, and made them glorious with sculptured colonnades and stately groups of statuary, and has adorned the eternal walls with noble paintings. She has built fortresses of such magnitude that the show-strongholds of the rest of the world are but modest little things by comparison; palaces that are wonders for rarity of materials, delicacy and beauty of workmanship, and for cost; and one tomb which men go around the globe to see. It takes eighty nations, speaking eighty languages, to people her, and they number three hundred millions.

On top of all this she is the mother and home of that wonder of wonders—*caste*—and of that mystery of mysteries, the satanic brotherhood of the Thugs.

India had the start of the whole world in the beginning of things. She had the first civilization; she had the first accumulation of material wealth; she was populous with deep thinkers and subtle intellects; she had mines, and woods, and a fruitful soil. It would seem as if she should have kept the lead, and should be to-day not the meek dependent of an alien master, but mistress of the world, and delivering law and command to every tribe and nation in it. But, in truth, there was never any possibility of such supremacy for her. If there had been but one India and one language—but there were eighty of them! Where there are eighty nations and several hundred governments, fighting and quarreling must be the common business of life; unity of purpose and policy are impossible;

out of such elements supremacy in the world cannot come. Even caste itself could have had the defeating effect of a multiplicity of tongues, no doubt; for it separates a people into layers, and layers, and still other layers, that have no community of feeling with each other; and in such a condition of things as that, patriotism can have no healthy growth.

It was the division of the country into so many States and nations that made Thuggee possible and prosperous. It is difficult to realize the situation. But perhaps one may approximate it by imagining the States of our Union peopled by separate nations, speaking separate languages, with guards and custom-houses strung along all frontiers, plenty of interruptions for travelers and traders, interpreters able to handle all the languages very rare or non-existent, and a few wars always going on here and there and yonder as a further embarrassment to commerce and excursioning. It would make intercommunication in a measure ungeneral. India had eighty languages, and more custom-houses than cats. No clever man with the instinct of a highway robber could fail to notice what a chance for business was here offered. India was full of clever men with the highwayman instinct, and so, quite naturally, the brotherhood of the Thugs came into being to meet the long-felt want.

How long ago that was nobody knows — centuries, it is supposed. One of the chiefest wonders connected with it was the success with which it kept its secret. The English trader did business in India two hundred years and more before he ever heard of it; and yet it was assassinating its thousands all around him every year, the whole time.

CHAPTER XLIV.

The old saw says, " Let a sleeping dog lie." Right. Still, when there is much at stake it is better to get a newspaper to do it. — *Pudd'nhead Wilson's New Calendar.*

FROM DIARY:

JANUARY 28. I learned of an *official* Thug-book the other day. I was not aware before that there was such a thing. I am allowed the temporary use of it. We are making preparations for travel. Mainly the preparations are purchases of bedding. This is to be used in sleeping berths in the trains; in private houses sometimes; and in nine-tenths of the hotels. It is not realizable; and yet it is true. It is a survival; an apparently unnecessary thing which in some strange way has outlived the conditions which once made it necessary. It comes down from a time when the railway and the hotel did not exist; when the occasional white traveler went horseback or by bullock-cart, and stopped over night in the small dak-bungalow provided at easy distances by the government — a shelter, merely, and nothing more. He had to carry bedding along, or do without. The dwellings of the English residents are spacious and comfortable and commodiously furnished, and surely it must be an odd sight to see half a dozen guests come filing into such a place and dumping blankets and pillows here and there and everywhere. But custom makes incongruous things congruous.

One buys the bedding, with waterproof hold-all for it at almost any shop — there is no difficulty about it.

January 30. What a spectacle the railway station was, at train-time! It was a very large station, yet when we arrived

A RAILWAY STATION.

it seemed as if the whole world was present — half of it inside, the other half outside, and both halves, bearing mountainous head-loads of bedding and other freight, trying simultaneously to pass each other, in opposing floods, in one narrow door. These opposing floods were patient, gentle, long-suffering natives, with whites scattered among them at rare intervals; and wherever a white man's native servant appeared, *that* native seemed to have put aside his natural gentleness for the time and invested himself with the white man's privilege of making a way for himself by promptly shoving all intervening black things out of it. In these exhibitions of authority Satan was scandalous. He was probably a Thug in one of his former incarnations.

Inside the great station, tides upon tides of rainbow-costumed natives swept along, this way and that, in massed and bewildering confusion, eager, anxious, belated, distressed; and washed up to the long trains and flowed into them with their packs and bundles, and disappeared, followed at once by the next wash, the next wave. And here and there, in the midst of this hurly-burly, and seemingly undisturbed by it, sat great groups of natives on the bare stone floor, — young, slender brown women, old, gray wrinkled women, little soft brown babies, old men, young men, boys; all poor people, but all the females among them, both big and little, bejeweled with cheap and showy nose-rings, toe-rings, leglets, and armlets, these things constituting all their wealth, no doubt. These silent crowds sat there with their humble bundles and baskets and small household gear about them, and patiently waited — for what? A train that was to start at some time or other during the day or night! They hadn't timed themselves well, but that was no matter — the thing had been so ordered from on high, therefore why worry? There was plenty of time, hours and hours of it, and the thing that was to happen would happen — there was no hurrying it.

The natives traveled third class, and at marvelously cheap rates. They were packed and crammed into cars that held each about fifty; and it was said that often a Brahmin of the highest caste was thus brought into personal touch, and consequent defilement, with persons of the lowest castes — no doubt a very shocking thing if a body could understand it and properly appreciate it. Yes, a Brahmin who didn't own a rupee and couldn't borrow one, might have to touch elbows with a rich hereditary lord of inferior caste, inheritor of an ancient title a couple of yards long, and he would just have to stand it; for if either of the two was allowed to go in the cars where the sacred white people were, it probably wouldn't be the august poor Brahmin. There was an immense string of those third-class cars, for the natives travel by hordes; and a weary hard night of it the occupants would have, no doubt.

When we reached our car, Satan and Barney had already arrived there with their train of porters carrying bedding and parasols and cigar boxes, and were at work. We named him Barney for short; we couldn't use his real name, there wasn't time.

It was a car that promised comfort; indeed, luxury. Yet the cost of it — well, economy could no further go; even in France; not even in Italy. It was built of the plainest and cheapest partially-smoothed boards, with a coating of dull paint on them, and there was nowhere a thought of decoration. The floor was bare, but would not long remain so when the dust should begin to fly. Across one end of the compartment ran a netting for the accommodation of hand-baggage; at the other end was a door which would shut, upon compulsion, but wouldn't stay shut; it opened into a narrow little closet which had a wash-bowl in one end of it, and a place to put a towel,

in case you had one with you — and you would be sure to
have towels, because you buy them with the bedding, knowing
that the railway doesn't furnish them. On each side of the
car, and running fore and aft, was a broad leather-covered sofa
— to sit on in the day and sleep on at night. Over each sofa
hung, by straps, a wide, flat, leather-covered shelf — to sleep
o 1. In the daytime you can hitch it up against the wall, out
of the way — and then you have a big unencumbered and most
comfortable room to spread out in. No car in any country is
quite its equal for comfort (and privacy) I think. For usually
there are but two persons in it ; and even when there are four
there is but little sense of impaired privacy. Our own cars at
home can surpass the railway world in all details but that one:
they have no cosiness ; there are too many people together.

At the foot of each sofa was a side-door, for entrance and
exit.

Along the whole length of the sofa on each side of the car
ran a row of large single-plate windows, of a blue tint — blue
to soften the bitter glare of the sun and protect one's eyes
from torture. These could be let down out of the way when
one wanted the breeze. In the roof were two oil lamps which
gave a light strong enough to read by ; each had a green-cloth
attachment by which it could be covered when the light
should be no longer needed.

While we talked outside with friends, Barney and Satan
placed the hand-baggage, books, fruits, and soda-bottles in the
racks, and the hold-alls and heavy baggage in the closet, hung
the overcoats and sun-helmets and towels on the hooks, hoisted
the two bed-shelves up out of the way, then shouldered their
bedding and retired to the third class.

Now then, you see what a handsome, spacious, light, airy,
homelike place it was, wherein to walk up and down, or sit
and write, or stretch out and read and smoke. A central

door in the forward end of the compartment opened into a similar compartment. It was occupied by my wife and daughter. About nine in the evening, while we halted a while at a station, Barney and Satan came and undid the clumsy big hold-alls, and spread the bedding on the sofas in both compartments — mattresses, sheets, gay coverlets, pillows, all complete; there are no chambermaids in India — apparently it was an office that was never heard of. Then they closed the communicating door, nimbly tidied up our place, put the night-clothing on the beds and the slippers under them, then returned to their own quarters.

January 31. It was novel and pleasant, and I stayed awake as long as I could, to enjoy it, and to read about those strange people the Thugs. In my sleep they remained with me, and tried to strangle me. The leader of the gang was that giant Hindoo who was such a picture in the strong light when we were leaving those Hindoo betrothal festivities at two o'clock in the morning — Rao Bahadur Baskirao Balinkanje Pitale, Vakeel to the Gaikwar of Baroda. It was he that brought me the invitation from his master to go to Baroda and lecture to that prince — and now he was misbehaving in my dreams. But all things can happen in dreams. It is indeed as the Sweet Singer of Michigan says — irrelevantly, of course, for the one and unfailing great quality which distinguishes her poetry from Shakespeare's and makes it precious to us is its stern and simple irrelevancy:

> My heart was gay and happy,
>   This was ever in my mind,
> There is better times a coming,
>   And I hope some day to find
> Myself capable of composing,
>   It was my heart's delight
> To compose on a sentimental subject
>   If it came in my mind just right.*

---

* "The Sentimental Song Book," p. 49; theme, "The Author's Early Life," 19th stanza.

Baroda. Arrived at 7 this morning. The dawn was just beginning to show. It was forlorn to have to turn out in a strange place at such a time, and the blinking lights in the station made it seem night still. But the gentlemen who had come to receive us were there with their servants, and they make quick work; there was no lost time. We were soon outside and moving swiftly through the soft gray light, and presently were comfortably housed — with more servants to help than we were used to, and with rather embarassingly important officials to direct them. But it was custom; they spoke Ballarat English, their bearing was charming and hospitable, and so all went well.

Breakfast was a satisfaction. Across the lawns was visible in the distance through the open window an Indian well, with two oxen tramping leisurely up and down long inclines, drawing water; and out of the stillness came the suffering screech of the machinery — not quite musical, and yet soothingly melancholy and dreamy and reposeful — a wail of lost spirits, one might imagine. And commemorative and reminiscent, perhaps; for of course the Thugs used to throw people down that well when they were done with them.

After breakfast the day began, a sufficiently busy one. We were driven by winding roads through a vast park, with noble forests of great trees, and with tangles and jungles of lovely growths of a humbler sort; and at one place three large gray apes came out and pranced across the road — a good deal of a surprise and an unpleasant one, for such creatures belong in the menagerie, and they look artificial and out of place in a wilderness.

We came to the city, by and by, and drove all through it. Intensely Indian, it was, and crumbly, and mouldering, and immemorially old, to all appearance. And the houses — oh, indescribably quaint and curious they were, with their fronts an elaborate lace-work of intricate and beautiful wood-carving, and now and then further adorned with rude pictures of elephants and princes and gods done in shouting colors; and all the ground floors along these cramped and narrow lanes occupied as shops — shops unbelievably small and impossibly packed with merchantable rubbish, and with nine-tenths-naked natives squatting at their work of hammering, pounding, brazing, soldering, sewing, designing, cooking, measuring out grain, grinding it, repairing idols — and then the swarm of ragged and noisy humanity under the horses' feet and everywhere, and the pervading reek and fume and smell! It was all wonderful and delightful.

Imagine a file of elephants marching through such a crevice of a street and scraping the paint off both sides of it with their hides. How big they must look, and how little they must make the houses look; and when the elephants are in their glittering court costume, what a contrast they must make with the humble and sordid surroundings. And when a mad elephant goes raging through, belting right and left with his trunk, how do these swarms of people get out of the way? I suppose it is a thing which happens now and then in the mad season (for elephants have a mad season).

I wonder how old the town is. There are patches of building — massive structures, monuments, apparently — that are so battered and worn, and seemingly so tired and so burdened with the weight of age, and so dulled and stupefied with trying to remember things they forgot before history began, that they give one the feeling that they must have been a part of original Creation. This is indeed one of the oldest of the princedoms of India, and has always been celebrated for its barbaric pomps and splendors, and for the wealth of its princes.

A MAD SEASON.

CHAPTER XLV.

It takes your enemy and your friend, working together, to hurt you to the heart ; the one to slander you and the other to get the news to you.—*Pudd'nhead Wilson's New Calendar.*

OUT of the town again ; a long drive through open country, by winding roads among secluded villages nestling in the inviting shade of tropic vegetation, a Sabbath stillness everywhere, sometimes a pervading sense of solitude, but always barefoot natives gliding by like spirits, without sound of footfall, and others in the distance dissolving away and vanishing like the creatures of dreams. Now and then a string of stately camels passed by —  always interesting things to look at — and they were velvet-shod by nature, and made no noise. Indeed, there were no noises of any sort in

ROADSIDE VIEWS.

this paradise. Yes, once there was one, for a moment : a file of native convicts passed along in charge of an officer, and we

(410)

caught the soft clink of their chains. In a retired spot, resting himself under a tree, was a holy person — a naked black fakeer, thin and skinny, and whitey-gray all over with ashes.

By and by to the elephant stables, and I took a ride; but it was by request — I did not ask for it, and didn't want it; but I took it, because otherwise they would have thought I was afraid, which I was. The elephant kneels down, by command — one end of him at a time — and you climb the ladder and get into the howdah, and then he gets up, one end at a time, just as a ship gets up over a wave; and after that, as he strides monstrously about, his motion is much like a ship's motion. The mahout bores into the back of his head with a great iron prod and you wonder at his temerity and at the elephant's patience, and you think that perhaps the patience will not last; but it does, and nothing happens. The mahout talks to the elephant in a low voice all the time, and the elephant seems to understand it all and to be pleased with it; and he obeys every order in the most contented and docile way. Among these twenty-five elephants were two which were larger than any I had ever seen before, and if I had thought I could learn to not be afraid, I would have taken one of them while the police were not looking.

In the howdah-house there were many howdahs that were made of silver, one of gold, and one of old ivory, and equipped with cushions and canopies of rich and costly stuffs. The wardrobe of the elephants was there, too; vast velvet covers stiff and heavy with gold embroidery; and bells of silver and gold; and ropes of these metals for fastening the things on — harness, so to speak; and monster hoops of massive gold for the elephant to wear on his ankles when he is out in procession on business of state.

But we did not see the treasury of crown jewels, and that was a disappointment, for in mass and richness it ranks only second in India. By mistake we we taken to see the new palace instead, and we used up the last remnant of our spare time there. It was a pity, too; for the new palace is mixed modern American-European, and has not a merit except costliness. It is wholly foreign to India, and impudent and out of place. The architect has escaped. This comes of overdoing the suppression of the Thugs; they had their merits. The old palace is oriental and charming, and in consonance with the country. The old palace would still be great if there were nothing of it but the spacious and lofty hall where the durbars are held. It is not a good place to lecture in, on account of the echoes, but it is a good place to hold durbars in and regulate the affairs of a kingdom, and that is what it is for. If I had it I would have a durbar every day, instead of once or twice a year.

The prince is an educated gentleman. His culture is European. He has been in Europe five times. People say that this is costly amusement for him, since in crossing the sea he must sometimes be obliged to drink water from vessels that are more or less public, and thus damage his caste. To get it purified again he must make pilgrimage to some renowned Hindoo temples and contribute a fortune or two to them. His people are like the other Hindoos, profoundly religious; and they could not be content with a master who was impure.

We failed to see the jewels, but we saw the gold cannon and the silver one — they seemed to be six-pounders. They were not designed for business, but for salutes upon rare and particularly important state occasions. An ancestor of the present Gaikwar had the silver one made, and a subsequent ancestor had the gold one made, in order to outdo him.

This sort of artillery is in keeping with the traditions of Baroda, which was of old famous for style and show. It used to entertain visiting rajahs and viceroys with tiger-fights, elephant-fights, illuminations, and elephant-processions of the most glittering and gorgeous character.

It makes the circus a pale, poor thing.

In the train, during a part of the return journey from Baroda, we had the company of a gentleman who had with him a remarkable looking dog. I had not seen one of its kind before, as far as I could remember; though of course I might have seen one and not noticed it, for I am not acquainted with dogs, but only with cats. This dog's coat was smooth and shiny and black, and I think it had tan trimmings around the edges of the dog, and perhaps underneath. It was a long, low dog, with very short, strange legs — legs that curved inboard, something like parentheses turned the wrong way (. Indeed, it was

A PRIZE WINNER.

made on the plan of a bench for length and lowness. It seemed to be satisfied, but I thought the plan poor, and structurally weak, on account of the distance between the forward supports and those abaft. With age the dog's back was likely to sag; and it seemed to me that it would have been a stronger and more practicable dog if it had had some more legs. It had not begun to sag yet, but the shape of the legs showed that the undue weight imposed upon them was beginning to tell. It had a long nose, and floppy ears that hung down, and a resigned expression of countenance. I did not like to ask what kind of a dog it was, or how it came to be deformed, for it was plain that the gentleman was very fond of it, and naturally he could be sensitive about it. From delicacy I thought it best not to seem to notice it too much. No doubt a man with a dog like that feels just as a person does who has a child that is out of true. The gentleman was not merely fond of the dog, he was also proud of it — just the same again, as a mother feels about her child when it is an idiot. I could see that he was proud of it, notwithstanding it was such a long dog and looked so resigned and pious. It had been all over the world with him, and had been pilgriming like that for years and years. It had traveled 50,000 miles by sea and rail, and had ridden in front of him on his horse 8,000. It had a silver medal from the Geographical Society of Great Britain for its travels, and I saw it. It had won prizes in dog shows, both in India and in England — I saw them. He said its pedigree was on record in the Kennel Club, and that it was a well-known dog. He said a great many people in London could recognize it the moment they saw it. I did not say anything, but I did not think it anything strange; I should know that dog again, myself, yet I am not careful about noticing dogs. He said that when he walked along in London, people often stopped and looked at the dog. Of course I did not say anything, for I did not want to hurt his feelings, but I could have explained to him that if you take a great long low dog like that and waddle it along the street anywhere in the world and not charge anything, people will stop and look. He was gratified because the dog took prizes. But that was nothing; if I were built like that I could take prizes myself. I wished I knew what kind of a dog it was, and what it was for, but I could not very well ask, for that would show that I did not know. Not that I want a dog like that, but only to know the secret of its birth.

I think he was going to hunt elephants with it, because I know, from remarks dropped by him, that he has hunted large game in India and Africa, and likes it. But I think that if he tries to hunt elephants with it, he is going to be disappointed.

I do not believe that it is suited for elephants. It lacks energy, it lacks force of character, it lacks bitterness. These things all show in the meekness and resignation of its expression. It would not attack an elephant, I am sure of it. It might not run if it saw one coming, but it looked to me like a dog that would sit down and pray.

I wish he had told me what breed it was, if there are others; but I shall know the dog next time, and then if I can bring myself to it I will put delicacy aside and ask. If I seem strangely interested in dogs, I have a reason for it; for a dog saved me from an embarrassing position once, and that has made me grateful to these animals; and if by study I could learn to tell some of the kinds from the others, I should be greatly pleased. I only know one kind apart, yet, and that is the kind that saved me that time. I always know that kind when I meet it, and if it is hungry or lost I take care of it. The matter happened in this way:

It was years and years ago. I had received a note from Mr. Augustin Daly of the Fifth Avenue Theatre, asking me to call the next time I should be in New York. I was writing plays, in those days, and he was admiring them and trying to get me a chance to get them played in Siberia. I took the first train — the early one — the one that leaves Hartford at 8.29 in the morning. At New Haven I bought a paper, and found it filled with glaring display-lines about a "bench-show" there. I had often heard of bench-shows, but had never felt any interest in them, because I supposed they were lectures that were not well attended. It turned out, now, that it was not that, but a dog-show. There was a double-leaded column about the king-feature of this one, which was called a Saint Bernard, and was worth $10,000, and was known to be the largest and finest of his species in the world. I read all this with interest, because out of my school-boy readings I dimly remembered how the priests and pilgrims of St. Bernard used to go out in the storms and dig these dogs out of the snowdrifts when lost and exhausted, and give them brandy and save their lives, and drag them to the monastery and restore them with gruel.

Also, there was a picture of this prize-dog in the paper, a noble great creature with a benignant countenance, standing by a table. He was placed in that way so that one could get a right idea of his great dimensions. You could see that he was just a shade higher than the table — indeed, a huge fellow for a dog. Then there was a description which went into the details. It gave his enormous weight — 150½ pounds, and his length — 4 feet 2 inches, from stem to stern-post; and his height — 3 feet 1 inch, to the top of his back. The pictures and the figures so impressed me, that I could see the beautiful colossus before me, and I kept on thinking about him for the next two hours; then I reached New York, and he dropped out of my mind.

In the swirl and tumult of the hotel lobby I ran across Mr. Daly's comedian, the late James Lewis, of beloved memory, and I casually mentioned that I was going to call upon Mr. Daly in the evening at 8. He looked surprised, and said he reckoned not. For answer I handed him Mr. Daly's note. Its substance was: "Come to my private den, over the theater, where we cannot be interrupted. And come by the back way, not the front. No. 642 Sixth Avenue is a cigar shop; pass through it and you are in a paved court, with high buildings all around; enter the second door on the left, and come up stairs."

" Is this all ? "

" Yes," I said.

" Well, you'll never get in."

" Why ? "

" Because you won't. Or if you do you can draw on me

(Owen Gormlay.)

"WELL, SOR, WHAT WILL *YOU* HAVE?"

for a hundred dollars; for you will be the first man that has accomplished it in twenty-five years.   I can't think what Mr. Daly can have been absorbed in.   He has forgotten a most important detail, and he will feel humiliated in the morning when he finds that you tried to get in and couldn't."

" Why, what is the trouble ? "

" I'll tell you.   You see — "

At that point we were swept apart by the crowd, somebody detained me with a moment's talk, and we did not get together again.   But it did not matter; I believed he was joking, anyway.

At eight in the evening I passed through the cigar shop and into the court and knocked at the second door.

" Come in ! "

I entered.   It was a small room, carpetless, dusty, with a naked deal table, and two cheap wooden chairs for furniture. A giant Irishman was standing there, with shirt collar and vest unbuttoned, and no coat on.   I put my hat on the table, and was about to say something, when the Irishman took the innings himself.   And not with marked courtesy of tone:

" Well, sor, what will *you* have ? "

I was a little disconcerted, and my easy confidence suffered a shrinkage.   The man stood as motionless as Gibraltar, and kept his unblinking eye upon me.   It was very embarrassing, very humiliating.   I stammered at a false start or two; then —

" I have just run down from — "

"Av ye plaze, ye'll not smoke here, ye understand."

I laid my cigar on the window-ledge; chased my flighty thoughts a moment, then said in a placating manner : —

" I — I have come to see Mr. Daly."

" Oh, ye *have*, have ye ? "

" Yes."

" Well, ye'll not see him."

" But he *asked* me to come."

" Oh, he *did*, did he ? "

" Yes, he sent me this note, and — "

" Lemme see it."

For a moment I fancied there would be a change in the atmosphere, now ; but this idea was premature.   The big man

" HE HAD IT UPSIDE DOWN."

was examining the note searchingly under the gas-jet. A glance showed me that he had it upside down — disheartening evidence that he could not read.

" Is ut his own hand-write ? "

" Yes — he wrote it himself."

" He did, did he ? "

" Yes."

" H'm. Well, then, why ud he write it like that ? "

" How do you mean ? "

" I mane, why wudn't he put his name to ut ? "

" His name *is* to it.   *That's* not it — you are looking at *my* name."

I thought that that was a home shot, but he did not betray that he had been hit.   He said :

" It's not an aisy one to spell ; how do you pronounce ut ? "

" Mark Twain."

" H'm. H'm. Mike Train. H'm. I don't remember ut. What is it ye want to see him about ? "

" It isn't I that want to see *him*, he wants to see *me*."

" Oh, he does, does he ? "

" Yes."

" What does he want to see ye about ? "

" I don't know."

" Ye don't *know*! And ye confess it, becod ! Well, I can tell ye wan thing — ye'll not see him. Are ye in the business ? "

" What business ? "

" The show business."

A fatal question. I recognized that I was defeated. If I answered no, he would cut the matter short and wave me to the door without the grace of a word — I saw it in his uncompromising eye ; if I said I was a lecturer, he would despise me, and dismiss me with opprobrious words ; if I said I was a dramatist, he would throw me out of the window. I saw that my case was hopeless, so I chose the course which seemed least humiliating : I would pocket my shame and glide out without answering. The silence was growing lengthy.

" I'll ask ye again. Are ye in the show business yerself ? "

" Yes ! "

I said it with splendid confidence ; for in that moment the very twin of that grand New Haven dog loafed into the room, and I saw that Irishman's eye light eloquently with pride and affection.

" Ye are ? And what is it ? "

" I've got a bench-show in New Haven."

The weather *did* change then.

" You don't *say*, sir ! And that's *your* show, sir ! Oh, it's a grand show, it's a wonderful show, sir, and a proud man I am to see your honor this day. And ye'll be an expert, sir, and ye'll know all about dogs — more than ever they know theirselves, I'll take me oath to ut."

I said, with modesty —

" I believe I have some reputation that way. In fact, my business requires it."

"Ye have *some* reputation, your honor! Bedad I believe you! There's not a jintleman in the worrld that can lay over ye in the judgmint of a dog, sir. Now I'll vinture that your honor'll know that dog's dimensions there better than he knows them his own self, and just by the casting of your educated eye upon him. Would you mind giving a guess, if ye'll be so good?"

I knew that upon my answer would depend my fate. If I made this dog bigger than the prize-dog, it would be bad diplomacy, and suspicious; if I fell too far short of the prize-dog, that would be equally damaging. The dog was standing by the table, and I believed I knew the difference between him and the one whose picture I had seen in the newspaper to a shade. I spoke promptly up and said —

"It's no trouble to guess this noble creature's figures: height, three feet; length, four feet and three-quarters of an inch; weight, a hundred and forty-eight and a quarter."

The man snatched his hat from its peg and danced on it with joy, shouting: —

"Ye've hardly missed it the hair's breadth, hardly the shade of a shade, your honor! Oh, it's the miraculous eye ye've got, for the judgmint of a dog!"

And still pouring out his admiration of my capacities, he snatched off his vest and scoured off one of the wooden chairs with it, and scrubbed it and polished it, and said —

"There, sit down, your honor, I'm ashamed of meself that I forgot ye were standing all this time; and do put on your hat, ye mustn't take cold, it's a drafty place; and here is your cigar, sir, a getting cold, I'll give ye a light. There. The place is all yours, sir, and if ye'll just put your feet on the table and make yourself at home, I'll stir around and get a candle and light ye up the ould crazy stairs and see that ye don't come to anny harm, for be this time Mr. Daly'll be that impatient to see your honor that he'll be taking the roof off."

"YE DON'T SAY, SIR!"

He conducted me cautiously and tenderly up the stairs, lighting the way and protecting me with friendly warnings, then pushed the door open and bowed me in and went his way, mumbling hearty things about my wonderful eye for points of a dog. Mr. Daly was writing and had his back to me. He glanced over his shoulder presently, then jumped up and said—

HE LIGHTED ME UP THE STAIRS.

"Oh, dear me, I forgot all about giving instructions. I was just writing you to beg a thousand pardons. But how is it you are here? How did you get by that Irishman? You are the first man that's done it in five and twenty years. You didn't bribe him, I know that; there's not money enough in New York to do it. And you didn't persuade him; he is all ice and iron: there isn't a soft place nor a warm one in him anywhere. What is your secret? Look here; you owe me a hundred dollars for unintentionally giving you a chance to perform a miracle—for it *is* a miracle that you've done."

"That is all right," I said, "collect it of Jimmy Lewis."

That good dog not only did me that good turn in the time of my need, but he won for me the envious reputation among all the theatrical people from the Atlantic to the Pacific of being the only man in history who had ever run the blockade of Augustin Daly's back door.

CHAPTER XLVI.

ON *the Train.* Fifty years ago, when I was a boy in the then remote and sparsely peopled Mississippi valley, vague tales and rumors of a mysterious body of professional murderers came wandering in from a country which was constructively as far from us as the constellations blinking in space — India; vague tales and rumors of a sect called Thugs, who waylaid travelers in lonely places and killed them for the contentment of a god whom they worshiped; tales which everybody liked to listen to and nobody believed — except with reservations. It was considered that the stories had gathered bulk on their travels. The matter died down and a lull followed. Then Eugene Sue's " Wandering Jew " appeared, and made great talk for a while. One character in it was a chief of Thugs — " Feringhea "—a mysterious and terrible Indian who was as slippery and sly as a serpent, and as deadly; and he stirred up the Thug interest once more. But it did not last. It presently died again — this time to stay dead.

At first glance it seems strange that this should have happened; but really it was not strange — on the contrary, it was natural; I mean on our side of the water. For the source whence the Thug tales mainly came was a Government Report, and without doubt was not republished in America; it was probably never even seen there. Government Reports have no general circulation. They are distributed to the few, and are not always read by those few. I heard of this Report for the first time a day or two ago, and borrowed it. It is full of fascinations; and it turns those dim, dark fairy tales of my boyhood days into realities.

The Report was made in 1839 by Major Sleeman, of the Indian Service, and was printed in Calcutta in 1840. It is a clumsy, great, fat, poor sample of the printer's art, but good enough for a government printing-office in that old day and in that remote region, perhaps. To Major Sleeman was given the general superintendence of the giant task of ridding India of Thuggee, and he and his seventeen assistants accomplished it. It was the Augean Stables over again. Captain Vallancey, writing in a Madras journal in those old times, makes this remark:

"The day that sees this far-spread evil eradicated from India and known only in name, will greatly tend to immortalize British rule in the East."

He did not overestimate the magnitude and difficulty of the work, nor the immensity of the credit which would justly be due to British rule in case it was accomplished.

Thuggee became known to the British authorities in India about 1810, but its wide prevalence was not suspected; it was not regarded as a serious matter, and no systematic measures were taken for its suppression until about 1830. About that time Major Sleeman captured Eugene Sue's Thug-chief, "Feringhea," and got him to turn King's evidence. The revelations were so stupefying that Sleeman was not able to believe them. Sleeman thought he knew every criminal within his jurisdiction, and that the worst of them were merely thieves; but Feringhea told him that he was in reality living in the midst of a swarm of professional murderers; that they had been all about him for many years, and that they buried their dead close by. These seemed insane tales; but Feringhea said come and see — and he took him to a grave and dug up a hundred bodies, and told him all the circumstances of the killings, and named the Thugs who had done the work. It was a staggering business. Sleeman captured some of these Thugs and proceeded to examine them separately, and with proper precautions against collusion; for he would not believe any Indian's unsupported word. The evidence gathered proved the truth of what Feringhea had said, and also revealed the fact that gangs of Thugs were plying their trade all over India. The astonished government now took hold of Thuggee, and for ten years made systematic and relentless war upon it, and finally destroyed it. Gang after gang was captured, tried, and punished. The Thugs were harried and hunted from one end of India to the other. The government got all their secrets out of them; and also got the names of the members of the bands, and recorded them in a book, together with their birthplaces and places of residence.

The Thugs were worshipers of Bhowanee; and to this god they sacrificed anybody that came handy; but they kept the dead man's things themselves, for the god cared for nothing but the corpse. Men were initiated into the sect with solemn ceremonies. Then they were taught how to strangle a person with the sacred choke-cloth, but were not allowed to perform officially with it until after long practice. No half-educated strangler could choke a man to death quickly enough to keep him from uttering a sound — a muffled scream, gurgle, gasp, moan, or something of the sort; but the expert's work was instantaneous: the cloth was whipped around the victim's neck, there was a sudden twist, and the head fell silently forward, the eyes starting from the sockets; and all was over. The Thug carefully guarded against resistance. It was usual to to get the victims to sit down, for that was the handiest position for business.

If the Thug had planned India itself it could not have been more conveniently arranged for the needs of his occupation.

There were no public conveyances. There were no conveyances
for hire. The traveler went on foot or in a bullock cart or on a
horse which he bought for the purpose. As soon as he was out
of his own little State or principality he was among strangers;
nobody knew him, nobody took note of him, and from that time
his movements could no longer be traced. He did not stop in
towns or villages, but camped outside of them and sent his servants in to buy provisions. There were no habitations between
villages. Whenever he was between villages he was an easy prey,
particularly as he usually traveled by night, to avoid the heat.
He was always being overtaken by strangers who offered him
the protection of their company, or asked for the protection of
his — and these strangers were often Thugs, as he presently
found out to his cost. The landholders, the native police, the
petty princes, the village officials, the customs officers were in
many cases protectors and harborers of the Thugs, and
betrayed travelers to them for a share of the spoil. At first
this condition of things made it next to impossible for the
government to catch the marauders; they were spirited away
by these watchful friends. All through a vast continent, thus
infested, helpless people of every caste and kind moved along
the paths and trails in couples and groups silently by night,
carrying the commerce of the country — treasure, jewels,
money, and petty batches of silks, spices, and all manner of
wares. It was a paradise for the Thug.

When the autumn opened, the Thugs began to gather together by pre-concert. Other people had to have interpreters
at every turn, but not the Thugs; *they* could talk together,
no matter how far apart they were born, for they had a
language of their own, and they had secret signs by which
they knew each other for Thugs; and they were always friends.
Even their diversities of religion and caste were sunk in
devotion to their calling, and the Moslem and the high-caste

and low-caste Hindoo were staunch and affectionate brothers in Thuggery.

When a gang had been assembled, they had religious worship, and waited for an omen. They had definite notions about the omens. The cries of certain animals were good omens, the cries of certain other creatures were bad omens. A bad omen would stop proceedings and send the men home.

The sword and the strangling-cloth were sacred emblems. The Thugs worshiped the sword at home before going out to the assembling-place; the strangling-cloth was worshiped at the place of assembly. The chiefs of most of the bands performed the religious ceremonies themselves; but the *Kaets* delegated them to certain official stranglers (Chaurs). The rites of the Kaets were so holy that no one but the Chaur was allowed to touch the vessels and other things used in them.

Thug methods exhibit a curious mixture of caution and the absence of it; cold business calculation and sudden, unreflecting impulse; but there were two details which were constant, and not subject to caprice: patient persistence in following up the prey, and pitilessness when the time came to act.

Caution was exhibited in the strength of the bands. They never felt comfortable and confident unless their strength exceeded that of any party of travelers they were likely to meet by four or fivefold. Yet it was never their purpose to attack openly, but only when the victims were off their guard. When they got hold of a party of travelers they often moved along in their company several days, using all manner of arts to win their friendship and get their confidence. At last, when this was accomplished to their satisfaction, the real business began. A few Thugs were privately detached and sent forward in the dark to select a good killing-place and *dig the graves.* When the rest reached the spot a halt was called, for a rest or a smoke. The travelers were invited to sit. By signs, the chief appointed certain Thugs to sit down in front of the travelers as if to wait upon them, others to sit down beside them and engage them in conversation, and certain expert stranglers to stand behind the travelers and be ready when the signal was given. The signal was usually some commonplace remark, like " Bring the tobacco." Sometimes a considerable wait ensued after all the actors were in their places — the chief was biding his time, in order to make everything sure. Meantime, the talk droned on, dim figures moved about in the dull light, peace and tranquility reigned, the travelers resigned themselves to the pleasant reposefulness and comfort of the situation, unconscious of the death-angels standing motionless at their backs. The time was ripe, now, and the signal came : " Bring the tobacco." There was a mute swift movement, all in the same instant the men at each victim's sides seized his hands, the man in front seized his feet, and pulled, the man at his back whipped the cloth around his neck and gave it a twist — the head sunk forward, the tragedy was over. The bodies were stripped and covered up in the graves, the spoil packed for transportation, then the Thugs gave pious thanks to Bhowanee, and departed on further holy service.

The Report shows that the travelers moved in exceedingly small groups — twos, threes, fours, as a rule; a party with a dozen in it was rare. The Thugs themselves seem to have been the only people who moved in force. They went about in gangs of 10, 15, 25, 40, 60, 100, 150, 200, 250, and one gang of 310 is mentioned. Considering their numbers, their catch was not extraordinary — particularly when you consider that they were not in the least fastidious, but took anybody they could get, whether rich or poor, and sometimes even killed children. Now and then they killed women, but it was considered sinful to do it, and unlucky. The "season" was six or eight months long. One season the half dozen Bundelkand and

Gwalior gangs aggregated 712 men, and they murdered 210 people. One season the Malwa and Kandeish gangs aggregated 702 men, and they murdered 232. One season the Kandeish and Berar gangs aggregated 963 men, and they murdered 385 people.

Here is the tally-sheet of a gang of *sixty* Thugs for a whole season — gang under two noted chiefs, " Chotee and Sheik Nungoo from Gwalior " :

" Left Poora, in Jhansee, and on arrival at Sarora murdered a traveler.

" On nearly reaching Bhopal, met 3 Brahmins, and murdered them.

" Cross the Nerbudda ; at a village called Hutteea, murdered a Hindoo.

" Went through Aurungabad to Walagow; there met a Havildar of the barber caste and 5 sepoys (native soldiers) ; in the evening came to Jokur, and in the morning killed them near the place where the treasure-bearers were killed the year before.

" Between Jokur and Dholeea met a sepoy of the shepherd caste ; killed him in the jungle.

" Passed through Dholeea and lodged in a village ; two miles beyond, on the road to Indore, met a Byragee (beggar — holy mendicant); murdered him at the Thapa.

" In the morning, beyond the Thapa, fell in with 3 Marwarie travelers ; murdered them.

" Near a village on the banks of the Taptee met 4 travelers and killed them.

" Between Choupra and Dhoreea met a Marwarie ; murdered him.

" At Dhoreea met 3 Marwaries ; took them two miles and murdered them.

" Two miles further on, overtaken by three treasure-bearers ; took them two miles and murdered them in the jungle.

" Came on to Khurgore Bateesa in Indore, divided spoil, and dispersed.

" A total of 27 men murdered on one expedition."

Chotee (to save his neck) was informer, and furnished these facts. Several things are noticeable about his resumé. 1. Business brevity ; 2, absence of emotion ; 3, smallness of the parties encountered by the 60 ; 4, variety in character and quality of the game captured ; 5, Hindoo and Mohammedan chiefs in business together for Bhowanee ; 6, the sacred caste of the Brahmins not respected by either ; 7, nor yet the character of that mendicant, that Byragee.

A beggar is a holy creature, and some of the gangs spared him on that account, no matter how slack business might be; but other gangs slaughtered not only him, but even that sacredest of sacred creatures, the *fakeer* — that repulsive skin-and-bone thing that goes around naked and mats his bushy hair with dust and dirt, and so beflours his lean body with ashes that he looks like a specter. Sometimes a fakeer trusted a shade too far in the protection of his sacredness. In the middle of a tally-sheet of Feringhea's, who had been out with forty Thugs, I find a case of the kind. After the killing of thirty-nine men and one woman, the fakeer appears on the scene:

"Approaching Doregow, met 3 pundits; also a fakeer, mounted on a pony; he was plastered over with sugar to collect flies, and was covered with them. Drove off the fakeer, and killed the other three.

"Leaving Doregow, the fakeer joined again, and went on in company to Raojana; met 6 Khutries on their way from Bombay to Nagpore. Drove off the fakeer with stones, and killed the 6 men in camp, and buried them in the grove.

"Next day the fakeer joined again; made him leave at Mana. Beyond there, fell in with two Kahars and a sepoy, and came on towards the place selected for the murder. When near it, the fakeer came again. Losing all patience with him, gave Mithoo, one of the gang, 5 rupees ($2.50) to murder him, and take the sin upon himself. All four were strangled, including the fakeer. Surprised to find among the fakeer's effects 30 pounds of coral, 350 strings of small pearls, 15 strings of large pearls, and a gilt necklace."

It it curious, the little effect that time has upon a really interesting circumstance. This one, so old, so long ago gone down into oblivion, reads with the same freshness and charm that attach to the news in the morning paper; one's spirits go up, then down, then up again, following the chances which the fakeer is running; now you hope, now you despair, now you hope again; and at last everything comes out right, and you feel a great wave of personal satisfaction go weltering through you, and without thinking, you put out your hand to pat Mithoo on the back, when — puff! the whole thing has vanished away, there is nothing there; Mithoo and all the

28

crowd have been dust and ashes and forgotten, oh, so many, many, *many* lagging years! And then comes a sense of injury: you don't know whether Mithoo got the swag, along with the sin, or had to divide up the swag and keep all the sin himself. There is no literary art about a government report. It stops a story right in the most interesting place.

These reports of Thug expeditions run along interminably in one monotonous tune: "Met a sepoy — killed him; met 5 pundits — killed them; met 4 Rajpoots and a woman — killed them"— and so on, till the statistics get to be pretty dry. But this small trip of Feringhea's Forty had some little variety about it. Once they came across a man hiding in a *grave* — a thief; he had stolen 1,100 rupees from Dhunroj Seith of Parowtee. They strangled him and took the money. They had no patience with thieves. They killed two treasure-bearers, and got 4,000 rupees. They came across two bullocks "laden with copper pice," and killed the four drivers and took the money. There must have been half a ton of it. I think it takes a double handful of pice to make an anna, and 16 annas to make a rupee; and even in those days the rupee was worth only half a dollar. Coming back over their tracks from Baroda, they had another picturesque stroke of luck: "The Lohars of Oodeypore" put a traveler in their charge "for safety." Dear, dear, across this abyssmal gulf of time we still see Feringhea's lips uncover his teeth, and through the dim haze we catch the incandescent glimmer of his smile. He accepted that trust, good man; and so we know what went with the traveler.

Even Rajahs had no terrors for Feringhea; he came across an elephant-driver belonging to the Rajah of Oodeypore and promptly strangled him.

"A total of 100 men and 5 women murdered on this expedition."

Among the reports of expeditions we find mention of victims of almost every quality and estate:

Native soldiers.	Chuprassies.	Women servants seeking
Fakeers.	Treasure-bearers.	work.
Mendicants.	Children.	Shepherds.
Holy-water carriers.	Cowherds.	Archers.
Carpenters.	Gardeners.	Table-waiters.
Peddlers.	Shopkeepers.	Weavers.
Tailors.	Palanquin-bearers.	Priests.
Blacksmiths.	Farmers.	Bankers.
Policemen (native).	Bullock-drivers.	Boatmen.
Pastry cooks.	Male servants seeking	Merchants.
Grooms.	work.	Grass-cutters.
Mecca pilgrims.		

Also a prince's cook; and even the water-carrier of that sublime lord of lords and king of kings, the Governor-General of India! How broad they were in their tastes! They also murdered actors — poor wandering barn-stormers. There are two instances recorded; the first one by a gang of Thugs under a chief who soils a great name borne by a better man — Kipling's deathless " Gungadin ":

"After murdering 4 sepoys, going on toward Indore, met 4 strolling players, and persuaded them to come with us, on the pretense that we would see their performance at the next stage. Murdered them at a temple near Bhopal."

Second instance:

" At Deohuttee, joined by comedians. Murdered them eastward of that place."

But this gang was a particularly bad crew. On that expedition they murdered a fakeer and twelve beggars. And yet Bhowanee protected them; for once when they were strangling a man in a wood when a crowd was going by close at hand and the noose slipped and the man screamed, Bhowanee made a camel burst out at the same moment with a roar that drowned the scream; and before the man could repeat it the breath was choked out of his body.

The cow is so sacred in India that to kill her keeper is an awful sacrilege, and even the Thugs recognized this; yet now and then the lust for blood was too strong, and so they did kill a few cow-keepers. In one of these instances the witness who killed the cowherd said, " In Thuggee this is strictly forbidden, and is an act from which no good can come. I was ill of a fever for ten days afterward. I do believe that evil will follow the murder of a man with a cow. If there be no cow it does not signify." Another Thug said he held the cowherd's feet while this witness did the strangling. He felt no concern, " because the bad fortune of such a deed is upon the strangler and not upon the assistants; even if there should be a hundred of them."

There were thousands of Thugs roving over India constantly, during many generations. They made Thuggee a hereditary vocation and taught it to their sons and to their son's sons. Boys were in full membership as early as 16 years of age; veterans were still at work at 70. What was the fascination, what was the impulse? Apparently, it was partly piety, largely gain, and there is reason to suspect that the *sport* afforded was the chiefest fascination of all. Meadows Taylor makes a Thug in one of his books claim that the pleasure of killing men was the white man's beast-hunting instinct enlarged, refined, ennobled. I will quote the passage:

CHAPTER XLVII.

Simple rules for saving money : To save half, when you are fired by an eager impulse to contribute to a charity, wait, and count forty. To save three-quarters, count sixty. To save it all, count sixty-five.— *Pudd'nhead Wilson's New Calendar.*

THE Thug said :

"How many of you English are passionately devoted to sporting ! Your days and months are passed in its excitement. A tiger, a panther, a buffalo or a hog rouses your utmost energies for its destruction — you even risk your lives in its pursuit. How much higher game is a Thug's !"

That must really be the secret of the rise and development of Thuggee. The joy of killing! the joy of seeing killing done — these are traits of the human race at large. We white people are merely modified Thugs; Thugs fretting under the restraints of a not very thick skin of civilization; Thugs who long ago enjoyed the slaughter of the Roman arena, and later the burning of doubtful Christians by authentic Christians in the public squares, and who now, with the Thugs of Spain and Nimes, flock to enjoy the blood and misery of the bull-ring. We have no tourists of either sex or any religion who are able to resist the delights of the bull-ring when opportunity offers; and we are gentle Thugs in the hunting-season, and love to chase a tame rabbit and kill it. Still, we have made some progress — microscopic, and in truth scarcely worth mentioning, and certainly nothing to be proud of — still, it is progress: we no longer take pleasure in slaughtering or burning helpless men. We have reached a little altitude where we may look down upon the Indian Thugs with a complacent shudder ; and we may even hope for a day, many centuries hence, when our posterity will look down upon us in the same way.

There are many indications that the Thug often hunted men for the mere sport of it; that the fright and pain of the quarry were no more to him than are the fright and pain of the rabbit or the stag to us; and that he was no more ashamed of beguiling his game with deceits and abusing its trust than are we when we have imitated a wild animal's call and shot it when it honored us with its confidence and came to see what we wanted:

"Madara, son of Nihal, and I, Ramzam, set out from Kotdee in the cold weather and followed the high road for about twenty days in search of travelers, until we came to Selempore, where we met a very old man going to the east. We won his confidence in this manner: he carried a load which was too heavy for his old age; I said to him, ' You are an old man, I will aid you in carrying your load, as you are from my part of the country.' He said, ' Very well, take me with you.' So we took him with us to Selempore, where we slept that night. We woke him next morning before dawn and set out, and at the distance of three miles we seated him to rest while it was still very dark. Madara was ready behind him, and strangled him. He never spoke a word. He was about 60 or 70 years of age."

Another gang fell in with a couple of barbers and persuaded them to come along in their company by promising them the job of shaving the whole crew — 30 Thugs. At the place appointed for the murder 15 got shaved, and actually paid the barbers for their work. Then killed them and took back the money.

A gang of forty-two Thugs came across two Brahmins and a shopkeeper on the road, beguiled them into a grove and got up a *concert* for their entertainment. While these poor fellows were listening to the music the stranglers were standing behind them; and at the proper moment for dramatic effect they applied the noose.

The most devoted fisherman must have a bite at least as often as once a week or his passion will cool and he will put up his tackle. The tiger-sportsman must find a tiger at least once a fortnight or he will get tired and quit. The elephant-hunter's enthusiasm will waste away little by little, and his zeal will perish at last if he plod around a month without find-ing a member of that noble family to assassinate.

But when the lust in the hunter's heart is for the noblest of all quarries, man, how different is the case! and how watery and poor is the zeal and how childish the endurance of those other hunters by comparison. Then, neither hunger, nor thirst, nor fatigue, nor deferred hope, nor monotonous disap-pointment, nor leaden-footed lapse of time can conquer the hunter's patience or weaken the joy of his quest or cool the splendid rage of his desire. Of all the hunting-passions that burn in the breast of man, there is none that can lift him superior to discouragements like these but the one — the royal sport, the supreme sport, whose quarry is his brother. By comparison, tiger-hunting is a colorless poor thing, for all it has been so bragged about.

Why, the Thug was content to tramp patiently along, afoot, in the wasting heat of India, week after week, at an average of nine or ten miles a day, if he might but hope to find game some time or other and refresh his longing soul with blood. Here is an instance:

"I (Ramzam) and Hyder set out, for the purpose of strangling travelers, from Guddapore, and proceeded via the Fort of Julalabad, Newulgunge, Bangermow, on the banks of the Ganges (upwards of 100 miles), from whence we returned by another route. Still no travelers! till we reached Bowaneegunge, where we fell in with a traveler, a boatman; we inveigled him and about two miles east of there Hyder strangled him as he stood — for he was troubled and afraid, and would not sit. We then made a long journey (about 130 miles) and reached Hussunpore Bundwa, where at the tank we fell in with a traveler — he slept there that night; next morning we followed him and tried to win his confidence; at the distance of two miles we endeavored to induce him to sit down — but he would not, having become aware of us. I attempted to strangle him as he walked along, but did not succeed; both of us then fell upon him, he made a great outcry, 'They are murdering me!' at length we strangled him and flung his body into a well. After this we returned to our homes, having been out a month and traveled about 260 miles. A total of two men murdered on the expedition."

And here is another case — related by the terrible Futty Khan, a man with a tremendous record, to be re-mentioned by and by;

"I, with three others, traveled for about 45 days a distance of about 200 miles in search of victims along the highway to Bundwa and returned by Davodpore (another 200 miles) during which journey we had only one murder, which happened in this manner. Four miles to the east of Noubustaghat we fell in with a traveler, an old man. I, with Koshal and Hyder, inveigled him and accompanied him that day within 3 miles of Rampoor, where, after dark, in a lonely place, we got him to sit down and rest ; and while I kept him in talk, seated before him, Hyder behind strangled him : he made no resistance. Koshal stabbed him under the arms and in the throat, and we flung the body into a running stream. We got about 4 or 5 rupees each ($2 or $2.50). We then proceeded homewards. A total of one man murdered on this expedition."

There. They tramped 400 miles, were gone about three months, and harvested two dollars and a half apiece. But the mere pleasure of the hunt was sufficient. That was pay enongh. They did no grumbling.

Every now and then in this big book one comes across that pathetic remark : " We tried to get him to sit down but he would not." It tells the whole story. Some accident had awakened the suspicion in him that these smooth friends who had been petting and coddling him and making him feel so safe and so fortunate after his forlorn and lonely wanderings were the dreaded Thugs; and now their ghastly invitation to " sit and rest " had confirmed its truth. He knew there was no help for him, and that he was looking his last upon earthly things, but " he would not sit." No, not that — it was too awful to think of !

There are a number of instances which indicate that when a man had once tasted the regal joys of man-hunting he could not be content with the dull monotony of a crimeless life after ward. Example, from a Thug's testimony :

" We passed through to Kurnaul, where we found a former Thug named Junooa, an old comrade of ours, who had turned religious mendicant and become a disciple and holy. He came to us in the serai and weeping with joy returned to his old trade."

Neither wealth nor honors nor dignities could satisfy a reformed Thug for long. He would throw them all away, some day, and go back to the lurid pleasures of hunting men, and being hunted himself by the British.

Ramzam was taken into a great native grandee's service and given authority over five villages. "My authority extended over these people to summons them to my presence, to make them stand or sit. I dressed well, rode my pony, and had two sepoys, a scribe and a village guard to attend me. During three years I used to pay each village a monthly visit, and no one suspected that I was a Thug! The chief man used to wait on me to transact business, and as I passed along, old and young made their salaam to me."

And yet during that very three years he got leave of absence "to attend a wedding," and instead went off on a Thugging lark with six other Thugs and hunted the highway for fifteen days! — with satisfactory results.

Afterwards he held a great office under a Rajah. There he had ten miles of country under his command and a military guard of fifteen men, with authority to call out 2,000 more upon occasion. But the British got on his track, and they crowded him so that he had to give himself up. See what a figure he was when he was gotten up for style and had all his things on: " I was fully armed — a sword, shield, pistols, a matchlock musket and a flint gun, for I was fond of being thus arrayed, and when so armed feared not though forty men stood before me."

He gave himself up and proudly proclaimed himself a Thug. Then by request he agreed to betray his friend and pal, Buhram, a Thug with the most tremendous record in India. " I went to the house where Buhram slept (often has he led our gangs!) I woke him, he knew me well, and came outside to me. It was a cold night, so under pretence of warming myself, but in reality to have light for his seizure by the guards, I lighted some straw and made a blaze. We were warming our hands. The guards drew around us. I said to them, "This is Buhram," and he was seized just as a cat seizes a mouse. Then Buhram said, "I am a Thug! my father was a Thug, my grandfather was a Thug, and I have thugged with many!"

So spoke the mighty hunter, the mightiest of the mighty, the Gordon Cumming of his day. Not much regret noticeable in it.*

So many many times this Official Report leaves one's curiosity unsatisfied. For instance, here is a little paragraph out of the record of a certain band of 193 Thugs, which has that defect:

"Fell in with Lall Sing Subahdar and his family, consisting of nine persons. Traveled with them two days, and the third put them all to death except the two children, little boys of one and a half years old."

There it stops. What did they do with those poor little fellows? What was their subsequent history? Did they purpose training them up as Thugs? How could they take care of such little creatures on a march which stretched over several months? No one seems to have cared to ask any questions about the babies. But I do wish I knew.

One would be apt to imagine that the Thugs were utterly callous, utterly destitute of human feelings, heartless toward

---

* "Having planted a bullet in the shoulder-bone of an elephant, and caused the agonized creature to lean for support against a tree, I proceeded to brew some coffee. Having refreshed myself, taking observations of the elephant's spasms and writhings between the sips, I resolved to make experiments on vulnerable points, and, approaching very near, I fired several bullets at different parts of his enormous skull. He only acknowledged the shots by a salaam-like movement of his trunk, with the point of which he gently touched the wounds with a striking and peculiar action. Surprised and shocked to find that I was only prolonging the suffering of the noble beast, which bore its trials with such dignified composure, I resolved to finish the proceeding with all possible despatch, and accordingly opened fire upon him from the left side. Aiming at the shoulder, I fired six shots with the two-grooved rifle, which must have eventually proved mortal, after which I fired six shots at the same part with the Dutch six-pounder. Large tears now trickled down from his eyes, which he slowly shut and opened, his colossal frame shivered convulsively, and falling on his side he expired." — *Gordon Cumming.*

their own families as well as toward other people's; but this
was not so. Like all other Indians, they had a passionate love
for their kin. A shrewd British officer who knew the Indian
character, took that characteristic into account in laying his
plans for the capture of Eugene Sue's famous Feringhea. He
found out Feringhea's hiding-place, and sent a guard by night
to seize him, but the squad was awkward and he got away.
However, they got the rest of the family — the mother, wife,
child, and brother — and brought them to the officer, at Jubbulpore; the officer did not fret, but bided his time: " I knew
Feringhea would not go far while links so dear to him were in
my hands." He was right. Feringhea knew all the danger
he was running by staying in the neighborhood, still he could
not tear himself away. The officer found that he divided his time
between five villages where he had relatives and friends who
could get news for him from his family in Jubbulpore jail;
and that he never slept two consecutive nights in the same
village. The officer traced out his several haunts, then pounced
upon all the five villages on the one night and at the same
hour, and got his man.

Another example of family affection. A little while
previously to the capture of Feringhea's family, the British
officer had captured Feringhea's foster-brother, leader of a
gang of ten, and had tried the eleven and condemned them to
be hanged. Feringhea's captured family arrived at the jail
the day before the execution was to take place. The foster-
brother, Jhurhoo, entreated to be allowed to see the aged
mother and the others. The prayer was granted, and this is
what took place — it is the British officer who speaks:

"In the morning, just before going to the scaffold, the interview took
place before me. He fell at the old woman's feet and begged that she would
relieve him from the obligations of the milk with which she had nourished
him from infancy, as he was about to die before he could fulfill any of them.
She placed her hands on his head, and he knelt, and she said she forgave him
all, and bid him die like a man."

If a capable artist should make a picture of it, it would be full of dignity and solemnity and pathos; and it could touch you. You would imagine it to be anything but what it was. There is reverence there, and tenderness, and gratefulness, and compassion, and resignation, and fortitude, and self-respect — and no sense of disgrace, no thought of dishonor. Everything is there that goes to make a noble parting, and give it a moving grace and beauty and dignity. And yet one of these people is a Thug and the other a mother of Thugs! The incongruities of our human nature seem to reach their limit here.

I wish to make note of one curious thing while I think of it. One of the very commonest remarks to be found in this bewildering array of Thug confessions is this:

"Strangled him and *threw him in a well!*" In one case they threw sixteen into a well — and they had thrown others in the same well before. It makes a body thirsty to read about it.

And there is another very curious thing. The bands of Thugs had *private graveyards*. They did not like to kill and bury at random, here and there and everywhere. They preferred to wait, and toll the victims along, and get to one of their regular burying-places (bheels) if they could. In the little kingdom of Oude, which was about half as big as Ireland and about as big as the State of Maine, they had *two hundred and seventy-four bheels*. They were scattered along *fourteen hundred miles of road*, at an average of only *five miles apart*, and the British government traced out and located each and every one of them and set them down on the map.

The Oude bands seldom went out of their own country, but they did a thriving business within its borders. So did outside bands who came in and helped. Some of the Thug leaders of Oude were noted for their successful careers. Each of four of them confessed to above 300 murders; another to nearly 400;

our friend Ramzam to 604 — he is the one who got leave of absence to attend a wedding and went thugging instead; and he is also the one who betrayed Buhram to the British.

But the biggest records of all were the murder-lists of Futty Khan and Buhram. Futty Khan's number is smaller than Ramzam's, but he is placed at the head because his *average* is the best in Oude-Thug history per year of service. His slaughter was 508 men in twenty years, and he was still a young man when the British stopped his industry. Buhram's list was 931 murders, but it took him forty years. His average was one man and nearly all of another man per month for forty years, but Futty Khan's average was *two* men and a little of another man per month during his twenty years of usefulness.

There is one very striking thing which I wish to call attention to. You have surmised from the listed callings followed by the victims of the Thugs that nobody could travel the Indian roads unprotected and live to get through; that the Thugs respected no quality, no vocation, no religion, nobody; that they killed every unarmed man that came in their way. That is wholly true — with one reservation. In all the long file of Thug confessions *an English traveler is mentioned but once* — and this is what the Thug says of the circumstance:

"He was on his way from Mhow to Bombay. *We studiously avoided him.* He proceeded next morning with a number of travelers *who had sought his protection*, and they took the road to Baroda."

We do not know who he was; he flits aross the page of this rusty old book and disappears in the obscurity beyond; but he is an impressive figure, moving through that valley of death serene and unafraid, clothed in the might of the English name.

We have now followed the big official book through, and we understand what Thuggee was, what a bloody terror it was, what a desolating scourge it was. In 1830 the English found this cancerous organization imbedded in the vitals of the empire, doing its devastating work in secrecy, and assisted, protected, sheltered, and hidden by innumerable confederates — big and little native chiefs, customs officers, village officials, and native police, all ready to lie for it, and the mass of the people, through fear, persistently pretending to know nothing about its doings; and this condition of things had existed for generations, and was formidable with the sanctions of age and old custom. If ever there was an unpromising task, if ever there was a hopeless task in the world, surely it was offered here — the task of conquering Thuggee. But that little handful of English officials in India set their sturdy and confident grip upon it, and ripped it out, root and branch! How modest do Captain Vallancey's words sound now, when we read them again, knowing what we know:

" The day that sees this far-spread evil completely eradicated from India, and known only in name, will greatly tend to immortalize British rule in the East."

It would be hard to word a claim more modestly than that for this most noble work.

CHAPTER XLVIII.

Grief can take care of itself; but to get the full value of a joy you must have somebody to divide it with.--*Pudd'nhead Wilson's New Calendar.*

WE left Bombay for Allahabad by a night train. It is the custom of the country to avoid day travel when it can conveniently be done. But there is one trouble: while you can seemingly "secure" the two lower berths by making early application, there is no ticket as witness of it, and no other producible evidence in case your proprietorship shall chance to be challenged. The word "engaged" appears on the window, but it doesn't state who the compartment is engaged *for*. If your Satan and your Barney arrive before somebody else's servants, and spread the bedding on the two sofas and then stand guard till you come, all will be well; but if they step aside on an errand, they may find the beds promoted to the two shelves, and somebody else's demons standing guard over their master's beds, which in the meantime have been spread upon your sofas.

You do not pay anything extra for your sleeping place; that is where the trouble lies. If you buy a fare-ticket and fail to use it, there is room thus made available for someone else; but if the place were secured to you it would remain vacant, and yet your ticket would secure you *another* place when you were presently ready to travel.

However, no explanation of such a system can make it seem quite rational to a person who has been used to a more rational system. If our people had the arranging of it, we should charge extra for securing the place, and then the road would suffer no loss if the purchaser did not occupy it.

(447)

The present system encourages good manners — and also discourages them. If a young girl has a lower berth and an elderly lady comes in, it is usual for the girl to offer her place to this late comer; and it is usual for the late comer to thank her courteously and take it. But the thing happens differently sometimes. When we were ready to leave Bombay my daughter's satchels were holding possession of her berth — a lower one. At the last moment, a middle-aged American lady swarmed into the compartment, followed by native porters laden with her baggage. She was growling and snarling and scolding, and trying to make herself phenomenally disagreeable; and succeeding. Without a word, she hoisted the satchels into the hanging shelf, and took possession of that lower berth.

On one of our trips Mr. Smythe and I got out at a station to walk up and down, and when we came back Smythe's bed was in the hanging shelf and an English cavalry officer was in bed on the sofa which he had lately been occupying. It was mean to be glad about it, but it is the way we are made; I could not have been gladder if it had been my enemy that had suffered this misfortune. We all like to see people in trouble, if it doesn't cost us anything. I was so happy over Mr. Smythe's chagrin that I couldn't go to sleep for thinking of it and enjoying it. I knew he supposed the officer had committed the robbery himself, whereas without a doubt the officer's servant had done it without his knowledge. Mr. Smythe kept this incident warm in his heart, and longed for a chance to get even with somebody for it. Sometime afterward the opportunity came, in Calcutta. We were leaving on a 24-hour journey to Darjeeling. Mr. Barclay, the general superintendent, has made special provision for our accommodation, Mr. Smythe said; so there was no need to hurry about getting to the train; consequently, we were a little late.

When we arrived, the usual immense turmoil and confusion of a great Indian station were in full blast. It was an immoderately long train, for all the natives of India were going by it somewhither, and the native officials were being pestered to frenzy by belated and anxious people. They didn't know where our car was, and couldn't remember having received any orders about it. It was a deep disappointment; moreover, it looked as if our half of our party would be left behind altogether. Then Satan came running and said he had found a compartment with one shelf and one sofa unoccupied, and had made our beds and had stowed our baggage. We rushed to the place, and just as the train was ready to pull out and the porters were slamming the doors to, all down the line, an officer of the Indian Civil Service, a good friend of ours, put his head in and said : —

"I have been hunting for you everywhere. What are you doing here? Don't you know —"

The train started before he could finish. Mr. Smythe's opportunity was come. His bedding, on the shelf, at once changed places with the bedding — a stranger's — that was occupying the sofa that was opposite to mine. About ten o'clock we stopped somewhere, and a large Englishman of official military bearing stepped in. We pretended to be asleep. The lamps were covered, but there was light enough for us to note his look of surprise. He stood there, grand and fine, peering down at Smythe, and wondering in silence at the situation. After a bit he said : —

"Well!" And that was all.

But that was enough. It was easy to understand. It meant: "This is extraordinary. This is high-handed. I haven't had an experience like this before."

He sat down on his baggage, and for twenty minutes we watched him through our eyelashes, rocking and swaying there

29

to the motion of the train.   Then we came to a station, and he
got up and went out, muttering: " I *must* find a lower berth,
or wait over."   His servant came presently and carried away
his things.

Mr. Smythe's sore place was healed, his hunger for revenge
was satisfied.   But he couldn't sleep, and neither could I ; for
this was a venerable old car, and nothing about it was taut.
The closet door slammed all night, and defied every fastening

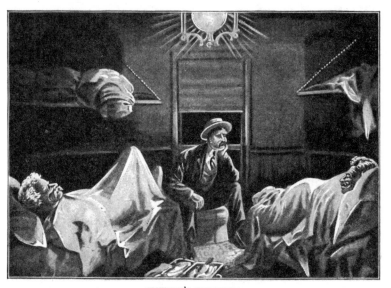

SMYTHE'S REVENGE.

we could invent.   We got up very much jaded, at dawn, and
stepped out at a way station ; and, while we were taking a cup
of coffee, that Englishman ranged up alongside, and somebody
said to him :

" So you didn't stop off, after all ? "

" No.   The guard found a place for me that had been
engaged and not occupied.   I had a whole saloon car all to
myself — oh, quite palatial !   I never had such luck in my
life."

That was our car, you see. We moved into it, straight off, the family and all. But I asked the English gentleman to remain, and he did. A pleasant man, an infantry colonel; and doesn't know, yet, that Smythe robbed him of his berth, but thinks it was done by Smythe's servant without Smythe's knowledge. He was assisted in gathering this impression.

The Indian trains are manned by natives exclusively. The Indian stations — except very large and important ones — are manned entirely by natives, and so are the posts and telegraphs. The rank and file of the police are natives. All these people are pleasant and accommodating. One day I left an express train to lounge about in that perennially ravishing show, the ebb and flow and whirl of gaudy natives, that is always surging up and down the spacious platform of a great Indian station; and I lost myself in the ecstasy of it, and when I turned, the train was moving swiftly away. I was going to sit down and wait for another train, as I would have done at home; I had no thought of any other course. But a native official, who had a green flag in his hand, saw me, and said politely:

" Don't you belong in the train, sir?"

" Yes," I said.

He waved his flag, and the train came back! And he put me aboard with as much ceremony as if I had been the General Superintendent. They are kindly people, the natives. The face and the bearing that indicate a surly spirit and a bad heart seemed to me to be so rare among Indians — so nearly non-existent, in fact — that I sometimes wondered if Thuggee wasn't a dream, and not a reality. The bad hearts *are* there, but I believe that they are in a small, poor minority. One thing is sure: They are much the most *interesting* people in the world — and the nearest to being incomprehensible. At any rate, the hardest to account for. Their character and

their history, their customs and their religion, confront you
with riddles at every turn — riddles which are a trifle more
perplexing after they are explained than they were before.
You can get the *facts* of a custom — like caste, and Suttee,
and Thuggee, and so on — and with the facts a theory which
tries to explain, but never quite does it to your satisfaction.
You can never quite understand *how* so strange a thing could
have been born, nor *why*.

For instance — the *Suttee*. This is the explanation of it:
A woman who throws away her life when her husband dies is
instantly joined to him again, and is forever afterward happy
with him in heaven; her family will build a little monument
to her, or a temple, and will hold her in honor, and, indeed,
worship her memory always; they will themselves be held
in honor by the public; the woman's self-sacrifice has
conferred a noble and lasting distinction upon her posterity.
And, besides, see what she has escaped: If she had elected
to live, she would be a disgraced person; she could not re-
marry; her family would despise her and disown her; she
would be a friendless outcast, and miserable all her days.

Very well, you say, but the explanation is not complete
yet. *How* did people come to drift into such a strange custom? What was the origin of the idea? "Well, nobody
knows; it was probably a revelation sent down by the
gods." One more thing: Why was such a cruel death chosen
— why wouldn't a gentle one have answered? "Nobody
knows; maybe that was a revelation, too."

No — you can never understand it. It all seems impossible. You resolve to believe that a widow never burnt herself
willingly, but went to her death because she was afraid to defy
public opinion. But you are not able to keep that position.
History drives you from it. Major Sleeman has a convincing
case in one of his books. In his government on the Nerbudda

he made a brave attempt on the 28th of March, 1828, to put down Suttee on his own hook and without warrant from the Supreme Government of India. He could not foresee that the Government would put it down itself eight months later. The only backing he had was a bold nature and a compassionate heart. He issued his proclamation abolishing the Suttee in his district. On the morning of Tuesday — note the day of the week — the 24th of the following November, Ummed Singh Upadhya, head of the most respectable and most extensive Brahmin family in the district, died, and presently came a deputation of his sons and grandsons to beg that his old widow might be allowed to burn herself upon his pyre. Sleeman threatened to enforce his order, and punish severely any man who assisted; and he placed a police guard to see that no one did so. From the early morning the old widow of sixty-five had been sitting on the bank of the sacred river by her dead, waiting through the long hours for the permission; and at last the refusal came instead. In one little sentence Sleeman gives you a pathetic picture of this lonely old gray figure: all day and all night "she remained sitting by the edge of the water without eating or drinking." The next morning the body of the husband was burned to ashes in a pit eight feet square and three or four feet deep, in the view of several thousand spectators. Then the widow waded out to a bare rock in the river, and everybody went away but her sons and other relations. All day she sat there on her rock in the blazing sun without food or drink, and with no clothing but a sheet over her shoulders.

The relatives remained with her and all tried to persuade her to desist from her purpose, for they deeply loved her. She steadily refused. Then a part of the family went to Sleeman's house, ten miles away, and tried again to get him to let her burn herself. He refused, hoping to save her yet.

All that day she scorched in her sheet on the rock, and all that night she kept her vigil there in the bitter cold. Thursday morning, in the sight of her relatives, she went through a ceremonial which said more to them than any words could have done; she put on the *dhaja* (a course red turban) and broke her bracelets in pieces. By these acts she became a dead person in the eye of the law, and excluded from her caste forever. By the iron rule of ancient custom, if she should now choose to live she could never return to her family. Sleeman was in deep trouble. If she starved herself to death her family would be disgraced; and, moreover, starving would be a more lingering misery than the death by fire. He went back in the evening thoroughly worried. The old woman remained on her rock, and there in the morning he found her with her *dhaja* still on her head. "She talked very collectedly, telling me that she had determined to mix her ashes with those of her departed husband, and should patiently wait my permission to do so, assured that God would enable her to sustain life till that was given, though she dared not eat or drink. Looking at the sun, then rising before her over a long and beautiful reach of the river, she said calmly, ' My soul has been for five days with my husband's near that sun; nothing but my earthly frame is left; and this, I know, you will in time suffer to be mixed with his ashes in yonder pit, because it is not in your nature or usage wantonly to prolong the miseries of a poor old woman.' "

He assured her that it was his desire and duty to save her, and to urge her to live, and to keep her family from the disgrace of being thought her murderers. But she said she was not afraid of their being thought so; that they had all, like good children, done everything in their power to induce her to live, and to abide with them; "and if I should consent I know they would love and honor me, but my duties to them have now ended. I commit them all to your care, and I go to attend my husband, Ummed Singh Upadhya, with whose ashes on the funeral pile mine have been already three times mixed."

She believed that she and he had been upon the earth three several times as wife and husband, and that she had burned herself to death three times upon his pyre. That is why she said that strange thing. Since she had broken her bracelets and put on the red turban she regarded herself as a corpse; otherwise she would not have allowed herself to do her husband the irreverence of pronouncing his name. "This was the first time in her long life that she had ever uttered her husband's name, for in India no woman, high or low, ever pronounces the name of her husband."

Major Sleeman still tried to shake her purpose. He promised to build her a fine house among the temples of her ancestors upon the bank of the river and make handsome provision for her out of rent-free lands if she would consent to live; and if she wouldn't he would allow no stone or brick to ever mark the place where she died. But she only smiled and said, "My pulse has long ceased to beat, my spirit has departed; I shall suffer nothing in the burning; and if you wish proof, order some fire and you shall see this arm consumed without giving me any pain."

Sleeman was now satisfied that he could not alter her purpose. He sent for all the chief members of the family and said he would suffer her to burn herself if they would enter into a written engagement to abandon the suttee in their family thenceforth. They agreed; the papers were drawn out and signed, and at noon, Saturday, word was sent to the poor old woman. She seemed greatly pleased. The ceremonies of bathing were gone through with, and by three o'clock she was ready and the fire was briskly burning in the pit. She had now gone without food or drink during more than four days

and a half. She came ashore from her rock, first wetting her
sheet in the waters of the sacred river, for without that safe-
guard any shadow which might fall upon her would convey
impurity to her; then she walked to the pit, leaning upon
one of her sons and a nephew — the distance was a hundred
and fifty yards.

"I had sentries placed all around, and no other person was allowed to
approach within five paces. She came on with a calm and cheerful countenance, stopped once, and casting her eyes upwards, said, 'Why have they
kept me five days from thee, my husband?' On coming to the sentries her
supporters stopped and remained standing; she moved on, and walked once
around the pit, paused a moment, and while muttering a prayer, threw some
flowers into the fire. She then walked up deliberately and steadily to the
brink, stepped into the centre of the flame, sat down, and leaning back in the
midst as if reposing upon a couch, was consumed without uttering a shriek or
betraying one sign of agony."

It is fine and beautiful. It compels one's reverence and
respect — no, has it freely, and without compulsion. We see
how the custom, once started, could continue, for the soul of it
is that stupendous power, Faith; faith brought to the pitch of
effectiveness by the cumulative force of example and long use
and custom; but we cannot understand how the first widows
came to take to it. That is a perplexing detail.

Sleeman says that it was usual to play music at the suttee,
but that the white man's notion that this was to drown the
screams of the martyr is not correct; that it had a quite different purpose. It was believed that the martyr died prophe-
cying; that the prophecies sometimes foretold disaster, and it
was considered a kindness to those upon whom it was to fall to
drown the voice and keep them in ignorance of the misfortune
that was to come.

HE STOLE BLANKETS.

CHAPTER XLIX.

He had had much experience of physicians, and said "the only way to keep your health is to eat what you don't want, drink what you don't like, and do what you'd druther not."— *Pudd'nhead Wilson's New Calendar.*

IT was a long journey — two nights, one day, and part of another day, from Bombay eastward to Allahabad; but it was always interesting, and it was not fatiguing. At first the night travel promised to be fatiguing, but that was on account of *pyjamas*. This foolish night-dress consists of jacket and drawers. Sometimes they are made of silk, sometimes of a raspy, scratchy, slazy woolen material with a sandpaper surface. The drawers are loose elephant-legged and elephant-waisted things, and instead of buttoning around the body there is a draw-string to produce the required shrinkage. The jacket is roomy, and one buttons it in front. Pyjamas are hot on a hot night and cold on a cold night — defects which a nightshirt is free from. I tried the pyjamas in order to be in the fashion; but I was obliged to give them up, I couldn't stand them. There was no sufficient change from day-gear to night-gear. I missed the refreshing and luxurious sense, induced by the night-gown, of being undressed, emancipated, set free from restraints and trammels. In place of that, I had the worried, confined, oppressed, suffocated sense of being abed with my clothes on. All through the warm half of the night the coarse surfaces irritated my skin and made it feel baked and feverish, and the dreams which came in the fitful flurries of slumber were such as distress the sleep of the damned, or ought to; and all through the cold other half of the night I could get no time for sleep because I had to employ it all in stealing blankets. But blankets are of no value at such a time; the higher they are piled the more effectively they cork the cold in and keep it from getting out. The result is that your legs are ice, and you know how you will feel by and by when you are buried. In a sane interval I discarded the pyjamas, and led a rational and comfortable life thenceforth.

Out in the country in India, the day begins early. One sees a plain, perfectly flat, dust-colored and brick-yardy, stretching limitlessly away on every side in the dim gray light, striped everywhere with hard-beaten narrow paths, the vast flatness broken at wide intervals by bunches of spectral trees that mark where villages are; and along all the paths are slender women and the black forms of lanky naked men moving to their work, the women with brass water-jars on their heads, the men carrying hoes. The man is not entirely naked; always there is a bit of white rag, a loin-cloth; it amounts to a bandage, and is a white accent on his black person, like the silver band around the middle of a pipe-stem. Sometimes he also wears a fluffy and voluminous white turban, and this adds a second accent. He then answers properly to Miss Gordon Cumming's flash-light picture of him—as a person who is dressed in "a turban and a pocket handkerchief."

All day long one has this monotony of dust-colored dead levels and scattering bunches of trees and mud villages. You soon realize that India is not beautiful; still there is an enchantment about it that is beguiling, and which does not pall. You cannot tell just what it is that makes the spell, perhaps, but you feel it and confess it, nevertheless. Of course, at bottom, you know in a vague way that it is *history*; it is that that affects you, a haunting sense of the myriads of human lives that have blossomed, and withered, and perished here, repeating and repeating and repeating, century after century, and age after age, the barren and meaningless process; it is this sense that gives to this forlorn, uncomely land power to speak to the spirit and make friends with it; to speak to it with a voice bitter with satire, but eloquent with melancholy. The deserts of Australia and the ice-barrens of Greenland have no speech, for they have no venerable history; with nothing to tell of man and his vanities, his fleeting glories and his miseries, they have nothing wherewith to spiritualize their ugliness and veil it with a charm.

There is nothing pretty about an Indian village — a mud one — and I do not remember that we saw any but mud ones on that long flight to Allahabad. It is a little bunch of dirt-colored mud hovels jammed together within a mud wall. As a rule, the rains had beaten down parts of some of the houses, and this gave the village the aspect of a mouldering and hoary ruin. I believe the cattle and the vermin live inside the wall; for I saw cattle coming out and cattle going in; and whenever I saw a villager, he was scratching. This last is only circumstantial evidence, but I think it has value. The village has a battered little temple or two, big enough to hold an idol, and with custom enough to fat-up a priest and keep him comfortable. Where there are Mohammedans there are generally a few sorry tombs outside the village that have a decayed and neglected look. The villages interested me because of things which Major Sleeman says about them in his books — particularly what he says about the division of labor in them. He says that the whole face of India is parceled out into estates of villages; that nine-tenths of the vast population of the land consist of cultivators of the soil; that it is these cultivators who inhabit the villages; that there are certain "established" village servants — mechanics and others who are apparently paid a wage by the village at large, and whose callings remain in certain families and are handed down from father to son, like an estate. He gives a list of these established servants:

Priest, blacksmith, carpenter, accountant, washerman, basket-maker, potter, watchman, barber, shoemaker, brazier, confectioner, weaver, dyer, etc. In his day witches abounded, and it was not thought good business wisdom for a man to marry his daughter into a family that hadn't a witch in it, for she would need a witch on the premises to protect her children from the evil spells which would certainly be cast upon them by the witches connected with the neighboring families.

The office of midwife was hereditary in the family of the basket-maker. It belonged to his wife. She might not be competent, but the office was hers, anyway. Her pay was not high — 25 cents for a boy, and half as much for a girl. The girl was not desired, because she would be a disastrous expense by and by. As soon as she should be old enough to begin to wear clothes for propriety's sake, it would be a disgrace to the family if she were not married; and to marry her meant financial ruin; for by custom the father must spend upon feasting and wedding-display everything he had and all he could borrow — in fact, reduce himself to a condition of poverty which he might never more recover from.

It was the dread of this prospective ruin which made the killing of girl-babies so prevalent in India in the old days before England laid the iron hand of her prohibitions upon the piteous slaughter. One may judge of how prevalent the custom was, by one of Sleeman's casual electrical remarks, when he speaks of children at play in villagès — *where girl-voices were never heard!*

The wedding-display folly is still in full force in India, and by consequence the destruction of girl-babies is still furtively practiced; but not largely, because of the vigilance of the government and the sternness of the penalties it levies.

In some parts of India the village keeps in its pay three other servants: an astrologer to tell the villager when he may plant his crop, or make a journey, or marry a wife, or strangle a child, or borrow a dog, or climb a tree, or catch a rat, or swindle a neighbor, without offending the alert and solicitous heavens; and what his dream means, if he has had one and was not bright enough to interpret it himself by the details of his dinner; the two other established servants were the tiger-persuader and the hailstorm discourager. The one kept away the tigers if he could, and collected the wages anyway, and the other kept off the hailstorms, or explained why he failed. He charged the same for explaining a failure that he did for scoring a success. A man is an idiot who can't earn a living in India.

Major Sleeman reveals the fact that the trade union and the boycott are antiquities in India. India seems to have originated everything. The "sweeper" belongs to the bottom caste; he is the lowest of the low — all other castes despise him and scorn his office. But that does not trouble him. His caste is a caste, and that is sufficient for him, and so he is proud of it, not ashamed. Sleeman says:

A STREET SPRINKLER.

" It is perhaps not known to many of my countrymen, even in India, that in every town and city in the country the right of sweeping the houses and streets is a monopoly, and is supported entirely by the pride of castes among the scavengers, who are all of the lowest class. The right of sweeping within a certain range is recognized by the caste to belong to a certain member; and if any other member presumes to sweep within that range, he is excommunicated — no other member will smoke out of his pipe or drink out of his jug; and he can get restored to caste only by a feast to the whole body of sweepers. If any housekeeper within a particular circle happens to offend the sweeper of that range, none of his filth will be removed till he pacifies him, because no other sweeper will dare to touch it; and the people of a town are often more tyrannized over by these people than by any other."

A footnote by Major Sleeman's editor, Mr. Vincent Arthur Smith, says that in our day this tyranny of the sweepers' guild is one of the many difficulties which bar the progress of Indian sanitary reform. Think of this:

"The sweepers cannot be readily coerced, because no Hindoo or Mussul-man would do their work to save his life, nor will he pollute himself by beating the refractory scavenger."

They certainly do seem to have the whip-hand; it would be difficult to imagine a more impregnable position. "The vested rights described in the text are so fully recognized in practice that *they are frequently the subject of sale or mortgage.*" Just like a milk-route; or like a London crossing-sweepership. It is said that the London crossing-sweeper's right to his cross-ing is recognized by the rest of the guild; that they protect

him in its possession; that certain choice cross-ings are valuable property, and are saleable at high figures. I have noticed that the man who sweeps in front of the Army and Navy Stores has a wealthy South African aristocratic style about him; and when he is off his guard, he has exactly that look on his face which you always see in the face of a man who has is saving up his daughter to marry her to a duke.

A SOUTH AFRICAN STYLE.

It appears from Sleeman that in India the occupation of elephant-driver is confined to Mohammedans. I wonder why that is. The water-carrier (*bheestie*) is a Mohammedan, but it is said that the reason of that is, that the Hindoo's religion does not allow him to touch the skin of dead kine, and that is what the water-sack is made of; it would defile him. And it doesn't allow him to eat meat; the animal that fur-nished the meat was murdered, and to take any creature's life is a sin. It is a good and gentle religion, but inconvenient.

A great Indian river, at low water, suggests the familiar anatomical picture of a skinned human body, the intricate mesh of interwoven muscles and tendons to stand for water-channels, and the archipelagoes of fat and flesh inclosed by them to stand for the sandbars. Somewhere on this journey we passed such a river, and on a later journey we saw in the Sutlej the duplicate of that river. Curious rivers they are; low shores a dizzy distance apart, with nothing between but an enormous acreage of sand-flats with sluggish little veins of water dribbling around amongst them; Saharas of sand, small-pox-pitted with footprints punctured in belts as straight as the equator clear from the one shore to the other (barring the channel-interruptions) — a dry-shod ferry, you see. Long railway bridges are required for this sort of rivers, and India has them. You approach Allahabad by a very long one. It was now carrying us across the bed of the Jumna, a bed which did not seem to have been slept in for one while or more. It wasn't all river-bed — most of it was overflow ground.

Allahabad means "City of God." I get this from the books. From a printed curiosity — a letter written by one of those brave and confident Hindoo strugglers with the English tongue, called a "babu"— I got a more compressed transla-tion: "Godville." It is perfectly correct, but that is the most that can be said for it.

We arrived in the forenoon, and short-handed; for Satan got left behind somewhere that morning, and did not overtake us until after nightfall. It seemed very peaceful without him. The world seemed asleep and dreaming.

I did not see the native town, I think. I do not remember why; for an incident connects it with the Great Mutiny, and that is enough to make any place interesting. But I saw the English part of the city. It is a town of wide avenues and

30

noble distances, and is comely and alluring, and full of suggestions of comfort and leisure, and of the serenity which a
good conscience buttressed by a sufficient bank account gives.
The bungalows (dwellings) stand well back in the seclusion
and privacy of large enclosed compounds (private grounds, as
we should say) and in the shade and shelter of trees. Even
the photographer and the prosperous merchant ply their industries in the elegant reserve of big compounds, and the citizens
drive in there upon their business occasions. And not in cabs —
no; in the Indian cities cabs are for the drifting stranger; all the
white citizens have private carriages; and each carriage has a
flock of white-turbaned black footmen and drivers all over it.
The vicinity of a lecture-hall looks like a snowstorm, and
makes the lecturer feel like an opera. India has many names,
and they are correctly descriptive. It is the Land of Contradictions, the Land of Subtlety and Superstition, the Land of
Wealth and Poverty, the Land of Splendor and Desolation, the
Land of Plague and Famine, the Land of the Thug and the
Poisoner, and of the Meek and the Patient, the Land of the
Suttee, the Land of the Unreinstatable Widow, the Land where
All Life is Holy, the Land of Cremation, the Land where the
Vulture is a Grave and a Monument, the Land of the Mul-
titudinous Gods; and if signs go for anything, it is the Land
of the Private Carriage.

In Bombay the forewoman of a millinery shop came to the
hotel in her private carriage to take the measure for a gown —
not for me, but for another. She had come out to India to
make a temporary stay, but was extending it indefinitely; indeed, she was purposing to end her days there. In London, she
said, her work had been hard, her hours long; for economy's
sake she had had to live in shabby rooms and far away from
the shop, watch the pennies, deny herself many of the common
comforts of life, restrict herself in effect to its bare necessities,

eschew cabs, travel third-class by underground train to and from her work, swallowing coal-smoke and cinders all the way, and sometimes troubled with the society of men and women who were less desirable than the smoke and the cinders. But in Bombay, on almost any kind of wages, she could live in comfort, and keep her carriage, and have six servants in place of the woman-of-all-work she had had in her English home. Later, in Calcutta, I found that the Standard Oil clerks had small one-horse vehicles, and did no walking; and I was told that the clerks of the other large concerns there had the like equipment. But to return to Allahabad.

I was up at dawn, the next morning. In India the tourist's servant does not sleep in a room in the hotel, but rolls himself up head and ears in his blanket and stretches himself on the veranda, across the front of his master's door, and spends the night there. I don't believe anybody's servant occupies a room. Apparently, the bungalow servants sleep on the veranda; it is roomy, and goes all around the house. I speak of men-servants; I saw none of the other sex. I think there are none, except child-nurses. I was up at dawn, and walked around the veranda, past the rows of sleepers. In front of one door a Hindoo servant was squatting, waiting for his master to call him. He had polished the yellow shoes and placed them by the door, and now he had nothing to do but wait. It was freezing cold, but there he was, as motionless as a sculptured image, and as patient. It troubled me. I wanted to say to him, Don't crouch there like that and freeze; nobody requires it of you; stir around and get warm." But I hadn't the words. I thought of saying *jeldy jow*, but I couldn't remember what it meant, so I didn't say it. I knew another phrase, but it wouldn't come to my mind. I moved on, purposing to dismiss him from my thoughts, but his bare legs and bare feet kept him there. They kept drawing me back from the sunny

side to a point whence I could see him. At the end of an
hour he had not changed his attitude in the least degree. It
was a curious and impressive exhibition of meekness and
patience, or fortitude or indifference, I did not know which.
But it worried me, and it was spoiling my morning.   In fact,
it spoiled two hours of it quite thoroughly.   I quitted this
vicinity, then, and left him to punish himself as much as he
might want to.   But up to that time the man had not changed
his attitude a hair.   He will always remain with me, I suppose;

"IT WORRIED ME"

his figure never grows vague in my memory.   Whenever I
read of Indian resignation, Indian patience under wrongs,
hardships, and misfortunes, he comes before me.   He becomes
a personification, and stands for India in trouble.   And for
untold ages India in trouble has been pursued with the very
remark which I was going to utter but didn't, because its
meaning had slipped me: *Jeldy jow!* ("Come, shove along!")
Why, it was the very thing.

In the early brightness we made a long drive out to the

Fort. Part of the way was beautiful. It led under stately
trees and through groups of native houses and by the usual
village well, where the picturesque gangs are always flocking
to and fro and laughing and chattering; and this time brawny
men were deluging their bronze bodies with the limpid water,
and making a refreshing and enticing show of it; enticing, for
the sun was already transacting business, firing India up for
the day. There was plenty of this early bathing going on, for
it was getting toward breakfast time, and with an unpurified
body the Hindoo must not eat.

Then we struck into the hot plain, and found the roads
crowded with pilgrims of both sexes, for one of the great
religious fairs of India was being held, just beyond the Fort,
at the junction of the sacred rivers, the Ganges and the Jumna.
Three sacred rivers, I should have said, for there is a subterranean one. Nobody has seen it, but that doesn't signify.
The fact that it is there is enough. These pilgrims had come
from all over India; some of them had been months on the
way, plodding patiently along in the heat and dust, worn,
poor, hungry, but supported and sustained by an unwavering
faith and belief; they were supremely happy and content,
now; their full and sufficient reward was at hand; they were
going to be cleansed from every vestige of sin and corruption
by these holy waters which make utterly pure whatsoever thing
they touch, even the dead and rotten. It is wonderful, the
power of a faith like that, that can make multitudes upon
multitudes of the old and weak and the young and frail enter
without hesitation or complaint upon such incredible journeys
and endure the resultant miseries without repining. It is done
in love, or it is done in fear; I do not know which it is. No
matter what the impulse is, the act born of it is beyond imagination marvelous to our kind of people, the cold whites.
There are choice great natures among us that could exhibit

the equivalent of this prodigious self-sacrifice, but the rest of us know that we should not be equal to anything approaching it. Still, we all talk self-sacrifice, and this makes me hope that we are large enough to honor it in the Hindoo.

Two millions of natives arrive at this fair every year. How many start, and die on the road, from age and fatigue and disease and scanty nourishment, and how many die on the return, from the same causes, no one knows; but the tale is great, one may say enormous. Every twelfth year is held to be a year of peculiar grace; a greatly augmented volume of pilgrims results then. The twelfth year has held this distinction since the remotest times, it is said. It is said also that there is to be but one more twelfth year — for the Ganges. After that, that holiest of all sacred rivers will cease to be holy, and will be abandoned by the pilgrim for many centuries; how many, the wise men have not stated. At the end of that interval it will become holy again. Meantime, the data will be arranged by those people who have charge of all such matters, the great chief Brahmins. It will be like shutting down a mint. At a first glance it looks most unbrahminically uncommercial, but I am not disturbed, being soothed and tranquilized by their reputation. "Brer fox he lay low," as Uncle Remus says; and at the judicious time he will spring something on the Indian public which will show that he was not financially asleep when he took the Ganges out of the market.

Great numbers of the natives along the roads were bringing away holy water from the rivers. They would carry it far and wide in India and sell it. Tavernier, the French traveler (17th century), notes that Ganges water is often given at weddings, "each guest receiving a cup or two, according to the liberality of the host; sometimes 2,000 or 3,000 rupees' worth of it is consumed at a wedding."

The Fort is a huge old structure, and has had a large

RELIGIOUS FAIR AT ALLAHABAD.

experience in religions.   In its great court stands a monolith which was placed there more than 2,000 years ago to preach Budhism by its pious inscription; the Fort was built three centuries ago by a Mohammedan Emperor — a resanctification of the place in the interest of *that* religion.   There is a Hindoo temple, too, with subterranean ramifications stocked with shrines and idols; and now the Fort belongs to the English, it contains a Christian Church.   Insured in all the companies.

From the lofty ramparts one has a fine view of the sacred rivers.   They join at that point — the pale blue Jumna, apparently clean and clear, and the muddy Ganges, dull yellow and not clean.   On a long curved spit between the rivers, towns of tents were visible, with a multitude of fluttering pennons, and a mighty swarm of pilgrims.   It was a troublesome place to get down to, and not a quiet place when you arrived; but it was interesting.   There was a world of activity and turmoil and noise, partly religious, partly commercial;   for the Mohammedans were there to curse and sell, and the Hindoos to buy and pray.   It is a fair as well as a religious festival.   Crowds were bathing, praying, and drinking the purifying waters, and many sick pilgrims had come long journeys in palanquins to be healed of their maladies by a bath; or if that might not be, then to die on the blessed banks and so make sure of heaven. There were fakeers in plenty, with their bodies dusted over with ashes and their long hair caked together with cow-dung; for the cow is holy and so is the rest of it;   so holy that the good Hindoo peasant frescoes the walls of his hut with this refuse, and also constructs ornamental figures out of it for the gracing of his dirt floor.   There were seated families, fearfully and wonderfully painted, who by attitude and grouping represented the families of certain great gods.   There was a holy man who sat naked by the day and by the week on a cluster of iron spikes, and did not seem to mind it; and another holy man, who stood all day holding his withered arms motionless aloft, and was said to have been doing it for years. All of these performers have a cloth on the ground beside them for the reception of contributions, and even the poorest of the people give a trifle and hope that the sacrifice will be blessed to him. At last came a procession of naked holy people marching by and chanting, and I wrenched myself away.

CHAPTER L.

THE journey to Benares was all in daylight, and occupied but a few hours. It was admirably dusty. The dust settled upon you in a thick ashy layer and turned you into a fakeer, with nothing lacking to the rôle but the cow manure and the sense of holiness. There was a change of cars about mid-afternoon at Moghul-serai — if that was the name — and a wait of two hours there for the Benares train. We could have found a carriage and driven to the sacred city, but we should have lost the wait. In other countries a long wait at a station is a dull thing and tedious, but one has no right to have that feeling in India. You have the monster crowd of bejeweled natives, the stir, the bustle, the confusion, the shifting splendors of the costumes — dear me, the delight of it, the charm of it are beyond speech. The two-hour wait was over too soon. Among other satisfying things to look at was a minor native prince from the backwoods somewhere, with his guard of honor, a ragged but wonderfully gaudy gang of fifty dark barbarians armed with rusty flint-lock muskets. The general show came so near to exhausting variety that one would have said that no addition to it could be conspicuous, but when this Falstaff and his motleys marched through it one saw that that seeming impossibility had happened.

We got away by and by, and soon reached the outer edge of Benares; then there was another wait; but, as usual, with something to look at. This was a cluster of little canvas-boxes — palanquins. A canvas-box is not much of a sight — when empty; but when there is a lady in it, it is an object of interest. These boxes were grouped apart, in the full blaze of the terrible sun during the three-quarters of an hour that we tarried there. They contained zenana ladies. The had to sit up; there was not room enough to stretch out. They probably did not mind it. They are used to the close captivity of their dwellings all their lives; when they go a journey they are carried to the train in these boxes; in the train they have to be secluded from inspection. Many people pity them, and I always did it myself and never charged anything; but it is doubtful if this compassion is valued. While we were in India some good-hearted Europeans in one of the cities proposed to restrict a large park to the use of zenana ladies, so that they could go there and in assured privacy go about unveiled and enjoy the sunshine and air as they had never enjoyed them before. The good intentions back of the proposition were recognized, and sincere thanks returned for it, but the proposition itself met with a prompt declination at the hands of those who were authorized to speak for the zenana ladies. Apparently, the idea was shocking to the ladies — indeed, it was quite manifestly shocking. Was that proposition the equivalent of inviting European ladies to assemble scantily and scandalously clothed in the seclusion of a private park? It seemed to be about that.

Without doubt modesty is nothing less than a holy feeling; and without doubt the person whose rule of modesty has been trangressed feels the same sort of wound that he would feel if something made holy to him by his religion had suffered a desecration. I say "rule of modesty" because there are about a million rules in the world, and this makes a million standards to be looked out for. Major Sleeman mentions the case of some high-caste veiled ladies who were profoundly scandalized when some English young ladies passed by with faces bare to the world; so scandalized that they spoke out with strong indignation and wondered that people could be so shameless as to expose their persons like that. And yet "the legs of the objectors were naked to mid-thigh." Both parties were clean-minded and irreproachably modest, while abiding by their separate rules, but they couldn't have traded rules for a change without suffering considerable discomfort. All human rules are more or less idiotic, I suppose. It is best so, no doubt. The way it is now, the asylums can hold the sane people, but if we tried to shut up the insane we should run out of building materials.

You have a long drive through the outskirts of Benares before you get to the hotel. And all the aspects are melancholy. It is a vision of dusty sterility, decaying temples, crumbling tombs, broken mud walls, shabby huts. The whole region seems to ache with age and penury. It must take ten thousand years of want to produce such an aspect. We were still outside of the great native city when we reached the hotel. It was a quiet and homelike house, inviting, and manifestly comfortable. But we liked its annex better, and went thither. It was a mile away, perhaps, and stood in the midst of a large compound, and was built bungalow fashion, everything on the ground floor, and a veranda all around. They have doors in India, but I don't know why. They don't fasten, and they stand open, as a rule, with a curtain hanging in the doorspace to keep out the glare of the sun. Still, there is plenty of privacy, for no white person will come in without notice, of course. The native men servants will, but they don't seem to count. They glide in, barefoot and noiseless, and are in the midst before one knows it. At first this is a shock, and sometimes it is an embarrassment; but one has to get used to it, and does.

There was one tree in the compound, and a monkey lived
in it. At first I was strongly interested in the tree, for I was
told that it was the renowned *peepul* — the tree in whose
shadow you cannot tell a lie. This one failed to stand the test,
and I went away from it disappointed. There was a softly
creaking well close by, and a couple of oxen drew water from
it by the hour, superintended by two natives dressed in the
usual "turban and pocket-handkerchief." The tree and the
well were the only scenery, and so the compound was a soothing and lonesome and satisfying place; and very restful after
so many activities. There was nobody in our bungalow but
ourselves; the other guests were in the next one, where the
table d'hote was furnished. A body could not be more pleasantly situated. Each room had the customary bath attached —
a room ten or twelve feet square, with a roomy stone-paved
pit in it and abundance of water. One could not easily
improve upon this arrangement, except by furnishing it with
cold water and excluding the hot, in deference to the fervency
of the climate; but that is forbidden. It would damage the
bather's health. The stranger is warned against taking cold
baths in India, but even the most intelligent strangers are
fools, and they do not obey, and so they presently get laid up.
I was the most intelligent fool that passed through, that year.
But I am still more intelligent now. Now that it is too late.

I wonder if the *dorian*, if that is the name of it, is another
superstition, like the peepul tree. There was a great abundance and variety of tropical fruits, but the dorian was never
in evidence. It was never the season for the dorian. It was
always going to arrive from Burma sometime or other, but it
never did. By all accounts it was a most strange fruit, and
incomparably delicious to the taste, but not to the smell. Its
rind was said to exude a stench of so atrocious a nature that
when a dorian was in the room even the presence of a polecat

was a refreshment. We found many who had eaten the dorian, and they all spoke of it with a sort of rapture. They said that if you could hold your nose until the fruit was in your mouth a sacred joy would suffuse you from head to foot that would make you oblivious to the smell of the rind, but that if your grip slipped and you caught the smell of the rind before the fruit was in your mouth, you would faint. There is a fortune in that rind. Some day somebody will import it into Europe and sell it for cheese.

Benares was not a disappointment. It justified its reputation as a curiosity. It is on high ground, and overhangs a grand curve of the Ganges. It is a vast mass of building, compactly crusting a hill, and is cloven in all directions by an intricate confusion of cracks which stand for streets. Tall, slim minarets and beflagged temple-spires rise out of it and give it picturesqueness,

THEY SPOKE OF IT WITH RAPTURE.

viewed from the river. The city is as busy as an ant-hill, and the hurly-burly of human life swarming along the web of narrow streets reminds one of the ants. The sacred cow swarms along, too, and goes whither she pleases, and takes toll of the grain-shops, and is very much in the way, and is a good deal of a nuisance, since she must not be molested.

Benares is older than history, older than tradition, older even than legend, and looks twice as old as all of them put together. From a Hindoo statement quoted in Rev. Mr. Parker's compact and lucid Guide to Benares, I find that the site of the town was the beginning-place of the Creation. It was merely an upright "lingam," at first, no larger than a stove-pipe, and stood in the midst of a shoreless ocean. This was the work of the God Vishnu. Later he spread the lingam out till its surface was ten miles across. Still it was not large enough for the business; therefore he presently built the globe around it. Benares is thus the center of the earth. This is considered an advantage.

It has had a tumultuous history, both materially and spiritually. It started Brahminically, many ages ago; then by and by Buddha came in recent times 2,500 years ago, and after that it was Buddhist during many centuries — twelve, perhaps — but the Brahmins got the upper hand again, then, and have held it ever since. It is unspeakably sacred in Hindoo eyes, and is as unsanitary as it is sacred, and smells like the rind of the dorian. It is the headquarters of the Brahmin faith, and one-eighth of the population are priests of that church. But it is not an overstock, for they have all India as a prey. All India flocks thither on pilgrimage, and pours its savings into the pockets of the priests in a generous stream, which never fails. A priest with a good stand on the shore of the Ganges is much better off than the sweeper of the best crossing in London. A good stand is worth a world of money. The holy proprietor of it sits under his grand spectacular umbrella and blesses people all his life, and collects his commission, and grows fat and rich; and the stand passes from father to son, down and down and down through the ages, and remains a permanent and lucrative estate in the family. As Mr. Parker suggests, it can become a subject of dispute, at one time or another, and then the matter will be settled, not by prayer and fasting and consultations with Vishnu, but by the intervention of a much more puissant power — an English court. In Bombay I was told by an American missionary that in India there are 640 Protestant missionaries at work. At first it seemed an immense force, but of course that was a thoughtless idea. One missionary to 500,000 natives — no, that is not a force; it is the reverse of it; 640 marching against an intrenched camp of 300,000,000 — the odds are too great. A force of 640 in Benares alone would have its hands over-full with 8,000 Brahmin priests for adversary. Missionaries need to be well equipped with hope and confidence, and this equipment they seem to have always had in all parts of the world. Mr. Parker has it. It enables him to get a favorable outlook out of statistics which might add up differently with other mathematicians. For instance:

"During the past few years competent observers declare that the number of pilgrims to Benares has increased."

And then he adds up this fact and gets this conclusion:

"But the revival, if so it may be called, has in it the marks of death. It is a spasmodic struggle before dissolution."

In this world we have seen the Roman Catholic power dying, upon these same terms, for many centuries. Many a time we have gotten all ready for the funeral and found it postponed again, on account of the weather or something. Taught by experience, we ought not to put on our things for this Brahminical one till we see the procession move. Apparently one of the most uncertain things in the world is the funeral of a religion.

I should have been glad to acquire some sort of idea of Hindoo theology, but the difficulties were too great, the matter was too intricate. Even the mere A, B, C of it is baffling.

31

There is a trinity — Brahma, Shiva, and Vishnu — independent powers, apparently, though one cannot feel quite sure of that, because in one of the temples there is an image where an attempt has been made to concentrate the three in one person. The three have other names and plenty of them, and this makes confusion in one's mind. The three have wives and the wives have several names, and this increases the confusion. There are children, the children have many names, and thus the confusion goes on and on. It is not worth while to try to get any grip upon the cloud of minor gods, there are too many of them.

It is even a justifiable economy to leave Brahma, the chiefest god of all, out of your studies, for he seems to cut no great figure in India. The vast bulk of the national worship is lavished upon Shiva and Vishnu and their families. Shiva's symbol — the "lingam" with which Vishnu began the Creation — is worshiped by everybody, apparently. It is the commonest object in Benares. It is on view everywhere, it is garlanded with flowers, offerings are made to it, it suffers no neglect. Commonly it is an upright stone, shaped like a thimble — sometimes like an elongated thimble. This priapus-worship, then, is older than history. Mr. Parker says that the lingams in Benares "*outnumber the inhabitants.*"

In Benares there are many Mohammedan mosques. There are Hindoo temples without number — these quaintly shaped and elaborately sculptured little stone jugs crowd all the lanes. The Ganges itself and every individual drop of water in it are temples. Religion, then, is the *business* of Benares, just as gold-production is the business of Johannesburg. Other industries count for nothing as compared with the vast and all-absorbing rush and drive and boom of the town's specialty. Benares is the sacredest of sacred cities. The moment you step across the sharply-defined line which separates it from the rest of the globe, you stand upon ineffably and unspeakably holy ground. Mr. Parker says : " It is impossible to convey any adequate idea of the intense feelings of veneration and affection with which the pious Hindoo regards 'Holy Kashi' (Benares)." And then he gives you this vivid and moving picture :

" Let a Hindoo regiment be marched through the district, and as soon as they cross the line and enter the limits of the holy place they rend the air with cries of ' Kashi ji ki jai — jai — jai ! (Holy Kashi ! Hail to thee ! Hail ! Hail ! Hail)'. The weary pilgrim scarcely able to stand, with age and weakness, blinded by the dust and heat, and almost dead with fatigue, crawls out of the oven-like railway carriage and as soon as his feet touch the ground he lifts up his withered hands and utters the same pious exclamation. Let a European in some distant city in casual talk in the bazar mention the fact that he has lived at Benares, and at once voices will be raised to call down blessings on his head, for a dweller in Benares is of all men most blessed."

It makes our own religious enthusiasm seem pale and cold. Inasmuch as the life of religion is in the heart, not the head, Mr. Parker's touching picture seems to promise a sort of indefinite postponement of that funeral.

CHAPTER LI.

Let me make the superstitions of a nation and I care not who makes its laws or its songs either. — *Pudd'nhead Wilson's New Calendar.*

YES, the city of Benares is in effect just a big church, a religious hive, whose every cell is a temple, a shrine or a mosque, and whose every conceivable earthly and heavenly good is procurable under one roof, so to speak — a sort of Army and Navy Stores, theologically stocked.

I will make out a little itinerary for the pilgrim; then you will see how handy the system is, how convenient, how comprehensive. If you go to Benares with a serious desire to spiritually benefit yourself, you will find it valuable. I got some of the facts from conversations with the Rev. Mr. Parker and the others from his Guide to Benares; they are therefore trustworthy.

1. *Purification.* At sunrise you must go down to the Ganges and bathe, pray, and drink some of the water. This is for your general purification.

2. *Protection against Hunger.* Next, you must fortify yourself against the sorrowful earthly ill just named. This you will do by worshiping for a moment in the Cow Temple. By the door of it you will find an image of Ganesh, son of Shiva; it has the head of an elephant on a human body; its face and hands are of silver. You will worship it a little, and pass on, into a covered veranda, where you will find devotees reciting from the sacred books, with the help of instructors. In this place are groups of rude and dismal idols. You may contribute something for their support; then pass into the tem-

ple, a grim and stenchy place, for it is populous with sacred cows and with beggars. You will give something to the beggars, and "reverently kiss the tails" of such cows as pass along, for these cows are peculiarly holy, and this act of worship will secure you from hunger for the day.

DO IT REVERENTLY.

3. "*The Poor Man's Friend.*" You will next worship this god. He is at the bottom of a stone cistern in the temple of Dalbhyeswar, under the shade of a noble peepul tree on the bluff overlooking the Ganges, so you must go back to the river. The Poor Man's Friend is the god of *material prosperity* in general, and the god of the *rain* in particular. You will secure material prosperiy, tor both, by worshiping him. He is Shiva, under a new alias, and he abides in the bottom of that cistern, in the form of a stone lingam. You pour Ganges water over him, and in return for this homage you get the promised benefits. If there is any delay about the rain, you must pour water in until the cistern is full; the rain will then be sure to come.

4. *Fever.* At the Kedar Ghat you will find a long flight of stone steps leading down to the river. Half way down is a tank filled with sewage. Drink as much of it as you want. It is for fever.

5. *Smallpox.* Go straight from there to the central Ghat. At its upstream end you will find a small whitewashed building, which is a temple sacred to Sitala, goddess of smallpox. Her under-study is there — a rude human figure behind a brass screen. You will worship this for reasons to be furnished presently.

6. *The Well of Fate.* For certain reasons you will next go and do homage at this well. You will find it in the Dandpan Temple, in the city. The sunlight falls into it from a square hole in the masonry above. You will approach it with awe, for your life is now at stake. You will bend over and look. If the fates are propitious, you will see your face pictured in the water far down in the well. If matters have been otherwise ordered, a sudden cloud will mask the sun and you will see nothing. This means that you have not six months to live. If you are already at the point of death, your circumstances are now serious. There is no time to lose. Let this world go, arrange for the next one. Handily situated, at your very elbow, is opportunity for this. You turn and worship the image of Maha Kal, the Great Fate, and happiness in the life to come is secured. If there is breath in your body yet, you should now make an effort to get a further lease of the present life. You have a chance. There is a chance for everything in this admirably stocked and wonderfully systemized Spiritual and Temporal Army and Navy Store. You must get yourself carried to the

7. *Well of Long Life.* This is within the precincts of the mouldering and venerable Briddhkal Temple, which is one of the oldest in Benares. You pass in by a stone image of the monkey god, Hanuman, and there, among the ruined courtyards, you will find a shallow pool of stagnant sewage. It smells like the best limburger cheese, and is filthy with the washings of rotting lepers, but that is nothing, bathe in it;

THE WELL OF FATE.

bathe in it gratefully and worshipfully, for this is the Fountain of Youth; these are the Waters of Long Life. Your gray hairs will disappear, and with them your wrinkles and your rheumatism, the burdens of care and the weariness of age, and you will come out young, fresh, elastic, and full of eagerness for the new race of life. Now will come flooding upon you the manifold desires that haunt the dear dreams of the morning of life. You will go whither you will find

8. *Fulfillment of Desire.* To wit, to the Kameshwar Temple, sacred to Shiva as the Lord of Desires. Arrange for yours there. And if you like to look at idols among the pack and jam of temples, there you will find enough to stock a museum. You will begin to commit sins now with a fresh, new vivacity; therefore, it will be well to go frequently to a place where you can get

9. *Temporary Cleansing from Sin.* To wit, to the Well of the Earring. You must approach this with the profoundest reverence, for it is unutterably sacred. It is, indeed, the most sacred place in Benares, the very Holy of Holies, in the estimation of the people. It is a railed tank, with stone stairways leading down to the water. The water is not clean. Of course it could not be, for people are always bathing in it. As long as you choose to stand and look, you will see the files of sinners descending and ascending — descending soiled with sin, ascending purged from it. "The liar, the thief, the murderer, and the adulterer may here wash and be clean," says the Rev. Mr. Parker, in his book. Very well. I know Mr. Parker, and I believe it; but if anybody else had said it, I should consider him a person who had better go down in the tank and take another wash. The god Vishnu dug this tank. He had nothing to dig with but his "discus." I do not know what a discus is, but I know it is a poor thing to dig tanks with, because, by the time this one was finished, it was full of

sweat — Vishnu's sweat.   He constructed the site that
Benares stands on, and afterward built the globe around it,
and thought nothing of it, yet sweated like that over a little
thing like this tank.   One of these statements is doubtful.   I
do not know which one it is, but I think it difficult not to
believe that a god who could build a world around Benares
would not be intelligent enough to build it around the tank
too, and not have to dig it.   Youth, long life, temporary purification from sin, salvation through propitiation of the Great
Fate — these are all good.   But you must do something more.
You must

10. *Make Salvation Sure.*   There are several ways.   To
get drowned in the Ganges is one, but that is not pleasant.
To die within the limits of Benares is another; but that is a
risky one, because you might be out of town when your time
came.   The best one of all is the Pilgrimage Around the City.
You must walk; also, you must go barefoot.   The tramp is
forty-four miles, for the road winds out into the country a
piece, and you will be marching five or six days.   But you
will have plenty of company.   You will move with throngs
and hosts of happy pilgrims whose radiant costumes will
make the spectacle beautiful and whose glad songs and holy
pæans of triumph will banish your fatigues and cheer your
spirit; and at intervals there will be temples where you may
sleep and be refreshed with food.   The pilgrimage completed,
you have purchased salvation, and paid for it.   But you may
not get it unless you

11. *Get Your Redemption Recorded.*   You can get this
done at the Sakhi Binayak Temple, and it is best to do it, for
otherwise you might not be able to prove that you had made
the pilgrimage in case the matter should some day come to be
disputed.   That temple is in a lane back of the Cow Temple.
Over the door is a red image of Ganesh of the elephant head,

son and heir of Shiva, and Prince of Wales to the Theological
Monarchy, so to speak. Within is a god whose office it is to
record your pilgrimage and be responsible for you. You will
not see him, but you will see a Brahmin who will attend to the
matter and take the money. If he should forget to collect the
money, you can remind him. *He* knows that your salvation
is now secure, but of course you would like to know it yourself. You have nothing to do but go and pray, and pay at the
   12. *Well of the Knowledge of Salvation.* It is close to
the Golden Temple. There you will see, sculptured out of a
single piece of black marble, a bull which is much larger than
any living bull you have ever seen, and yet is not a good likeness after all. And there also you will see a very uncommon
thing — an image of Shiva. You have seen his lingam fifty
thousand times already, but this is Shiva himself, and said to
be a good likeness. It has three eyes. He is the only god in
the firm that has three. "The well is covered by a fine
canopy of stone supported by forty pillars," and around it you
will find what you have already seen at almost every shrine
you have visited in Benares, a mob of devout and eager pilgrims. The sacred water is being ladled out to them; with it
comes to them the knowledge, clear, thrilling, absolute, that
they are saved; and you can see by their faces that there is
one happiness in this world which is supreme, and to which no
other joy is comparable. You receive your water, you make
your deposit, and now what more would you have? Gold,
diamonds, power, fame? All in a single moment these things
have withered to dirt, dust, ashes. The world has nothing to
give you now. For you it is bankrupt.

   I do not claim that the pilgrims do their acts of worship in
the order and sequence above charted out in this Itinerary of
mine, but I think logic suggests that they ought to do so.
Instead of a helter-skelter worship, we then have a definite

starting-place, and a march which carries the pilgrim steadily forward by reasoned and logical progression to a definite goal. Thus, his Ganges bath in the early morning gives him an appetite; he kisses the cow-tails, and that removes it. It is now business hours, and longings for material prosperity rise in his mind, and he goes and pours water over Shiva's symbol; this insures the prosperity, but also brings on a rain, which gives him a fever. Then he drinks the sewage at the Kedar Ghat to cure the fever; it cures the fever but gives him the smallpox. He wishes to know how it is going to turn out; he goes to the Dandpan Temple and looks down the well. A clouded sun shows him that death is near. Logically his best course for the present, since he cannot tell at what moment he may die, is to secure a happy hereafter; this he does, through the agency of the Great Fate. He is safe, now, for heaven; his next move will naturally be to keep out of it as long as he can. Therefore he goes to the Briddhkal Temple and secures Youth and long life by bathing in a puddle of leper-pus which would kill a microbe. Logically, Youth has re-equipped him for sin and with the disposition to commit it; he will naturally go to the fane which is consecrated to the Fulfillment of Desires, and make arrangements. Logically, he will now go to the Well of the Earring from time to time to unload and freshen up for further banned enjoyments. But first and last and all the time he is human, and therefore in his reflective intervals he will always be speculating in "futures." He will make the Great Pilgrimage around the city and so make his salvation absolutely sure; he will also have record made of it, so that it may remain absolutely sure and not be forgotten or repudiated in the confusion of the Final Settlement. Logically, also, he will wish to have satisfying and tranquilizing *personal* knowledge that that salvation is secure; therefore he goes to the Well of the Knowledge of Salvation, adds that completing detail, and then goes about his affairs serene and content; serene and content, for he is now royally endowed with an advantage which no religion in this world could give him but his own; for henceforth he may commit as many million sins as he wants to and nothing can come of it.

Thus the system, properly and logically ordered, is neat, compact, clearly defined, and covers the whole ground. I desire to recommend it to such as find the other systems too difficult, exacting, and irksome for the uses of this fretful brief life of ours.

However, let me not deceive any one. My Itinerary lacks a detail. I must put it in. The truth is, that after the pilgrim has faithfully followed the requirements of the Itinerary through to the end and has secured his salvation and also the personal knowledge of that fact, there is still an accident possible to him which can annul the whole thing. If he should ever cross to the other side of the Ganges and get caught out and die there he would at once come to life again in the form

CONSIDERING THE MATTER.

of an ass. Think of that, after all this trouble and expense. You see how capricious and uncertain salvation is there. The Hindoo has a childish and unreasoning aversion to being turned into an ass. It is hard to tell why. One could properly expect an ass to have an aversion to being turned into a Hindoo. One could understand that he could lose dignity by it; also self-respect, and nine-tenths of his intelligence. But the Hindoo changed into an ass wouldn't lose anything, unless you count his religion. And he would gain much — release from his slavery to two million gods and twenty million priests, fakeers, holy mendicants, and other sacred bacilli; he would escape the Hindoo hell; he would also escape the Hindoo heaven. These are advantages which the Hindoo ought to consider; then he would go over and die on the other side.

Benares is a religious Vesuvius. In its bowels the theological forces have been heaving and tossing, rumbling, thundering and quaking, boiling, and weltering and flaming and smoking for ages. But a little group of missionaries have taken post at its base, and they have hopes. There are the Baptist Missionary Society, the Church Missionary Society, the London Missionary Society, the Wesleyan Missionary Society, and the Zenana Bible and Medical Mission. They have schools, and the principal work seems to be among the children. And no doubt that part of the work prospers best, for grown people everywhere are always likely to cling to the religion they were brought up in.

BENARES.

CHAPTER LII.

Wrinkles should merely indicate where smiles have been.
— *Pudd'nhead Wilson's New Calendar.*

IN one of those Benares temples we saw a devotee working for salvation in a curious way. He had a huge wad of clay beside him and was making it up into little wee gods no bigger than carpet tacks. He stuck a grain of rice into each — to represent the lingam, I think. He turned them out nimbly, for he had had long practice and had acquired great facility. Every day he made 2,000 gods, then threw them into the holy Ganges. This act of homage brought him the profound homage of the pious — also their coppers. He had a sure living here, and was earning a high place in the hereafter.

The Ganges front is the supreme show-place of Benares. Its tall bluffs are solidly caked from water to summit, along a stretch of three miles, with a splendid jumble of massive and picturesque masonry, a bewildering and beautiful confusion of stone platforms, temples, stair-flights, rich and stately palaces — nowhere a break, nowhere a glimpse of the bluff itself; all the long face of it is compactly walled from sight by this crammed perspective of platforms, soaring stairways, sculptured temples, majestic palaces, softening away into the distances; and there is movement, motion, human life everywhere, and brilliantly costumed — streaming in rainbows up and down the lofty stairways, and massed in metaphorical flower-gardens on the miles of great platforms at the river's edge.

All this masonry, all this architecture represents piety. The palaces were built by native princes whose homes, as a rule, are far from Benares, but who go there from time to time to refresh their souls with the sight and touch of the Ganges, the river of their idolatry. The stairways are records of acts of piety; the crowd of costly little temples are tokens of money spent by rich men for present credit and hope of future reward. Apparently, the rich Christian who spends large sums upon his religion is conspicuous with us, by his rarity, but the rich Hindoo who doesn't spend large sums upon his religion is seemingly non-existent. With us the poor spend money on their religion, but they keep back some to live on. Apparently, in India, the poor bankrupt themselves daily for their religion. The rich Hindoo can afford his pious outlays; he gets much glory for his spendings, yet keeps back a sufficiency of his income for temporal purposes; but the poor Hindoo is entitled to compassion, for his spendings keep him poor, yet get him no glory.

We made the usual trip up and down the river, seated in chairs under an awning on the deck of the usual commodious hand-propelled ark; made it two or three times, and could have made it with increasing interest and enjoyment many times more; for, of course, the palaces and temples would grow more and more beautiful every time one saw them, for that happens with all such things; also, I think one would not get tired of the bathers, nor their costumes, nor of their ingenuities in getting out of them and into them again without exposing too much bronze, nor of their devotional gesticulations and absorbed bead-tellings.

But I should get tired of seeing them wash their mouths with that dreadful water and drink it. In fact, I did get tired of it, and very early, too. At one place where we halted for a while, the foul gush from a sewer was making the water turbid and murky all around, and there was a random corpse slopping around in it that had floated down from up country. Ten steps

32

THE PURIFYING WATERS OF THE GANGES.

below that place stood a crowd of men, women, and comely young maidens waist deep in the water — and they were scooping it up in their hands and drinking it. Faith can certainly do wonders, and this is an instance of it. Those people were not drinking that fearful stuff to assuage thirst, but in order to purify their souls and the interior of their bodies. According to their creed, the Ganges water makes everything pure that it touches — instantly and utterly pure. The sewer water was not an offence to them, the corpse did not revolt them; the sacred water had touched both, and both were now snow-pure, and could defile no one. The memory of that sight will always stay by me; but not by request.

A word further concerning the nasty but all-purifying Ganges water. When we went to Agra, by and by, we happened there just in time to be in at the birth of a marvel — a memorable scientific discovery — the discovery that in certain ways the foul and derided Ganges water *is* the most puissant purifier in the world! This curious fact, as I have said, had just been added to the treasury of modern science. It had long been noted as a strange thing that while Benares is often afflicted with the cholera she does not spread it beyond her borders. This could not be accounted for. Mr. Henkin, the scientist in the employ of the government of Agra, concluded to examine the water. He went to Benares and made his tests. He got water at the mouths of the sewers where they empty into the river at the bathing ghats; a cubic centimetre of it contained millions of germs; at the end of six hours they were *all dead*. He caught a floating corpse, towed it to the shore, and from beside it he dipped up water that was swarming with cholera germs; at the end of six hours they were *all dead*. He added swarm after swarm of cholera germs to this water; within the six hours *they always died*, to the last sample. Repeatedly, he took pure well water which was barren of animal life, and put into it a few cholera germs; they always began to propagate at once, and always within six hours they swarmed — and were *numberable by millions upon millions.*

For ages and ages the Hindoos have had absolute faith that the water of the Ganges was absolutely pure, could not be defiled by any contact whatsoever, and infallibly made pure and clean whatsoever thing touched it. They still believe it, and that is why they bathe in it and drink it, caring nothing for its *seeming* filthiness and the floating corpses. The Hindoos have been laughed at, these many generations, but the laughter will need to modify itself a little from now on. How did they find out the water's secret in those ancient ages? Had they germ-scientists then? We do not know. We only know that they had a civilization long before we emerged from savagery. But to return to where I was before; I was about to speak of the burning-ghat.

They do not burn fakeers — those revered mendicants. They are so holy that they can get to their place without that sacrament, provided they be consigned to the consecrating river. We saw one carried to mid-stream and thrown overboard. He was sandwiched between two great slabs of stone.

We lay off the cremation-ghat half an hour and saw nine corpses burned. I should not wish to see any more of it, unless I might select the parties. The mourners follow the bier through the town and down to the ghat; then the bier-bearers deliver the body to some low-caste natives — Doms — and the mourners turn about and go back home. I heard no crying and saw no tears, there was no ceremony of parting. Apparently, these expressions of grief and affection are reserved for the privacy of the home. The dead women came draped in red, the men in white. They are laid in the water at the river's edge while the pyre is being prepared.

The first subject was a man. When the Doms unswathed

him to wash him, he proved to be a sturdily built, well-nourished and handsome old gentleman, with not a sign about him to suggest that he had ever been ill. Dry wood was brought and built up into a loose pile; the corpse was laid upon it and covered over with fuel. Then a naked holy man who was sitting on high ground a little distance away began to talk and shout with great energy, and he kept up this noise right along. It may have been the funeral sermon, and probably was. I forgot to say that one of the mourners remained behind when the others went away. This was the dead man's son, a boy of ten or twelve, brown and handsome, grave and self-possessed, and clothed in flowing white.

EXTRA EXPENSE.

He was there to burn his father. He was given a torch, and while he slowly walked seven times around the pyre the naked black man on the high ground poured out his sermon more clamorously than ever. The seventh circuit completed, the boy applied the torch at his father's head, then at his feet; the flames sprang briskly up with a sharp crackling noise, and the lad went away. Hindoos do not want daughters, because their weddings make such a ruinous expense; but they want sons, so that at death they may have honorable exit from the world; and there is no honor equal to the honor of having one's pyre lighted by one's son. The father who dies sonless is in a grievous situation indeed, and is pitied. Life being uncertain, the Hindoo marries while he is still a boy, in the hope that he will have a son ready when the day of his need shall come. But if he have no son, he will adopt one. This answers every purpose.

Meantime the corpse is burning, also several others. It is a dismal business. The stokers did not sit down in idleness, but moved briskly about, punching up the fires with long poles, and now and then adding fuel. Sometimes they hoisted the half of a skeleton into the air, then slammed it down and beat it with the pole, breaking it up so that it would burn better. They hoisted skulls up in the same way and banged and battered them. The sight was hard to bear; it would have been harder if the mourners had stayed to witness it. I had but a moderate desire to see a cremation, so it was soon satisfied. For sanitary reasons it would be well if cremation were universal; but this form is revolting, and not to be recommended.

The fire used is sacred, of course — for there is money in it. Ordinary fire is forbidden; there is no money in it. I was told that this sacred fire is all furnished by one person, and that he has a monopoly of it and charges a good price for it. Sometimes a rich mourner pays a thousand rupees for it. To get to paradise from India is an expensive thing. Every detail connected with the matter costs something, and helps to fatten a priest. I suppose it is quite safe to conclude that that fire-bug is in holy orders.

Close to the cremation-ground stand a few time-worn stones which are remembrances of the suttee. Each has a rough carving upon it, representing a man and a woman standing or walking hand in hand, and marks the spot where a widow went to her death by fire in the days when the suttee flourished. Mr. Parker said that widows would burn themselves now if the government would allow it. The family that can point

to one of these little memorials and say:
"She who burned herself there was an
ancestress of ours," is envied.

It is a curious people. With them, all
life seems to be sacred except human life.
Even the life of vermin is sacred, and must
not be taken. The good Jain wipes off a
seat before using it, lest he cause the death
of some valueless insect by sitting down on
it. It grieves him to have to drink water,
because the provisions in his stomach may
not agree with the microbes. Yet
India invented Thuggery and the
Suttee. India is a hard country to
understand.

We went to the temple of the
Thug goddess, Bhowanee, or Kali, or
Durga. She has these names and
others. She is the only god to whom
living sacrifices are made. Goats are
sacrificed to her. Monkeys would
be cheaper. There are plenty of them
about the place. Being sacred, they
make themselves very free, and scramble around wherever they please. The
temple and its porch are beautifully
carved, but this is not the case with
the idol. Bho-
wanee is not
pleasant to look
at. She has a
silver face, and

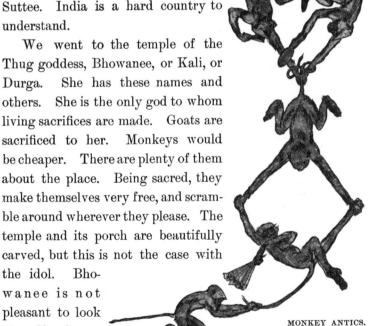

MONKEY ANTICS.

a projecting swollen
tongue painted a deep red. She wears a necklace of skulls.

In fact, none of the idols in Benares are handsome or attractive. And what a swarm of them there is! The town is a vast museum of idols — and all of them crude, misshapen, and ugly. They flock through one's dreams at night, a wild mob of nightmares. When you get tired of them in the temples and take a trip on the river, you find idol giants, flashily painted, stretched out side by side on the shore. And apparently wherever there is room for one more lingam, a lingam is there. If Vishnu had foreseen what his town was going to be, he would have called it Idolville or Lingamburg.

The most conspicuous feature of Benares is the pair of slender white minarets which tower like masts from the great Mosque of Aurangzeb. They seem to be always in sight, from everywhere, those airy, graceful, inspiring things. But masts is not the right word, for masts have a perceptible taper, while these minarets have not. They are 142 feet high, and only $8\frac{1}{2}$ feet in diameter at the base, and $7\frac{1}{2}$ at the summit — scarcely any taper at all. These are the proportions of a candle; and fair and fairy-like candles these are. Will be, anyway, some day, when the Christians inherit them and top them with the electric light. There is a great view from up there — a wonderful view. A large gray monkey was part of it, and damaged it. A monkey has no judgment. This one was skipping about the upper great heights of the mosque — skipping across empty yawning intervals which were almost too wide for him, and which he only just barely cleared, each time, by the skin of his teeth. He got me so nervous that I couldn't look at the view. I couldn't look at anything but him. Every time he went sailing over one of those abysses my breath stood still, and when he grabbed for the perch he was going for, I grabbed too, in sympathy. And he was perfectly indifferent, perfectly unconcerned, and I did all the panting myself. He came within an ace of losing his life a dozen times, and I was so troubled about him that I would have shot him if I had had anything to do it with. But I strongly recommend the view. There is more monkey than view, and there is always going to be more monkey while that idiot survives, but what view you get is superb. All Benares, the river, and the region round about are spread before you. Take a gun, and look at the view.

The next thing I saw was more reposeful. It was a new kind of art. It was a picture painted on water. It was done by a native. He sprinkled fine dust of various colors on the still surface of a basin of water, and out of these sprinklings a dainty and pretty picture gradually grew, a picture which a breath could destroy. Somehow it was impressive, after so much browsing among massive and battered and decaying fanes that rest upon ruins, and those ruins upon still other ruins, and those upon still others again. It was a sermon, an allegory, a symbol of Instability. Those creations in stone were only a kind of water pictures, after all.

A prominent episode in the Indian career of Warren Hastings had Benares for its theater. Wherever that extraordinary man set his foot, he left his mark. He came to Benares in 1781 to collect a fine of £500,000 which he had levied upon its Rajah, Cheit Singh, on behalf of the East India Company. Hastings was a long way from home and help. There were, probably, not a dozen Englishmen within reach; the Rajah was in his fort with his myriads around him. But no matter. From his little camp in a neighboring garden, Hastings sent a party to arrest the sovereign. He sent on this daring mission a couple of hundred native soldiers — sepoys — under command of three young English lieutenants. The Rajah submitted without a word. The incident lights up the Indian situation electrically, and gives one a vivid sense of the strides which the English had made and the mastership they had acquired in the land since the date of Clive's great victory. In a quarter of a century, from being nobodies, and feared by none, they were become confessed lords and masters, feared by all, sovereigns included, and served by all, sovereigns included. It makes the fairy tales sound true. The English had not been afraid to enlist native soldiers to fight against their own people and keep them obedient. And now Hastings was not afraid to come away out to this remote place with a handful of such soldiers and send them to arrest a native sovereign.

The lieutenants imprisoned the Rajah in his own fort. It was beautiful, the pluckiness of it, the impudence of it. The arrest enraged the Rajah's people, and all Benares came storming about the place and threatening vengeance. And yet, but for an accident, nothing important would have resulted, perhaps. The mob found out a most strange thing, an almost incredible thing — that this handful of soldiers had come on this hardy errand with empty guns and no ammunition. This has been attributed to thoughtlessness, but it could hardly have been that, for in such large emergencies as this, intelligent people *do* think. It must have been indifference, an over-confidence born of the proved submissiveness of the native character, when confronted by even one or two stern Britons in their war paint. But, however that may be, it was a fatal discovery that the mob had made. They were full of courage, now, and they broke into the fort and massacred the helpless soldiers and their officers. Hastings escaped from Benares by night and got safely away, leaving the principality in a state of wild insurrection ; but he was back again within the month, and quieted it down in his prompt and virile way, and took the Rajah's throne away from him and gave it to another man. He was a capable kind of person was Warren Hastings. This was the only time he was ever out of ammunition. Some of his acts have left stains upon his name which can never be washed away, but he saved to England the Indian Empire, and that was the best service that was ever done to the Indians themselves, those wretched heirs of a hundred centuries of pitiless oppression and abuse.

CHAPTER LIII.

True irreverence is disrespect for another man's god.
> — *Pudd'nhead Wilson's New Calendar.*

IT was in Benares that I saw another living god. That makes two. I believe I have seen most of the greater and lesser wonders of the world, but I do not remember that any of them interested me so overwhelmingly as did that pair of gods.

When I try to account for this effect I find no difficulty about it. I find that, as a rule, when a thing is a wonder to us it is not because of what *we* see in it, but because of what *others* have seen in it. We get almost all our wonders at second hand. We are eager to see any celebrated thing — and we never fail of our reward; just the deep privilege of gazing upon an object which has stirred the enthusiasm or evoked the reverence or affection or admiration of multitudes of our race is a thing which we value; we are profoundly glad that we have seen it, we are permanently enriched from having seen it, we would not part with the memory of that experience for a great price. And yet that very spectacle may be the *Taj.* You cannot keep your enthusiasms down, you cannot keep your emotions within bounds when that soaring bubble of marble breaks upon your view. But these are not *your* enthusiasms and emotions — they are the accumulated emotions and enthusiasms of a thousand fervid writers, who have been slowly and steadily storing them up in your heart day by day and year by year all your life; and now they burst out in a flood and overwhelm you; and you could not be a whit happier if

they were your very own. By and by you sober down, and then
you perceive that you have been drunk on the smell of somebody else's cork. For ever and ever the memory of my distant first glimpse of the Taj will compensate me for creeping
around the globe to have that great privilege.

But the Taj — with all your inflation of delusive emotions,
acquired at second-hand from people to whom in the majority
of cases they were also delusions acquired at second-hand — a
thing which you fortunately did not think of or it might have
made you doubtful of what you imagined were your own —
what is the Taj as a marvel, a spectacle and an uplifting and
overpowering wonder, compared with a living, breathing,
speaking personage whom several millions of human beings
devoutly and sincerely and unquestioningly believe to be a
God, and humbly and gratefully worship as a God?

He was sixty years old when I saw him. He is called Sri
108 Swami Bhaskarananda Saraswati. That is one form of it.
I think that that is what you would call him in speaking to
him — because it is short. But you would use more of his name
in addressing a letter to him ; courtesy would require this. Even
then you would not have to use all of it, but only this much :

*Sri 108 Matparamahansapaɩ ivᴠajakacharyaswamibhaskar-
anandasaraswati.*

You do not put " Esq." after it, for that is not necessary.
The word which opens the volley is itself a title of honor —
" Sri." The " 108 " stands for the rest of his names, I believe.
Vishnu has 108 names which he does not use in business, and
no doubt it is a custom of gods and a privilege sacred to their
order to keep 108 extra ones in stock. Just the restricted
name set down above is a handsome property, without the 108.
By my count it has 58 letters in it. This removes the long
German words from competition ; they are permanently out of
the race.

Sri 108 S. B. Saraswati has attained to what among the Hindoos is called the "state of perfection." It is a state which other Hindoos reach by being born again and again, and over and over again into this world, through one re-incarnation after another — a tiresome long job covering centuries and decades of centuries, and one that is full of risks, too, like the accident of dying on the wrong side of the Ganges some time or other and waking up in the form of an ass, with a fresh start necessary and the numerous trips to be made all over again. But in reaching perfection, Sri 108 S. B. S. has escaped all that. He is no longer a part or a feature of this world; his substance has changed, all earthiness has departed out of it; he is utterly holy, utterly pure; nothing can desecrate this holiness or stain this purity; he is no longer of the earth, its concerns are matters foreign to him, its pains and griefs and troubles cannot reach him. When he dies, Nirvana is his; he will be absciced into the substance of the Supreme Deity and be at peace forever.

The Hindoo Scriptures point out how this state is to be reached, but it is only once in a thousand years, perhaps, that candidate accomplishes it. This one has traversed the course required, stage by stage, from the beginning to the end, and now has nothing left to do but wait for the call which shall release him from a world in which he has now no part nor lot. First, he passed through the student stage, and became learned in the holy books. Next he became citizen, householder, husband, and father. That was the required second stage. Then — like John Bunyan's Christian — he bade perpetual good-bye to his family, as required, and went wandering away. He went far into the desert and served a term as hermit. Next, he became a beggar, "in accordance with the rites laid down in the Scriptures," and wandered about India eating the bread of mendicancy. A quarter of a century ago he reached the stage of purity. This needs no garment; its symbol is nudity; he discarded the waist-cloth which he had previously worn. He could resume it now if he chose, for neither that nor any other contact can defile him; but he does not choose.

There are several other stages, I believe, but I do not remember what they are. But he has been through them. Throughout the long course he was perfecting himself in holy learning, and writing commentaries upon the sacred books. He was also meditating upon Brahma, and he does that now.

White marble relief-portraits of him are sold all about India. He lives in a good house in a noble great garden in Benares, all meet and proper to his stupendous rank. Necessarily he does not go abroad in the streets. Deities would never be able to move about handily in any country. If one whom we recognized and adored as a god should go abroad in our streets, and the day it was to happen were known, all traffic would be blocked and business would come to a standstill.

This god is comfortably housed, and yet modestly, all things considered, for if he wanted to live in a palace he would only need to speak and his worshipers would gladly build it. Sometimes he sees devotees for a moment, and comforts them and blesses them, and they kiss his feet and go away happy. Rank is nothing to him, he being a god. To him all men are alike. He sees whom he pleases and denies himself to whom he pleases. Sometimes he sees a prince and denies himself to a pauper; at other times he receives the pauper and turns the prince away. However, he does not receive many of either class. He has to husband his time for his meditations. I think he would receive Rev. Mr. Parker at any time. I think he is sorry for Mr. Parker, and I think Mr. Parker is sorry for him; and no doubt this compassion is good for both of them.

When we arrived we had to stand around in the garden a

little while and wait, and the outlook was not good, for he had
been turning away Maharajas that day and receiving only the
riff-raff, and we belonged in between, somewhere. But presently, a servant came out saying it was all right, he was coming.

And sure enough, he came, and I saw him — that object of
the worship of millions. It was a strange sensation, and
thrilling. I wish I could feel it stream through my veins
again. And yet, to me he was not a god, he was only a Taj.
The thrill was not my thrill, but had come to me secondhand
from those invisible millions of believers. By a hand-shake
with their god I had ground-circuited their wire and got their
monster battery's whole charge.

He was tall and slender, indeed emaciated. He had a clean
cut and conspicuously intellectual face, and a deep and kindly
eye. He looked many years older than he really was, but
much study and meditation and fasting and prayer, with the
arid life he had led as hermit and beggar, could account for
that. He is wholly nude when he receives natives, of whatever
rank they may be, but he had white cloth around his loins now,
a concession to Mr. Parker's Europe prejudices, no doubt.

As soon as I had sobered down a little we got along very
well together, and I found him a most pleasant and friendly
deity. He had heard a deal about Chicago, and showed a
quite remarkable interest in it, for a god. It all came of the
World's Fair and the Congress of Religions. If India knows
about nothing else American, she knows about those, and will
keep them in mind one while.

He proposed an exchange of autographs, a delicate attention which made me believe in him, but I had been having my
doubts before. He wrote his in his book, and I have a reverent regard for that book, though the words run from right to
left, and so I can't read it. It was a mistake to print in that
way. It contains his voluminous comments on the Hindoo

holy writings, and if I could make them out I would try for
perfection myself. I gave him a copy of Huckleberry Finn.
I thought it might rest him up a little to mix it in along with
his meditations on Brahma, for he looked tired, and I knew
that if it didn't do him any good it wouldn't do him any harm.

श्री १०८ मत्परमहंसपरिव्राजकाचार्य्य
स्वामिभास्करानन्दसरस्वति शिष्य
मीनवाहादुरराणानैपाळीमहाशय ।

Sri 108 matparamahansaparivra-
jakacharyaswamibhaskara-
nandasaraswati, sishya
mina Bahadur Rana

PAGE FROM BAHADUR'S BOOK.

He has a scholar meditating under him — Mina Bahadur
Rana — but we did not see him. He wears clothes and is
very imperfect. He has written a little pamphlet about his
master, and I have that. It contains a wood-cut of the master
and himself seated on a rug in the garden. The portrait of
the master is very good indeed. The posture is exactly that
which Brahma himself affects, and it requires long arms and

limber legs, and can be accumulated only by gods and the india-rubber man. There is a life-size marble relief of Shri 108, S.B.S. in the garden. It represents him in this same posture.

Dear me! It is a strange world. Particularly the Indian division of it. This pupil, Mina Bahadur Rana, is not a commonplace person, but a man of distinguished capacities and attainments, and, apparently, he had a fine worldly career in front of him. He was serving the Nepal Government in a high capacity at the Court of the Viceroy of India, twenty years ago. He was an able man, educated, a thinker, a man of property. But the longing

TITLE PAGE OF BAHADUR'S BOOK.

to devote himself to a religious life came upon him, and he resigned his place, turned his back upon the vanities and comforts of the world, and went away into the solitudes to live in a hut and study the sacred writings and meditate upon virtue and holiness and seek to attain them. This sort of religion resembles ours. Christ recommended the rich to give away all their property and follow Him in poverty, not in worldly comfort. American and English millionaires do it every day, and thus verify and confirm to the world the tremendous forces that lie in religion. Yet many people scoff at them for this loyalty to duty, and many will scoff at Mina Bahadur Rana and call him a crank. Like many Christians of great character and intellect, he has made the study of his Scriptures and the writing of books of commentaries

33

upon them the loving labor of his life. Like them, he has believed that his was not an idle and foolish waste of his life, but a most worthy and honorable employment of it. Yet, there are many people who will see in those others, men worthy of homage and deep reverence, but in him merely a crank. But I shall not. He has my reverence. And I don't offer it as a common thing and poor, but as an unusual thing and of value. The ordinary reverence, the reverence defined and explained by the dictionary costs nothing. Reverence for one's own sacred things — parents, religion, flag, laws, and respect for one's own beliefs — these are feelings which we cannot even help. They come natural to us; they are involuntary, like breathing. There is no personal merit in breathing. But the reverence which is difficult, and which has personal merit in it, is the respect which you pay, without compulsion, to the political or religious attitude of a man whose beliefs are not yours. You can't revere his gods or his politics, and no one expects you to do that, but you could respect his belief in them if you tried hard enough; and you could respect *him*, too, if you tried hard enough. But it is very, very difficult ; it is next to impossible, and so we hardly ever try. If the man doesn't believe as we do, we say he is a crank, and that settles it. I mean it does nowadays, because now we can't burn him.

We are always canting about people's "irreverence," always charging this offense upon somebody or other, and thereby intimating that we are better than that person and do not commit that offense ourselves. Whenever we do this we are in a lying attitude, and our speech is cant; for none of us are reverent — in a meritorious way; deep down in our hearts we are all irreverent. There is probably not a single exception to this rule in the earth. There is probably not one person whose reverence rises higher than respect for his *own* sacred things;

and therefore, it is not a thing to boast about and be proud of, since the most degraded savage has that — and, like the best of us, has nothing higher. To speak plainly, we despise all reverences and all objects of reverence which are outside the pale of our own list of sacred things. And yet, with strange inconsistency, we are shocked when other people despise and defile the things which are holy to us. Suppose we should meet with a paragraph like the following, in the newspapers:

" Yesterday a visiting party of the British nobility had a picnic at Mount Vernon, and in the tomb of Washington they ate their luncheon, sang popular songs, played games, and danced waltzes and polkas."

Should we be shocked? Should we feel outraged? Should we be amazed? Should we call the performance a desecration? Yes, that would all happen. We should denounce those people in round terms, and call them hard names.

And suppose we found this paragraph in the newspapers:

" Yesterday a visiting party of American pork-millionaires had a picnic in Westminster Abbey, and in that sacred place they ate their luncheon, sang popular songs, played games, and danced waltzes and polkas."

Would the English be shocked? Would they feel outraged? Would they be amazed? Would they call the performance a desecration? That would all happen. The pork-millionaires would be denounced in round terms; they would be called hard names.

In the tomb at Mount Vernon lie the ashes of America's most honored son; in the Abbey, the ashes of England's greatest dead; the tomb of tombs, the costliest in the earth, the wonder of the world, the Taj, was built by a great Emperor to honor the memory of a perfect wife and perfect mother, one in whom there was no spot or blemish, whose love was his stay and support, whose life was the light of the world to him ; in it her ashes lie, and to the Mohammedan millions of India it is a holy place; to them it is what Mount Vernon is to Americans, it is what the Abbey is to the English.

A PICNIC IN A SEPULCHRE.

Major Sleeman wrote forty or fifty years ago (the italics are mine):

"I would here enter my humble protest against the *quadrille and lunch parties* which are sometimes given to European ladies and gentlemen of the station at this imperial tomb; drinking and dancing are no doubt very good things in their season, but they are sadly out of place *in a sepulchre.*"

Were there any Americans among those lunch parties? If they were invited, there were.

If my imagined lunch-parties in Westminster and the tomb of Washington should take place, the incident would cause a vast outbreak of bitter eloquence about Barbarism and Irreverence; and it would come from two sets of people who would go next day and dance in the Taj if they had a chance.

As we took our leave of the Benares god and started away we noticed a group of natives waiting respectfully just within the gate — a Rajah from somewhere in India, and some people of lesser consequence. The god beckoned them to come, and as we passed out the Rajah was kneeling and reverently kissing his sacred feet.

If Barnum — but Barnum's ambitions are at rest. This god will remain in the holy peace and seclusion of his garden, undisturbed. Barnum could not have gotten him, anyway. Still, he would have found a substitute that would answer.

CHAPTER LIV.

Do not undervalue the headache. While it is at its sharpest it seems a bad investment; but when relief begins, the unexpired remainder is worth $4 a minute.
                                            —*Pudd'nhead Wilson's New Calendar.*

A COMFORTABLE railway journey of seventeen and a half hours brought us to the capital of India, which is likewise the capital of Bengal — Calcutta. Like Bombay, it has a population of nearly a million natives and a small gathering of white people. It is a huge city and fine, and is called the City of Palaces. It is rich in historical memories; rich in British achievement — military, political, commercial; rich in the results of the miracles done by that brace of mighty magicians, Clive and Hastings. And has a cloud-kissing monument to one Ochterlony.

It is a fluted candlestick 250 feet high. This lingam is the only large monument in Calcutta, I believe. It is a fine ornament, and will keep Ochterlony in mind.

Wherever you are, in Calcutta, and for miles around, you can see it; and always when you see it you think of Ochterlony. And so there is not an hour in the day that you do not think of Ochterlony and wonder who he was. It is good that Clive cannot come back, for he would think it was for Plassey; and then that great spirit would be wounded when the revelation came that it was not. Clive would find out that it was for Ochterlony; and he would think Ochterlony was a battle. And he would think it was a great one, too, and he would say, " With three thousand I whipped sixty thousand and founded the Empire — and there is no monument; this other soldier must have whipped a billion with a dozen and saved the world."

(517)

But he would be mistaken. Ochterlony was a man, not a battle. And he did good and honorable service, too; as good and honorable service as has been done in India by seventy-five or a hundred other Englishmen of courage, rectitude, and distinguished capacity. For India has been a fertile breeding-ground of such men, and remains so; great men, both in war and in the civil service, and as modest as great. But they have no monuments, and were not expecting any. Ochterlony could not have been expecting one, and it is not at all likely that he desired one — certainly not until Clive and Hastings should be supplied. Every day Clive and Hastings lean on the battlements of heaven and look down and wonder which of the two the monument is for; and they fret and worry because they cannot find out, and so the peace of heaven is spoiled for them and lost. But not for Ochterlony. Ochterlony is not troubled. He doesn't suspect that it is his monument. Heaven is sweet and peaceful to him. There is a sort of unfairness about it all.

Indeed, if monuments were always given in India for high achievements, duty straitly performed, and smirchless records, the landscape would be monotonous with them. The handful of English in India govern the Indian myriads with apparent ease, and without noticeable friction, through tact, training, and distinguished administrative ability, reinforced by just and liberal laws — and by keeping their word to the native whenever they give it.

England is far from India and knows little about the eminent services performed by her servants there, for it is the newspaper correspondent who makes fame, and he is not sent to India but to the continent, to report the doings of the princelets and the dukelets, and where they are visiting and whom they are marrying. Often a British official spends thirty or forty years in India, climbing from grade to grade by services which would make him celebrated anywhere else, and finishes as a vice-sovereign, governing a great realm and millions of subjects; then he goes home to England substantially unknown and unheard of, and settles down in some modest corner, and is as one extinguished. Ten years later there is a twenty-line obituary in the London papers, and the reader is paralyzed by the splendors of a career which he is not sure that he had ever heard of before. But meanwhile he has learned all about the continental princelets and dukelets.

The average man is profoundly ignorant of countries that lie remote from his own. When they are mentioned in his presence one or two facts and maybe a couple of names rise like torches in his mind, lighting up an inch or two of it and leaving the rest all dark. The mention of Egypt suggests some Biblical facts and the Pyramids — nothing more. The mention of South Africa suggests Kimberly and the diamonds and there an end. Formerly the mention, to a Hindoo, of America suggested a name — George Washington — with that his familiarity with our country was exhausted. Latterly his familiarity with it has doubled in bulk; so that when America is mentioned now, two torches flare up in the dark caverns of his mind and he says, "Ah, the country of the great man — Washington; and of the Holy City — Chicago." For he knows about the Congress of Religion, and this has enabled him to get an erroneous impression of Chicago.

When India is mentioned to the citizen of a far country it suggests Clive, Hastings, the Mutiny, Kipling, and a number of other great events; and the mention of Calcutta infallibly brings up the Black Hole. And so, when that citizen finds himself in the capital of India he goes first of all to see the Black Hole of Calcutta — and is disappointed.

The Black Hole was not preserved; it is gone, long, long ago. It is strange. Just as it stood, it was itself a monument ; a ready-made one.   It was finished, it was complete, its materials were strong and lasting, it needed no furbishing up, no repairs; it merely needed to be let alone.   It was the first brick, the Foundation Stone, upon which was reared a mighty Empire — the Indian Empire of Great Britain.   It was the ghastly episode of the Black Hole that maddened the British and brought Clive, that young military marvel, raging up from Madras; it was the seed from which sprung Plassey ; and it was that extraordinary battle, whose like had not been seen in the earth since Agincourt, that laid deep and strong the foundations of England's colossal Indian sovereignty.

And yet within the time of men who still live, the Black Hole was torn down and thrown away as carelessly as if its bricks were common clay, not ingots of historic gold.   There is no accounting for human beings.

The supposed site of the Black Hole is marked by an en-graved plate.   I saw that ; and better that than nothing. The Black Hole was a prison — a *cell* is nearer the right word — eighteen feet square, the dimensions of an ordinary bed-chamber ; and into this place the victorious Nabob of Bengal packed 146 of his English prisoners.   There was hardly standing room for them ; scarcely a breath of air was to be got ; the time was night, the weather sweltering hot.   Before the dawn came, the captives were all dead but twenty-three.   Mr. Holwell's long account of the awful episode was familiar to the world a hundred years ago, but one seldom sees in print even an extract from it in our day.   Among the striking things in it is this.   Mr. Holwell, perishing with thirst, kept himself alive by sucking the perspiration from his sleeves.   It gives one a vivid idea of the situation.   He presently found that while he was busy drawing life from one of his sleeves a young English gentleman was stealing supplies from the other one.   Holwell was an unselfish man, a man of the most generous impulses; he lived and died famous for these fine and rare qualities; yet when he found out what was happening to that unwatched sleeve, he took the precaution to suck that one dry first. The miseries of the Black Hole were able to change even a nature like his. But that young gentleman was one of the twenty-three survivors, and he said it was the stolen perspiration that saved his life. From the middle of Mr. Holwell's narrative I will make a brief excerpt:

"Then a general prayer to Heaven, to hasten the approach of the flames to the right and left of us, and put a period to our misery. But these failing, they whose strength and spirits were quite exhausted laid themselves down and expired quietly upon their fellows: others who had yet some strength and vigor left made a last effort at the windows, and several succeeded by leaping and scrambling over the backs and heads of those in the first rank, and got hold of the bars, from which there was no removing them. Many to the right and left sunk with the violent pressure, and were soon suffocated; for now a steam arose from the living and the dead, which affected us in all its circumstances as if we were forcibly held with our heads over a bowl full of strong volatile spirit of hartshorn, until suffocated; nor could the effluvia of the one be distinguished from the other, and frequently, when I was forced by the load upon my head and shoulders to hold my face down, I was obliged, near as I was to the window, instantly to raise it again to avoid suffocation. I need not, my dear friend, ask your commiseration, when I tell you, that in this plight, from half an hour past eleven till near two in the morning, I sustained the weight of a heavy man, with his knees in my back and the pressure of his whole body on my head. A Dutch surgeon who had taken his seat upon my left shoulder, and a Topaz (a black Christian soldier) bearing on my right; all which nothing could have enabled me to support but the props and pressure equally sustaining me all around. The two latter I frequently dislodged by shifting my hold on the bars and driving my knuckles into their ribs; but my friend above stuck fast, held immovable by two bars.

"I exerted anew my strength and fortitude; but the repeated trials and efforts I made to dislodge the insufferable incumbrances upon me at last quite exhausted me; and towards two o'clock, finding I must quit the window or sink where I was, I resolved on the former, having bore, truly for the sake of others, infinitely more for life than the best of it is worth. In the rank close behind me was an officer of one of the ships, whose name was Cary, and who had behaved with much bravery during the siege (his wife, a fine woman, though country born, would not quit him, but accompanied him into the prison, and was one who survived). This poor wretch had been long raving for water and air; I told him I was determined to give up life, and recommended his gaining my station. On my quitting it he made a fruitless attempt to get my place; but the Dutch surgeon, who sat on my shoulder, supplanted him. Poor Cary expressed his thankfulness, and said he would give up life too ; but it was with the utmost labor we forced our way from the window (several in the inner ranks appearing to me dead standing, unable to fall by the throng and equal pressure around). He laid himself down to die ; and his death, I believe, was very sudden ; for he was a short, full, sanguine man. His strength was great ; and, I imagine, had he not retired with me, I should never have been able to force my way. I was at this time sensible of no pain, and little uneasiness ; I can give you no better idea of my situation than by repeating my simile of the bowl of spirit of hartshorn. I found a stupor coming on apace, and laid myself down by that gallant old man, the Rev. Mr. Jervas Bellamy, who laid dead with his son, the lieutenant, hand in hand, near the southernmost wall of the prison. When I had lain there some little time, I still had reflection enough to suffer some uneasiness in the thought that I should be trampled upon, when dead, as I myself had done to others. With some difficulty I raised myself, and gained the platform a second time, where I presently lost all sensation; the last trace of sensibility that I have been able to recollect after my laying down, was my sash being uneasy about my waist, which I untied, and threw from me. Of what passed in this interval, to the time of my resurrection from this hole of horrows, I can give you no account."

There was plenty to see in Calcutta, but there was not plenty of time for it. I saw the fort that Clive built ; and the place where Warren Hastings and the author of the Junius Letters fought their duel ; and the great botanical gardens ; and the fashionable afternoon turnout in the Maidan ; and a grand review of the garrison in a great plain at sunrise ; and a military tournament in which great bodies of native soldiery exhibited the perfection of their drill at all arms, a spectacular and beautiful show occupying several nights and closing with the mimic storming of a native fort which was as good as the reality for thrilling and accurate detail, and better than the reality for security and comfort ; we had a pleasure excursion on the *Hoogly* by courtesy of friends, and devoted the rest of the time to social life and the Indian museum. One should spend a month in the museum, an enchanted palace of Indian antiquities. Indeed, a person might spend half a year among the beautiful and wonderful things without exhausting their interest.

It was winter. We were of Kipling's " hosts of tourists who travel up and down India in the cold weather showing how things ought to be managed." It is a common expression there, " the cold weather," and the people think there is such a thing. It is because they have lived there half a lifetime, and their perceptions have become blunted. When a person is accustomed to 138 in the shade, his ideas about cold weather are not valuable. I had read, in the histories, that the June marches made between Lucknow and Cawnpore by the British forces in the time of the Mutiny were made in that kind of weather — 138 in the shade — and had taken it for historical embroidery. I had read it again in Serjeant-Major Forbes-Mitchell's account of his military experiences in the Mutiny — at least I thought I had — and in Calcutta I asked him if it was true, and he said it was. An officer of high rank who had been in the thick of the Mutiny said the same. As long as those men were talking about what they knew, they were trustworthy, and I believed them ; but when they said it was now " cold weather," I saw that they had traveled outside of their sphere of knowledge and were floundering. I believe that in India " cold weather " is merely a conventional phrase and has come into use through the necessity of having some way to distinguish between weather which will melt a brass door-knob and weather which will only make it mushy. It was

MUSHY WEATHER.

observable that brass ones were in use while I was in Calcutta, showing that it was not yet time to change to porcelain ; I was told the change to porcelain was not usually made until May. But this cold weather was too warm for us ; so we started to Darjeeling, in the Himalayas — a twenty-four hour journey.

CHAPTER LV.

There are 869 different forms of lying, but only one of them has been squarely forbidden. Thou shalt not bear false witness against thy neighbor.
— *Pudd'nhead Wilson's New Calendar.*

FROM DIARY:

February 14. We left at 4:30 P.M. Until dark we moved through rich vegetation, then changed to a boat and crossed the Ganges.

February 15. Up with the sun. A brilliant morning, and frosty. A double suit of flannels is found necessary. The plain is perfectly level, and seems to stretch away and away and away, dimming and softening, to the uttermost bounds of nowhere. What a soaring, strenuous, gushing fountain spray of delicate greenery a bunch of bamboo is! As far as the eye can reach, these grand vegetable geysers grace the view, their spoutings refined to steam by distance. And there are fields of bananas, with the sunshine glancing from the varnished surface of their drooping vast leaves. And there are frequent groves of palm; and an effective accent is given to the landscape by isolated individuals of this picturesque family, towering, clean-stemmed, their plumes broken and hanging ragged, Nature's imitation of an umbrella that has been out to see what a cyclone is like and is trying not to look disappointed. And everywhere through the soft morning vistas we glimpse the villages, the countless villages, the myriad villages, thatched, built of clean new matting, snuggling among grouped palms and sheaves of bamboo; villages, villages, no end of villages, not three hundred yards apart, and dozens and dozens

(524)

of them in sight all the time; a mighty City, hundreds of miles long, hundreds of miles broad, made all of villages, the biggest city in the earth, and as populous as a European kingdom. I have seen no such city as this before. And there is a continuously repeated and replenished multitude of naked men in view on both sides and ahead. We fly through it mile after mile, but still it is always there, on both sides and ahead — brown-bodied, naked men and boys, plowing in the fields. But *not a woman*. In these two hours I have not seen a woman or a girl working in the fields.

" From Greenland's icy mountains,
   From India's coral strand,
 Where Afric's sunny fountains
   Roll down their golden sand.
 From many an ancient river,
   From many a palmy plain,
 They call us to deliver
   Their land from error's chain."

INDIA.

Those are beautiful verses, and they have remained in my memory all my life. But if the closing lines are true, let us hope that when we come to answer the call and deliver the land from its errors, we shall secrete from it some of our high-civilization ways, and at the same time borrow some of its pagan ways to enrich our high system with. We have a right to do this.   If we lift those people up, we have a right to lift ourselves up nine or ten grades or so, at their expense.   A few years ago I spent several weeks at Tölz, in Bavaria.   It is a Roman Catholic region, and not even Benares is more deeply or pervasively or intelligently devout.   In my diary of those days I find this:

"We took a long drive yesterday around about the lovely country roads. But it was a drive whose pleasure was damaged in a couple of ways: by the dreadful shrines and by the shameful spectacle of gray and venerable old grandmothers toiling in the fields.   The shrines were frequent along the roads — figures of the Saviour nailed to the cross and streaming with blood from the wounds of the nails and the thorns.

"When missionaries go from here do they find fault with the pagan idols?   I saw many women seventy and even eighty years old mowing and binding in the fields, and pitchforking the loads into the wagons."

I was in Austria later, and in Munich.   In Munich I saw gray old women pushing trucks up hill and down, long distances, trucks laden with barrels of beer, incredible loads.   In my Austrian diary I find this:

AUSTRIA.

"In the fields I often see a woman and a cow harnessed to the plow, and a man driving.

"In the public street of Marienbad to-day, I saw an old, bent, gray-headed woman, *in harness with a dog*, drawing a laden sled over bare dirt roads and bare pavements; and at his ease walked the driver, smoking his pipe, a hale fellow not thirty years old."

Five or six years ago I bought an open boat, made a kind of a canvas wagon-roof over the stern of it to shelter me from sun and rain; hired a courier and a boatman, and made a twelve-day floating voyage down the Rhone from Lake Bourget to Marseilles.   In my diary of that trip I find this entry.   I was far down the Rhone then:

"Passing St. Etienne, 2:15 P.M. On a distant ridge inland, a tall open-work structure commandingly situated, with a statue of the Virgin standing on it A devout country. All down this river, wherever there is a crag there is a statue of the Virgin on it. I believe I have seen a hundred of them. And yet, in many respects, the peasantry seem to be mere pagans, and destitute of any considerable degree of civilization.

" . . . . We reached a not very promising looking village about 4 o'clock, and I concluded to tie up for the day ; munching fruit and fogging the hood with pipe-smoke had grown monotonous ; I could not have the hood furled, because the floods of rain fell unceasingly. The tavern was on the river bank, as is the custom. It was dull there, and melancholy — nothing to do but look out of the window into the drenching rain, and shiver ; one could do that, for it was bleak and cold and windy, and country France furnishes no fire. Winter overcoats did not help me much ; they had to be supplemented with rugs. The raindrops were so large and struck the river with such force that they knocked up the water like pebble-splashes.

"With the exception of a very occasional wooden-shod peasant, nobody was abroad in this bitter weather — I mean nobody of our sex. But all weathers are alike to the women in these continental countries. To them and the other animals, life is serious ; nothing interrupts their slavery. Three of them were washing clothes in the river under the window when I arrived, and they continued at it as long as there was light to work by. One was apparently thirty ; another — the mother ! — above fifty ; the third — grandmother ! — so old and worn and gray she could have passed for eighty ; I took

FRANCE.

her to be that old. They had no waterproofs nor rubbers, of course ; over their shoulders they wore gunny-sacks — simply conductors for rivers of water ; some of the volume reached the ground ; the rest soaked in on the way.

"At last a vigorous fellow of thirty-five arrived, dry and comfortable, smoking his pipe under his big umbrella in an open donkey-cart — husband, son, and grandson of those women ! He stood up in the cart, sheltering himself, and began to superintend, issuing his orders in a masterly tone of command, and showing temper when they were not obeyed swiftly enough.

Without complaint or murmur the drowned women patiently carried out the orders, lifting the immense baskets of soggy, wrung-out clothing into the cart and stowing them to the man's satisfaction. There were six of the great baskets, and a man of mere ordinary strength could not have lifted any one of them. The cart being full now, the Frenchman descended, still sheltered by his umbrella, entered the tavern, and the women went drooping homeward, trudging in the wake of the cart, and soon were blended with the deluge and lost to sight.

"When I went down into the public room, the Frenchman had his bottle of wine and plate of food on a bare table black with grease, and was "chomping" like a horse. He had the little religious paper which is in everybody's hands on the Rhone borders, and was enlightening himself with the histories of French saints who used to flee to the desert in the Middle Ages to escape the contamination of woman. For two hundred years France has been sending missionaries to other savage lands. To spare to the needy from poverty like hers is fine and true generosity."

But to get back to India — where, as my favorite poem says —

"Every prospect pleases,
And only man is vile."

It is because Bavaria and Austria and France have not introduced their civilization to him yet. But Bavaria and Austria and France are on their way. They are coming. They will rescue him; they will refine the vileness out of him.

Some time during the forenoon, approaching the mountains, we changed from the regular train to one composed of little canvas-sheltered cars that skimmed along within a foot of the ground and seemed to be going fifty miles an hour when they were really making about twenty. Each car had seating capacity for half-a-dozen persons; and when the curtains were up one was substantially out of doors, and could see everywhere, and get all the breeze, and be luxuriously comfortable. It was not a pleasure excursion in name only, but in fact.

After a while we stopped at a little wooden coop of a station just within the curtain of the sombre jungle, a place with a deep and dense forest of great trees and scrub and vines all about it. The royal Bengal tiger is in great force there, and is very bold and unconventional. From this lonely little station a message once went to the railway manager in Calcutta: " Tiger eating station-master on front porch; telegraph instructions."

It was there that I had my first tiger hunt. I killed thirteen. We were presently away again, and the train began to climb the mountains. In one place seven wild elephants crossed the track, but two of them got away before I could overtake them. The railway journey up the mountain is forty miles, and it takes eight hours to make it. It is so wild and interesting and exciting and enchanting that it ought to take a week. As for the vegetation, it is a museum. The jungle seemed to contain samples of every rare and curious tree and bush that we had ever seen or heard of. It is from that museum, I think, that the globe must have been supplied with the trees and vines and shrubs that it holds precious.

The road is infinitely and charmingly crooked. It goes winding in and out under lofty cliffs that are smothered in vines and foliage, and around the edges of bottomless chasms; and all the way one glides by files of picturesque natives, some carrying burdens up, others going down from their work in the tea-gardens; and once there was a gaudy wedding procession, all bright tinsel and color, and a bride, comely and girlish, who peeped out from the curtains of her palanquin, exposing her face with that pure delight which the young and happy take in sin for sin's own sake.

By and by we were well up in the region of the clouds, and from that breezy height we looked down and afar over a wonderful picture — the Plains of India, stretching to the horizon, soft and fair, level as a floor, shimmering with heat, mottled with cloud-shadows, and cloven with shining rivers. Immediately below us, and receding down, down, down, toward the valley, was a shaven confusion of hilltops, with ribbony roads

34

and paths squirming and snaking cream-yellow all over them and about them, every curve and twist sharply distinct.

At an elevation of 6,000 feet we entered a thick cloud, and it shut out the world and kept it shut out. We climbed 1,000 feet higher, then began to descend, and presently got down to Darjeeling, which is 6,000 feet above the level of the Plains.

We had passed many a mountain village on the way up, and seen some new kinds of natives, among them many samples of the fighting Ghurkas. They are not large men, but they are strong and resolute. There are no better soldiers among Britain's native troops. And we had passed shoals of their women climbing the forty miles of steep road from the valley to their mountain homes, with tall baskets on their backs hitched to their foreheads by a band, and containing a freightage weighing — I will not say how many hundreds of pounds, for the sum is unbelievable. These were young women, and they strode smartly along under these astonishing burdens with the air of people out for a holiday. I was told that a woman will carry a piano on her back all the way up the mountain ; and that more than once a woman had done it. If these were old women I should regard the Ghurkas as no more civilized than the Europeans.

At the railway station at Darjeeling you find plenty of cab-substitutes — open coffins, in which you sit, and are then borne on men's shoulders up the steep roads into the town.

Up there we found a fairly comfortable hotel, the property of an indiscriminate and incoherent landlord, who looks after nothing, but leaves everything to his army of Indian servants. No, he does look after the bill — to be just to him — and the tourist cannot do better than follow his example. I was told by a resident that the summit of Kinchinjunga is often hidden in the clouds, and that sometimes a tourist has waited twenty-two days and then been obliged to go away without a sight of

A CAB SUBSTITUTE.

it. And yet went not disappointed ; for when he got his hotel bill he recognized that he was now seeing the highest thing in the Himalayas. But this is probably a lie.

After lecturing I went to the Club that night, and that was a comfortable place. It is loftily situated, and looks out over a vast spread of scenery ; from it you can see where the boundaries of three countries come together, some thirty miles away ; Thibet is one of them, Nepaul another, and I think Herzegovina was the other. Apparently, in every town and city in India the gentlemen of the British civil and military service have a club; sometimes it is a palatial one, always it is pleasant and homelike. The hotels are not always as good as they might be, and the stranger who has access to the Club is grateful for his privilege and knows how to value it.

Next day was Sunday. Friends came in the gray dawn with horses, and my party rode away to a distant point where Kinchinjunga and Mount Everest show up best, but I stayed at home for a private view ; for it was very old, and I was not acquainted with the horses, any way. I got a pipe and a few blankets and sat for two hours at the window, and saw the sun drive away the veiling gray and touch up the snow-peaks one after another with pale pink splashes and delicate washes of gold, and finally flood the whole mighty convulsion of snow-mountains with a deluge of rich splendors.

Kinchinjunga's peak was but fitfully visible, but in the between times it was vividly clear against the sky — away up there in the blue dome more than 28,000 feet above sea level — the loftiest land I had ever seen, by 12,000 feet or more. It was 45 miles away. Mount Everest is a thousand feet higher, but it was not a part of that sea of mountains piled up there before me, so I did not see it ; but I did not care, because I think that mountains that are as high as that are disagreeable.

I changed from the back to the front of the house and

MOUNT KINCHINJUNGA.

spent the rest of the morning there, watching the swarthy
strange tribes flock by from their far homes in the Himalayas.
All ages and both sexes were represented, and the breeds were

THE PRAYER WHEEL.

quite new to me, though the
costumes of the Thibetans made
them look a good deal like
Chinamen.   The prayer-wheel
was a frequent feature.   It
brought me near to these people,
and made them seem kinfolk of
mine.   Through our preacher we
do much of our praying by
proxy.   We do not whirl him
around a stick, as they do, but
that is merely a detail.   The
swarm swung briskly by, hour
after hour, a strange and striking
pageant.   It was wasted there,
and it seemed a pity.   It should
have been sent streaming through
the cities of Europe or America, to refresh eyes weary of the
pale monotonies of the circus-pageant.   These people were
bound for the bazar, with things to sell.   We went down there,
later, and saw that novel congress of the wild peoples, and
plowed here and there through it, and concluded that it would
be worth coming from Calcutta to see, even if there were no
Kinchinjunga and Everest.

CHAPTER LVI.

ON Monday and Tuesday at sunrise we again had fair-to-middling views of the stupendous mountains; then, being well cooled off and refreshed, we were ready to chance the weather of the lower world once more.

We traveled up hill by the regular train five miles to the summit, then changed to a little canvas-canopied hand-car for the 35-mile descent. It was the size of a sleigh, it had six seats and was so low that it seemed to rest on the ground. It had no engine or other propelling power, and needed none to help it fly down those steep inclines. It only needed a strong brake, to modify its flight, and it had that. There was a story of a disastrous trip made down the mountain once in this little car by the Lieutenant-Governor of Bengal, when the car jumped the track and threw its passengers over a precipice. It was not true, but the story had value for me, for it made me nervous, and nervousness wakes a person up and makes him alive and alert, and heightens the thrill of a new and doubtful experience. The car could really jump the track, of course; a pebble on the track, placed there by either accident or malice, at a sharp curve where one might strike it before the eye could discover it, could derail the car and fling it down into India; and the fact that the lieutenant-governor had escaped was no proof that I would have the same luck. And standing there, looking down upon the Indian Empire from the airy altitude of 7,000 feet, it seemed unpleasantly far, dangerously far, to be flung from a hand-car.

But after all, there was but small danger — for me. What there was, was for Mr. Pugh, inspector of a division of the Indian police, in whose company and protection we had come from Calcutta. He had seen long service as an artillery officer, was less nervous than I was, and so he was to go ahead of us in a pilot hand-car, with a Ghurka and another native; and the plan was that when we should see his car jump over a precipice we must put on our break and send for another pilot. It was a good arrangement. Also Mr. Barnard, chief engineer of the mountain-division of the road, was to take personal charge of our car, and he had been down the mountain in it many a time.

Everything looked safe. Indeed, there was but one questionable detail left: the regular train was to follow us as soon as we should start, and it might run over us. Privately, I thought it would.

The road fell sharply down in front of us and went corkscrewing in and out around the crags and precipices, down, down, forever down, suggesting nothing so exactly or so uncomfortably as a crooked toboggan slide with no end to it. Mr. Pugh waved his flag and started, like an arrow from a bow, and before I could get out of the car we were gone too. I had previously had but one sensation like the shock of that departure, and that was the gaspy shock that took my breath away the first time that I was discharged from the summit of a toboggan slide. But in both instances the sensation was pleasurable — intensely so; it was a sudden and immense exaltation, a mixed ecstasy of deadly fright and unimaginable joy. I believe that this combination makes the perfection of human delight.

The pilot car's flight down the mountain suggested the swoop of a swallow that is skimming the ground, so swiftly and smoothly and gracefully it swept down the long straight

WE PLAYED WITH THE TRAIN.

reaches and soared in and out of the bends and around the corners. We raced after it, and seemed to flash by the capes and crags with the speed of light; and now and then we almost overtook it — and had hopes; but it was only playing with us; when we got near, it released its brake, make a spring around a corner, and the next time it spun into view, a few seconds later, it looked as small as a wheelbarrow, it was so far away. We played with the train in the same way. We often got out to gather flowers or sit on a precipice and look at the scenery, then presently we would hear a dull and growing roar, and the long coils of the train would come into sight behind and above us; but we did not need to start till the locomotive was close down upon us — then we soon left it far behind. It had to stop at every station, therefore it was not an embarrassment to us. Our brake was a good piece of machinery; it could bring the car to a standstill on a slope as steep as a house-roof.

The scenery was grand and varied and beautiful, and there was no hurry; we could always stop and examine it. There was abundance of time. We did not need to hamper the train; if it wanted the road, we could switch off and let it go by, then overtake it and pass it later. We stopped at one place to see the Gladstone Cliff, a great crag which the ages and the weather have sculptured into a recognizable portrait of the venerable statesman. Mr. Gladstone is a stockholder in the road, and Nature began this portrait ten thousand years ago, with the idea of having the compliment ready in time for the event.

We saw a banyan tree which sent down supporting stems from branches which were sixty feet above the ground. That is, I suppose it was a banyan; its bark resembled that of the great banyan in the botanical gardens at Calcutta, that spider-legged thing with its wilderness of vegetable columns. And there were frequent glimpses of a totally leafless tree upon whose innumerable twigs and branches a cloud of crimson butterflies had lighted — apparently. In fact these brilliant red butterflies were flowers, but the illusion was good. Afterward in South Africa, I saw another splendid effect made by red flowers. This flower was probably called the torchplant — should have been so named, anyway. It had a slender stem several feet high, and from its top stood up a single tongue of flame, an intensely red flower of the size and shape of a small corn-cob. The stems stood three or four feet apart all over a great hill-slope that was a mile long, and make one think of what the Place de la Concorde would be if its myriad lights were red instead of white and yellow.

A few miles down the mountain we stopped half an hour to see a Thibetan dramatic performance. It was in the open air on the hillside. The audience was composed of Thibetans, Ghurkas, and other unusual people. The costumes of the actors were in the last degree outlandish, and the performance was in keeping with the clothes. To an accompaniment of barbarous noises the actors stepped out one after another and began to spin around with immense swiftness and vigor and violence, chanting the while, and soon the whole troupe would be spinning and chanting and raising the dust. They were performing an ancient and celebrated historical play, and a Chinaman explained it to me in pidjin English as it went along. The play was obscure enough without the explanation; with the explanation added, it was opake. As a drama this ancient historical work of art was defective, I thought, but as a wild and barbarous spectacle t'-- representation was beyond criticism.

Far down the mountain we got out to look at a piece of remarkable loop-engineering — a spiral where the road curves upon itself with such abruptness that when the regular train came down and entered the loop, we stood over it and saw the locomotive disappear under our bridge, then in a few moments appear again, chasing its own tail; and we saw it gain on it, overtake it, draw ahead past the rear cars, and run a race with that end of the train. It was like a snake swallowing itself.

Half-way down the mountain we stopped about an hour at Mr. Barnard's house for refreshments, and while we were sitting on the veranda looking at the distant panorama of hills through a gap in the forest, we came very near seeing a leopard kill a calf.* It is a wild place and lovely. From the woods all about came the songs of birds,— among them the contributions of a couple of birds which I was not then acquainted with: the brain-fever bird and the coppersmith. The song of the brain-fever demon starts on a low but steadily rising key, and is a spiral twist which augments in intensity and severity with each added spiral, growing sharper and sharper, and more and more painful, more and more agonizing, more and more maddening, intolerable, unendurable, as it bores deeper and deeper and deeper into the listener's brain, until at last the brain fever comes as a relief and the man dies. I am bringing some of these birds home to America. They will be a great curiosity there, and it is believed that in our climate they will multiply like rabbits.

The coppersmith bird's note at a certain distance away has the ring of a sledge on granite; at a certain other distance the hammering has a more metallic ring, and you might think that the bird was mending a copper kettle; at another distance it has a more woodeny thump, but it is a thump that is full of energy, and sounds just like starting a bung. So he is a hard bird to name with a single name; he is a stone-breaker, coppersmith, and bung-starter, and even then he is not completely named, for when he is close by you find that there is a soft,

---

* It killed it the day before.

THE LOOP.

deep, melodious quality in his thump, and for that no satisfying name occurs to you. You will not mind his other notes,
but when he camps near enough for you to hear that one, you
presently find that his measured and monotonous repetition of
it is beginning to disturb you; next it will weary you, soon it
will distress you, and before long each thump will hurt your
head; if this goes on, you will lose your mind with the pain
and misery of it, and go crazy. I am bringing some of these
birds home to America. There is nothing like them there.
They will be a great surprise, and it is said that in a climate
like ours they will surpass expectation for fecundity.

I am bringing some nightingales, too, and some cue-owls.
I got them in Italy. The song of the nightingale is the deadliest known to ornithology. That demoniacal shriek can kill
at thirty yards. The note of the cue-owl is infinitely soft and
sweet — soft and sweet as the whisper of a flute. But pene-
trating — oh, beyond belief; it can bore through boiler-iron.
It is a lingering note, and comes in triplets, on the one un-
changing key: *hoo-o-o, hoo-o-o, hoo-o-o ;* then a silence of fifteen
seconds, then the triplet again; and so on, all night. At first
it is divine; then less so; then trying; then distressing; then
excruciating; then agonizing, and at the end of two hours the
listener is a maniac.

And so, presently we took to the hand-car and went flying
down the mountain again; flying and stopping, flying and
stopping, till at last we were in the plain once more and stowed
for Calcutta in the regular train. That was the most enjoyable day I have spent in the earth. For rousing, tingling,
rapturous pleasure there is no holiday trip that approaches the
bird-flight down the Himalayas in a hand-car. It has no
fault, no blemish, no lack, except that there are only thirty-
five miles of it instead of five hundred.

CHAPTER LVII.

She was not quite what you would call refined. She was not quite what you would call unrefined. She was the kind of person that keeps a parrot.
— *Pudd'nhead Wilson's New Calendar*.

SO far as I am able to judge, nothing has been left undone, either by man or Nature, to make India the most extraordinary country that the sun visits on his round. Nothing seems to have been forgotten, nothing overlooked. Always, when you think you have come to the end of her tremendous specialties and have finished hanging tags upon her as the Land of the Thug, the Land of the Plague, the Land of Famine, the Land of Giant Illusions, the Land of Stupendous Mountains, and so forth, another specialty crops up and another tag is required. I have been overlooking the fact that India is by an unapproachable supremacy — the Land of Murderous Wild Creatures. Perhaps it will be simplest to throw away the tags and generalize her with one all-comprehensive name, as the Land of Wonders.

For many years the British Indian Government has been trying to destroy the murderous wild creatures, and has spent a great deal of money in the effort. The annual official returns show that the undertaking is a difficult one.

These returns exhibit a curious annual uniformity in results; the sort of uniformity which you find in the annual output of suicides in the world's capitals, and the proportions of deaths by this, that, and the other disease. You can always come close to foretelling how many suicides will occur in Paris, London, and New York, next year, and also how many deaths will result from cancer, consumption, dog-bite, falling out of the window, getting run over by cabs, etc., if you know the statistics of those matters for the present year. In the same way, with one year's Indian statistics before you, you can guess closely at how many people were killed in that Empire by tigers during the previous year, and the year before that, and the year before that, and at how many were killed in each of those years by bears, how many by wolves, and how many by snakes ; and you can also guess closely at how many people are going to be killed each year for the coming five years by each of those agencies. You can also guess closely at how many of each agency the government is going to kill each year for the next five years.

I have before me statistics covering a period of six consecutive years. By these, I know that in India the tiger kills something over 800 persons every year, and that the government responds by killing about double as many tigers every year. In four of the six years referred to, the tiger got 800 odd ; in one of the remaining two years he got only 700, but in the other remaining year he made his average good by scoring 917. He is always sure of his average. Anyone who bets that the tiger will kill 2,400 people in India in any three consecutive years has invested his money in a certainty ; anyone who bets that he will kill 2,600 in any three consecutive years, is absolutely sure to lose.

As strikingly uniform as are the statistics of suicide, they are not any more so than are those of the tiger's annual output of slaughtered human beings in India. The government's work is quite uniform, too ; it about doubles the tiger's average. In six years the tiger killed 5,000 persons, minus 50 ; in the same six years 10,000 tigers were killed, minus 400.

The wolf kills nearly as many people as the tiger — 700 a year to the tiger's 800 odd — but while he is doing it, more than 5,000 of his tribe fall.

35

The leopard kills an average of 230 people per year, but loses 3,300 of his own mess while he is doing it.

The bear kills 100 people per year at a cost of 1,250 of his own tribe.

The tiger, as the figures show, makes a very handsome fight against man. But it is nothing to the elephant's fight. The king of beasts, the lord of the jungle, loses four of his mess per year, but he kills *forty-five* persons to make up for it.

But when it comes to killing cattle, the lord of the jungle is not interested. He kills but 100 in six years — horses of hunters, no doubt — but in the same six the tiger kills more than 84,000, the leopard 100,000, the bear 4,000, the wolf 70,000, the hyena more than 13,000, other wild beasts 27,000, and the snakes 19,000, a grand total of more than 300,000 ; an average of 50,000 head per year.

In response, the government kills, in the six years, a total of 3,201,232 wild beasts and snakes. Ten for one.

It will be perceived that the snakes are not much interested in cattle ; they kill only 3,000 odd per year. The snakes are much more interested in man. India swarms with deadly snakes. At the head of the list is the cobra, the deadliest known to the world, a snake whose bite kills where the rattle-snake's bite merely entertains.

In India, the annual man-killings by snakes are as uniform, as regular, and as forecastable as are the tiger-average and the suicide-average. Anyone who bets that in India, in any three consecutive years the snakes will kill 49,500 persons, will win his bet ; and anyone who bets that in India in any three con-secutive years, the snakes will kill 53,500 persons, will lose his bet. In India the snakes kill 17,000 people a year ; they hardly ever fall short of it ; they as seldom exceed it. An insurance actuary could take the Indian census tables and the government's snake tables and tell you within sixpence how much it would be worth to insure a man against death by snake-bite there. If I had a dollar for every person killed per year in India, I would rather have it than any other property, as it is the only property in the world not subject to shrinkage.

I should like to have a royalty on the government-end of the snake business, too, and am in London now trying to get it; but when I get it it is not going to be as regular an income as the other will be if I get that; I have applied for it. The snakes transact their end of the business in a more orderly and systematic way than the government transacts its end of it, because the snakes have had a long experience and know all about the traffic. You can make sure that the government will never kill fewer than 110,000 snakes in a year, and that it will never quite reach 300,000 — too much room for oscillation; good speculative stock, to bear or bull, and buy and sell long and short, and all that kind of thing, but not eligible for investment like the other. The man that speculates in the government's snake crop wants to go carefully. I would not advise a man to buy a single crop at all — I mean a crop of futures — for the possible wobble is something quite extraordinary. If he can buy *six* future crops in a bunch, seller to deliver 1,500,000 altogether, that is another matter. I do not know what snakes are worth now, but I know what they would be worth then, for the statistics show that the seller could not come within 427,000 of carrying out his contract. However, I think that a person who speculates in snakes is a fool, anyway. He always regrets it afterwards.

To finish the statistics. In six years the wild beasts kill 20,000 persons, and the snakes kill 103,000. In the same six the government kills 1,073,546 snakes. Plenty left.

There are narrow escapes in India. In the very jungle where I killed sixteen tigers and all those elephants, a cobra bit me but it got well; everyone was surprised. This could not happen twice in ten years, perhaps. Usually death would result in fifteen minutes.

We struck out westward or northwestward from Calcutta on an itinerary of a zig-zag sort, which would in the course of time carry us across India to its northwestern corner and the border of Afghanistan. The first part of the trip carried us through a great region which was an endless garden — miles and miles of the beautiful flower from whose juices comes the opium, and at Muzaffurpore we were in the midst of the indigo culture; thence by a branch road to the Ganges at a point near Dinapore, and by a train which would have missed the connection by a week but for the thoughtfulness of some British officers who were along, and who knew the ways of trains that are run by natives without white supervision. This train stopped at every village; for no purpose connected with business, apparently. We put out nothing, we took nothing aboard. The train hands stepped ashore and gossiped with friends a quarter of an hour, then pulled out and repeated this at the succeeding villages. We had thirty-five miles to go and six hours to do it in, but it was plain that we were not going to make it. It was then that the English officers said it was now necessary to turn this gravel train into an express. So they gave the engine-driver a rupee and told him to fly. It was a simple remedy. After that we made ninety miles an hour. We crossed the Ganges just at dawn, made our connection, and went to Benares, where we stayed twenty-four hours and inspected that strange and fascinating piety-hive again; then left for Lucknow, a city which is perhaps the most conspicuous of the many monuments of British fortitude and valor that are scattered about the earth.

The heat was pitiless, the flat plains were destitute of grass, and baked dry by the sun they were the color of pale dust, which was flying in clouds. But it was much hotter than this when the relieving forces marched to Lucknow in the time of the Mutiny. Those were the days of 138° in the shade.

CHAPTER LVIII.

Make it a point to do something every day that you don't want to do. This is the golden rule for acquiring the habit of doing your duty without pain.
— *Pudd'nhead Wilson's New Calendar.*

IT seems to be settled, now, that among the many causes from which the Great Mutiny sprang, the main one was the annexation of the kingdom of Oudh by the East India Company — characterized by Sir Henry Lawrence as "the most unrighteous act that was ever committed." In the spring of 1857, a mutinous spirit was observable in many of the native garrisons, and it grew day by day and spread wider and wider. The younger military men saw something very serious in it, and would have liked to take hold of it vigorously and stamp it out promptly; but they were not in authority. Old men were in the high places of the army — men who should have been retired long before, because of their great age — and they regarded the matter as a thing of no consequence. They loved their native soldiers, and would not believe that anything could move them to revolt. Everywhere these obstinate veterans listened serenely to the rumbling of the volcanoes under them, and said it was nothing.

And so the propagators of mutiny had everything their own way. They moved from camp to camp undisturbed, and painted to the native soldier the wrongs his people were suffering at the hands of the English, and made his heart burn for revenge. They were able to point to two facts of formidable value as backers of their persuasions: In Clive's day, native armies were incoherent mobs, and without effective arms;

therefore, they were weak against Clive's organized handful of well-armed men, but the thing was the other way, now. The British forces were native; they had been trained by the British, organized by the British, armed by the British, all the power was in their hands — they were a club made by British hands to beat out British brains with. There was nothing to oppose their mass, nothing but a few weak battalions of British soldiers scattered about India, a force not worth speaking of. This argument, taken alone, might not have succeeded, for the bravest and best Indian troops had a wholesome dread of the white soldier, whether he was weak or strong; but the agitators backed it with their second and best point — *prophecy* — a prophecy a hundred years old. The Indian is open to prophecy at all times; argument may fail to convince him, but not prophecy. There was a prophecy that a hundred years from the year of that battle of Clive's which founded the British Indian Empire, the British power would be overthrown and swept away by the natives.

The Mutiny broke out at Meerut on the 10th of May, 1857, and fired a train of tremendous historical explosions. Nana Sahib's massacre of the surrendered garrison of Cawnpore occurred in June, and the long siege of Lucknow began. The military history of England is old and great, but I think it must be granted that the crushing of the Mutiny is the greatest chapter in it. The British were caught asleep and unprepared. They were a few thousands, swallowed up in an ocean of hostile populations. It would take months to inform England and get help, but they did not falter or stop to count the odds, but with English resolution and English devotion they took up their task, and went stubbornly on with it, through good fortune and bad, and fought the most unpromising fight that one may read of in fiction or out of it, and won it thoroughly.

The Mutiny broke out so suddenly, and spread with such rapidity that there was but little time for occupants of weak outlying stations to escape to places of safety. Attempts were made, of course, but they were attended by hardships as bitter as death in the few cases which were successful; for the heat ranged between 120 and 138 in the shade; the way led through hostile peoples, and food and water were hardly to be had. For ladies and children accustomed to ease and comfort and plenty, such a journey must have been a cruel experience. Sir G. O. Trevelyan quotes an example:

"This is what befell Mrs. M——, the wife of the surgeon at a certain station on the southern confines of the insurrection. 'I heard,' she says, 'a number of shots fired, and, looking out, I saw my husband driving furiously from the mess-house, waving his whip. I ran to him, and, seeing a bearer with my child in his arms, I caught her up, and got into the buggy. At the mess-house we found all the officers assembled, together with sixty sepoys, who had remained faithful. We went off in one large party, amidst a general conflagration of our late homes. We reached the caravanserai at Chattapore the next morning, and thence started for Callinger. At this point our sepoy escort deserted us. We were fired upon by matchlockmen, and one officer was shot dead. We heard, likewise, that the people had risen at Callinger, so we returned and walked back ten miles that day. M—— and I carried the child alternately. Presently Mrs. Smalley died of sunstroke. We had no food amongst us. An officer kindly lent us a horse. We were very faint. The Major died, and was buried; also the Sergeant-major and some women. The bandsmen left us on the nineteenth of June. We were fired at again by match-lockmen, and changed direction for Allahabad. Our party consisted of nine gentlemen, two children, the sergeant and his wife. On the morning of the twentieth, Captain Scott took Lottie on to his horse. I was riding behind my husband, and she was so crushed between us. She was two years old on the first of the month. We were both weak through want of food and the effect of the sun. Lottie and I had no head covering. M—— had a sepoy's cap I found on the ground. Soon after sunrise we were followed by villagers armed with clubs and spears. One of them struck Captain Scott's horse on the leg. He galloped off with Lottie, and my poor husband never saw his child again. We rode on several miles, keeping away from villages, and then crossed the river. Our thirst was extreme. M—— had dreadful cramps, so that I had to hold him on the horse. I was very uneasy about him. The day before I saw the drummer's wife eating chupatties, and asked her to give a piece to the child, which she did. I now saw water in a ravine. The descent was steep, and our only drinking-vessel was M——'s cap. Our horse got water, and I bathed my neck. I had no stockings, and my feet were torn and blistered. Two peasants came in sight, and we were frightened and rode off. The sergeant held our horse, and M—— put me up and mounted. I think he must have got suddenly faint, for I fell and he over me, on the road, when the horse started off. Some time before he said, and Barber, too, that he could not live many hours. I felt he was dying before we came to the ravine. He told me his wishes about his children and myself, and took leave. My brain seemed burnt up. No tears came. As soon as we fell, the sergeant let go the horse, and it went off; so that escape was cut off. We sat down on the ground waiting for death. Poor fellow! he was very weak; his thirst was frightful, and I went to get him water. Some villagers came, and took my rupees and watch. I took off my wedding-ring, and twisted it in my hair, and replaced the guard. I tore off the skirt of my dress to bring water in, but was no use, for when I returned my beloved's eyes were fixed, and, though I called and tried to restore him, and poured water into his mouth, it only rattled in his throat. He never spoke to me again. I held him in my arms till he sank gradually down. I felt frantic, but could not cry. I was alone. I bound his head and face in my dress, for there was no earth to bury him. The pain in my hands and feet was dreadful. I went down to the ravine, and sat in the water on a stone, hoping to get off at night and look for Lottie. When I came back from the water, I saw that they had not taken her little watch, chain, and seals, so I tied them under my petticoat. In an hour, about thirty villagers came, they dragged me out of the ravine, and took off my jacket, and found the little chain. They then dragged me to a village, mocking me all the way, and disputing as to whom I was to belong to. The whole population came to look at me   I asked for a bedstead, and lay down outside the door of a hut. They had a dozen of cows, and yet refused me milk. When night came, and the village was quiet, some old woman brought me a leafful of rice. I was too parched to eat, and they gave me water. The morning after a neighboring Rajah sent a palanquin and a horseman to fetch me, who told me that a little child and three Sahibs had come to his master's house.   And so the poor mother found her lost one, 'greatly blistered,' poor little creature.   It is not for Europeans in India to pray that their flight be not in the winter."

In the first days of June the aged general, Sir Hugh Wheeler commanding the forces at Cawnpore, was deserted by his native troops; then he moved out of the fort and into an exposed patch of open flat ground and built a 'our-foot mud wall around it. He had with him a few hundred white soldiers and officers, and apparently more women and children than soldiers. He was short of provisions, short of arms, short of ammunition, short of military wisdom, short of everything but courage and devotion to duty. The defense of that open lot through twenty-one days and nights of hunger, thirst, Indian heat, and a never-ceasing storm of bullets, bombs, and cannon-balls — a defense conducted, not by the aged and infirm general, but by a young officer named Moore — is one of the most heroic episodes in history. When at last the Nana found it impossible to conquer these starving men and women with powder and ball, he resorted to treachery, and that succeeded. He agreed to supply them with food and send them to Allahabad in boats. Their mud wall and their barracks were in ruins, their provisions were at the point of exhaustion, they had done all that the brave could do, they had conquered an honorable compromise, their forces had been fearfully reduced by casualties and by disease, they were not able to continue the contest longer. They came forth helpless but suspecting no treachery, the Nana's host closed around them, and at a signal from a trumpet the massacre began. About two hundred women and children were spared — for the present — but all the men except three or four were killed. Among the incidents of the massacre quoted by Sir G. O. Trevelyan, is this:

"When, after the lapse of some twenty minutes, the dead began to out-number the living; — when the fire slackened, as the marks grew few and far between; then the troopers who had been drawn up to the right of the temple plunged into the river, sabre between teeth, and pistol in hand. Thereupon two half-caste Christian women, the wives of musicians in the band of the Fifty-sixth, witnessed a scene which should not be related at second-hand. 'In the boat where I was to have gone,' says Mrs. Bradshaw, confirmed throughout by Mrs. Setts, 'was the school-mistress and twenty-two misses. General Wheeler came last in a palkee. They carried him into the water near the boat. I stood close by. He said, 'Carry me a little further towards the boat.' But a trooper said, 'No, get out here.' As the General got out of the palkee, headforemost, the trooper gave him a cut with his sword into the neck, and he fell into the water. My son was killed near him. I saw it; alas! alas! Some were stabbed with bayonets; others cut down. Little in-fants were torn in pieces. We saw it; we did; and tell you only what we saw. Other children were stabbed and thrown into the river. The school-girls were burnt to death. I saw their clothes and hair catch fire. In the water, a few paces off, by the next boat, we saw the youngest daughter of Colonel Williams. A sepoy was going to kill her with his bayonet. She said, 'My father was always kind to sepoys.' He turned away, and just then a villager struck her on the head with a club, and she fell into the water These people likewise saw good Mr. Moncrieff, the clergyman, take a book

from his pocket that he never had leisure to open, and heard him commence
a prayer for mercy which he was not permitted to conclude. Another de-
ponent observed an European making for a drain like a scared water-rat,
when some boatmen, armed with cudgels, cut off his retreat, and beat him
down dead into the mud."

The women and children who had been reserved from the
massacre were imprisoned during a fortnight in a small building, one story high — a cramped place, a slightly modified
Black Hole of Calcutta. They were waiting in suspense;
there was none who could forecaste their fate. Meantime the
news of the massacre had traveled far and an army of rescuers
with Havelock at its head was on its way — at least an army
which hoped to be rescuers. It was crossing the country by
forced marches, and strewing its way with its own dead —
men struck down by cholera, and by a heat which reached
135°. It was in a vengeful fury, and it stopped for nothing —
neither heat, nor fatigue, nor disease, nor human opposition.
It tore its impetuous way through hostile forces, winning victory after victory, but still striding on and on, not halting to
count results. And at last, after this extraordinary march, it
arrived before the walls of Cawnpore, met the Nana's massed
strength, delivered a crushing defeat, and entered.

But too late — only a few hours too late. For at the last
moment the Nana had decided upon the massacre of the
captive women and children, and had commissioned three
Mohammedans and two Hindoos to do the work. Sir G. O.
Trevelyan says:

"Thereupon the five men entered. It was the short gloaming of Hindo-
stan — the hour when ladies take their evening drive. She who had accosted
the officer was standing in the doorway. With her were the native doctor
and two Hindoo menials. That much of the business might be seen from the
veranda, but all else was concealed amidst the interior gloom. Shrieks and
scuffling acquainted those without that the journeymen were earning their
hire. Survur Khan soon emerged with his sword broken off at the hilt. He
procured another from the Nana's house, and a few minutes after appeared
again on the same errand. The third blade was of better temper; or perhaps
the thick of the work was already over. By the time darkness had closed in,

the men came forth and locked up the house for the night.   Then the screams ceased, but the groans lasted till morning.

"The sun rose as usual.   When he had been up nearly three hours the five repaired to the scene of their labors over night.   They were attended by a few sweepers, who proceeded to transfer the contents of the house to a dry well situated behind some trees which grew hard by.   'The bodies,' says one who was present throughout, 'were dragged out, most of them by the hair of the head.   Those who had clothing worth taking were stripped.   Some of the women were alive.   I cannot say how many ; but *three could speak*.   They prayed for the sake of God that an end might be put to their sufferings.   I remarked one very stout woman, a half-caste, who was severely wounded in both arms, who entreated to be killed.   She and two or three others were placed against the bank of the cut by which bullocks go down in drawing water.   The dead were first thrown in.   Yes : there was a great crowd looking on ; they were standing along the walls of the compound.   They were principally city people and villagers.   Yes : there were also sepoys.   *Three boys were alive*.   They were fair children.   The eldest, I think, must have been six or seven, and the youngest five years.   They were running around the well (where else could they go to ?), and there was none to save them.   No : none said a word or tried to save them.'

"At length the smallest of them made an infantile attempt to get away.   The little thing had been frightened past bearing by the murder of one of the surviving ladies.   He thus attracted the observation of a native who flung him and his companions down the well."

The soldiers had made a march of eighteen days, almost without rest, to save the women and the children, and now they were too late — all were dead and the assassin had flown. What happened then, Trevelyan hesitated to put into words. "Of what took place, the less said is the better."

Then he continues :

"But there was a spectacle to witness which might excuse much.   Those who, straight from the contested field, wandered sobbing through the rooms of the ladies' house, saw what it were well could the outraged earth have straightway hidden.   The inner apartment was ankle-deep in blood.   The plaster was scored with sword-cuts ; not high up as where men have fought, but low down, and about the corners, as if a creature had crouched to avoid the blow.   Strips of dresses, vainly tied around the handles of the doors, signified the contrivance to which feminine despair had resorted as a means of keeping out the murderers.   Broken combs were there, and the frills of children's trousers, and torn cuffs and pinafores, and little round hats, and one or two shoes with burst latchets, and one or two daguerreotype cases with cracked glasses.   An officer picked up a few curls, preserved in a bit of cardboard, and marked 'Ned's hair, with love'; but around were strewn locks, some near a yard in length, dissevered, not as a keepsake, by quite other scissors."

The battle of Waterloo was fought on the 18th of June, 1815. I do not state this fact as a reminder to the reader, but as news to him. For a forgotten fact *is* news when it comes again. Writers of books have the fashion of whizzing by vast and renowned historical events with the remark, " The details of this tremendous episode are too familiar to the reader to need repeating here." They know that that is not true. It is a low kind of flattery. They know that the reader has forgotten every detail of it, and that nothing of the tremendous event is left in his mind but a vague and formless luminous smudge. Aside from the desire to flatter the reader, they have another reason for making the remark — two reasons, indeed. They do not remember the details themselves, and do not want the trouble of hunting them up and copying them out; also, they are afraid that if they search them out and print them they will be scoffed at by the book-reviewers for retelling those worn old things which are familiar to everybody. They should not mind the reviewer's jeer; *he* doesn't remember any of the worn old things until the book which he is reviewing has retold them to him.

I have made the quoted remark myself, at one time and another, but I was not doing it to flatter the reader; I was merely doing it to save work. If I had known the details without brushing up, I would have put them in; but I didn't, and I did not want the labor of posting myself; so I said, " The details of this tremendous episode are too familiar to the reader to need repeating here." I do not like that kind of a lie; still, it does save work.

I am not trying to get out of repeating the details of the Siege of Lucknow in fear of the reviewer; I am not leaving them out in fear that they would not interest the reader; I am leaving them out partly to save work; mainly for lack of room. It is a pity, too; for there is not a dull place anywhere in the great story.

Ten days before the outbreak (May 10th) of the Mutiny, all was serene at Lucknow, the huge capital of Oudh, the kingdom which had recently been seized by the India Company. There was a great garrison, composed of about 7,000 native troops and between 700 and 800 whites. These white soldiers and their families were probably the only people of their race there; at their elbow was that swarming population of warlike natives, a race of born soldiers, brave, daring, and fond of fighting. On high ground just outside the city stood the palace of that great personage, the Resident, the representative of British power and authority. It stood in the midst of spacious grounds, with its due complement of outbuildings, and the grounds were enclosed by a wall — a wall not for defense, but for privacy. The mutinous spirit was in the air, but the whites were not afraid, and did not feel much troubled.

Then came the outbreak at Meerut, then the capture of Delhi by the mutineers; in June came the three-weeks leaguer of Sir Hugh Wheeler in his open lot at Cawnpore — 40 miles distant from Lucknow — then the treacherous massacre of that gallant little garrison; and now the great revolt was in full flower, and the comfortable condition of things at Lucknow was instantly changed.

There was an outbreak there, and Sir Henry Lawrence marched out of the Residency on the 30th of June to put it down, but was defeated with heavy loss, and had difficulty in getting back again. That night the memorable siege of the Residency — called the siege of Lucknow — began. Sir Henry was killed three days later, and Brigadier Inglis succeeded him in command.

Outside of the Residency fence was an immense host of hostile and confident native besiegers; inside it were 480 loyal native soldiers, 730 white ones, and 500 women and children.

In those days the English garrisons always managed to hamper themselves sufficiently with women and children.

The natives established themselves in houses close at hand and began to rain bullets and cannon-balls into the Residency; and this they kept up, night and day, during four months and a half, the little garrison industriously replying all the time. The women and children soon became so used to the roar of the guns that it ceased to disturb their sleep. The children imitated siege and defense in their play. The women — with any pretext, or with none — would sally out into the storm-swept grounds.

The defense was kept up week after week, with stubborn fortitude, in the midst of death, which came in many forms — by bullet, small-pox, cholera, and by various diseases induced by unpalatable and insufficient food, by the long hours of wearying and exhausting overwork in the daily and nightly battle in the oppressive Indian heat, and by the broken rest caused by the intolerable pest of mosquitoes, flies, mice, rats, and fleas.

Six weeks after the beginning of the siege more than one-half of the original force of white soldiers was dead, and close upon three-fifths of the original native force.

But the fighting went on just the same. The enemy mined, the English counter-mined, and, turn about, they blew up each other's posts. The Residency grounds were honey-combed with the enemy's tunnels. Deadly courtesies were constantly exchanged — sorties by the English in the night; rushes by the enemy in the night — rushes whose purpose was to breach the walls or scale them; rushes which cost heavily, and always failed.

The ladies got used to all the horrors of war — the shrieks of mutilated men, the sight of blood and death. Lady Inglis makes this mention in her diary:

WHERE THE WOMEN AND CHILDREN LODGED.

"Mrs. Bruere's nurse was carried past our door to-day, wounded in the eye. To extract the bullet it was found necessary to take out the eye—a fearful operation. Her mistress held her while it was performed."

The first relieving force failed to relieve. It was under Havelock and Outram, and arrived when the siege had been going on for three months. It fought its desperate way to Lucknow, then fought its way through the city against odds of a hundred to one, and entered the Residency; but there was not enough left of it, then, to do any good. It lost more men in its last fight than it found in the Residency when it got in. It became captive itself.

The fighting and starving and dying by bullets and disease went steadily on. Both sides fought with energy and industry. Captain Birch puts this striking incident in evidence. He is speaking of the third month of the siege:

"As an instance of the heavy firing brought to bear on our position this month may be mentioned the cutting down of the upper story of a brick building simply by *musketry firing*. This building was in a most exposed position. All the shots which just missed the top of the rampart cut into the dead wall pretty much in a straight line, and at length cut right through and brought the upper story tumbling down. The upper structure on the top of the brigade-mess also fell in. The Residency house was a wreck. Captain Anderson's post had long ago been knocked down, and Innes' post also fell in. These two were riddled with round shot. As many as 200 were picked up by Colonel Masters."

The exhausted garrison fought doggedly on all through the next month — October. Then, November 2d, news came — Sir Colin Campbell's relieving force would soon be on its way from Cawnpore.

On the 12th the boom of his guns was heard.

On the 13th the sounds came nearer — he was slowly, but steadily, cutting his way through, storming one stronghold after another.

On the 14th he captured the Martiniere College, and ran up the British flag there. It was seen from the Residency.

Next he took the Dilkoosha.

36

On the 17th he took the former mess-house of the 32d regiment — a fortified building, and very strong. "A most exciting, anxious day," writes Lady Inglis in her diary. "About 4 P. M., two strange officers walked through our yard, leading their horses "— and by that sign she knew that communication was established between the forces, that the relief was real, this time, and that the long siege of Lucknow was ended.

The last eight or ten miles of Sir Colin Campbell's march was through seas of blood. The weapon mainly used was the bayonet, the fighting was desperate. The way was milestoned with detached strong buildings of stone, fortified, and heavily garrisoned, and these had to be taken by assault. Neither side asked for quarter, and neither gave it. At the Secundrabagh, where nearly two thousand of the enemy occupied a great stone house in a garden, the work of slaughter was continued until every man was killed. That is a sample of the character of that devastating march.

There were but few trees in the plain at that time, and from the Residency the progress of the march, step by step, victory by victory, could be noted; the ascending clouds of battle-smoke marked the way to the eye, and the thunder of the guns marked it to the ear.

Sir Colin Campbell had not come to Lucknow to hold it, but to save the occupants of the Residency, and bring them away. Four or five days after his arrival the secret evacuation by the troops took place, in the middle of a dark night, by the principal gate (the Bailie Guard). The two hundred women and two hundred and fifty children had been previously removed. Captain Birch says:

"And now commenced a movement of the most perfect arrangement and successful generalship — the withdrawal of the whole of the various forces, a combined movement requiring the greatest care and skill. First, the garrison in immediate contact with the enemy at the furthest extremity of the

THE BAILIE GUARD GATE.

Residency position was marched out. Every other garrison in turn fell in behind it, and so passed out through the Bailie Guard gate, till the whole of our position was evacuated. Then Havelock's force was similarly withdrawn, post by post, marching in rear of our garrison. After them in turn came the forces of the Commander-in-Chief, which joined on in the rear of Havelock's force. Regiment by regiment was withdrawn with the utmost order and regularity. The whole operation resembled the movement of a telescope. Stern silence was kept, and the enemy took no alarm."

Lady Inglis, referring to her husband and to General Sir James Outram, sets down the closing detail of this impressive midnight retreat, in darkness and by stealth, of this shadowy host through the gate which it had defended so long and so well:

"At twelve precisely they marched out, John and Sir James Outram remaining till all had passed, and then they took off their hats to the Bailie Guard, the scene of as noble a defense as I think history will ever have to relate."

RUINS OF THE RESIDENCY.

CHAPTER LIX.

Don't part with your illusions. When they are gone you may still exist but you have ceased to live. — *Pudd'nhead Wilson's New Calendar.*

Often, the surest way to convey misinformation is to tell the strict truth.
— *Pudd'nhead Wilson's New Calendar.*

WE were driven over Sir Colin Campbell's route by a British officer, and when I arrived at the Residency I was so familiar with the road that I could have led a retreat over it myself; but the compass in my head has been out of order from my birth, and so, as soon as I was within the battered Bailie Guard and turned about to review the march and imagine the relieving forces storming their way along it, everything was upside down and wrong end first in a moment, and I was never able to get straightened out again. And now, when I look at the battle-plan, the confusion remains. In me the east was born west, the battle-plans which have the east on the right-hand side are of no use to me.

The Residency ruins are draped with flowering vines, and are impressive and beautiful. They and the grounds are sacred now, and will suffer no neglect nor be profaned by any sordid or commercial use while the British remain masters of India. Within the grounds are buried the dead who gave up their lives there in the long siege.

After a fashion, I was able to imagine the fiery storm that raged night and day over the place during so many months, and after a fashion I could imagine the men moving through it, but I could not satisfactorily place the 200 women, and I could do nothing at all with the 250 children. I knew by

THE RESIDENCY GATEWAY.

Lady Inglis' diary that the children carried on their small affairs very much as if blood and carnage and the crash and thunder of a siege were natural and proper features of nursery life, and I tried to realize it; but when her little Johnny came rushing, all excitement, through the din and smoke, shouting, "Oh, mamma, the white hen has laid an egg!" I saw that I could not do it. Johnny's place was under the bed. I could imagine him there, because I could imagine myself there; and I think I should not have been interested in a hen that was laying an egg; my interest would have been with the parties that were laying the bombshells. I sat at dinner with one of those children in the Club's Indian palace, and I knew that all through the siege he was perfecting his teething and learning to talk; and while to me he was the most impressive object in Lucknow after the Residency ruins, I was not able to imagine what his life had been during that tempestuous infancy of his, nor what sort of a curious surprise it must have been to him to be marched suddenly out into a strange dumb world where there wasn't any noise, and nothing going on. He was only forty-one when I saw him, a strangely youthful link to connect the present with so ancient an episode as the Great Mutiny.

By and by we saw Cawnpore, and the open lot which was the scene of Moore's memorable defense, and the spot on the shore of the Ganges where the massacre of the betrayed garrison occurred, and the small Indian temple whence the bugle-signal notified the assassins to fall on. This latter was a lonely spot, and silent. The sluggish river drifted by, almost currentless. It was dead low water, narrow channels with vast sandbars between, all the way across the wide bed; and the only living thing in sight was that grotesque and solemn bald-headed bird, the Adjutant, standing on his six-foot stilts, solitary on a distant bar, with his head sunk between his shoulders, thinking; thinking of his prize, I suppose — the dead Hindoo that lay awash at his feet, and whether to eat him alone or invite friends. He and his prey were a proper accent to that mournful place. They were in keeping with it, they emphasized its loneliness and its solemnity.

And we saw the scene of the slaughter of the helpless women and children, and also the costly memorial that is built over the well which contains their remains. The Black Hole of Calcutta is gone, but a more reverent age is come, and whatever remembrancer still exists of the moving and heroic sufferings and achievements of the garrisons of Lucknow and Cawnpore will be guarded and preserved.

In Agra and its neighborhood, and afterwards at Delhi, we saw forts, mosques, and tombs, which were built in the great days of the Mohammedan emperors, and which are marvels of cost, magnitude, and richness of materials and ornamentation, creations of surpassing grandeur, wonders which do indeed make the like things in the rest of the world seem tame and inconsequential by comparison. I am not purposing to describe them. By good fortune I had not read too much about them, and therefore was able to get a natural and rational focus upon them, with the result that they thrilled, blessed, and exalted me. But if I had previously overheated my imagination by drinking too much pestilential literary hot Scotch, I should have suffered disappointment and sorrow.

I mean to speak of only one of these many world-renowned buildings, the Taj Mahal, the most celebrated construction in the earth. I had read a great deal too much about it. I saw it in the daytime, I saw it in the moonlight, I saw it near at hand, I saw it from a distance; and I knew all the time, that of its kind it was *the* wonder of the world, with no competitor now and no possible future competitor; and yet, it was not *my* Taj. My Taj had been built by excitable literary people; it was solidly lodged in my head, and I could not blast it out.

I wish to place before the reader some of the usual descriptions of the Taj, and ask him to take note of the impressions left in his mind. These descriptions do really state the truth —as nearly as the limitations of language will allow. But language is a treacherous thing, a most unsure vehicle, and it can seldom arrange descriptive words in such a way that they will not inflate the facts — by help of the reader's imagination, which is always ready to take a hand, and work for nothing, and do the bulk of it at that.

I will begin with a few sentences from the excellent little local guide-book of Mr. Satya Chandra Mukerji. I take them from here and there in his description :

"The inlaid work of the Taj and the flowers and petals that are to be found on all sides on the surface of the marble evince a most delicate touch."

That is true.

"The inlaid work, the marble, the flowers, the buds, the leaves, the petals, and the lotus stems are almost without a rival in the whole of the civilized world."

"The work of inlaying with stones and gems is found in the highest perfection in the Taj."

Gems, inlaid flowers, buds, and leaves to be found on all sides. What do you see before you? Is the fairy structure growing? Is it becoming a jewel casket?

"The whole of the Taj produces a wonderful effect that is equally sublime and beautiful."

Then Sir William Wilson Hunter :

"The Taj Mahal with its beautiful domes, 'a dream of marble,' rises on the river bank."

"The materials are white marble and red sandstone."

"The complexity of its design and the delicate intricacy of the workmanship baffle description."

Sir William continues. I will italicize some of his words :

"The mausoleum stands on a raised marble platform at each of whose corners rises a tall and slender minaret of graceful proportions and of ex-quisite beauty. Beyond the platform stretch the two wings, one of which is itself a mosque of great architectural merit. In the center of the whole design the mausoleum occupies a square of 186 feet, with the angles deeply truncated so as to form an unequal octagon. The main feature in this central pile is the great dome, which swells upward to nearly two-thirds of a sphere and tapers at its extremity into a pointed spire crowned by a crescent. Beneath it an enclosure of marble trellis-work surrounds the tomb of the princess and of her husband, the Emperor. Each corner of the mausoleum is covered by a similar though much smaller dome erected on a pediment pierced with graceful Saracenic arches. Light is admitted into the interior through a double screen of pierced marble, which tempers the glare of an Indian sky while its whiteness prevents the mellow effect from degenerating into gloom. The internal decorations consist of inlaid work in *precious stones, such as agate, jasper, etc., with which every squandril or salient point in the architecture is richly fretted.* Brown and violet marble is also freely employed in wreaths, scrolls, and lintels to relieve the monotony of white wall. *In regard to color and design, the interior of the Taj may rank first in the world for purely decorative workmanship;* while the perfect symmetry of its exterior, once seen can never be forgotten, nor the aërial grace of its domes, rising like marble bubbles into the clear sky. The Taj represents the most highly elaborated stage of ornamentation reached by the Indo-Mohammedan builders, the stage in which the architect ends and the *jeweler* begins. In its magnificent gateway the diagonal ornamentation at the corners, which satisfied the designers of the gateways of Itimad-ud-doulah and Sikandra mausoleums is superseded by fine marble cables, in bold twists, strong and handsome. The triangular insertions of white marble and large flowers have in like manner given place to *fine inlaid work.* Firm perpendicular lines in black marble with well proportioned panels of the same material are effectively used in the interior of the gateway. On its top the Hindu brackets and monolithic architraves of Sikandra are replaced by Moorish carped arches, usually single blocks of red sandstone, in the Kiosks and pavilions which adorn the roof. From the pillared pavilions a magnificent view is obtained of the Taj gardens below, with the noble Jumna river at their farther end, and the city and fort of Agra in the distance. From this beautiful and splendid gateway one passes up a straight alley shaded by evergreen trees cooled by a broad shallow piece of water running along the middle of the path to the Taj itself. *The Taj is entirely of marble and gems.* The red sandstone of the other Mohammedan buildings has entirely disappeared, or rather the red sandstone which used to form the thickness of the walls, is in the Taj itself overlaid completely with white marble, and the white marble is itself *inlaid with precious stones arranged in lovely patterns of flowers.* A feeling of purity impresses itself on the eye and the mind from the absence of the coarser material which forms so invariable a material in Agra architecture. The lower wall and panels are covered with tulips, oleanders, and full-blown lilies, in flat carving on the white marble ; and *although the inlaid work of flowers done in gems* is very *brilliant* when looked at closely, there is on the whole but little color, and the all-prevailing sentiment is one of whiteness, silence, and calm. The whiteness is broken only by the fine color of the inlaid gems, by lines in black marble, and by delicately written inscriptions, also in black, from the Koran. Under the dome of *the vast mausoleum* a high and beautiful screen of open tracery in white marble rises around the two tombs, or rather cenotaphs of the emperor and his princess ; and in this *marvel of marble* the carving has advanced from the old geometrical patterns to a trellis-work of flowers and foliage, handled with great freedom and spirit. The two cenotaphs in the center of the *exquisite* enclosure have no carving except the plain *Kalamdan* or oblong pen-box on the tomb of Emperor Shah Jehan. But both cenotaphs are *inlaid with flowers made of costly gems,* and with the ever graceful oleander scroll."

Bayard Taylor, after describing the details of the Taj, goes on to say :

"On both sides the palm, the banyan, and the feathery bamboo mingle their foliage ; the song of birds meets your ears, and the odor of roses and lemon flowers sweetens the air. Down such a vista and over such a foreground rises the Taj. There is no mystery, no sense of partial failure about the Taj. *A thing of perfect beauty and of absolute finish* in every detail, it might pass for the work of genii who knew naught of the weaknesses and ills with which mankind are beset."

All of these details are true. But, taken together, they state a falsehood — to *you.* You cannot add them up correctly. Those writers know the values of their words and phrases, but to you the words and phrases convey other and uncertain values. To those writers their phrases have values which I think I am now acquainted with ; and for the help of the reader I will here repeat certain of those words and phrases, and follow them with numerals which shall represent those values — then we shall see the difference between a writer's ciphering and a mistaken reader's :

Precious stones, such as agate, jasper, etc. — 5.

With which every salient point is richly fretted — 5.

First in the world for purely decorative workmanship — 9.

The Taj represents the stage where the architect ends and the jeweler begins — 5.

The Taj is entirely of marble and gems — 7.

Inlaid with precious stones in lovely patterns of flowers — 5.

The inlaid work of flowers done in gems is very brilliant — (followed by a most important modification which the reader is sure to read too carelessly) — 2.

The vast mausoleum — 5.

This marvel of marble — 5.

The exquisite enclosure — 5.

Inlaid with flowers made of costly gems — 5.

A thing of perfect beauty and absolute finish — 5.

Those details are correct; the figures which I have placed after them represent quite fairly their individual values. Then why, as a whole, do they convey a false impression to the reader? It is because the reader — beguiled by his heated imagination — masses them in the wrong way. The *writer* would mass the first three figures in the following way, and they would speak the truth:

5

5

9

Total — 19

But the reader masses them thus — and then they tell a lie — 559.

The writer would add all of his twelve numerals together, and then the sum would express the whole truth about the Taj, and the truth only — 63.

But the reader — always helped by his imagination — would put the figures in a row one after the other, and get this sum, which would tell him a noble big lie:

559575255555.

You must put in the commas yourself; I have to go on with my work.

The reader will always be sure to put the figures together in that wrong way, and then as surely before him will stand, sparkling in the sun, a gem-crusted Taj tall as the Matterhorn.

I had to visit Niagara fifteen times before I succeeded in getting my imaginary Falls gauged to the actuality and could begin to sanely and wholesomely wonder at them for what they were, not what I had expected them to be. When I first ap-

THE TAJ MAHAL.

AN EXAGGERATED NIAGARA.

proached them it was with my face lifted toward the sky, for I thought I was going to see an Atlantic ocean pouring down thence over cloud-vexed Himalayan heights, a sea-green wall of water sixty miles front and six miles high, and so, when the toy reality came suddenly into view — that beruffled little wet apron hanging out to dry — the shock was too much for me, and I fell with a dull thud.

Yet slowly, surely, steadily, in the course of my fifteen visits, the proportions adjusted themselves to the facts, and I came at last to realize that a waterfall a hundred and sixty-five feet high and a quarter of a mile wide was an impressive thing. It was not a dipperful to my vanished great vision, but it would answer.

I know that I ought to do with the Taj as I was obliged to do with Niagara — see it fifteen times, and let my mind gradually get rid of the Taj built in it by its describers, by help of my imagination, and substitute for it the Taj of fact. It would be noble and fine, then, and a marvel; not the marvel which it replaced, but still a marvel, and fine enough. I am a careless reader, I suppose — an *impressionist* reader; an impressionist reader of what is *not* an impressionist picture; a reader who overlooks the informing details or masses their sum improperly, and gets only a large splashy, general effect — an effect which is not correct, and which is not warranted by the particulars placed before me — particulars which I did not examine, and whose meanings I did not cautiously and carefully estimate. It is an effect which is some thirty-five or forty times finer than the reality, and is therefore a great deal better and more valuable than the reality; and so, I ought never to hunt up the reality, but stay miles away from it, and thus preserve undamaged my own private mighty Niagara tumbling out of the vault of heaven, and my own ineffable Taj, built of tinted mists upon jeweled arches of rainbows supported by colonnades of moonlight. It is a mistake for a person with an unregulated imagination to go and look at an illustrious world's wonder.

I suppose that many, many years ago I gathered the idea that the Taj's place in the achievements of man was exactly the place of the ice-storm in the achievements of Nature; that the Taj represented man's supremest possibility in the creation of grace and beauty and exquisiteness and splendor, just as the ice-storm represents Nature's supremest possibility in the combination of those same qualities. I do not know how long ago that idea was bred in me, but I know that I cannot remember back to a time when the thought of either of these symbols of gracious and unapproachable perfection did not at once suggest the other. If I thought of the ice-storm, the Taj rose before me divinely beautiful; if I thought of the Taj, with its encrustings and inlayings of jewels, the vision of the ice-storm rose. And so, to me, all these years, the Taj has had no rival among the temples and palaces of men, none that even remotely approached it — it was man's architectural ice-storm.

Here in London the other night I was talking with some Scotch and English friends, and I mentioned the ice-storm, using it as a figure — a figure which failed, for none of them had heard of the ice-storm. One gentleman, who was very familiar with American literature, said he had never seen it mentioned in any book. That is strange. And I, myself, was not able to say that I had seen it mentioned in a book; and yet the autumn foliage, with all other American scenery, has received full and competent attention.

The oversight is strange, for in America the ice-storm is an event. And it is not an event which one is careless about. When it comes, the news flies from room to room in the house, there are bangings on the doors, and shoutings, "The

ice-storm! the ice-storm!" and even the laziest sleepers throw
off the covers and join the rush for the windows. The ice-
storm occurs in mid-winter, and usually its enchantments are
wrought in the silence and the darkness of the night. A
fine drizzling rain falls hour after hour upon the naked twigs
and branches of the trees, and as it falls it freezes. In time
the trunk and every branch and twig are incased in hard pure
ice; so that the tree looks like a skeleton tree made all of
glass — glass that is crystal-clear. All along the under side of
every branch and twig is a comb of little icicles — the frozen
drip. Sometimes these pendants do not quite amount to
icicles, but are round beads — frozen tears.

The weather clears, toward dawn, and leaves a brisk pure
atmosphere and a sky without a shred of cloud in it — and
everything is still, there is not a breath of wind. The dawn
breaks and spreads, the news of the storm goes about the
house, and the little and the big, in wraps and blankets, flock
to the window and press together there, and gaze intently out
upon the great white ghost in the grounds, and nobody says a
word, nobody stirs. All are waiting; they know what is
coming, and they are waiting — waiting for the miracle. The
minutes drift on and on and on, with not a sound but the
ticking of the clock; at last the sun fires a sudden sheaf of
rays into the ghostly tree and turns it into a white splendor of
glittering diamonds. Everybody catches his breath, and feels
a swelling in his throat and a moisture in his eyes — but waits
again; for he knows what is coming; there is more yet. The
sun climbs higher, and still higher, flooding the tree from its
loftiest spread of branches to its lowest, turning it to a glory of
white fire; then in a moment, without warning, comes the
great miracle, the supreme miracle, the miracle without its fellow in the earth; a gust of wind sets every branch and twig
to swaying, and in an instant turns the whole white tree into a

spouting and spraying explosion of flashing gems of every conceivable color; and there it stands and sways this way and that, flash! flash! flash! a dancing and glancing world of rubies, emeralds, diamonds, sapphires, the most radiant spectacle, the most blinding spectacle, the divinest, the most exquisite, the most intoxicating vision of fire and color and intolerable and unimaginable splendor that ever any eye has rested upon in this world, or will ever rest upon outside of the gates of heaven.

By all my senses, all my faculties, I know that the ice-storm is Nature's supremest achievement in the domain of the superb and the beautiful; and by my reason, at least, I know that the Taj is man's ice-storm.

In the ice-storm every one of the myriad ice-beads pendant from twig and branch is an individual gem, and changes color with every motion caused by the wind; each tree carries a million, and a forest-front exhibits the splendors of the single tree multiplied by a thousand.

It occurs to me now that I have never seen the ice-storm put upon canvas, and have not heard that any painter has tried to do it. I wonder why that is. Is it that paint cannot counterfeit the intense blaze of a sun-flooded jewel? There should be, and must be, a reason, and a good one, why the most enchanting sight that Nature has created has been neglected by the brush.

Often, the surest way to convey misinformation is to tell the strict truth. The describers of the Taj have used the word *gem* in its strictest sense — its scientific sense. In that sense it is a mild word, and promises but little to the eye — nothing bright, nothing brilliant, nothing sparkling, nothing splendid in the way of color. It *accurately* describes the sober and unobtrusive gem-work of the Taj; that is, to the very highly-educated one person in a thousand; but it most falsely describes it to the 999.   But the 999 are the people who ought to be especially taken care of, and to them it does not mean quiet-colored designs wrought in carnelians, or agates, or such things; they know the word in its wide and ordinary sense only, and so to them it means diamonds and rubies and opals and their kindred, and the moment their eyes fall upon it in print they see a vision of glorious colors clothed in fire.

These describers are writing for the "general," and so, in order to make sure of being understood, they ought to use words in their ordinary sense, or else explain.   The word *fountain* means one thing in Syria, where there is but a handful of people; it means quite another thing in North America, where there are 75,000,000.   If I were describing some Syrian scenery, and should exclaim, "Within the narrow space of a quarter of a mile square I saw, in the glory of the flooding moonlight, two hundred noble fountains — imagine the spectacle!" the North American would have a vision of clustering columns of water soaring aloft, bending over in graceful arches, bursting in beaded spray and raining white fire in the moonlight — and he would be deceived.   But the Syrian would not be deceived; he would merely see two hundred fresh-water springs — two hundred drowsing puddles, as level and unpretentious and unexcited as so many door-mats, and even with the help of the moonlight he would not lose his grip in the presence of the exhibition.   My word "fountain" would be correct; it would speak the strict truth; and it would convey the strict truth to the handful of Syrians, and the strictest misinformation to the North American millions.   With their gems — and gems — and more gems — and gems again — and still other gems — the describers of the Taj are within their legal but not their moral rights; they are dealing in the strictest scientific truth; and in doing it they succeed to admiration in telling "what ain't so."

CHAPTER LX.

SATAN (impatiently) to NEW-COMER. The trouble with you Chicago people is, that you think you are the best people down here; whereas you are merely the most numerous. — *Pudd'nhead Wilson's New Calendar.*

WE wandered contentedly around here and there in India; to Lahore, among other places, where the Lieutenant-Governor lent me an elephant. This hospitality stands out in my experiences in a stately isolation. It was a fine elephant, affable, gentlemanly, educated, and I was not afraid of it. I even rode it with confidence through the crowded lanes of the native city, where it scared all the horses out of their senses, and where children were always just escaping its feet. It took the middle of the road in a fine independent way, and left it to the world to get out of the way or take the consequences. I am used to being afraid of collisions when I ride or drive, but when one is on top of an elephant that feeling is absent. I could have ridden in comfort through a regiment of runaway teams. I could easily learn to prefer an elephant to any other vehicle, partly because of that immunity from collisions, and partly because of the fine view one has from up there, and partly because of the dignity one feels in that high place, and partly because one can look in at the windows and see what is going on privately among the family. The Lahore horses were used to elephants, but they were rapturously afraid of them just the same. It seemed curious. Perhaps the better they know the elephant the more they respect him in that peculiar way. In our own case we are not afraid of dynamite till we get acquainted with it.

We drifted as far as Rawal Pindi, away up on the Afghan frontier — I think it was the Afghan frontier, but it may have been Hertzegovina — it was around there somewhere — and down again to Delhi, to see the ancient architectural wonders there and in Old Delhi and not describe them, and also to see the scene of the illustrious assault, in the Mutiny days, when the British carried Delhi by storm, one of the marvels of history for impudent daring and immortal valor.

We had a refreshing rest, there in Delhi, in a great old mansion which possessed historical interest. It was built by a rich Englishman who had become orientalized — so much so that he had a zenana. But he was a broad-minded man, and remained so. To please his harem he built a mosque ; to please himself he built an English church. That kind of a man will arrive, somewhere. In the Mutiny days the mansion was the British general's headquarters. It stands in a great garden — oriental fashion — and about it are many noble trees. The trees harbor monkeys ; and they are monkeys of a watchful and enterprising sort, and not much troubled with fear. They invade the house whenever they get a chance, and carry off everything they don't want. One morning the master of the house was in his bath, and the window was open. Near it stood a pot of yellow paint and a brush. Some monkeys appeared in the window ; to scare them away, the gentleman threw his sponge at them. They did not scare at all ; they jumped into the room and threw yellow paint all over him from the brush, and drove him out ; then they painted the walls and the floor and the tank and the windows and the furniture yellow, and were in the dressing-room painting that when help arrived and routed them.

Two of these creatures came into my room in the early morning, through a window whose shutters I had left open, and when I woke one of them was before the glass brushing his hair, and the other one had my note-book, and was reading a page of humorous notes and crying. I did not mind the one with the hair-brush, but the conduct of the other one hurt me; it hurts me yet. I threw something at him, and that was wrong, for my host had told me that the monkeys were best left alone. They threw everything at me that they could lift, and then went into the bathroom to get some more things, and I shut the door on them.

At Jeypore, in Rajputana, we made a considerable stay. We were not in the native city, but several miles from it, in the small European official suburb. There were but few Europeans — only fourteen — but they were all kind and hospitable, and it amounted to being at home. In Jeypore we found again what we had found all about India — that while the Indian servant is in his way a very real treasure, he will sometimes bear watching, and the Englishman watches him. If he sends him on an errand, he wants more than the man's word for it that he did the errand. When fruit and vegetables were sent to us, a "chit" came with them — a receipt for us to sign; otherwise the things might not arrive. If a gentleman sent up his carriage, the chit stated "from" such-and-such an hour "to" such-and-such an hour — which made it unhandy for the coachman and his two or three subordinates to put us off with a part of the allotted time and devote the rest of it to a lark of their own.

We were pleasantly situated in a small two-storied inn, in an empty large compound which was surrounded by a mud wall as high as a man's head. The inn was kept by nine Hindoo brothers, its owners. They lived, with their families, in a one-storied building within the compound, but off to one side, and there was always a long pile of their little comely brown children loosely stacked in its veranda, and a detachment of the parents wedged among them, smoking the hookah or the

AN HONEST CRITIC.

howdah, or whatever they call it. By the veranda stood a palm, and a monkey lived in it, and led a lonesome life, and always looked sad and weary, and the crows bothered him a good deal.

The inn cow poked about the compound and emphasized the secluded and country air of the place, and there was a dog of no particular breed, who was always present in the compound, and always asleep, always stretched out baking in the sun and adding to the deep tranquility and reposefulness of the place, when the crows were away on business. White-draperied servants were coming and going all the time, but they seemed only spirits, for their feet were bare and made no sound. Down the lane a piece lived an elephant in the shade of a noble tree, and rocked and rocked, and reached about with his trunk, begging of his brown mistress or fumbling the children playing at his feet. And there were camels about, but they go on velvet feet, and were proper to the silence and serenity of the surroundings.

The Satan mentioned at the head of this chapter was not our Satan, but the other one. Our Satan was lost to us. In these later days he had passed out of our life — lamented by me, and sincerely. I was missing him; I am missing him yet, after all these months. He was an astonishing creature to fly around and do things. He didn't always do them quite right, but he *did* them, and did them suddenly. There was no time wasted. You would say:

"Pack the trunks and bags, Satan."

"Wair good" (very good).

Then there would be a brief sound of thrashing and slashing and humming and buzzing, and a spectacle as of a whirlwind spinning gowns and jackets and coats and boots and things through the air, and then — with bow and touch —

"Awready, master."

It was wonderful. It made one dizzy. He crumpled dresses a good deal, and he had no particular plan about the work —at first—except to put each article into the trunk it didn't belong in. But he soon reformed, in this matter. Not entirely; for, to the last, he would cram into the satchel sacred to literature any odds and ends of rubbish that he couldn't find a handy place for elsewhere. When threatened with death for this, it did not trouble him; he only looked pleasant, saluted with soldierly grace, said "Wair good," and did it again next day.

He was always busy; kept the rooms tidied up, the boots polished, the clothes brushed, the wash-basin full of clean water, my dress clothes laid out and ready for the lecture-hall an hour ahead of time; and he dressed me from head to heel in spite of my determination to do it myself, according to my lifelong custom.

He was a born boss, and loved to command, and to jaw and dispute with inferiors and harry them and bullyrag them He was fine at the railway station — yes, he was at his finest there. He would shoulder and plunge and paw his violent way through the packed multitude of natives with nineteen coolies at his tail, each bearing a trifle of luggage — one a trunk, another a parasol, another a shawl, another a fan, and so on; one article to each, and the longer the procession, the better he was suited — and he was sure to make for some engaged sleeper and begin to hurl the owner's things out of it, swearing that it was ours and that there had been a mistake. Arrived at our own sleeper, he would undo the bedding-bundles and make the beds and put everything to rights and shipshape in two minutes; then put his head out at a window and have a restful good time abusing his gang of coolies and disputing their bill until we arrived and made him pay them and stop his noise.

Speaking of noise, he certainly was the noisest little devil in India — and that is saying much, very much, indeed. I loved him for his noise, but the family detested him for it. They could not abide it; they could not get reconciled to it. It humiliated them. As a rule, when we got within six hundred yards of one of those big railway stations, a mighty racket of screaming and shrieking and shouting and storming would break upon us, and I would be happy to myself, and the family would say, with shame:

"There — that's Satan. Why *do* you keep him?"

And, sure enough, there in the whirling midst of fifteen hundred wondering people we would find that little scrap of a creature gesticulating like a spider with the colic, his black eyes snapping, his fez-tassel dancing, his jaws pouring out floods of billingsgate upon his gang of beseeching and astonished coolies.

I loved him; I couldn't help it; but the family — why, they could hardly speak of him with patience. To this day I regret his loss, and wish I had him back; but they — it is different with them. He was a native, and came from Surat. Twenty degrees of latitude lay between his birthplace and Manuel's, and fifteen hundred between their ways and characters and dispositions. I only liked Manuel, but I loved Satan. This latter's real name was intensely Indian. I could not quite get the hang of it, but it sounded like Bunder Rao Ram Chunder Clam Chowder. It was too long for handy use, anyway; so I reduced it.

When he had been with us two or three weeks, he began to make mistakes which I had difficulty in patching up for him. Approaching Benares one day, he got out of the train to see if be could get up a misunderstanding with somebody, for it had been a weary, long journey and he wanted to freshen up. He found what he was after, but kept up his pow-wow a shade too long and got left. So there we were in a strange city and no chambermaid. It was awkward for us, and we told him he must not do so any more. He saluted and said in his dear, pleasant way, " Wair good." Then at Lucknow he got drunk. I said it was a fever, and got the family's compassion and solicitude aroused ; so they gave him a teaspoonful of liquid quinine and it set his vitals on fire. He made several grimaces which gave me a better idea of the Lisbon earthquake than any I have ever got of it from paintings and descriptions. His drunk was still portentously solid next morning, but I could have pulled him through with the family if he would only have taken another spoonful of that remedy ; but no, although he was stupefied, his memory still had flickerings of life ; so he smiled a divinely dull smile and said, fumblingly saluting :

" Scoose me, mem Saheb, scoose me, Missy Saheb ; Satan not prefer it, please."

Then some instinct revealed to them that he was drunk. They gave him prompt notice that next time this happened he must go. He got out a maudlin and most gentle " Wair good," and saluted indefinitely.

Only one short week later he fell again. And oh, sorrow ! not in a hotel this time, but in an English gentleman's private house. And in Agra, of all places. So he had to go. When I told him, he said patiently, " Wair good," and made his parting salute, and went out from us to return no more forever. Dear me ! I would rather have lost a hundred angels than that one poor lovely devil. What style he used to put on, in a swell hotel or in a private house — snow-white muslin from his chin to his bare feet, a crimson sash embroidered with gold thread around his waist, and on his head a great sea-green turban like to the turban of the Grand Turk.

He was not a liar, but he will become one if he keeps on. He told me once that he used to crack cocoanuts with his

CITY GATE. — JEYPORE.

teeth when he was a boy; and when I asked how he got them into his mouth, he said he was upward of six feet high at that time, and had an unusual mouth. And when I followed him up and asked him what had become of that other foot, he said a house fell on him and he was never able to get his stature back again. Swervings like these from the strict line of fact often beguile a truthful man on and on until he eventually becomes a liar.

His successor was a Mohammedan, Sahadat Mohammed Khan; very dark, very tall, very grave. He went always in flowing masses of white, from the top of his big turban down to his bare feet. His voice was low. He glided about in a noiseless way, and looked like a ghost. He was competent and satisfactory. But where he was, it seemed always Sunday. It was not so in Satan's time.

SAHADAT KHAN.

Jeypore is intensely Indian, but it has two or three features which indicate the presence of European science and European interest in the weal of the common public, such as the liberal water-supply furnished by great works built at the State's expense; good sanitation, resulting in a degree of healthfulness unusually high for India; a noble pleasure garden, with privileged days for women; schools for the instruction of native youth in advanced art, both ornamental and utilitarian; and a new and beautiful palace stocked with a museum of extraordinary interest and value. Without the Maharaja's sympathy and purse these beneficences could not have been created; but he is a man of wide views and large generosities, and all such matters find hospitality with him.

We drove often to the city from the hotel Kaiser-i-Hind, a journey which was always full of interest, both night and day, for that country road was never quiet, never empty, but was always India in motion, always a streaming flood of brown people clothed in smouchings from the rainbow, a tossing and moiling flood, happy, noisy, a charming and satisfying confusion of strange human and strange animal life and equally strange and outlandish vehicles.

And the city itself is a curiosity. Any Indian city is that, but this one is not like any other that we saw. It is shut up in a lofty turreted wall; the main body of it is divided into six parts by perfectly straight streets that are more than a hundred feet wide; the blocks of houses exhibit a long frontage of the most taking architectural quaintnesses, the straight lines being broken everywhere by pretty little balconies, pillared and highly ornamented, and other cunning and cozy and inviting perches and projections, and many of the fronts are curiously pictured by the brush, and the whole of them have the soft rich tint of strawberry ice-cream. One cannot look down the far stretch of the chief street and persuade himself that these are real houses, and that it is all out of doors — the impression that it is an unreality, a picture, a scene in a theater, is the only one that will take hold.

Then there came a great day when this illusion was more pronounced than ever. A rich Hindoo had been spending a fortune upon the manufacture of a crowd of idols and accompanying paraphernalia whose purpose was to illustrate scenes in the life of his especial god or saint, and this fine show was to be brought through the town in processional state at ten in the morning. As we passed through the great public pleasure garden on our way to the city we found it crowded with natives. That was one sight. Then there was another. In the midst of the spacious lawns stands the palace which contains the museum — a beautiful construction of stone which shows arched colonnades, one above another, and receding,

STREET SCENE IN JEYPORE.

terrace-fashion, toward the sky. Every one of these terraces, all the way to the top one, was packed and jammed with natives. One must try to imagine those solid masses of splendid color, one above another, up and up, against the blue sky, and the Indian sun turning them all to beds of fire and flame.

Later, when we reached the city, and glanced down the chief avenue, smouldering in its crushed-strawberry tint, those splendid effects were repeated; for every balcony, and every fanciful bird-cage of a snuggery countersunk in the house-fronts, and all the long lines of roofs were crowded with people, and each crowd was an explosion of brilliant color.

Then the wide street itself, away down and down and down into the distance, was alive with gorgeously-clothed people — not still, but moving, swaying, drifting, eddying, a delirious display of all colors and all shades of color, delicate, lovely, pale, soft, strong, stunning, vivid, brilliant, a sort of storm of sweet-pea blossoms passing on the wings of a hurricane; and presently, through this storm of color, came swaying and swinging the majestic elephants, clothed in their Sunday best of gaudinesses, and the long procession of fanciful trucks freighted with their groups of curious and costly images, and then the long rear-guard of stately camels, with their picturesque riders.

For color, and picturesqueness, and novelty, and outlandishness, and sustained interest and fascination, it was the most satisfying show I had ever seen, and I suppose I shall not have the privilege of looking upon its like again.

CHAPTER LXI.

In the first place God made idiots. This was for practice. Then He made School Boards. — *Pudd'nhead Wilson's New Calendar.*

SUPPOSE we applied no more ingenuity to the instruction of deaf and dumb and blind children than we sometimes apply in our American public schools to the instruction of children who are in possession of all their faculties? The result would be that the deaf and dumb and blind would acquire nothing. They would live and die as ignorant as bricks and stones. The methods used in the asylums are rational. The teacher exactly measures the child's capacity, to begin with; and from thence onwards the tasks imposed are nicely gauged to the gradual development of that capacity; the tasks keep pace with the steps of the child's progress, they don't jump miles and leagues ahead of it by irrational caprice and land in vacancy — according to the average public-school plan. In the public school, apparently, they teach the child to spell cat, then ask it to calculate an eclipse; when it can read words of two syllables, they require it to explain the circulation of the blood; when it reaches the head of the infant class they bully it with conundrums that cover the domain of universal knowledge. This sounds extravagant — and is; yet it goes no great way beyond the facts.

I received a curious letter one day, from the Punjab (you must pronounce it Pun*jawb*). The handwriting was excellent, and the wording was English — English, and yet not exactly English. The style was easy and smooth and flowing, yet there was something subtly foreign about it — something tropically ornate and sentimental and rhetorical. It turned out to be the work of a Hindoo youth, the holder of a humble clerical billet in a railway office. He had been educated in one of the numerous colleges of India. Upon inquiry I was told that the country was full of young fellows of his like. They had been educated away up to the snow-summits of learning — and the market for all this elaborate cultivation was minutely out of proportion to the vastness of the product. This market consisted of some thousands of small clerical posts under the government — the supply of material for it was multitudinous. If this youth with the flowing style and the blossoming English was occupying a small railway clerkship, it meant that there were hundreds and hundreds as capable as he, or he would be in a high place; and it certainly meant that there were thousands whose education and capacity had fallen a little short, and that they would have to go *without* places. Apparently, then, the colleges of India were doing what our high schools have long been doing — richly over-supplying the market for highly-educated service; and thereby doing a damage to the scholar, and through him to the country.

At home I once made a speech deploring the injuries inflicted by the high school in making handicrafts distasteful to boys who would have been willing to make a living at trades and agriculture if they had but had the good luck to stop with the common school. But I made no converts. Not one, in a community overrun with educated idlers who were above following their fathers' mechanical trades, yet could find no market for their book-knowledge. The same mail that brought me the letter from the Punjab, brought also a little book published by Messrs. Thacker, Spink & Co., of Calcutta, which interested me, for both its preface and its contents treated of this matter of over-education. In the preface occurs this paragraph from the *Calcutta Review*. For "Government office" read "dry-goods clerkship" and it will fit more than one region of America:

"The education that we give makes the boys a little less clownish in their manners, and more intelligent when spoken to by strangers. On the other hand, it has made them less contented with their lot in life, and less willing to work with their hands. The form which discontent takes in this country is not of a healthy kind; for, the Natives of India consider that the only occupation worthy of an educated man is that of a writership in some office, and especially in a Government office. The village schoolboy goes back to the plow with the greatest reluctance; and the town schoolboy carries the same discontent and inefficiency into his father's workshop. Sometimes these ex-students positively refuse at first to work; and more than once parents have openly expressed their regret that they ever allowed their sons to be inveigled to school."

The little book which I am quoting from is called " Indo-Anglian Literature," and is well stocked with " baboo " English — clerkly English, booky English, acquired in the schools. Some of it is very funny, — almost as funny, perhaps, as what you and I produce when we try to write in a language not our own; but much of it is surprisingly correct and free. If I were going to quote *good* English — but I am not. India is well stocked with natives who speak it and write it as well as the best of us. I merely wish to show some of the quaint imperfect attempts at the use of our tongue. There are many letters in the book; poverty imploring help — bread, money, kindness, office — generally an office, a clerkship, some way to get food and a rag out of the applicant's unmarketable education; and food not for himself alone, but sometimes for a dozen helpless relations in addition to his own family; for those people are astonishingly unselfish, and admirably faithful to their ties of kinship. Among us I think there is nothing approaching it. Strange as some of these wailing and supplicating letters are, humble and even groveling as some of them are, and quaintly funny and confused as a goodly number of them are, there is still a pathos about them, as a rule, that checks the rising laugh

and reproaches it. In the following letter "father" is not to
be read literally.   In Ceylon a little native beggar-girl embarrassed me by calling me father, although I knew she was mistaken.   I was so new that I did not know that she was merely
following the custom of the dependent and the supplicant.

"Sir,
    "I pray please to give me some action (work) for I am very poor boy I
have no one to help me even so father for it so it seemed in thy good sight,
you give the Telegraph Office, and another work what is your wish I am very
poor boy, this understand what is your wish you my father I am your son
this understand what is your wish.
                                   "Your Sirvent, P. C. B."

Through ages of debasing oppression suffered by these
people at the hands of their native rulers, they come legiti-
mately by the attitude and language of fawning and flattery,
and one must remember this in mitigation when passing judgment upon the native character.   It is common in these letters
to find the petitioner furtively trying to get at the white man's
soft religious side ; even this poor boy baits his hook with a
macerated Bible-text in the hope that it may catch something
if all else fail.

Here is an application for the post of instructor in English
to some children :

    "My Dear Sir or Gentleman, that your Petitioner has much qualification
in the Language of English to instruct the young boys ; I was given to
understand that your of suitable children has to acquire the knowledge of
English language."

As a sample of the flowery Eastern style, I will take a
sentence or two from a long letter written by a young native
to the Lieutenant-Governor of Bengal — an application for
employment :

"Honored and Much Respected Sir,
    "I hope your honor will condescend to hear the tale of this poor creature.
I shall overflow with gratitude at this mark of your royal condescension.
The bird-like happiness has flown away from my nest-like heart and has not
hitherto returned from the period whence the rose of my father's life suffered
the autumnal breath of death, in plain English he passed through the
gates of Grave, and from that hour the phantom of delight has never danced
before me."

"I WAS EMBARRASSED."

It is all school-English, book-English, you see; and good enough, too, all things considered. If the native boy had but that one study he would shine, he would dazzle, no doubt. But that is not the case. He is situated as are our public-school children — loaded down with an over-freightage of other studies; and frequently they are as far beyond the actual point of progress reached by him and suited to the stage of development attained, as could be imagined by the insanest fancy. Apparently — like our public-school boy — he must work, work, work, in school and out, and play but little. Apparently — like our public-school boy — his "education" consists in learning *things*, not the meaning of them; he is fed upon the husks, not the corn. From several essays written by native schoolboys in answer to the question of how they spend their day, I select one — the one which goes most into detail:

"66. At the break of day I rises from my own bed and finish my daily duty, then I employ myself till 8 o'clock, after which I employ myself to bathe, then take for my body some sweet meat, and just at 9½ I came to school to attend my class duty, then at 2½ P. M. I return from school and engage myself to do my natural duty, then, I engage for a quarter to take my tiffin, then I study till 5 P. M., after which I began to play anything which comes .n my head. After 8½ half pass to eight we are began to sleep, before sleeping I told a constable just 11 o' he came and rose us from half pass eleven we began to read still morning."

It is not perfectly clear, now that I come to cipher upon it. He gets up at about 5 in the morning, or along there somewhere, and goes to bed about fifteen or sixteen hours afterward — that much of it seems straight; but why he should rise again three hours later and resume his studies till morning is puzzling.

I think it is because he is studying history. History requires a world of time and bitter hard work when your "education" is no further advanced than the cat's; when you are merely stuffing yourself with a mixed-up mess of empty names and random incidents and elusive dates, which no one teaches you how to interpret, and which, uninterpreted, pay you not a farthing's value for your waste of time. Yes, I think he had to get up at half-past 11 P. M. in order to be sure to be perfect with his history lesson by noon. With results as follows — from a Calcutta school examination :

" Q. *Who was Cardinal Wolsey?*

" Cardinal Wolsey was an Editor of a paper named *North Briton.* No. 45 of his publication he charged the King of uttering a lie from the throne. He was arrested and cast into prison ; and after releasing went to France.

" 3. As Bishop of York but died in disentry in a church on his way to be blockheaded.

" 8. Cardinal Wolsey was the son of Edward IV, after his father's death he himself ascended the throne at the age of (10) ten only, but when he surpassed or when he was fallen in his twenty years of age at that time he wished to make a journey in his countries under him, but he was opposed by his mother to do journey, and according to his mother's example he remained in the home, and then became King. After many times obstacles and many confusion he become King and afterwards his brother."

There is probably not a word of truth in that.

" Q. *What is the meaning of Ich Dien?*

" 10. An honor conferred on the first or eldest sons of English Sovereigns. It is nothing more than some feathers.

"11. Ich Dien was the word which was written on the feathers of the blind King who came to fight, being interlaced with the bridles of the horse.

" 13. Ich Dien is a title given to Henry VII by the Pope of Rome, when he forwarded the Reformation of Cardinal Wolsy to Rome, and for this reason he was called Commander of the faith."

A dozen or so of this kind of insane answers are quoted in the book from that examination. Each answer is sweeping proof, all by itself, that the person uttering it was pushed ahead of where he belonged when he was put into history ; proof that he had been put to the task of acquiring history before he had had a single lesson in the *art* of acquiring it, which is the equivalent of dumping a pupil into geometry before he has learned the progressive steps which lead up to it and make its acquirement possible. Those Calcutta novices had no business with history. There was no excuse for examing them in it, no excuse for exposing them and their teachers. They were totally empty ; there was nothing to " examine."

Helen Kellar has been dumb, stone deaf, and stone blind, ever since she was a little baby a year-and-a-half old ; and now at sixteen years of age this miraculous creature, this wonder of all the ages, passes the Harvard University examination in Latin, German, French history, *belles lettres*, and such things, and does it brilliantly, too, not in a commonplace fashion. She doesn't know merely *things*, she is splendidly familiar with the *meanings* of them.   When she writes an essay on a Shakespearean character, her English is fine and strong, her grasp of the subject is the grasp of one who *knows*, and her page is electric with light.   Has Miss Sullivan taught her by the methods of India and the American public school ? No, oh, no ; for then she would be deafer and dumber and blinder than she was before.   It is a pity that we can't educate all the children in the asylums.

To continue the Calcutta exposure :

" What is the meaning of a Sheriff ? "

" 25.   Sheriff is a post opened in the time of John.   The duty of Sheriff here in Calcutta, to look out and catch those carriages which is rashly driven out by the coachman ; but it is a high post in England.

" 26.   Sheriff was the English bill of common prayer.

" 27.   The man with whom the accusative persons are placed is çalled Sheriff.

" 28.   Sheriff — Latin term for 'shrub,' we called broom, worn by the first earl of Enjue, as an emblem of humility when they went to the pilgrimage, and from this their hairs took their crest and sur name.

" 29.   Sheriff is a kind of titlous sect of people, as Barons, Nobles, etc.

" 30.   *Sheriff*, a tittle given on those persons who were respective and pious in England."

The students were examined in the following bulky matters : Geometry, the Solar Spectrum, the Habeas Corpus Act, the British Parliament, and in Metaphysics they were asked to trace the progress of skepticism from Descartes to Hume.   It is within bounds to say that some of the results were astonishing.   Without doubt, there were students present who justified their teacher's wisdom in introducing them to these studies ; but the fact is also evident that others had been pushed into these studies to waste their time over them when they could have been profitably employed in hunting smaller game. Under the head of Geometry, one of the answers is this:

"49. The whole BD = the whole CA, and so-so-so-so-so-so — so.

To me this is cloudy, but I was never well up in geometry. That was the only effort made among the five students who appeared for examination in geometry; the other four wailed and surrendered without a fight. They are piteous wails, too, wails of despair; and one of them is an eloquent reproach; it comes from a poor fellow who has been laden beyond his strength by a stupid teacher, and is eloquent in spite of the poverty of its English. The poor chap finds himself required to explain riddles which even Sir Isaac Newton was not able to understand:

"50. Oh my dear father examiner you my father and you kindly give a number of pass you my great father.

"51. I am a poor boy and have no means to support my mother and two brothers who are suffering much for want of food. I get four rupees monthly from charity fund of this place, from which I send two rupees for their support, and keep two for my own support. Father, if I relate the unlucky circumstance under which we are placed, then, I think, you will not be able to suppress the tender tear.

"52. Sir which Sir Isaac Newton and other experienced mathematicians cannot understand I being third of Entrance Class can understand these which is too impossible to imagine. And my examiner also has put very tiresome and very heavy propositions to prove."

We must remember that these pupils had to do their thinking in one language, and express themselves in another and alien one. It was a heavy handicap. I have by me "English as She is Taught"—a collection of American examinations made in the public schools of Brooklyn by one of the teachers, Miss Caroline B. Le Row. An extract or two from its pages will show that when the American pupil is using but one language, and that one his own, his performance is no whit better than his Indian brother's:

' ON HISTORY.

"Christopher Columbus was called the father of his Country. Queen

Isabella of Spain sold her watch and chain and other millinery so that Columbus could discover America.

"The Indian wars were very desecrating to the country.

"The Indians pursued their warfare by hiding in the bushes and then scalping them.

"Captain John Smith has been styled the father of his country. His life was saved by his daughter Pochahantas.

"The Puritans found an insane asylum in the wilds of America.

"The Stamp Act was to make everybody stamp all materials so they should be null and void.

"Washington died in Spain almost broken-hearted. His remains were taken to the cathedral in Havana.

"Gorilla warfare was where men rode on gorillas."

In Brooklyn, as in India, they examine a pupil, and when they find out he doesn't know anything, they put him into literature, or geometry, or astronomy, or government, or something like that, so that he can properly display the assification of the whole system:

"ON LITERATURE.

"'Bracebridge Hall' was written by Henry Irving.

"Edgar A. Poe was a very curdling writer.

"Beowulf wrote the Scriptures.

"Ben Johnson survived Shakespeare in some respects.

"In the 'Canterbury Tale' it gives account of King Alfred on his way to the shrine of Thomas Bucket.

"Chaucer was the father of English pottery.

"Chaucer was succeeded by H. Wads. Longfellow."

We will finish with a couple of samples of "literature," — one from America, the other from India. The first is a Brooklyn public-school boy's attempt to turn a few verses of the "Lady of the Lake" into prose. You will have to concede that he did it:

"The man who rode on the horse performed the whip and an instrument made of steel alone with strong ardor not diminishing, for, being tired from the time passed with hard labor overworked with anger and ignorant with weariness, while every breath for labor he drew with cries full of sorrow, the young deer made imperfect who worked hard filtered in sight."

The following paragraph is from a little book which is famous in India — the biography of a distinguished Hindoo judge, Onoocool Chunder Mookerjee; it was written by his nephew, and is unintentionally funny — in fact, exceedingly so. I offer here the closing scene. If you would like to sample the rest of the book, it can be had by applying to the publishers, Messrs. Thacker, Spink & Co., Calcutta:

"And having said these words he hermetically sealed his lips not to open them again. All the well-known doctors of Calcutta that could be procured for a man of his position and wealth were brought, — Doctors Payne, Fayrer, and Nilmadhub Mookerjee and others; they did what they could do, with their puissance and knack of medical knowledge, but it proved after all as if to milk the ram! His wife and children had not the mournful consolation to hear his last words; he remained *sotto voce* for a few hours, and then was taken from us at 6.12 P.M. according to the caprice of God which passeth understanding."

CHAPTER LXII.

There are no people who are quite so vulgar as the over-refined ones.
— Pudd'nhead Wilson's New Calendar.

WE sailed from Calcutta toward the end of March; stopped a day at Madras; two or three days in Ceylon; then sailed westward on a long flight for Mauritius. From my diary:

April 7. We are far abroad upon the smooth waters of the Indian Ocean, now; it is shady and pleasant and peaceful under the vast spread of the awnings, and life is perfect again — ideal.

The difference between a river and the sea is, that the river looks fluid, the sea solid — usually looks as if you could step out and walk on it.

The captain has this peculiarity — he cannot tell the truth in a plausible way. In this he is the very opposite of the austere Scot who sits midway of the table; *he* cannot tell a lie in an *un*plausible way. When the captain finishes a statement the passengers glance at each other privately, as who should say, " Do you believe that ? " When the Scot finishes one, the look says, " How strange and interesting." The whole secret is in the manner and method of the two men. The captain is a little shy and diffident, and he states the simplest fact as if he were a little afraid of it, while the Scot delivers himself of the most abandoned lie with such an air of stern veracity that one is forced to believe it although one knows it isn't so. For instance, the Scot told about a pet flying-fish he once owned, that lived in a little fountain in his conservatory, and supported

39 (609)

itself by catching birds and frogs and rats in the neighboring fields. It was plain that no one at the table doubted this statement.

By and by, in the course of some talk about custom-house annoyances, the captain brought out the following simple every-day incident, but through his infirmity of style managed to tell it in such a way that it got no credence. He said:

A FEMALE UNCLE.

"I went ashore at Naples one voyage when I was in that trade, and stood around helping my passengers, for I could speak a little Italian. Two or three times, at intervals, the officer asked me if I had anything dutiable about me, and seemed more and more put out and disappointed every time I told him no. Finally a passenger whom I had helped through asked me to come out and take something. I thanked him, but excused myself, saying I had taken a whisky just before I came ashore.

"It was a fatal admission. The officer at once made me pay sixpence import-duty on the whisky — just from ship to shore, you see; and he fined me £5 for not declaring the goods, another £5 for falsely denying that I had anything dutiable about me, also £5 for concealing the goods, and £50 for smuggling, which is the maximum penalty for unlawfully bringing in goods under the value of sevenpence ha'penny. Altogether, sixty-five pounds sixpence for a little thing like that."

The Scot is always believed, yet he never tells anything but lies; whereas the captain is never believed, although he never tells a lie, so far as I can judge. If he should say his uncle was a male person, he would probably say it in such a way that nobody would believe it; at the same time the Scot could claim that he had a female uncle and not stir a doubt in anybody's mind. My own luck has been curious all my literary life; I never could tell a lie that anybody would doubt, nor a truth that anybody would believe.

Lots of pets on board — birds and things. In these far

countries the white people do seem to run remarkably to pets. Our host in Cawnpore had a fine collection of birds — the finest we saw in a private house in India. And in Colombo, Dr. Murray's great compound and commodious bungalow were

well populated with domesticated company from the woods: frisky little squirrels; a Ceylon mina walking sociably about the house; a small green parrot that whistled a single urgent note of call without motion of its beak; also chuckled; a monkey in a cage on the back veranda, and some more out in the trees; also a number of beautiful macaws in the trees; and various and sundry birds and animals of breeds not known to me. But no cat. Yet a cat would have liked that place.

A CAT WOULD LIKE THAT PLACE.

April 9. Tea-planting is the great business in Ceylon, now. A passenger says it often pays 40 per cent. on the investment. Says there is a boom.

April 10. The sea is a Mediterranean blue; and I believe that that is about the divinest color known to nature.

It is strange and fine — Nature's lavish generosities to her creatures. At least to all of them except man. For those that fly she has provided a home that is nobly spacious — a home which is forty miles deep and envelops the whole globe, and has not an obstruction in it.   For those that swim she has provided a more than imperial domain — a domain which is miles deep and covers four-fifths of the globe.   But as for man, she has cut him off with the mere odds and ends of the creation.   She has given him the thin skin, the meagre skin which is stretched over the remaining one-fifth — the naked bones stick up through it in most places.   On the one-half of this domain he can raise snow, ice, sand, rocks, and nothing else.   So the valuable part of his inheritance really consists of but a single fifth of the family estate;   and out of it he has to grub hard to get enough to keep him alive and provide kings and soldiers and powder to extend the blessings of civilization with.   Yet man, in his simplicity and complacency and inability to cipher, thinks Nature regards him as the important member of the family — in fact, her favorite.   Surely, it must occur to even his dull head, sometimes, that she has a curious way of showing it.

Afternoon.   The captain has been telling how, in one of his Arctic voyages, it was so cold that the mate's shadow froze fast to the deck and had to be ripped loose by main strength. And even then he got only about two-thirds of it back.   Nobody said anything, and the captain went away.   I think he is becoming disheartened. . . .   Also, to be fair, there is another word of praise due to this ship's library: it contains no copy of the Vicar of Wakefield, that strange menagerie of complacent hypocrites and idiots, of theatrical cheap-john heroes and heroines, who are always showing off, of bad people who are not interesting, and good people who are fatiguing. A singular book.   Not a sincere line in it, and not a character that invites respect; a book which is one long waste-pipe discharge of goody-goody puerilities and dreary moralities; a book which is full of pathos which revolts, and humor which grieves the heart.   There are few things in literature that are

"The Mate's shadow froze
fast to the deck."

more piteous, more pathetic, than the celebrated "humorous" incident of Moses and the spectacles.

Jane Austen's books, too, are absent from this library. Just that one omission alone would make a fairly good library out of a library that hadn't a book in it.

"THE BARBER . . . FLAYS US ON THE BREEZY DECK."

Customs in tropic seas. At 5 in the morning they pipe to wash down the decks, and at once the ladies who are sleeping there turn out and they and their beds go below. Then one after another the men come up from the bath in their pyjamas, and walk the decks an hour or two with bare legs and bare feet. Coffee and fruit served. The ship cat and her kitten now appear and get about their toilets; next the barber comes and flays us on the breezy deck. Breakfast at 9.30, and the day begins. I do not know how a day could be more reposeful: no motion; a level blue sea; nothing in sight from horizon to horizon; the speed of the ship furnishes a cooling breeze; there is no mail to read and answer; no newspapers to excite you; no telegrams to fret you or fright you — the world is far, far away; it has ceased to exist for you — seemed a fading dream, along in the first days; has dissolved to an unreality now; it is gone from your mind with all its businesses and ambitions, its prosperities and disasters, its exultations and despairs, its joys and griefs and cares and worries. They are no concern of yours any more; they have gone out of your life; they are a storm which has passed and left a deep calm behind. The people group themselves about the decks in their snowy white linen, and read, smoke, sew, play cards, talk, nap, and so on. In other ships the passengers are always ciphering about when they are going to arrive; out in these seas it is rare, very rare, to hear that subject broached. In other ships there is always an eager rush to the bulletin board at noon to find out what the "run" has been; in these seas the bulletin seems to attract no interest; I have seen no one visit it; in thirteen days I have visited it only once. Then I happened to notice the figures of the day's run. On that day there happened to be talk, at dinner, about the speed of modern ships. I was the only passenger present who knew this ship's gait. Necessarily, the Atlantic custom of betting on the ship's run is not a custom here — nobody ever mentions it.

I myself am wholly indifferent as to when we are going to "get in"; if any one else feels interested in the matter he has not indicated it in my hearing. If I had my way we should never get in at all. This sort of sea life is charged with an indestructible charm. There is no weariness, no fatigue, no worry, no responsibility, no work, no depression of spirits. There is nothing like this serenity, this comfort, this peace, this deep contentment, to be found anywhere on land. If I had my way I would sail on for ever and never go to live on the solid ground again.

One of Kipling's ballads has delivered the aspect and sentiment of this bewitching sea correctly :

" The Injian Ocean sets an' smiles
So sof', so bright, so bloomin' blue ;
There aren't a wave for miles an' miles
Excep' the jiggle from the screw."

April 14. It turns out that the astronomical apprentice worked off a section of the Milky Way on me for the Magellan Clouds. A man of more experience in the business showed one of them to me last night. It was small and faint and delicate, and looked like the ghost of a bunch of white smoke left floating in the sky by an exploded bombshell.

Wednesday, April 15. Mauritius. Arrived and anchored off Port Louis 2 A. M. Rugged clusters of crags and peaks, green to their summits ; from their bases to the sea a green plain with just tilt enough to it to make the water drain off. I believe it is in 56° E. and 22° S.— a hot tropical country. The green plain has an inviting look; has scattering dwellings nestling among the greenery. Scene of the sentimental adventure of Paul and Virginia.

Island under French control — which means a community which depends upon quarantines, not sanitation, for its health.

Thursday, April 16. Went ashore in the forenoon at Port Louis, a little town, but with the largest variety of nationalities and complexions we have encountered yet. French, English, Chinese, Arabs, Africans with wool, blacks with straight hair, East Indians, half-whites, quadroons — and great varieties in costumes and colors.

Took the train for Curepipe at 1.30 — two hours' run, gradually uphill. What a contrast, this frantic luxuriance of vegetation, with the arid plains of India; these architecturally picturesque crags and knobs and miniature mountains, with the monotony of the Indian dead-levels.

A native pointed out a handsome swarthy man of grave and dignified bearing, and said in an awed tone, " That is so-and-so; has held office of one sort or another under this government for 37 years — he is known all over this whole island — and in the other countries of the world perhaps — who knows? One thing is certain; you can speak his name anywhere in this whole island, and you will find not one grown person that has not heard it. It is a wonderful thing to be so celebrated; yet look at him; it makes no change in him; he does not even seem to know it."

Curepipe (means Pincushion or Pegtown, probably). Sixteen miles (two hours) by rail from Port Louis. At each end of every roof and on the apex of every dormer window a wooden peg two feet high stands up; in some cases its top is blunt, in others the peg is sharp and looks like a toothpick. The passion for this humble ornament is universal.

Apparently, there has been only one prominent event in the history of Mauritius, and that one didn't happen. I refer to the romantic sojourn of Paul and Virginia here. It was that story that made Mauritius known to the world, made the name familiar to everybody, the geographical position of it to nobody.

A clergyman was asked to guess what was in a box on a table. It was a vellum fan painted with the shipwreck, and was " *one of Virginia's wedding gifts.*"

April 18. This is the only country in the world where the stranger is not asked " How do you like this place?" This is indeed a large distinction. Here the citizen does the talking

about the country himself; the stranger is not asked to help.
you get all sorts of information.   From one citizen you gather
the idea that Mauritius was made first, and then heaven; and
that heaven was copied after Mauritius.   Another one tells you
that this is an exaggeration; that the two chief villages, Port
Louis and Curepipe, fall short of heavenly perfection; that nobody lives in Port Louis
except upon compulsion;
and that Curepipe is the
wettest and rainiest place
in the world.   An English
citizen said :

"In the early part of this
century Mauritius was used by
the French as a basis from which
to operate against England's
Indian merchantmen ; so England captured the island and
also the neighbor, Bourbon, to
stop that annoyance. England
gave Bourbon back; the government in London did not want
any more possessions in the West
Indies. If the government had
had a better quality of geogra-
phy in stock it would not have
wasted Bourbon in that foolish
way. A big war will tempo-
rarily shut up the Suez Canal
some day and the English ships
will have to go to India around
the Cape of Good Hope again;
then England will have to have
Bourbon and will take it.

"Mauritius was a crown
colony until 20 years ago, with
a governor appointed by the
Crown and assisted by a Coun-

THE WETTEST PLACE ON EARTH.

cil appointed by himself ; but Pope Hennessey came out as Governor then,
and he worked hard to get a part of the council made elective, and succeeded. So now the whole council is French, and in all ordinary matters of
legislation they vote together and in the French interest, not the English.

The English population is very slender ; it has not votes enough to elect a
legislator.   Half a dozen rich French families elect the legislature.   Pope
Hennessey was an Irishman, a Catholic, a Home Ruler, M.P., a hater of England and the English, a very troublesome person and a serious incumbrance
at Westminster; so it was decided to send him out to govern unhealthy countries, in hope that something would happen to him.   But nothing did.   The
first experiment was not merely a failure, it was more than a failure.   He
proved to be more of a disease himself than any he was sent to encounter.
The next experiment was here.   The dark scheme failed again.   It was an
off-season and there was nothing but measles here at the time.   Pope Hennessey's health was not affected.   He worked with the French and for the
French and against the English, and he made the English very tired and the
French very happy, and lived to have the joy of seeing the flag he served
publicly hissed.   His memory is held in worshipful reverence and affection
by the French.

"It is a land of extraordinary quarantines.   They quarantine a ship for
anything or for nothing ; quarantine her for 20 and even 30 days.   They once
quarantined a ship because her captain had had the smallpox when he was a
boy.   That and because he was English.

"The population is very small ; small to insignificance.   The majority is
East Indian ; then mongrels ; then negroes (descendants of the slaves of the
French times) ; then French ; then English.   There was an American, but
he is dead or mislaid.   The mongrels are the result of all kinds of mixtures ;
black and white, mulatto and white, quadroon and white, octoroon and white.
And so there is every shade of complexion ; ebony, old mahogany, horse-
chestnut, sorrel, molasses-candy, clouded amber, clear amber, old-ivory white,
new-ivory white, fish-belly white — this latter the leprous complexion frequent
with the Anglo-Saxon long resident in tropical climates.

"You wouldn't expect a person to be proud of being a Mauritian, now
would you ?   But it is so.   The most of them have never been out of the
island, and haven't read much or studied much, and they think the world
consists of three principal countries — Judæa, France, and Mauritius ; so they
are very proud of belonging to one of the three grand divisions of the globe.
They think that Russia and Germany are in England, and that England does
not amount to much.   They have heard vaguely about the United States and
the equator, but they think both of them are monarchies.   They think Mount
Peter Botte is the highest mountain in the world, and if you show one of
them a picture of Milan Cathedral he will swell up with satisfaction and say
that the idea of that jungle of spires was stolen from the forest of peg-tops
and toothpicks that makes the roofs of Curepipe look so fine and prickly.

"There is not much trade in books.   The newspapers educate and entertain the people.   Mainly the latter.   They have two pages of large-print
reading-matter — one of them English, the other French.   The English page
is a translation of the French one.   The typography is super-extra primitive :
in this quality it has not its equal anywhere.   There is no proof-reader now ;
he is dead.

"Where do they get matter to fill up a page in this little island lost in the wastes of the Indian Ocean? Oh, Madagascar. They discuss Madagascar and France. That is the bulk. Then they chock up the rest with advice to the Government. Also, slurs upon the English administration. The papers are all owned and edited by creoles — French.

"The language of the country is French. Everybody speaks it — has to. You have to know French — particularly mongrel French, the patois spoken by Tom, Dick, and Harry of the multiform complexions — or you can't get along.

"This was a flourishing country in former days, for it made then and still makes the best sugar in the world; but first the Suez Canal severed it from the world and left it out in the cold, and next the beetroot sugar, helped by bounties, captured the European markets. Sugar is the life of Mauritius, and it is losing its grip. Its downward course was checked by the depreciation of the rupee — for the planter pays wages in rupees but sells his crop for gold — and the insurrection in Cuba and paralyzation of the sugar industry there have given our prices here a life-saving lift; but the outlook has noth-

The Commercial Gazette

Port-Louis, M rdi 14 Avril 1896

CABLOGRAMMES

LES AFFAIRES CUBAINES
NOUVELLE DEFAITE DET INSURGÉS— LES MANIFESTATIONS EN ESPAGNE

Madrid, 8 mars—*Par service spécial* — Une dépêche de la Havane annonce que le co onel Vienna a baitu une bande d'in surgés commandés par Maceo, Les insur gés ont eu 72 morts et ont abandonné de nombreux blessés sur le champ de bae taille ; les Espagnols ont eu 22 blessés. Ils ont dispersé l'enuemi et lui ont pris 210 chevaux et des armes.

La tranquillité e t rétablie à Valence. Une nouvelle manifestation a eu lieu à Barcelone où la gendarmerie à cheval a dû charger pour disperser la foule.

Au cours des démonstrations qui se sont pro laites, aujourd'hui, à Saragosse, Bar celone et Valence, la foule s'est portée levant les consulats de France et a vive ment réclamé les consuls.

Mal de têt•, courbature, état fiévreux, •ents, mal do go•ge; tels sont les •'un cemm ne

ANCIENT NEWS AT PORT LOUIS.

ing permanently favorable about it. It takes a year to mature the canes — on the high ground three and six months longer — and there is always a chance that the annual cyclone will rip the profit out of the crop. In recent times a cyclone took the whole crop, as you may say ; and the island never saw a finer one. Some of the noblest sugar estates in the island are in deep difficulties. A dozen of them are investments of English capital; and the companies that own them are at work now, trying to settle up and get out with a saving of half the money they put in. You know, in these days, when a country begins to introduce the tea culture, it means that its own specialty has gone back on it Look at Bengal; look at Ceylon. Well, they've begun to introduce the tea culture, *here*.

"Many copies of Paul and Virginia are sold every year in Mauritius. No other book is so popular here except the Bible. By many it is supposed to be a part of the Bible. All the missionaries work up their French on it when they come here to pervert the Catholic mongrel. It is the greatest story that was ever written about Mauritius, and the only one."

CHAPTER LXIII.

APRIL 20. — The cyclone of 1892 killed and crippled hundreds of people; it was accompanied by a deluge of rain, which drowned Port Louis and *produced a water famine.* Quite true; for it burst the reservoir and the water-pipes; and for a time after the flood had disappeared there was much distress from want of water.

This is the only place in the world where *no* breed of matches can stand the damp. Only one match in 16 will light.

The roads are hard and smooth; some of the compounds are spacious, some of the bungalows commodious, and the roadways are walled by tall bamboo hedges, trim and green and beautiful; and there are azalea hedges, too, both the white and the red; I never saw that before.

As to healthiness: I translate from to-day's (April 20) *Merchants' and Planters' Gazette*, from the article of a regular contributor, "Carminge," concerning the death of the nephew of a prominent citizen:

ONLY ONE IN SIXTEEN.

"Sad and lugubrious existence, this which we lead in Mauritius; I believe there is no other country in the world where one dies more easily than among us. The least indisposition becomes a mortal malady; a simple headache develops into meningitis; a cold into pneumonia, and presently, when we are least expecting it, death is a guest in our home."

This daily paper has a meteorological report which tells you what the weather was day before yesterday.

One is never pestered by a beggar or a peddler in this town, so far as I can see. This is pleasantly different from India.

April 22. To such as believe that the quaint product called French civilization would be an improvement upon the civilization of New Guinea and the like, the snatching of Madagascar and the laying on of French civilization there will be fully justified. But why did the English allow the French to have Madagascar? Did she respect a theft of a couple of centuries ago? Dear me, robbery by European nations of each other's territories has never been a sin, is not a sin to-day. To the several cabinets the several political establishments of the world are clothes-lines; and a large part of the official duty of these cabinets is to keep an eye on each other's wash and grab what they can of it as opportunity offers. All the territorial possessions of all the political establishments in the earth — including America, of course — consist of pilferings from other people's wash. No tribe, howsoever insignificant, and no nation, howsoever mighty, occupies a foot of land that was not stolen. When the English, the French, and the Spaniards reached America, the Indian tribes had been raiding each other's territorial clothes-lines for ages, and every acre of ground in the continent had been stolen and re-stolen 500 times. The English, the French, and the Spaniards went to work and stole it all over again; and when that was satisfactorily accomplished they went diligently to work and stole it from each other. In Europe and Asia and Africa every acre of ground has been stolen several millions of times. A crime persevered in a thousand centuries ceases to be a crime, and becomes a virtue. This is the law of custom, and custom supersedes all other forms of law. Christian governments are as frank to-day, as open and above-board, in discussing projects for raiding each other's clothes-lines as ever they were before the Golden Rule came smiling into this inhospitable world and couldn't get a night's lodging anywhere. In 150 years England has beneficently retired garment after garment from the Indian lines, until there is hardly a rag of the original wash left dangling anywhere. In 800 years an obscure tribe of Muscovite savages has risen to the dazzling position of Land-Robber-in-Chief; she found a quarter of the world hanging out to dry on a hundred parallels of latitude, and she scooped in the whole wash. She keeps a sharp eye on a multitude of little lines that stretch along the northern boundaries of India, and every now and then she snatches a hip-rag or a pair of pyjamas. It is England's prospective property, and Russia knows it; but Russia cares nothing for that. In fact, in our day land-robbery, claim-jumping, is become a European governmental frenzy. Some have been hard at it in the borders of China, in Burma, in Siam, and the islands of the sea; and *all* have been at it in Africa. Africa has been as coolly divided up and portioned out among the gang as if they had bought it and paid for it. And now straightway they are beginning the old game again — to steal each other's grabbings. Germany found a vast slice of Central Africa with the English flag and the English missionary and the English trader scattered all over it, but with certain formalities neglected — no signs up, " Keep off the grass," " Trespassers forbidden," etc. — and she stepped in with a cold calm smile and put up the signs herself, and swept those English pioneers promptly out of the country.

There is a tremendous point there. It can be put into the

form of a maxim: Get your formalities ' right — never mind about the moralities.

It was an impudent thing; but England had to put up with it. Now, in the case of Madagascar, the formalities had originally been observed, but by neglect they had fallen into desuetude ages ago. England should have snatched Madagascar from the French clothes-line. Without an effort she could have saved those harmless natives from the calamity of French civilization, and she did not do it. Now it is too late.

The signs of the times show plainly enough what is going to happen. All the savage lands in the world are going to be brought under subjection to the Christian governments of Europe. I am not sorry, but glad. This coming fate might have been a calamity to those savage peoples two hundred years ago; but now it will in some cases be a benefaction. The sooner the seizure is consummated, the better for the savages.

"THE THIRD YEAR THEY DO NOT GATHER SHELLS."

The dreary and dragging ages of bloodshed and disorder and

40

oppression will give place to peace and order and the reign of law. When one considers what India was under her Hindoo and Mohammedan rulers, and what she is now; when he remembers the miseries of her millions then and the protections and humanities which they enjoy now, he must concede that the most fortunate thing that has ever befallen that empire was the establishment of British supremacy there. The savage lands of the world are to pass to alien possession, their peoples to the mercies of alien rulers. Let us hope and believe that they will all benefit by the change.

April 23. "The first year they gather shells; the second year they gather shells and drink; the third year they do not gather shells." (Said of immigrants to Mauritius.)

Population 375,000. 120 sugar factories.

Population 1851, 185,000. The increase is due mainly to

A STEVEDORE.

the introduction of Indian coolies. They now apparently form the great majority of the population. They are admirable breeders; their homes are always hazy with children. Great savers of money. A British officer told me that in India he paid his servant 10 rupees a month, and he had 11 cousins, uncles, parents, etc., dependent upon him, and he supported them on his wages. These thrifty coolies are said to be acquiring land a trifle at a time, and cultivating it; and may own the island by and by.

The Indian women do very hard labor for wages running from $40\frac{1}{100}$ of a rupee for twelve hours' work, to $50\frac{1}{100}$. They carry mats of sugar on their heads

(70 pounds) all day lading ships, for half a rupee, and work at gardening all day for less.

The camaron is a fresh water creature like a cray-fish. It is regarded here as the world's chiefest delicacy — and certainly it is good. Guards patrol the streams to prevent poaching it. A fine of Rs. 200 or 300 (they say) for poaching. Bait is thrown in the water; the camaron goes for it; the fisher drops his loop in and works it around and about the camaron he has selected, till he gets it over its tail; then there's a jerk or something to certify the camaron that it is his turn now; he suddenly backs away, which moves the loop still further up his person and draws it taut, and his days are ended.

Another dish, called palmiste, is like raw turnip-shavings and tastes like green almonds; is very delicate and good. Costs the life of a palm tree 12 to 20 years old — for it is the pith.

Another dish — looks like greens or a tangle of fine seaweed — is a preparation of the deadly nightshade. Good enough.

The monkeys live in the dense forests on the flanks of the toy mountains, and they flock down nights and raid the sugarfields. Also on other estates they come down and destroy a sort of bean-crop — just for fun, apparently — tear off the pods and throw them down

The cyclone of 1892 tore down two great blocks of stone buildings in the center of Port Louis — the chief architectural feature — and left the uncomely and apparently frail blocks standing. Everywhere in its track it annihilated houses, tore off roofs, destroyed trees and crops. The men were in the towns, the women and children at home in the country getting crippled, killed, frightened to insanity; and the rain deluging them, the wind howling, the thunder crashing, the lightning glaring. This for an hour or so. Then a lull and sunshine;

many ventured out of safe shelter; then suddenly here it came again from the opposite point and renewed and completed the devastation.   It is said the Chinese fed the sufferers for days on free rice.

Whole streets in Port Louis were laid flat — wrecked. During a minute and a half the wind blew 123 miles an hour; no official record made after that, when it may have reached 150.   It cut down an obelisk.   It carried an American ship

A CYCLONE.

into the woods after breaking the chains of two anchors.   They now use four — two forward, two astern.   Common report says it killed 1,200 in Port Louis alone, in half an hour.   Then came the lull of the central calm — people did not know the barometer was still going down — then suddenly all perdition broke loose again while people were rushing around seeking friends and rescuing the wounded.   The noise was comparable to nothing; there is nothing resembling it but thunder and cannon, and these are feeble in comparison.

What there is of Mauritius is beautiful. You have undulating wide expanses of sugar-cane — a fine, fresh green and very pleasant to the eye; and everywhere else you have a ragged luxuriance of tropic vegetation of vivid greens of varying shades, a wild tangle of underbrush, with graceful tall palms lifting their crippled plumes high above it; and you have stretches of shady dense forest with limpid streams frolicking through them, continually glimpsed and lost and glimpsed again in the pleasantest hide-and-seek fashion; and you have some tiny mountains, some quaint and picturesque groups of toy peaks, and a dainty little vest-pocket Matterhorn; and here and there and now and then a strip of sea with a white ruffle of surf breaks into the view.

That is Mauritius; and pretty enough. The details are few, the massed result is charming, but not imposing; not riotous, not exciting; it is a Sunday landscape. Perspective, and the enchantments wrought by distance, are wanting. There are no distances; there is no perspective, so to speak. Fifteen miles as the crow flies is the usual limit of vision. Mauritius is a garden and a park combined. It affects one's emotions as parks and gardens affect them. The surfaces of one's spiritual deeps are pleasantly played upon, the deeps themselves are not reached, not stirred. Spaciousness, remote altitudes, the sense of mystery which haunts apparently inaccessible mountain domes and summits reposing in the sky — these are the things which exalt the spirit and move it to see visions and dream dreams.

The Sandwich Islands remain my ideal of the perfect thing in the matter of tropical islands. I would add another story to Mauna Loa's 16,000 feet if I could, and make it particularly bold and steep and craggy and forbidding and snowy; and I would make the volcano spout its lava-floods out of its summit instead of its sides; but aside from these non-essentials I have no corrections to suggest. I hope these will be attended to; I do not wish to have to speak of it again.

CHAPTER LXIV.

When your watch gets out of order you have choice of two things to do: throw it in the fire or take it to the watch-tinker. The former is the quickest.
— *Pudd'nhead Wilson's New Calendar.*

THE *Arundel Castle* is the finest boat I have seen in these seas. She is thoroughly modern, and that statement covers a great deal of ground. She has the usual defect, the common defect, the universal defect, the defect that has never been missing from any ship that ever sailed — she has imperfect beds. Many ships have good beds, but no ship has *very* good ones. In the matter of beds all ships have been badly edited, ignorantly edited, from the beginning. The selection of the beds is given to some hearty, strong-backed, self-made man, when it ought to be given to a frail woman accustomed from girlhood to backaches and insomnia. Nothing is so rare, on either side of the ocean, as a perfect bed; nothing is so difficult to make. Some of the hotels on both sides provide it, but no ship ever does or ever did. In Noah's Ark the beds were simply scandalous. Noah set the fashion, and it will endure in one degree of modification or another till the next flood.

8 A.M. Passing Isle de Bourbon. Broken-up sky-line of volcanic mountains in the middle. Surely it would not cost much to repair them, and it seems inexcusable neglect to leave them as they are.

It seems stupid to send tired men to Europe to rest. It is no proper rest for the mind to clatter from town to town in the dust and cinders, and examine galleries and architecture,

(630)

and be always meeting people and lunching and teaing and dining, and receiving worrying cables and letters. And a sea voyage on the Atlantic is of no use — voyage too short, sea too rough. The peaceful Indian and Pacific Oceans and the long stretches of time are the healing thing.

STUPIDITY IN EUROPE.

May 2, A.M. A fair, great ship in sight, almost the first we have seen in these weeks of lonely voyaging. We are now in the Mozambique Channel, between Madagascar and South Africa, sailing straight west for Delagoa Bay.

Last night, the burly chief engineer, middle-aged, was standing telling a spirited seafaring tale, and had reached the most exciting place, where a man overboard was washing swiftly astern on the great seas, and uplifting despairing cries, everybody racing aft in a frenzy of excitement and fading hope, when the band, which had been silent a moment, began impressively its closing piece, the English national anthem. As simply as if he was unconscious of what he was doing, he stopped his story, uncovered, laid his laced cap against his breast, and slightly bent his grizzled head. The few bars finished, he put on his cap and took up his tale again as naturally as if that interjection of music had been a part of it. There was something touching and fine about it, and it was moving to reflect that he was one of a myriad, scattered over every part of the globe, who by turn was doing as he was doing every hour of the twenty-four — those awake doing it while the others slept — those impressive bars forever floating up out of the various climes, never silent and never lacking reverent listeners.

All that I remember about Madagascar is that Thackeray's little Billie went up to the top of the mast and there knelt him upon his knee, saying, "I see

" Jerusalem and Madagascar,
And North and South Amerikee."

May 3. Sunday. Fifteen or twenty Africanders who will end their voyage to-day and strike for their several homes from Delagoa Bay to-morrow, sat up singing on the after-deck in the moonlight till 3 A.M. Good fun and wholesome. And the songs were clean songs, and some of them were hallowed by tender associations. Finally, in a pause, a man asked, "Have you heard about the fellow that kept a diary crossing the Atlantic?" It was a discord, a wet blanket. The men were not in the mood for humorous dirt. The songs

GOOD FUN AND WHOLESOME.

had carried them to their homes, and in spirit they sat by those far hearthstones, and saw faces and heard voices other than those that were about them. And so this disposition to drag in an old indecent anecdote got no welcome; nobody answered. The poor man hadn't wit enough to see that he had blundered, but asked his question again. Again there was no response. It was embarrassing for him. In his con‑ fusion he chose the wrong course, did the wrong thing — began the anecdote. Began it in a deep and hostile stillness, where had been such life and stir and warm comradeship before. He delivered himself of the brief details of the diary's first day, and did it with some confidence and a fair degree of eagerness. It fell flat. There was an awkward pause. The two rows of men sat like statues. There was no movement, no sound. He *had* to go on; there was no other way, at least none that an animal of his calibre could think of. At the close of each day's diary the same dismal silence followed. When at last he finished his tale and sprung the indelicate surprise which is wont to fetch a crash of laughter, not a ripple of sound resulted. It was as if the tale had been told to dead men. After what seemed a long, long time, somebody sighed, somebody else stirred in his seat; presently, the men dropped into a low murmur of confidential talk, each with his neighbor, and the incident was closed. There were indica‑ tions that that man was fond of his anecdote; that it was his pet, his standby, his shot that never missed, his reputation‑ maker. But he will never tell it again. No doubt he will think of it sometimes, for that cannot well be helped; and then he will see a picture, and always the same picture — the double rank of dead men; the vacant deck stretching away in dimming perspective beyond them, the wide desert of smooth sea all abroad; the rim of the moon spying from behind a rag of black cloud; the remote top of the mizzenmast shearing a zigzag path through the fields of stars in the deeps of space; and this soft picture will remind him of the time that he sat in the midst of it and told his poor little tale and felt *so* lonesome when he got through.

Fifty Indians and Chinamen asleep in a big tent in the waist of the ship forward; they lie side by side with no space between; the former wrapped up, head and all, as in the Indian

INDIANS AND CHINAMEN.

streets, the Chinamen uncovered; the lamp and things for opium smoking in the center.

A passenger said it was ten 2-ton truck loads of dynamite that lately exploded at Johannesburg. Hundreds killed; he doesn't know how many; limbs picked up for miles around. Glass shattered, and roofs swept away or collapsed 200 yards off; fragment of iron flung three and a half miles.

It occurred at 3 p. m.; at 6, £65,000 had been subscribed. When this passenger left, £35,000 had been voted by city and state governments and £100,000 by citizens and business corporations. When news of the disaster was telephoned to the Exchange £35,000 were subscribed in the first five minutes. Subscribing was still going on when he left; the papers had ceased the names, only the amounts — too many names; not enough room. £100,000 subscribed by companies and citizens; if this is true, it must be what they call in Australia " a record " — the biggest instance of a spontaneous outpour for charity in history, considering the size of the population it was drawn from, $8 or $10 for each white resident, babies at the breast included.

Monday, May 4. Steaming slowly in the stupendous Delagoa Bay, its dim arms stretching far away and disappearing on both sides. It could furnish plenty of room for all the ships in the world, but it is shoal. The lead has given us 3½ fathoms several times and we are drawing that, lacking 6 inches.

A bold headland — precipitous wall, 150 feet high, very strong, red color, stretching a mile or so. A man said it was Portuguese blood — battle fought here with the natives last year. I think this doubtful. Pretty cluster of houses on the tableland above the red — and rolling stretches of grass and groups of trees, like England.

The Portuguese have the railroad (one passenger train a day) to the border — 70 miles — then the Netherlands Company have it. Thousands of tons of freight on the shore — no cover. This is Portuguese all over — indolence, piousness, poverty, impotence.

Crews of small boats and tugs, all jet black woolly heads and very muscular.

Winter. The South African winter is just beginning now,

but nobody but an expert can tell it from summer. However, I am tired of summer; we have had it unbroken for eleven months. We spent the afternoon on shore, Delagoa Bay. A small town — no sights. No carriages. Three 'rickshas, but we couldn't get them — apparently private. These Portuguese are a rich brown, like some of the Indians. Some of the blacks have the long horse heads and very long chins of the negroes of the picture books; but most of them are exactly like the negroes of our Southern States — round faces, flat noses, good-natured, and easy laughers.

Flocks of black women passed along, carrying outrageously heavy bags of freight on their heads — the quiver of their leg as the foot was planted and the strain exhibited by their bodies showed what a tax upon their strength the load was. They were stevedores, and doing full stevedore's work. They were very erect when unladen — from carrying weights on their heads — just like the Indian women. It gives them a proud, fine carriage.

LIKE OUR SOUTHERN NEGROES.

Sometimes one saw a woman carrying on her head a laden and top-heavy basket the shape of an inverted pyramid — its top the size of a soup-plate, its base the diameter of a teacup. It required nice balancing — and got it.

No bright colors; yet there were a good many Hindoos.

The Second Class Passenger came over as usual at "lights out" (11) and we lounged along the spacious vague solitudes of the deck and smoked the peaceful pipe and talked. He told me an incident in Mr. Barnum's life which was evidently characteristic of that great showman in several ways:

This was Barnum's purchase of Shakespeare's birthplace, a quarter of a century ago. The Second Class Passenger was in Jamrach's employ at the time and knew Barnum well. He said the thing began in this way. One morning Barnum and Jamrach were in Jamrach's little private snuggery back of the wilderness of caged monkeys and snakes and other commonplaces of Jamrach's stock in trade, refreshing themselves after an arduous stroke of business, Jamrach with something orthodox, Barnum with something heterodox — for Barnum was a teetotaler. The stroke of business was in the elephant line. Jamrach had contracted to deliver to Barnum in New York 18 elephants for $360,000 in time for the next season's opening. Then it occurred to Mr. Barnum that he needed a "card." He suggested Jumbo. Jamrach said he would have to think of something else — Jumbo couldn't be had; the Zoo wouldn't part with that elephant. Barnum said he was willing to pay a fortune for Jumbo if he could get him. Jamrach said it was no use to think about it; that Jumbo was as popular as the Prince of Wales and the Zoo wouldn't dare to sell him; all England would be outraged at the idea; Jumbo was an English institution; he was part of the national glory; one might as well think of buying the Nelson monument. Barnum spoke up with vivacity and said —

"It's a first-rate idea. *I'll buy the Monument.*"

Jamrach was speechless for a second. Then he said, like one ashamed —

"You caught me. I was napping. For a moment I thought you were in earnest."

Barnum said pleasantly —

"I *was* in earnest. I know they won't sell it, but no matter, I will not throw away a good idea for all that. All I want is a big advertisement. I will keep the thing in mind, and if nothing better turns up I will offer to buy it. That will answer every purpose. It will furnish me a couple of columns of gratis advertising in every English and American paper for

BARNUM'S CHANCE

a couple of months, and give my show the biggest boom a show ever had in this world."

Jamrach started to deliver a burst of admiration, but was interrupted by Barnum, who said —

"Here is a state of things! England ought to blush."

His eye had fallen upon something in the newspaper. He read it through to himself, then read it aloud. It said that the house that Shakespeare was born in at Stratford-on-Avon was falling gradually to ruin through neglect; that the room where the poet first saw the light was now serving as a butcher's shop; that all appeals to England to contribute money (the requisite sum stated) to buy and repair the house and place it in the care of salaried and trustworthy keepers had fallen resultless. Then Barnum said—

"There's my chance. Let Jumbo and the Monument alone for the present— they'll keep. I'll buy Shakespeare's house. I'll set it up in my Museum in New York and put a glass case around it and make a sacred thing of it; and you'll see all America flock there to worship; yes, and pilgrims from the whole earth; and I'll make them take their hats off, too. In America we know how to value anything that Shakespeare's touch has made holy. You'll see."

In conclusion the S. C. P. said:

"That is the way the thing came about. Barnum did buy Shakespeare's house. He paid the price asked, and received the properly attested documents of sale. Then there was an explosion, I can tell you. England rose! What, the birthplace of the master-genius of all the ages and all the climes — that priceless possession of Britain — to be carted out of the country like so much old lumber and set up for sixpenny desecration in a Yankee show-shop — the idea was not to be tolerated for a moment. England rose in her indignation, and Barnum was glad to relinquish his prize and offer apologies. However, he stood out for a compromise; he claimed a concession — England must let him have Jumbo. And England consented, but not cheerfully."

It shows how, by help of time, a story can grow — even after Barnum has had the first innings in the telling of it. Mr. Barnum told me the story himself, years ago. He said that the permission to buy Jumbo was not a concession; the purchase was made and the animal delivered before the public knew anything about it. Also, that the securing of Jumbo was all the advertisement he needed. It produced many columns of newspaper talk, free of cost, and he was satisfied. He said that if he had failed to get Jumbo he would have caused his notion of buying the Nelson Monument to be treacherously smuggled into print by some trusty friend, and after he had gotten a few hundred pages of gratuitous advertising out of it, he would have come out with a blundering, obtuse, but warm-hearted letter of apology, and in a postscript to it would have naïvely proposed to let the Monument go, and take Stonehenge in place of it at the same price.

It was his opinion that such a letter, written with well-simulated asinine innocence and gush would have gotten his ignorance and stupidity an amount of newspaper abuse worth six fortunes to him, and not purchasable for twice the money.

I knew Mr. Barnum well, and I placed every confidence in the account which he gave me of the Shakespeare birthplace episode. He said he found the house neglected and going to decay, and he inquired into the matter and was told that many times earnest efforts had been made to raise money for its proper repair and preservation, but without success. He then proposed to buy it. The proposition was entertained, and a price named — $50,000, I think; but whatever it was, Barnum paid the money down, without remark, and the papers were drawn up and executed. He said that it had been his purpose to set up the house in his Museum, keep it in repair, protect it from name-scribblers and other desecrators, and leave it by bequest to the safe and perpetual guardianship of the Smithsonian Institute at Washington.

But as soon as it was found that Shakespeare's house had passed into foreign hands and was going to be carried across the ocean, England was stirred as no appeal from the custo-

dians of the relic had ever stirred England before, and protests
came flowing in — and money, too, to stop the outrage. Offers
of re-purchase were made — offers of double the money that
Mr. Barnum had paid for the house. He handed the house
back, but took only the sum which it had cost him — but on the
condition that an endowment sufficient for the future safe-
guarding and maintenance of the sacred relic should be raised.
This condition was fulfilled.

That was Barnum's account of the episode; and to the end
of his days he claimed with pride and satisfaction that not
England, but America — represented by him — saved the birthplace of Shakespeare from destruction.

At 3 P. M., May 6th, the ship slowed down, off the land, and
thoughtfully and cautiously picked her way into the snug
harbor of Durban, South Africa.

CHAPTER LXV.

In statesmanship get the formalities right, never mind about the moralities.
— *Pudd'nhead Wilson's New Calendar.*

FROM DIARY:

Royal Hotel. Comfortable, good table, good service of natives and Madrasis. Curious jumble of modern and ancient city and village, primitiveness and the other thing. Electric bells, but they don't ring. Asked why they didn't, the watchman in the office said he thought they must be out of order; he thought so because some of them rang, but most of them didn't. Wouldn't it be a good idea to put them in order? He hesitated — like one who isn't quite sure — then conceded the point.

May 7. A bang on the door at 6. Did I want my boots cleaned? Fifteen minutes later another bang. Did we want coffee? Fifteen later, bang again, my wife's bath ready; 15 later, my bath ready. Two other bangs; I forget what they were about. Then lots of shouting back and forth, among the servants just as in an Indian hotel.

Evening. At 4 P. M. it was unpleasantly warm. Half-hour after sunset one needed a spring overcoat; by 8 a winter one.

Durban is a neat and clean town. One notices that without having his attention called to it.

Rickshaws drawn by splendidly built black Zulus, so overflowing with strength, seemingly, that it is a pleasure, not a pain, to see them snatch a rickshaw along. They smile and laugh and show their teeth — a good-natured lot. Not allowed to drink; 2ˢ per hour for one person; 3ˢ for two; 3ᵈ for a course — one person.

The chameleon in the hotel court. He is fat and indolent and contemplative; but is business-like and capable when a fly comes about — reaches out a tongue like a teaspoon and takes him in. He gums his tongue first. He is always pious, in his looks. And pious and thankful both, when Providence or one of us sends him a fly. He has a froggy head, and a back like a new grave — for shape; and hands like a bird's toes that have been frost-

bitten. But his eyes are his exhibition feature. A couple of skinny cones project from the sides of his head, with a wee shiny bead of an eye set in the apex of each; and these cones turn bodily like pivot-guns and point every-which-way, and they are independent of each other; each has its own exclusive machinery. When I am behind him and C. in front of him, he whirls one eye rearwards and the other forwards — which gives him a most

A CONGRESSIONAL EXPRESSION.

Congressional expression (one eye on the constituency and one on the swag); and then if something happens above and below him he shoots out one eye upward like a telescope and the other downward — and this changes his expression, but does not improve it.

Natives must not be out after the curfew bell without a pass. In Natal there are ten blacks to one white.

Sturdy plump creatures are the women. They comb their wool up to a peak and keep it in position by stiffening it with brown-red clay — half of this tower colored, denotes engagement; the whole of it colored denotes marriage.

None but heathen Zulus on the police; Christian ones not allowed.

May 9. A drive yesterday with friends over the Berea. Very fine roads and lofty, overlooking the whole town, the harbor, and the sea — beautiful views. Residences all along,

A FASHION IN HAIR.

set in the midst of green lawns with shrubs and generally one or two intensely red outbursts of poinsettia — the flaming splotch of blinding red a stunning contrast with the world of surrounding green. The cactus tree — candelabrum - like; and one twisted like gray writhing serpents. The " flat-crown " (should be flat-roof) — half a dozen naked branches full of elbows, slant upward like artificial supports, and fling a roof of delicate foliage out in a horizontal platform as flat as a floor; and you look up through this thin floor as through a green cobweb or veil. The branches are japanesich. All about you is a bewildering variety of unfamiliar and beautiful trees; one sort wonderfully dense foliage and very dark green — so dark that you notice it at once, notwithstanding there are so many orange trees. The "flamboyant" — not in flower, now, but when in flower lives up to its name, we are told. Another tree with a lovely upright tassel scattered among its rich greenery, red and glowing as a fire-coal. Here and there a gum-tree; half a dozen lofty Norfolk

Island pines lifting their fronded arms skyward. Groups of tall bamboo.

Saw one bird. Not many birds here, and *they* have no music — and the flowers not much smell, they grow so fast.

Everything neat and trim and clean like the town. The loveliest trees and the greatest variety I have ever seen anywhere, except approaching Darjeeling. Have not heard anyone call Natal the garden of South Africa, but that is what it probably is.

It was when Bishop of Natal that Colenso raised such a storm in the religious world. The concerns of religion are a vital matter here yet. A vigilant eye is kept upon Sunday. Museums and other dangerous resorts are not allowed to be open. You may sail on the Bay, but it is wicked to play cricket. For a while a Sunday concert was tolerated, upon condition that it must be admission free and the money taken by collection. But the collection was alarmingly large and that stopped the matter. They are particular about babies. A clergyman would not bury a child according to the sacred rites because it had not been baptized. The Hindoo is more liberal. He burns no child under three, holding that it does not need purifying.

The King of the Zulus, a fine fellow of 30, was banished six years ago for a term of seven years. He is occupying Napoleon's old stand — St. Helena. The people are a little nervous about having him come back, and they may well be, for Zulu kings have been terrible people sometimes — like Tchaka, Dingaan, and Cetewayo.

There is a large Trappist monastery two hours from Durban, over the country roads, and in company with Mr. Milligan and Mr. Hunter, general manager of the Natal government railways, who knew the heads of it, we went out to see it.

There it all was, just as one reads about it in books and

cannot believe that it is so — I mean the rough, hard work, the impossible hours, the scanty food, the coarse raiment, the Maryborough beds, the *tabu* of human speech, of social inter-course, of relaxation, of amusement, of entertainment, of the presence of woman in the men's establishment. There it all was. It was not a dream, it was not a lie. And yet with the fact before one's face it was still incredible. It is such a sweeping suppression of human instincts, such an extinction of the man as an individual.

La Trappe must have known the human race well. The scheme which he invented hunts out everything that a man wants and values — and withholds it from him. Apparently there is no detail that can help make life worth living that has not been carefully ascertained and placed out of the Trap-pist's reach. La Trappe must have known that there were men who would enjoy this kind of misery, but how did he find it out?

If he had consulted you or me he would have been told that his scheme lacked too many attractions; that it was im-possible; that it could never be floated. But there in the monastery was proof that he knew the human race better than it knew itself. He set his foot upon every desire that a man has — yet he floated his project, and it has prospered for two hundred years, and will go on prospering forever, no doubt.

Man likes personal distinction — there in the monastery it is obliterated. He likes delicious food — there he gets beans and bread and tea, and not enough of it. He likes to lie softly — there he lies on a sand mattress, and has a pillow and a blanket, but no sheet. When he is dining, in a great com-pany of friends, he likes to laugh and chat — there a monk reads a holy book aloud during meals, and nobody speaks or laughs. When a man has a hundred friends about him, even-

THE SUPPRESSION OF HUMAN INSTINCTS.

ings, he likes to have a good time and run late — there he and the rest go silently to bed at 8; and in the dark, too; there is but a loose brown robe to discard, there are no night-clothes to put on, a light is not needed. Man likes to lie abed late — there he gets up once or twice in the night to perform some religious office, and gets up finally for the day at two in the morning. Man likes light work or none at all — there he labors all day in the field, or in the blacksmith shop or the other shops devoted to the mechanical trades, such as shoe-making, saddlery, carpentry, and so on. Man likes the society of girls and women — there he never has it. He likes to have his children about him, and pet them and play with them — there he has none. He likes billiards — there is no table there. He likes outdoor sports and indoor dramatic and musical and social entertainments — there are none there. He likes to bet on things — I was told that betting is forbidden there. When a man's temper is up he likes to pour it out upon somebody — there this is not allowed. A man likes animals — pets; there are none there. He likes to smoke — there he cannot do it. He likes to read the news — no papers or magazines come there. A man likes to know how his parents and brothers and sisters are getting along when he is away, and if they miss him — there he cannot know. A man likes a pretty house, and pretty furniture, and pretty things, and pretty colors — there he has nothing but naked aridity and sombre colors. A man likes — name it yourself: whatever it is, it is absent from that place.

From what I could learn, all that a man gets for this is merely the saving of his soul.

It all seems strange, incredible, impossible. But La Trappe knew the race. He knew the powerful attraction of unattract-iveness; he knew that no life could be imagined, howsoever comfortless and forbidding, but somebody would want to try it.

This parent establishment of Germans began its work fifteen years ago, strangers, poor, and unencouraged; it owns
15,000 acres of land now, and raises grain and fruit, and makes
wines, and manufactures all manner of things, and has native
apprentices in its shops, and sends them forth able to read and
write, and also well equipped to earn their living by their
trades. And this young establishment has set up eleven
branches in South Africa, and in them they are christianizing
and educating and teaching wage-yielding mechanical trades
to 1,200 boys and girls. Protestant Missionary work is coldly
regarded by the commercial white colonist all over the heathen
world, as a rule, and its product is nicknamed "rice-Christians"
(occupationless incapables who join the church for revenue
only), but I think it would be difficult to pick a flaw in the
work of these Catholic monks, and I believe that the disposition to attempt it has not shown itself.

Tuesday, May 12. Transvaal politics in a confused condition. First the sentencing of the Johannesburg Reformers
startled England by its severity; on the top of this came
Kruger's exposure of the cipher correspondence, which showed
that the invasion of the Transvaal, with the design of seizing
that country and adding it to the British Empire, was planned
by Cecil Rhodes and Beit — which made a revulsion in English feeling, and brought out a storm against Rhodes and the
Chartered Company for degrading British honor. For a good
while I couldn't seem to get at a clear comprehension of it, it
was so tangled. But at last by patient study I have managed
it, I believe. As I understand it, the Uitlanders and other
Dutchmen were dissatisfied because the English would not
allow them to take any part in the government except to pay
taxes. Next, as I understand it, Dr. Kruger and Dr. Jameson,
not having been able to make the medical business pay, made
a raid into Matabeleland with the intention of capturing the

capital, Johannesburg, and holding the women and children to ransom until the Uitlanders and the other Boers should grant to them and the Chartered Company the political rights which had been withheld from them. They would have succeeded in this great scheme, as I understand it, but for the interference of Cecil Rhodes and Mr. Beit, and other Chiefs of the Matabele, who persuaded their countrymen to revolt and throw off their allegiance to Germany. This, in turn, as I understand it, provoked the King of Abyssinia to destroy the Italian army and fall back upon Johannesburg; this at the instigation of Rhodes, to bull the stock market.

CHAPTER LXVI.

WHEN I scribbled in my note-book a year ago the para-graph which ends the preceding chapter, it was meant to indicate, in an extravagant form, two things : the conflicting nature of the information conveyed by the citizen to the stranger concerning South African politics, and the resulting confusion created in the stranger's mind thereby.

But it does not seem so very extravagant now. Nothing could in that disturbed and excited time make South African politics clear or quite rational to the citizen of the country be-cause his personal interest and his political prejudices were in his way; and nothing could make those politics clear or rational to the stranger, the sources of his information being such as they were.

I was in South Africa some little time. When I arrived there the political pot was boiling fiercely. Four months pre-viously, Jameson had plunged over the Transvaal border with about 600 armed horsemen at his back, to go to the "relief of the women and children" of Johannesburg; on the fourth day of his march the Boers had defeated him in battle, and carried him and his men to Pretoria, the capital, as prisoners; the Boer government had turned Jameson and his officers over to the British government for trial, and shipped them to England; next, it had arrested 64 important citizens of Johannesburg as raid-conspirators, condemned their four leaders to death, then

(654)

THE POLITICAL POT.

commuted the sentences, and now the 64 were waiting, in jail, for further results. Before midsummer they were all out excepting two, who refused to sign the petitions for release; 58 had been fined $10,000 each and enlarged, and the four leaders had gotten off with fines of $125,000 each — with permanent exile added, in one case.

Those were wonderfully interesting days for a stranger, and I was glad to be in the thick of the excitement. Everybody was talking, and I expected to understand the whole of *one side* of it in a very little while.

I was disappointed. There were singularities, perplexities, unaccountabilities about it which I was not able to master. I had no personal access to Boers — their side was a secret to me, aside from what I was able to gather of it from published statements. My sympathies were soon with the Reformers in the Pretoria jail, with their friends, and with their cause. By diligent inquiry in Johannesburg I found out — apparently — all the details of their side of the quarrel except one — *what they expected to accomplish by an armed rising.*

Nobody seemed to know.

The reason why the Reformers were discontented and wanted some changes made, seemed quite clear. In Johannesburg it was claimed that the Uitlanders (strangers, foreigners) paid thirteen-fifteenths of the Transvaal taxes, yet got little or nothing for it. Their city had no charter; it had no municipal government; it could levy no taxes for drainage, water-supply, paving, cleaning, sanitation, policing. There was a police force, but it was composed of Boers, it was furnished by the State Government, and the city had no control over it. Mining was very costly; the government enormously increased the cost by putting burdensome taxes upon the mines, the output, the machinery, the buildings; by burdensome imposts

42

upon incoming materials; by burdensome railway-freight-charges. Hardest of all to bear, the government reserved to itself a monopoly in that essential thing, dynamite, and burdened it with an extravagant price. The detested Hollander from over the water held all the public offices. The government was rank with corruption. The Uitlander had no vote, and must live in the State ten or twelve years before he could get one. He was not represented in the Raad (legislature) that oppressed him and fleeced him. Religion was not free. There were no schools where the teaching was in English, yet the great majority of the white population of the State knew no tongue but that. The State would not pass a liquor law; but allowed a great trade in cheap vile brandy among the blacks, with the result that 25 per cent. of the 50,000 blacks employed in the mines were usually drunk and incapable of working.

There — it was plain enough that the *reasons* for wanting some changes made were abundant and reasonable, if this statement of the existing grievances was correct.

What the Uitlanders wanted was reform — *under the existing Republic.*

What they proposed to do was to secure these reforms by *prayer, petition, and persuasion.*

They did petition. Also, they issued a Manifesto, whose very first note is a bugle-blast of loyalty: "We want the establishment of this Republic as a true Republic."

Could anything be clearer than the Uitlander's statement of the grievances and oppressions under which they were suffering? Could anything be more legal and citizenlike and law-respecting than their attitude as expressed by their Manifesto? No. Those things were perfectly clear, perfectly comprehensible.

But at this point the puzzles and riddles and confusions

begin to flock in. You have arrived at a place which you cannot quite understand.

For you find that as a preparation for this loyal, lawful,
and in every way unexceptionable attempt to persuade the government to right their grievances, the Uitlanders had smuggled a Maxim gun or two and 1,500 muskets into the town,
concealed in oil tanks and coal cars, and had begun to form
and drill military companies composed òf clerks, merchants,
and citizens generally.

What was their idea? Did they suppose that the Boers
would attack them for *petitioning for redress?* That could
not be.

Did they suppose that the Boers would attack them even
for issuing a Manifesto *demanding* relief under the existing
government?

Yes, they apparently believed so, because the air was full of
talk of *forcing* the government to grant redress if it were not
granted peacefully.

The Reformers were men of high intelligence. If they
were in earnest, they were taking extraordinary risks. They
had enormously valuable properties to defend ; their town was
full of women and children ; their mines and compounds were
packed with thousands upon thousands of sturdy blacks. If
the Boers attacked, the mines would close, the blacks would
swarm out and get drunk ; riot and conflagration and the Boers
together might lose the Reformers more in a day, in money,
blood, and suffering, than the desired political relief could compensate in ten years if they won the fight and secured the
reforms.

It is May, 1897, now ; a year has gone by, and the confusions of that day have been to a considerable degree cleared
away. Mr. Cecil Rhodes, Dr. Jameson, and others responsible
for the Raid, have testified before the Parliamentary Committee

of Inquiry in London, and so have Mr. Lionel Phillips and other Johannesburg Reformers, monthly-nurses of the Revolution which was born dead. These testimonies have thrown light. Three books have been added much to this light: "South Africa As It Is," by Mr. Statham, an able writer partial to the Boers; "The Story of an African Crisis," by Mr. Garrett, a brilliant writer partial to Rhodes; and "A Woman's Part in a Revolution," by Mrs. John Hays Hammond, a vigorous and vivid diarist, partial to the Reformers. By liquifying the evidence of the prejudiced books and of the prejudiced parliamentary witnesses and stirring the whole together and pouring it into my own (prejudiced) moulds, I have got at the truth of that puzzling South African situation, which is this:

1. The capitalists and other chief men of Johannesburg were fretting under various political and financial burdens imposed by the State (the South African Republic, sometimes called "the Transvaal") and desired to procure by peaceful means a modification of the laws.

2. Mr. Cecil Rhodes, Premier of the British Cape Colony, millionaire, creator and managing director of the territorially-immense and financially unproductive South Africa Company; projector of vast schemes for the unification and consolidation of all the South African States into one imposing commonwealth or empire under the shadow and general protection of the British flag, thought he saw an opportunity to make profitable use of the Uitlander discontent above mentioned — make the Johannesburg cat help pull out one of his consolidation chestnuts for him. With this view he set himself the task of warming the lawful and legitimate petitions and supplications of the Uitlanders into seditious talk, and their frettings into threatenings — the final outcome to be revolt and armed rebellion. If he could bring about a bloody collision between those people

and the Boer government, Great Britain would have to interfere; her interference would be resisted by the Boers; she
would chastise them and add the Transvaal to her South African possessions. It was not a foolish idea, but a rational and
practical one.

After a couple of years of judicious plotting, Mr. Rhodes
had his reward; the revolutionary kettle was briskly boiling in
Johannesburg, and the Uitlander leaders were backing their
appeals to the government — now hardened into demands —
by threats of force and bloodshed. By the middle of December, 1895, the explosion seemed imminent. Mr. Rhodes was
diligently helping, from his distant post in Cape Town. He
was helping to procure arms for Johannesburg; he was also
arranging to have Jameson break over the border and come to
Johannesburg with 600 mounted men at his back. Jameson —
as per instructions from Rhodes, perhaps — wanted a letter
from the Reformers requesting him to come to their aid. It
was a good idea. It would throw a considerable share of the
responsibility of his invasion upon the Reformers. He got the
letter — that famous one urging him to fly to the rescue of the
women and children. He got it *two months* before he flew.
The Reformers seem to have thought it over and concluded
that they had not done wisely; for the next day after giving
Jameson the implicating document they wanted to withdraw
it and leave the women and children in danger; but they were
told that it was too late. The original had gone to Mr. Rhodes
at the Cape. Jameson had kept a copy, though.

From that time until the 29th of December, a good deal of
the Reformers' time was taken up with energetic efforts to
keep Jameson from coming to their assistance. Jameson's invasion had been set for the 26th. The Reformers were not
ready. The town was not united. Some wanted a fight,
some wanted peace; some wanted a new government, some

wanted the existing one reformed; apparently very few wanted the revolution to take place in the interest and under the ultimate shelter of the Imperial flag — British; yet a report began to spread that Mr. Rhodes's embarrassing assistance had for its end this latter object.

Jameson was away up on the frontier tugging at his leash, fretting to burst over the border. By hard work the Reformers got his starting-date postponed a little, and wanted to get it postponed eleven days. Apparently, Rhodes's agents were seconding their efforts — in fact wearing out the telegraph wires trying to hold him back. Rhodes was himself the only man who could have effectively postponed Jameson, but that would have been a disadvantage to his scheme; indeed, it could spoil his whole two years' work.

Jameson endured postponement three days, then resolved to wait no longer. Without any orders — excepting Mr. Rhodes's significant silence — he cut the telegraph wires on the 29th, and made his plunge that night, to go to the rescue of the women and children, by urgent request of a letter now nine days old — as per date, — a couple of months old, in fact. He read the letter to his men, and it affected them. It did not affect all of them alike. Some saw in it a piece of piracy of doubtful wisdom, and were sorry to find that they had been assembled to violate friendly territory instead of to raid native kraals, as they had supposed.

Jameson would have to ride 150 miles. He knew that there were suspicions abroad in the Transvaal concerning him, but he expected to get through to Johannesburg before they should become general and obstructive. But a telegraph wire had been overlooked and not cut. It spread the news of his invasion far and wide, and a few hours after his start the Boer farmers were riding hard from every direction to intercept him.

PATROL — STATE ARTILLERY.

2 OF JAMESON'S GUNS IN FORT PRETORIA, MANNED BY STATE ARTILLERY.

BOERS ASSEMBLING — PRETORIA.

BOER WAR SCENES.

As soon as it was known in Johannesburg that he was on his way to rescue the women and children, the grateful people put the women and children in a train and rushed them for Australia. In fact, the approach of Johannesburg's saviour created panic and consternation there, and a multitude of males of peaceable disposition swept to the trains like a sand-storm. The early ones fared best; they secured seats — by sitting in them — eight hours before the first train was timed to leave.

Mr. Rhodes lost no time. He cabled the renowned Johannesburg letter of invitation to the London press — the grayheadedest piece of ancient history that ever went over a cable.

The new poet laureate lost no time. He came out with a rousing poem lauding Jameson's prompt and splendid heroism in flying to the rescue of the women and children; for the poet could not know that he did not fly until two months after the invitation. He was deceived by the false date of the letter, which was December 20th.

Jameson was intercepted by the Boers on New Year's Day, and on the next day he surrendered. He had carried his copy of the letter along, and if his instructions required him — in case of emergency — to see that it fell into the hands of the Boers, he loyally carried them out. Mrs. Hammond gives him a sharp rap for his supposed carelessness, and emphasizes her feeling about it with burning italics: " It was picked up on the battle-field in a leathern pouch, supposed to be Dr. Jameson's saddle-bag. *Why, in the name of all that is discreet and honorable, didn't he eat it !* "

She requires too much. He was not in the service of the Reformers — excepting ostensibly; he was in the service of Mr. Rhodes. It was the only plain English document, undarkened by ciphers and mysteries, and responsibly signed and authenticated, which squarely implicated the Reformers in the raid, and it was not to Mr. Rhodes's interest that it should be eaten. Besides, that letter was not the original, it was only a copy. Mr. Rhodes had the original — and didn't eat it. He cabled it to the London press. It had already been read in England and America and all over Europe before Jameson dropped it on the battlefield. If the subordinate's knuckles deserved a rap, the principal's deserved as many as a couple of them.

That letter is a juicily dramatic incident and is entitled to all its celebrity, because of the odd and variegated effects which it produced. All within the space of a single week it had made Jameson an illustrious hero in England, a pirate in Pretoria, and an ass without discretion or honor in Johannesburg; also it had produced a poet-laureatic explosion of colored fireworks which filled the world's sky with giddy splendors, and the knowledge that Jameson was coming with it to rescue the women and children emptied Johannesburg of that detail of the population. For an old letter, this was much. For a letter two months old, it did marvels; if it had been a year old it would have done miracles.

CHAPTER LXVII.

THOSE latter days were days of bitter worry and trouble for the harassed Reformers.

From Mrs. Hammond we learn that on the 31st (the day after Johannesburg heard of the invasion), "the Reform Committee repudiates Dr. Jameson's inroad."

It also publishes its intention to adhere to the Manifesto.

It also earnestly desires that the inhabitants shall refrain from overt acts against the Boer government.

It also "distributes arms" at the Court House, and furnishes horses "to the newly-enrolled volunteers."

It also brings a Transvaal flag into the committee-room, and the entire body swear allegiance to it "with uncovered heads and upraised arms."

Also "one thousand Lee-Metford rifles have been given out" — to rebels.

Also, in a speech, Reformer Lionel Phillips informs the public that the Reform Committee Delegation has "been received with courtesy by the Government Commission," and "been assured that their proposals shall be earnestly considered." That "while the Reform Committee regretted Jameson's precipitate action, they would stand by him."

Also the populace are in a state of "wild enthusiasm," and "can scarcely be restrained; they want to go out to meet Jameson and bring him in with triumphal outcry."

Also the British High Commissioner has issued a damnifying proclamation against Jameson and all British abettors of his game. It arrives January 1st.

It is a difficult position for the Reformers, and full of hindrances and perplexities. Their duty is hard, but plain:

1. They have to repudiate the inroad, and stand by the inroader.

2. They have to swear allegiance to the Boer government, and distribute cavalry horses to the rebels.

3. They have to forbid overt acts against the Boer government, and distribute arms to its enemies.

4. They have to avoid collision with the British government, but still stand by Jameson and their new oath of allegiance to the Boer government, taken, uncovered, in presence of its flag.

They did such of these things as they could; they tried to do them all; in fact, did do them all, but only in turn, not simultaneously. In the nature of things they could not be made to simultane.

In preparing for armed revolution and in talking revolution, were the Reformers "bluffing," or were they in earnest? If they were in earnest, they were taking great risks — as has been already pointed out. A gentleman of high position told me in Johannesburg that he had in his possession a printed document proclaiming a *new* government and naming its president — one of the Reform leaders. He said that this proclamation had been ready for issue, but was suppressed when the raid collapsed. Perhaps I misunderstood him. Indeed, I must have misunderstood him, for I have not seen mention of this large incident in print anywhere.

Besides, I hope I am mistaken; for, if I am, then there is argument that the Reformers were privately not serious, but were only trying to scare the Boer government into granting the desired reforms.

The Boer government *was* scared, and it had a right to be. For if Mr. Rhodes's plan was to provoke a collision that would compel the interference of England, that was a serious matter. If it could be shown that that was also the Reformers' plan and purpose, it would prove that they had marked out a feasible project, at any rate, although it was one which could hardly fail to cost them ruinously before England should arrive. But it seems clear that they had no such plan nor desire. If, when the worst should come to the worst, they meant to overthrow the government, they also meant to inherit the assets themselves, no doubt.

This scheme could hardly have succeeded. With an army of Boers at their gates and 50,000 riotous blacks in their midst, the odds against success would have been too heavy — even if the whole town had been armed. With only 2,500 rifles in the place, they stood really no chance.

To me, the military problems of the situation are of more interest than the political ones, because by disposition I have always been especially fond of war. No, I mean fond of discussing war; and fond of giving military advice. If I had been with Jameson the morning after he started, I should have advised him to turn back. That was Monday; it was then that he received his first warning from a Boer source not to violate the friendly soil of the Transvaal. It showed that his invasion was known. If I had been with him on Tuesday morning and afternoon, when he received further warnings, I should have repeated my advice. If I had been with him the next morning — New Year's — when he received notice that "a few hundred" Boers were waiting for him a few miles ahead, I should not have advised, but commanded him to go back. And if I had been with him two or three hours later — a thing not conceivable to me — I should have retired him by force; for at that time he learned that the few hundred had now grown to 800; and that meant that the growing would go on growing.

For, by authority of Mr. Garrett, one knows that Jameson's 600 were only 530 at most, when you count out his native drivers, etc.; and that the 530 consisted largely of "green" youths, "raw young fellows," not trained and war-worn British soldiers; and I would have told Jameson that those lads would not be able to shoot effectively from horseback in the scamper and racket of battle, and that there would not be anything for them to shoot at, anyway, but rocks; for the Boers

BOERS RECEIVING ARMS AND EQUIPMENTS.

would be behind the rocks, not out in the open. I would have told him that 300 Boer sharpshooters behind rocks would be an overmatch for his 500 raw young fellows on horseback.

If pluck were the only thing essential to battle-winning, the English would lose no battles. But discretion, as well as pluck, is required when one fights Boers and Red Indians. In South Africa the Briton has always insisted upon standing bravely up, unsheltered, before the hidden Boer, and taking the results. Jameson's men would follow the custom. Jameson would not have listened to me — he would have been intent upon repeating history, according to precedent. Americans are not acquainted with the British-Boer war of 1881; but its history is interesting, and could have been instructive to Jameson if he had been receptive. I will cull some details of it from trustworthy sources — mainly from "Russell's Natal." Mr. Russell is not a Boer, but a Briton. He is inspector of schools, and his history is a text-book whose purpose is the instruction of the Natal English youth.

After the seizure of the Transvaal and the suppression of the Boer government by England in 1877, the Boers fretted for three years, and made several appeals to England for a restoration of their liberties, but without result. Then they gathered themselves together in a great mass-meeting at Krugersdorp, talked their troubles over, and resolved to fight for their deliverance from the British yoke. (Krugersdorp — the place where the Boers interrupted the Jameson raid.) The little handful of farmers rose against the strongest empire in the world. They proclaimed martial law and the re-establishment of their Republic. They organized their forces and sent them forward to intercept the British battalions. This, although Sir Garnet Wolseley had but lately made proclamation that " so long as the sun shone in the heavens," the Transvaal would be and remain English territory. And also in spite of the fact that the commander of the 94th regiment — already on the march to suppress this rebellion — had been heard to say that " the Boers would turn tail at the first beat of the big drum."*

Four days after the flag-raising, the Boer force which had been sent forward to forbid the invasion of the English troops met them at Bronkhorst Spruit — 246 men of the 94th regiment, in command of a colonel, the big drum beating, the band

---

* "South Africa As It Is," by F. Reginald Statham, page 82. London: T. Fisher Unwin, 1897.

playing — and the first battle was fought.   It lasted ten minutes. Result:

*British loss, more than 150 officers and men, out of the 246.
Surrender of the remnant.*

Boer loss — if any — not stated.

They are fine marksmen, the Boers.   From the cradle up,
they live on horseback and hunt wild animals with the rifle.
They have a passion for liberty and the Bible, and care for
nothing else.

"General Sir George Colley, Lieutenant-Governor and
Commander-in-Chief in Natal, felt it his duty to proceed at
once to the relief of the loyalists and soldiers beleaguered in
the different towns of the Transvaal." He moved out with
1,000 men and some artillery.   He found the Boers encamped
in a strong and sheltered position on high ground at Laing's
Nek — every Boer behind a rock.   Early in the morning of
the 28th January, 1881, he moved to the attack " with the
58th regiment, commanded by Colonel Deane, a mounted
squadron of 70 men, the 60th Rifles, the Naval Brigade with
three rocket tubes, and the Artillery with six guns." He
shelled the Boers for twenty minutes, then the assault was delivered, the 58th marching up the slope *in solid column.*   The
battle was soon finished, with this result, according to Russell:

British loss in killed and wounded, 174.

Boer loss, " trifling."

Colonel Deane was killed, and apparently every officer above
the grade of lieutenant was killed or wounded, for the 58th
retreated to its camp *in command of a lieutenant.* ("Africa as
It Is.")

That ended the second battle.

On the 7th of February General Colley discovered that the
Boers were flanking his position.   The next morning he left
his camp at Mount Pleasant and marched out and crossed the

Ingogo river with 270 men, started up the Ingogo heights, and there fought a battle which lasted from noon till nightfall. He then retreated, leaving his wounded with his military chaplain, and in recrossing the now swollen river lost some of his men by drowning. That was the third Boer victory. Result, according to Mr. Russell:

British loss 150 out of 270 engaged.

Boer loss, 8 killed, 9 wounded — 17.

There was a season of quiet, now, but at the end of about three weeks Sir George Colley conceived the idea of climbing, with an infantry and artillery force, the steep and rugged mountain of Amajuba in the night — a bitter hard task, but he accomplished it. On the way he left about 200 men to guard a strategic point, and took about 400 up the mountain with him. When the sun rose in the morning, there was an unpleasant surprise for the Boers; yonder were the English troops visible on top of the mountain two or three miles away, and now their own position was at the mercy of the English artillery. The Boer chief resolved to retreat — up that mountain. He asked for volunteers, and got them.

The storming party crossed the swale and began to creep up the steeps, " and from behind rocks and bushes they shot at the soldiers on the sky-line as if they were stalking deer," says Mr. Russell. There was " continuous musketry fire, steady and fatal on the one side, wild and ineffectual on the other." The Boers reached the top, and began to put in their ruinous work. Presently the British " broke and fled for their lives down the rugged steep." The Boers had won the battle. Result in killed and wounded, including among the killed the British General:

British loss, 226, out of 400 engaged.

Boer loss, 1 killed, 5 wounded.

43

That ended the war. England listened to reason, and recognized the Boer Republic — a government which has never been in any really awful danger since, until Jameson started after it with his 500 "raw young fellows." To recapitulate:

The Boer farmers and British soldiers fought 4 battles, and the Boers won them all. Result of the 4, in killed and wounded:

British loss, 700 men.

Boer loss, so far as known, 23 men.

It is interesting, now, to note how loyally Jameson and his several trained British military officers tried to make their battles conform to precedent. Mr. Garrett's account of the Raid is much the best one I have met with, and my impressions of the Raid are drawn from that.

When Jameson learned that near Krugersdorp he would find 800 Boers waiting to dispute his passage, he was not in the least disturbed. He was feeling as he had felt two or three days before, when he had opened his campaign with a historic remark to the same purport as the one with which the commander of the 94th had opened the Boer-British war of fourteen years before. That Commander's remark was, that the Boers "would turn tail at the first beat of the big drum." Jameson's was, that with his "raw young fellows" he could kick the (persons) of the Boers "all round the Transvaal." He was keeping close to historic precedent.

Jameson arrived in the presence of the Boers. They — according to precedent — were not visible. It was a country of ridges, depressions, rocks, ditches, moraines of mining-tailings — not even as favorable for cavalry work as Laing's Nek had been in the former disastrous days. Jameson shot at the ridges and rocks with his artillery, just as General Colley had done at the Nek; and did them no damage and persuaded no Boer to show himself. Then about a hundred of his men

Outside Pretoria Jail.—
Just after arrival of Dr Jameson.

Prisoners at Racecourse.—
—Pretoria, Jan. 11. 1896.

Boers Coming into Pretoria.

—Some of the Boers—
—Who Brought Dr Jameson—
—To Pretoria.

AFTER THE FIGHTING.

formed up to charge the ridge—according to the 58th's precedent at the Nek; but as they dashed forward they opened out in a long line, which was a considerable improvement on the 58th's tactics; when they had gotten to within 200 yards of the ridge the concealed Boers opened out on them and emptied 20 saddles. The unwounded dismounted and fired at the rocks over the backs of their horses; but the return-fire was too hot, and they mounted again, "and galloped back or crawled away into a clump of reeds for cover, where they were shortly afterward taken prisoners as they lay among the reeds. Some thirty prisoners were so taken, and during the night which followed the Boers carried away another thirty killed and wounded—the wounded to Krugersdorp hospital." Sixty per cent. of the assaulted force disposed of—according to Mr. Garrett's estimate.

It was according to Amajuba precedent, where the British loss was 226 out of about 400 engaged.

Also, in Jameson's camp, that night, "there lay about 30 wounded or otherwise disabled" men. Also during the night "some 30 or 40 young fellows got separated from the command and straggled through into Johannesburg." Altogether a possible 150 men gone, out of his 530. His lads had fought valorously, but had not been able to get near enough to a Boer to kick him around the Transvaal.

At dawn the next morning the column of something short of 400 whites resumed its march. Jameson's grit was stubbornly good; indeed, it was always that. He still had hopes. There was a long and tedious zigzagging march through broken ground, with constant harassment from the Boers; and at last the column "walked into a sort of trap," and the Boers "closed in upon it." "Men and horses dropped on all sides. In the column the feeling grew that unless it could burst through the Boer lines at this point it was done for.

The Maxims were fired until they grew too hot, and, water failing for the cool jacket, five of them jammed and went out of action. The 7-pounder was fired until only half an hour's ammunition was left to fire with. One last rush was made, and failed, and then the Staats Artillery came up on the left flank, and the game was up."

Jameson hoisted a white flag and surrendered.

There is a story, which may not be true, about an ignorant Boer farmer there who thought that this white flag was the national flag of England. He had been at Bronkhorst, and Laing's Nek, and Ingogo and Amajuba, and supposed that the English did not run up their flag excepting at the end of a fight.

PRISONERS AT ROLL CALL.

The following is (as I understand it) Mr. Garrett's estimate of Jameson's total loss in killed and wounded for the two days:

"When they gave in they were minus some 20 per cent. of combatants. There were 76 casualties. There were 30 men hurt or sick in the wagons. There were 27 killed on the spot or mortally wounded."

Total, 133, out of the original 530. It is just 25 per cent.* This is a large improvement upon the precedents established at Bronkhorst, Laing's Nek, Ingogo, and Amajuba, and seems to indicate that Boer marksmanship is not so good now as it was in those days. But there is one detail in which the Raid-episode exactly repeats history. By surrender at Bronkhorst, the whole British force disappeared from the theater of war; this was the case with Jameson's force.

In the Boer loss, also, historical precedent is followed with sufficient fidelity. In the 4 battles named above, the Boer loss, so far as known, was an average of 6 men per battle, to the British average loss of 175. In Jameson's battles, as per Boer official report, the Boer loss in killed was 4. Two of these were killed by the Boers themselves, by accident, the other by Jameson's army — one of them intentionally, the other by a pathetic mischance. "A young Boer named Jacobz was moving forward to give a drink to one of the wounded troopers (Jameson's) after the first charge, when another wounded man, mistaking his intention, shot him." There were three or four wounded Boers in the Krugersdorp hospital, and apparently no others have been reported. Mr. Garrett, "on a balance of probabilities, fully accepts the official version, and thanks Heaven the killed was not larger."

As a military man, I wish to point out what seems to me to be military errors in the conduct of the campaign which we have just been considering. I have seen active service in the field, and it was in the actualities of war that I acquired my training and my right to speak. I served two weeks in the beginning of our Civil War, and during all that time com-

---

* However, I judge that the total was really 150; for the number of wounded carried to Krugersdorp hospital was 53; not 30, as Mr. Garrett reports it. The lady whose guest I was in Krugerdorp gave me the figures. She was head nurse from the beginning of hostilities (Jan. 1) until the professional nurses arrived, Jan. 8th. Of the 53, "Three or four were Boers"; I quote her words.

manded a battery of infantry composed of twelve men. General Grant knew the history of my campaign, for I told it him. I also told him the principle upon which I had conducted it; which was, to tire the enemy. I tired out and disqualified many battalions, yet never had a casualty myself nor lost a man. General Grant was not given to paying compliments, yet he said frankly that if I had conducted the whole war much bloodshed would have been spared, and that what the army might have lost through the inspiriting results of collision in the field would have been amply made up by the liberalizing influences of travel. Further endorsement does not seem to me to be necessary.

Let us now examine history, and see what it teaches. In the 4 battles fought in 1881 and the two fought by Jameson, the British loss in killed, wounded, and prisoners, was substantially 1,300 men; the Boer loss, as far as is ascertainable, was about 30 men. These figures show that there was a defect somewhere. It was not in the absence of courage. I think it lay in the absence of discretion. The Briton should have done one thing or the other: discarded British methods and fought the Boer with Boer methods, or augmented his own force until — using British methods — it should be large enough to equalize results with the Boer.

To retain the British method requires certain things, determinable by arithmetic. If, for argument's sake, we allow that the aggregate of 1,716 British soldiers engaged in the 4 early battles was opposed by the same aggregate of Boers, we have this result: the British loss of 700 and the Boer loss of 23 argues that in order to equalize results in future battles you must make the British force thirty times as strong as the Boer force. Mr. Garrett shows that the Boer force immediately opposed to Jameson was 2,000, and that there were 6,000 more on hand by the evening of the second day. Arithmetic shows

TIRING THE ENEMY.

that in order to make himself the equal of the 8,000 Boers, Jameson should have had 240,000 men, whereas he merely had 530 boys. From a military point of view, backed by the facts of history, I conceive that Jameson's military judgment was at fault.

Another thing. Jameson was encumbered by artillery, ammunition, and rifles. The facts of the battle show that he should have had none of those things along. They were heavy, they were in his way, they impeded his march. There was nothing to shoot at but rocks — he knew quite well that there would be nothing to shoot at but rocks — and he knew that artillery and rifles have no effect upon rocks. He was badly overloaded with unessentials. He had 8 Maxims — a Maxim is a kind of Gatling, I believe, and shoots about 500 bullets per minute; he had one $12\frac{1}{2}$-pounder cannon and two 7-pounders; also, 145,000 rounds of ammunition. He worked the Maxims so hard upon the rocks that five of them became disabled — five of the Maxims, not the rocks. It is believed that upwards of 100,000 rounds of ammunition of the various kinds were fired during the 21 hours that the battles lasted. *One man killed.* He must have been much mutilated. It was a pity to bring those futile Maxims along. Jameson should have furnished himself with a battery of Pudd'nhead Wilson maxims instead. They are much more deadly than those others, and they are easily carried, because they have no weight.

Mr. Garrett — not very carefully concealing a smile — excuses the presence of the Maxims by saying that they were of very substantial use because their sputtering disordered the aim of the Boers, and in that way saved lives.

Three cannon, eight Maxims, and five hundred rifles yielded a result which emphasized a fact which had already been established — that the British system of standing out in the open to fight Boers who are behind rocks is not wise, not excusable, and ought to be abandoned for something more efficacious. For the purpose of war is to kill, not merely to waste ammunition.

THE DOCUMENT IN EVIDENCE.

If I could get the management of one of those campaigns, I would know what to do, for I have studied the Boer. He values the Bible above every other thing. The most delicious edible in South Africa is "biltong." You will have seen it mentioned in Olive Schreiner's books. It is what our plainsmen call "jerked beef." It is the Boer's main standby. He has a passion for it, and he is right.

If I had the command of the campaign I would go with rifles only, no cumbersome Maxims and cannon to spoil good rocks with. I would move surreptitiously by night to a point about a quarter of a mile from the Boer camp, and there I would build up a pyramid of biltong and Bibles fifty feet high, and then conceal my men all about. In the morning the Boers would send out spies, and then the rest would·come with a rush. I would surround them, and they would have to fight my men on equal terms, in the open. There wouldn't be any Amajuba results.*

---

*Just as I am finishing this book an unfortunate dispute has sprung up between Dr. Jameson and his officers, on the one hand, and Colonel Rhodes on the other, concerning the wording of a note which Colonel Rhodes sent from Johannesburg by a cyclist to Jameson just before hostilities began on the memorable New Year's Day. Some of the fragments of this note were found on the battlefield after the fight, and these have been pieced together; the dispute is as to what words the lacking fragments contained. Jameson says the note promised him a reinforcement of 300 men from Johannesburg. Colonel Rhodes denies this, and says he merely promised to send out "some" men "to meet you."

It seems a pity that these friends should fall out over so little a thing. If the 300 had been sent, what good would it have done? In 21 hours of industrious fighting, Jameson's 530 men, with 8 Maxims, 3 cannon, and 145,000 rounds of ammunition, killed an aggregate of 1 Boer. These statistics show that a reinforcement of 300 Johannesburgers, armed merely with muskets, would have killed, at the outside, only a little over a half of another Boer. This would not have saved the day. It would not even have seriously affected the general result. The figures show clearly, and with mathematical violence, that the only way to save Jameson, or even give him a fair and equal chance with the enemy, was for Johannesburg to send him 240 Maxims, 90 cannon, 600 carloads of ammunition, and 240,000 men. Johannesburg was not in a position to do this. Johannesburg has been called very hard names for not reinforcing Jameson. But in every instance this has been done by two classes of persons — people who do not read history, and people, like Jameson, who do not understand what it means, after they have read it.

CHAPTER LXVIII.

THE Duke of Fife has borne testimony that Mr. Rhodes deceived him. That is also what Mr. Rhodes did with the Reformers. He got them into trouble, and then stayed out himself. A judicious man. He has always been that. As to this there was a moment of doubt, once. It was when he was out on his last pirating expedition in the Matabele country. The cable shouted out that he had gone unarmed, to visit a party of hostile chiefs. It was true, too; and this dare-devil thing came near fetching another indiscretion out of the poet laureate. It would have been too bad, for when the facts were all in, it turned out that there was a lady along, too, and she also was unarmed.

In the opinion of many people Mr. Rhodes is South Africa; others think he is only a large part of it. These latter consider that South Africa consists of Table Mountain, the diamond mines, the Johannesburg gold fields, and Cecil Rhodes. The gold fields are wonderful in every way. In seven or eight years they built up, in a desert, a city of a hundred thousand inhabitants, counting white and black together; and not the ordinary mining city of wooden shanties, but a city made out of lasting material. Nowhere in the world is there such a concentration of rich mines as at Johannesburg. Mr. Bonamici, my manager there, gave me a small gold brick with some statistics engraved upon it which record the output of gold from the early days to July, 1895, and exhibit the strides which have

been made in the development of the industry; in 1888 the
output was $4,162,440; the output of the next five and a half
years was (total) $17,585,894; for the single year ending with
June, 1895, it was $45,553,700.

The capital which has developed the mines came from England, the mining engineers from America. This is the case
with the diamond mines also. South Africa seems to be the
heaven of the American scientific mining engineer. He gets
the choicest places, and keeps them. His salary is not based
upon what he would get in America, but apparently upon what
a whole family of him would get there.

The successful mines pay great dividends, yet the rock is
not rich, from a Californian point of view. Rock which yields
ten or twelve dollars a ton is considered plenty rich enough.
It is troubled with base metals to such a degree that twenty
years ago it would have been only about half as valuable as
it is now; for at that time there was no paying way of getting
anything out of such rock but the coarser-grained "free" gold;
but the new cyanide process has changed all that, and the gold
fields of the world now deliver up fifty million dollars' worth
of gold per year which would have gone into the tailing-pile
under the former conditions.

The cyanide process was new to me, and full of interest;
and among the costly and elaborate mining machinery there
were fine things which were new to me, but I was already familiar with the rest of the details of the gold-mining industry.
I had been a gold miner myself, in my day, and knew substantially everything that those people knew about it, except how
to make money at it. But I learned a good deal about the
Boers there, and that was a fresh subject. What I heard there
was afterwards repeated to me in other parts of South Africa.
Summed up — according to the information thus gained — this
is the Boer:

He is deeply religious, profoundly ignorant, dull, obstinate, bigoted, uncleanly in his habits, hospitable, honest in his dealings with the whites, a hard master to his black servant, lazy, a good shot, good horseman, addicted to the chase, a lover of political independence, a good husband and father, not fond of herding together in towns, but liking the seclusion and remoteness and solitude and empty vastness and silence of the veldt; a man of a mighty appetite, and not delicate about what he appeases it with — well-satisfied with pork and Indian corn and biltong, requiring only that the quantity shall not be stinted; willing to ride a long journey to take a hand in a rude all-night dance interspersed with vigorous feeding and boisterous jollity, but ready to ride twice as far for a prayer-meeting; proud of his Dutch and Huguenot origin and its religious and military history; proud of his race's achievements in South Africa, its bold plunges into hostile and uncharted deserts in search of free solitudes unvexed by the pestering and detested English, also its victories over the natives and the British; proudest of all, of the direct and effusive personal interest which the Deity has always taken in its affairs. He cannot read, he cannot write; he has one or two newspapers, but he is, apparently, not aware of it; until latterly he had no schools, and taught his children nothing; news is a term which has no meaning to him, and the thing itself he cares nothing about. He hates to be taxed and resents it. He has stood stock still in South Africa for two centuries and a half, and would like to stand still till the end of time, for he has no sympathy with Uitlander notions of progress. He is hungry to be rich, for he is human; but his preference has been for riches in cattle, not in fine clothes and fine houses and gold and diamonds. The gold and the diamonds have brought the godless stranger within his gates, also contamination and broken repose, and he wishes that they had never been discovered.

I think that the bulk of those details can be found in Olive Schreiner's books, and she would not be accused of sketching the Boer's portrait with an unfair hand.

Now what would you expect from that unpromising material? What ought you to expect from it? Laws inimical to religious liberty? Yes. Laws denying representation and suffrage to the intruder? Yes. Laws unfriendly to educational institutions? Yes. Laws obstructive of gold production? Yes. Discouragement of railway expansion? Yes. Laws heavily taxing the intruder and overlooking the Boer? Yes.

The Uitlander seems to have expected something very different from all that. I do not know why. Nothing different from it was rationally to be expected. A round man cannot be expected to fit a square hole right away. He must have time to modify his shape. The modification had begun in a detail or two, before the Raid, and was making some progress. It has made further progress since. There are wise men in the Boer government, and that accounts for the modification; the modification of the Boer mass has probably not begun yet. If the heads of the Boer government had not been wise men they would have hanged Jameson, and thus turned a very commonplace pirate into a holy martyr. But even their wisdom has its limits, and they will hang Mr. Rhodes if they ever catch him. That will round him and complete him and make him a saint. He has already been called by all other titles that symbolize human grandeur, and he ought to rise to this one, the grandest of all. It will be a dizzy jump from where he is now, but that is nothing, it will land him in good company and be a pleasant change for him.

Some of the things demanded by the Johannesburgers'

44

Manifesto have been conceded since the days of the Raid, and the others will follow in time, no doubt. It was most fortunate for the miners of Johannesburg that the taxes which distressed them so much were levied by the Boer government, instead of by their friend Rhodes and his Chartered Company of high-waymen, for these latter take *half* of whatever their mining victims find, they do not stop at a mere percentage. If the Johannesburg miners were under their jurisdiction they would be in the poorhouse in twelve months.

I have been under the impression all along that I had an unpleasant paragraph about the Boers somewhere in my note-book, and also a pleasant one. I have found them now. The unpleasant one is dated at an interior village, and says:

"Mr. Z. called. He is an English Afrikander; is an old resident, and has a Boer wife. He speaks the language, and his professional business is with the Boers exclusively. He told me that the ancient Boer families in the great region of which this village is the commercial center are falling victims to their inherited indolence and dullness in the materialistic latter-day race and struggle, and are dropping one by one into the grip of the usurer — getting hopelessly in debt — and are losing their high place and retiring to second and lower. The Boer's farm does not go to another Boer when he loses it, but to a foreigner. Some have fallen so low that they sell their daughters to the blacks."

Under date of another South African town I find the note which is creditable to the Boers:

"Dr. X. told me that in the Kafir war 1,500 Kafirs took refuge in a great cave in the mountains about 90 miles north of Johannesburg, and the Boers blocked up the entrance and smoked them to death. Dr. X. has been in there and seen the great array of bleached skeletons — one a woman with the skeleton of a child hugged to her breast."

The great bulk of the savages must go. The white man wants their lands, and all must go excepting such percentage of them as he will need to do his work for him upon terms to be determined by himself. Since history has removed the element of guesswork from this matter and made it certainty, the humanest way of diminishing the black population should be adopted, not the old cruel ways of the past. Mr. Rhodes and his gang have been following the old ways. They are chartered to rob and slay, and they lawfully do it, but not in a compassionate and Christian spirit. They rob the Mashonas and the Matabeles of a portion of their territories in the hallowed old style of "purchase" for a song, and then they force a quarrel and take the rest by the strong hand. They rob the natives of their cattle under the pretext that all the cattle in the country belonged to the king whom they have tricked and assassinated. They issue "regulations" requiring the incensed and harassed natives to work for the white settlers, and neglect their own affairs to do it. This is slavery, and is several times worse than was the American slavery which used to pain England so much; for when this Rhodesian slave is sick, superannuated, or otherwise disabled, he must support himself or starve—his master is under no obligation to support him.

The reduction of the population by Rhodesian methods to the desired limit is a return to the old-time slow-misery and lingering-death system of a discredited time and a crude "civilization." We humanely reduce an overplus of dogs by swift chloroform; the Boer humanely reduced an overplus of blacks by swift suffocation; the nameless but right-hearted Australian pioneer humanely reduced his overplus of aboriginal neighbors by a sweetened swift death concealed in a poisoned pudding. All these are admirable, and worthy of praise; you and I would rather suffer either of these deaths thirty times over in thirty successive days than linger out one of the Rhodesian twenty-year deaths, with its daily burden of insult, humiliation, and forced labor for a man whose entire race the victim hates. Rhodesia is a happy name for that land of piracy and pillage, and puts the right stain upon it.

Several long journeys gave us experience of the Cape Colony railways; easy-riding, fine cars; all the conveniences;

thorough cleanliness; comfortable beds furnished for the night trains. It was in the first days of June, and winter; the daytime was pleasant, the nighttime nice and cold. Spinning along all day in the cars it was ecstasy to breathe the bracing air and gaze out over the vast brown solitudes of the velvet plains, soft and lovely near by, still softer and lovelier further away, softest and loveliest of all in the remote distances, where dim island-hills seemed afloat, as in a sea — a sea made of dream-stuff and flushed with colors faint and rich; and dear me, the depth of the sky, and the beauty of the strange new cloud-forms, and the glory of the sunshine, the lavishness, the wastefulness of it! The vigor and freshness and inspiration of the air and the sun — well, it was all just as Olive Schreiner had made it in her books.

To me the veldt, in its sober winter garb, was surpassingly beautiful. There were unlevel stretches where it was rolling and swelling, and rising and subsiding, and sweeping superbly on and on, and still on and on like an ocean, toward the faraway horizon, its pale brown deepening by delicately graduated shades to rich orange, and finally to purple and crimson where it washed against the wooded hills and naked red crags at the base of the sky.

Everywhere, from Cape Town to Kimberley, and from Kimberley to Port Elizabeth and East London, the towns were well populated with tamed blacks; tamed and Christianized too, I suppose, for they wore the dowdy clothes of our Christian civilization. But for that, many of them would have been remarkably handsome. These fiendish clothes, together with the proper lounging gait, good-natured face, happy air, and easy laugh, made them precise counterparts of our American blacks; often where all the other aspects were strikingly and harmoniously and thrillingly African, a flock of these natives would intrude, looking wholly out of place, and spoil it all, making the thing a grating discord, half African and half American.

One Sunday in King William's Town a score of colored women came mincing across the great barren square dressed — oh, in the last perfection of fashion, and newness, and expensiveness, and showy mixture of unrelated colors,— all just as I had seen it so often at home; and in their faces and their gait was that languishing, aristocratic, divine delight in their finery which was so familiar to me, and had always been such a satisfaction to my eye and my heart. I seemed among old, old friends; friends of fifty years, and I stopped and cordially greeted them. They broke into a good-fellowship laugh, flashing their white teeth upon me, and all answered at once. I

OLD ACQUAINTANCES.

did not understand a word they said. I was astonished; I was not dreaming that they would answer in anything but American.

The voices, too, of the African women, were familiar to me — sweet and musical, just like those of the slave women of my early days. I followed a couple of them all over the Orange

Free State — no, over its capital — Bloemfontein, to hear their liquid voices and the happy ripple of their laughter. Their language was a large improvement upon American. Also upon the Zulu. It had no Zulu clicks in it; and it seemed to have no angles or corners, no roughness, no vile *s*'s or other hissing sounds, but was very, very mellow and rounded and flowing.

In moving about the country in the trains, I had opportunity to see a good many Boers of the veldt. One day at a village station a hundred of them got out of the third-class cars to feed.

HE SAID "NO" RUDELY.

Their clothes were very interesting. For ugliness of shapes, and for miracles of ugly colors inharmoniously associated, they were a record.

The effect was nearly as exciting and interesting as that produced by the brilliant and beautiful clothes and perfect taste always on view at the Indian railway stations. One man had corduroy trousers of a faded chewing-gum tint. And they were new — showing that this tint did not come by calamity, but was intentional; the very ugliest color I have ever seen. A gaunt, shackly country lout six feet high, in battered gray slouched hat with wide brim, and old resin-colored breeches, had on a hideous

HE FELT STUFFY.

brand-new woolen coat which was imitation tiger skin —
wavy broad stripes of dazzling yellow and deep brown. I
thought he ought to be hanged, and asked the station-
master if it could be arranged. He said no; and not only
that, but said it rudely; said it with a quite unnecessary
show of feeling. Then he muttered something about my being
a jackass, and walked away and pointed me out to people, and
did everything he could to turn public sentiment against me.
It is what one gets for trying to do good.

In the train that day a passenger told me some more about
Boer life out in the lonely veldt. He said the Boer gets up
early and sets his "niggers" at their tasks (pasturing the cattle,
and watching them); eats, smokes, drowses, sleeps; toward
evening superintends the milking, etc.; eats, smokes, drowses;
goes to bed at early candlelight in the fragrant clothes he
(and she) have worn all day and every week-day for years. I
remember that last detail, in Olive Schreiner's "Story of an
African Farm." And the passenger told me that the Boers
were justly noted for their hospitality. He told me a story
about it. He said that his grace the Bishop of a certain See
was once making a business-progress through the tavernless
veldt, and one night he stopped with a Boer; after supper was
shown to bed; he undressed, weary and worn out, and was
soon sound asleep; in the night he woke up feeling crowded
and suffocated, and found the old Boer and his fat wife in bed
with him, one on each side, with all their clothes on, and snor-
ing. He had to stay there and stand it — awake and suffering
— until toward dawn, when sleep again fell upon him for an
hour. Then he woke again. The Boer was gone, but the wife
was still at his side.

Those Reformers detested that Boer prison; they were not
used to cramped quarters and tedious hours, and weary idleness, and early to bed, and limited movement, and arbitrary

and irritating rules, and the absence of the luxuries which wealth comforts the day and the night with. The confinement told upon their bodies and their spirits; still, they were superior men, and they made the best that was to be made of the circumstances. Their wives smuggled delicacies to them, which helped to smooth the way down for the prison fare.

In the train Mr. B. told me that the Boer jail-guards treated the black prisoners — even political ones — mercilessly. An African chief and his following had been kept there nine months without trial, and during all that time they had been without shelter from rain and sun. He said that one day the guards put a big black in the stocks for dashing his soup on the ground; they stretched his legs painfully wide apart, and set him with his back down hill; he could not endure it, and put back his hands upon the slope for a support. The guard ordered him to withdraw the support — and kicked him in the back. "Then," said Mr. B., "the powerful black wrenched the stocks asunder and went for the guard; a Reform prisoner pulled him off, and thrashed the guard himself."

CHAPTER LXIX.

The very ink with which all history is written is merely fluid prejudice.
— *Pudd'nhead Wilson's New Calendar.*
There isn't a Parallel of Latitude but thinks it would have been the Equator if
it had had its rights. — *Pudd'nhead Wilson's New Calendar.*

NEXT to Mr. Rhodes, to me the most interesting convulsion of nature in South Africa was the diamond-crater. The Rand gold fields are a stupendous marvel, and they make all other gold fields small, but I was not a stranger to gold-mining; the veldt was a noble thing to see, but it was only another and lovelier variety of our Great Plains; the natives were very far from being uninteresting, but they were not new; and as for the towns, I could find my way without a guide through the most of them because I had learned the streets, under other names, in towns just like them in other lands; but the diamond mine was a wholly fresh thing, a splendid and absorbing novelty. Very few people in the world have seen the diamond in its home. It has but three or four homes in the world, whereas gold has a million. It is worth while to journey around the globe to see anything which can truthfully be called a novelty, and the diamond mine is the greatest and most select and restricted novelty which the globe has in stock.

The Kimberley diamond deposits were discovered about 1869, I think. When everything is taken into consideration, the wonder is that they were not discovered five thousand years ago and made familiar to the African world for the rest of time. For this reason the first diamonds were found on the surface of the ground. They were smooth and limpid, and in

the sunlight they vomited fire. They were the very things
which an African savage of any era would value above every
other thing in the world excepting a glass bead. For two or
three centuries we have been buying his lands, his cattle, his
neighbor, and any other thing he had for sale, for glass beads:
and so it is strange that he was indifferent to the diamonds
—for he must have picked them up many and many a time.
It would not occur to him to try to sell them to whites, of
course, since the whites already had plenty of glass beads, and
more fashionably shaped, too, than these; but one would think
that the poorer sort of black, who could not afford real glass,
would have been humbly content to decorate himself with the
imitation, and that presently the white trader would notice the
things, and dimly suspect, and carry some of them home, and
find out what they were, and at once empty a multitude of
fortune-hunters into Africa. There are many strange things in
human history; one of the strangest is that the sparkling diamonds laid there so long without exciting any one's interest.

The revelation came at last by accident. In a Boer's hut
out in the wide solitude of the plains, a traveling stranger
noticed a child playing with a bright object, and was told it
was a piece of glass which had been found in the veldt. The
stranger bought it for a trifle and carried it away; and being
without honor, made another stranger believe it was a diamond,
and so got $125 out of him for it, and was as pleased with him-
self as if he had done a righteous thing. In Paris the wronged
stranger sold it to a pawnshop for $10,000, who sold it to a
countess for $90,000, who sold it to a brewer for $800,000, who
traded it to a king for a dukedom and a pedigree, and the king
"put it up the spout." †   I know these particulars to be correct.

—————————

† From the Greek δ͡ξ Χη
meaning "pawned it."

The news flew around, and the South African diamond-boom began. The original traveler — the dishonest one — now remembered that he had once seen a Boer teamster chocking his wagon-wheel on a steep grade with a diamond as large as a football, and he laid aside his occupations and started out to hunt for it, but not with the intention of cheating anybody out of $125 with it, for he had reformed.

We now come to matters more didactic. Diamonds are not imbedded in rock ledges fifty miles long, like the Johannesburg gold, but are distributed through the rubbish of a filled-up well, so to speak. The well is rich, its walls are sharply defined; outside of the walls are no diamonds. The well is a crater, and a large one. Before it had been meddled with, its surface was even with the level plain, and there was no sign to suggest that it was there. The pasturage covering the surface of the Kimberley crater was sufficient for the support of a cow, and the pasturage underneath was sufficient for the support of a kingdom; but the cow did not know it, and lost her chance.

The Kimberley crater is roomy enough to admit the Roman Coliseum; the bottom of the crater has not been reached, and no one can tell how far down in the bowels of the earth it goes. Originally, it was a perpendicular hole packed solidly full of blue rock or cement, and scattered through that blue mass, like raisins in a pudding, were the diamonds. As deep down in the earth as the blue stuff extends, so deep will the diamonds be found.

There are three or four other celebrated craters near by — a circle three miles in diameter would enclose them all. They are owned by the De Beers Company, a consolidation of diamond properties arranged by Mr. Rhodes twelve or fourteen years ago. The De Beers owns other craters; they are under the grass, but the De Beers knows where they are, and will open them some day, if the market should require it.

Originally, the diamond deposits were the property of the Orange Free State; but a judicious "rectification" of the boundary line shifted them over into the British territory of Cape Colony. A high official of the Free State told me that the sum of $400,000 was handed to his commonwealth as a compromise, or indemnity, or something of the sort, and that he thought his commonwealth did wisely to take the money and keep out of a dispute, since the power was all on the one side and the weakness all on the other. The De Beers Company dig out $400,000 worth of diamonds per week, now. The Cape got the territory, but no profit; for Mr. Rhodes and the Rothschilds and the other De Beers people own the mines, and they pay no taxes.

In our day the mines are worked upon scientific principles, un-

SEARCHING FOR GEMS.

der the guidance of the ablest mining-engineering talent procurable in America. There are elaborate works for reducing the blue rock and passing it through one process after another until every diamond it contains has been hunted down and secured. I watched the "concentrators" at work — big tanks containing mud and water and invisible diamonds — and was told that each could stir and churn and properly treat 300 car-loads of mud per day — 1,600 pounds to the car-load — and reduce it to 3 car-loads of slush. I saw the 3 car-loads of slush taken to the "pulsators" and there reduced to a quarter of a load of nice clean dark-colored sand. Then I followed it to the sorting tables and saw the men deftly and swiftly spread it out and brush it about and seize the diamonds as they showed up. I assisted, and once I found a diamond half as large as an almond. It is an exciting kind of fishing, and you feel a fine thrill of pleasure every time you detect the glow of one of those limpid pebbles through the veil of dark sand. I would like to spend my Saturday holidays in that charming sport every now and then. Of course there are disappointments. Sometimes you find a diamond which is not a diamond; it is only a quartz crystal or some such worthless thing. The expert can generally distinguish it from the precious stone which it is counterfeiting; but if he is in doubt he lays it on a flatiron and hits it with a sledge-hammer. If it is a diamond it holds its own; if it is anything else, it is reduced to powder. I liked that experiment very much, and did not tire of repetitions of it. It was full of enjoyable apprehensions, unmarred by any personal sense of risk. The De Beers concern treats 8,000 carloads — about 6,000 tons — of blue rock per day, and the result is three pounds of diamonds. Value, uncut, $50,000 to $70,000. After cutting, they will weigh considerably less than a pound, but will be worth four or five times as much as they were before.

All the plain around that region is spread over, a foot deep, with blue rock, placed there by the Company, and looks like a plowed field. Exposure for a length of time make the rock easier to work than it is when it comes out of the mine. If mining should cease now, the supply of rock spread over those fields would furnish the usual 8,000 car-loads per day to the separating works during three years. The fields are fenced and watched; and at night they are under the constant inspection of lofty electric searchlight. They contain fifty or sixty million dollars' worth of diamonds, and there is an abundance of enterprising thieves around.

In the dirt of the Kimberley streets there is much hidden wealth. Some time ago the people were granted the privilege of a free wash-up. There was a general rush, the work was done with thoroughness, and a good harvest of diamonds was gathered.

The deep mining is done by natives. There are many hundreds of them. They live in quarters built around the inside of a great compound. They are a jolly and good-natured lot,

INSIDE OF COMPOUND WITH NET COVERING TO PREVENT THOWING GEMS
OUT OF THE CRATER.

and accommodating. They performed a war-dance for us, which was the wildest exhibition I have ever seen. They are not allowed outside of the compound during their term of service — three months, I think it is, as a rule. They go down the shaft, stand their watch, come up again, are searched, and go to bed or to their amusements in the compound ; and this routine they repeat, day in and day out.

It is thought that they do not now steal many diamonds — successfully. They used to swallow them, and find other ways of concealing them, but the white man found ways of beating

their various games. One man cut his leg and shoved a diamond into the wound, but even that project did not succeed.
When they find a fine large diamond they are more likely to
report it than to steal it, for in the former case they get a reward, and in the latter they are quite apt to merely get into
trouble. Some years ago, in a mine not owned by the De
Beers, a black found what has been claimed to be the largest
diamond known to the world's history; and as a reward he
was released from service and given a blanket, a horse, and
five hundred dollars.
It made him a Van-
derbilt. He could
buy four wives, and
have money left.
Four wives are an
ample support for a
native. With four
wives he is wholly
independent, and
need never do a
stroke of work again.

That great dia-
mond weighs 971
carats. Some say it
is as big as a piece of
alum, others say it
is as large as a bite
of rock candy, but
the best authorities
agree that it is almost

JAGERSFONTIEN DIAMOND,

THE LARGEST DIAMOND IN THE WORLD,

WEIGHT 971 CARATS,

EXACT SIZE.

exactly the size of a chunk of ice. But those details are not
important; and in my opinion not trustworthy. It has a flaw

45

in it, otherwise it would be of incredible value. As it is, it is held to be worth $2,000,000. After cutting it ought to be worth from $5,000,000 to $8,000,000, therefore persons desiring to save money should buy it now. It is owned by a syndicate, and apparently there is no satisfactory market for it. It is earning nothing; it is eating its head off. Up to this time it has made nobody rich but the native who found it.

He found it in a mine which was being worked by contract. That is to say, a company had bought the privilege of taking from the mine 5,000,000 carloads of blue-rock, for a sum down and a royalty. Their speculation had not paid; but on the very day that their privilege ran out that native found the $2,000,000-diamond and handed it over to them. Even the diamond culture is not without its romantic episodes.

The Koh-i-Noor is a large diamond, and valuable; but it cannot compete in these matters with three which — according to legend — are among the crown trinkets of Portugal and Russia. One of these is held to be worth $20,000,000; another, $25,000,000, and the third something over $28,000,000.

Those are truly wonderful diamonds, whether they exist or not; and yet they are of but little importance by comparison with the one wherewith the Boer wagoner chocked his wheel on that steep grade as heretofore referred to. In Kimberley I had some conversation with the man who saw the Boer do that — an incident which had occurred twenty-seven or twenty-eight years before I had my talk with him. He assured me that that diamond's value could have been over a billion dollars, but not under it. I believed him, because he had devoted twenty-seven years to hunting for it, and was in a position to know.

A fitting and interesting finish to an examination of the tedious and laborious and costly processes whereby the diamonds are gotten out of the deeps of the earth and freed from the base stuffs which imprison them is the visit to the De Beers offices in the town of Kimberley, where the result of each day's mining is brought every day, and weighed, assorted, valued, and deposited in safes against shipping-day. An unknown and unaccredited person cannot get into that place; and it seemed apparent from the generous supply of warning and protective and prohibitory signs that were posted all about that not even the known and accredited can steal diamonds there without inconvenience.

NATIVE MINERS GAMBLING.

We saw the day's output — shining little nests of diamonds, distributed a foot apart, along a counter, each nest reposing upon a sheet of white paper. That day's catch was about $70,000 worth. In the course of a year half a ton of diamonds pass under the scales there and sleep on that counter; the resulting money is $18,-000,000 or $20,000,000. Profit, about $12,000,000.

Young girls were doing the sorting — a nice, clean, dainty, and probably distressing employment. Every day ducal incomes sift and sparkle through the fingers of those young girls; yet they go to bed at night as poor as they were when they got up in the morning. The same thing next day, and all the days.

They are beautiful things, those diamonds, in their native state. They are of various shapes; they have flat surfaces, rounded borders, and never a sharp edge. They are of all colors and shades of color, from dewdrop white to actual black;

and their smooth and rounded surfaces and contours, variety of color, and transparent limpidity make them look like piles of assorted candies. A very light straw color is their commonest tint. It seemed to me that these uncut gems must be more beautiful than any cut ones could be; but when a collection of cut ones was brought out, I saw my mistake. Nothing is so beautiful as a rose diamond with the light playing through it, except that uncostly thing which is just like it — wavy sea-water with the sunlight playing through it and striking a white-sand bottom.

Before the middle of July we reached Cape Town, and the end of our African journeyings. And well satisfied; for, towering above us was Table Mountain — a reminder that we had now seen each and all of the great features of South Africa except Mr. Cecil Rhodes. I realize that that is a large exception. I know quite well that whether Mr. Rhodes is the lofty and worshipful patriot and statesman that multitudes believe him to be, or Satan come again, as the rest of the world account him, he is still the most imposing figure in the British empire outside of England. When he stands on the Cape of Good Hope, his shadow falls to the Zambesi. He is the only colonial in the British dominions whose goings and comings are chronicled and discussed under all the globe's meridians, and whose speeches, unclipped, are cabled from the ends of the earth; and he is the only unroyal outsider whose arrival in London can compete for attention with an eclipse.

That he is an extraordinary man, and not an accident of fortune, not even his dearest South African enemies were willing to deny, so far as I heard them testify. The whole South African world seemed to stand in a kind of shuddering awe of him, friend and enemy alike. It was as if he were deputy-God on the one side, deputy-Satan on the other, proprietor of the people, able to make them or ruin them by his breath, worshiped by many, hated by many, but blasphemed by none among the judicious, and even by the indiscreet in guarded whispers only.

What is the secret of his formidable supremacy? One says it is his prodigious wealth — a wealth whose drippings in salaries and in other ways support multitudes and make them his interested and loyal vassals; another says it is his personal magnetism and his persuasive tongue, and that these hypnotize and make happy slaves of all that drift within the circle of their influence; another says it is his majestic ideas, his vast schemes for the territorial aggrandizement of England, his patriotic and unselfish ambition to spread her beneficent protection and her just rule over the pagan wastes of Africa and make luminous the African darkness with the glory of her name; and another says he wants the earth and wants it for his own, and that the belief that he will get it and let his friends in on the ground floor is *the* secret that rivets so many eyes upon him and keeps him in the zenith where the view is unobstructed.

One may take his choice. They are all the same price. One fact is sure: he keeps his prominence and a vast following, no matter what he does. He "deceives" the Duke of Fife — it is the Duke's word — but that does not destroy the Duke's loyalty to him. He tricks the Reformers into immense trouble with his Raid, but the most of them believe he meant well. He weeps over the harshly-taxed Johannesburgers and makes them his friends; at the same time he taxes his Charter-settlers 50 per cent., and so wins their affection and their confidence that they are squelched with despair at every rumor that the Charter is to be annulled. He raids and robs and slays and enslaves the Matabele and gets worlds of Charter-Christian applause for it. He has beguiled England into buying Charter waste paper for Bank of England notes, ton for ton, and the ravished still burn incense to him as the Eventual God of Plenty. He has done everything he could think of to pull himself down to the ground; he has done more than enough to pull sixteen common-run great men down; yet there he stands, to this day, upon his dizzy summit under the dome of the sky, an apparent permanency, the marvel of the time, the mystery of the age, an Archangel with wings to half the world, Satan with a tail to the other half.

I admire him, I frankly confess it; and when his time comes I shall buy a piece of the rope for a keepsake.

CONCLUSION.

I have traveled more than anyone else, and I have noticed that even the angels speak English with an accent. — Pudd'nhead Wilson's New Calendar.

I SAW Table Rock, anyway — a majestic pile. It is 3,000 feet high. It is also 17,000 feet high. These figures may be relied upon. I got them in Cape Town from the two best-informed citizens, men who had made Table Rock the study of their lives. And I saw Table Bay, so named for its levelness. I saw the Castle — built by the Dutch East India Company three hundred years ago — where the Commanding General lives; I saw St. Simon's Bay, where the Admiral lives. I saw the Government, also the Parliament, where they quarreled in two languages when I was there, and agreed in none. I saw the club. I saw and explored the beautiful sea-girt drives that wind about the mountains and through the paradise where the villas are. Also I saw some of the fine old Dutch mansions, pleasant homes of the early times, pleasant homes to-day, and enjoyed the privilege of their hospitalities.

And just before I sailed I saw in one of them a quaint old picture which was a link in a curious romance — a picture of a pale, intellectual young man in a pink coat with a high black collar. It was a portrait of Dr. James Barry, a military surgeon who came out to the Cape fifty years ago with his regiment. He was a wild young fellow, and was guilty of various kinds of misbehavior. He was several times reported to headquarters in England, and it was in each case expected that orders would come out to deal with him promptly and severely, but for some mysterious reason no orders of any kind ever came back — nothing came but just an impressive silence. This made him an imposing and uncanny wonder to the town.

Next, he was promoted — away up. He was made Medical Superintendent General, and transferred to India. Presently he was back at the Cape again and at his escapades once more. There were plenty of pretty girls, but none of them caught him, none of them could get hold of his heart; evidently he was not a marrying man. And that was another marvel, another puzzle, and made no end of perplexed talk. Once he was called in the night, an obstetric service, to do what he could for a woman who was believed to be dying. He was prompt and scientific, and saved both mother and child. There are other instances of record which testify to his mastership of his profession ; and many which testify to his love of it and his devotion to it. Among other adventures of his was a duel of a desperate sort, fought with swords, at the Castle. He killed his man.

The child heretofore mentioned as having been saved by Dr. Barry so long ago, was named for him, and still lives in Cape Town. He had Dr. Barry's portrait painted, and gave it to the gentleman in whose old Dutch house I saw it — the quaint figure in pink coat and high black collar.

The story seems to be arriving nowhere. But that is because I have not finished. Dr. Barry died in Cape Town 30 years ago. It was then discovered that he was *a woman*.

The legend goes that enquiries — soon silenced — developed the fact that she was a daughter of a great English house, and that that was why her Cape wildnesses brought no punishment and got no notice when reported to the government at home. Her name was an *alias*. She had disgraced herself with her people; so she chose to change her name and her sex and take a new start in the world.

We sailed on the 15th of July in the *Norman*, a beautiful ship, perfectly appointed. The voyage to England occupied a short fortnight, without a stop except at Madeira. A good and restful voyage for tired people, and there were several of us. I seemed to have been lecturing a thousand years, though it was only a twelvemonth, and a considerable number of the others were Reformers who were fagged out with their five months of seclusion in the Pretoria prison.

Our trip around the earth ended at the Southampton pier, where we embarked thirteen months before. It seemed a fine and large thing to have accomplished — the circumnavigation of this great globe in that little time, and I was privately proud of it. For a moment. Then came one of those vanity-snubbing astronomical reports from the Observatory-people, whereby it appeared that another great body of light had lately flamed up in the remotenesses of space which was traveling at a gait which would enable it to do all that I had done in *a minute and a half*. Human pride is not worth while; there is always something lying in wait to take the wind out of it.

TO THE PERSON

SITTING IN DARKNESS

TO THE PERSON
SITTING
IN DARKNESS

———

— BY —

MARK TWAIN

———

REPRINTED BY PERMISSION FROM THE NORTH AMERICAN
REVIEW, FEBRUARY, 1901

TO THE PERSON SITTING IN DARKNESS.

BY MARK TWAIN.

Extending the Blessings of Civilization to our Brother who Sits in Darkness has been a good trade and has paid well, on the whole; and there is money in it yet, if carefully worked—but not enough, in my judgment, to make any considerable risk advisable. The People that Sit in Darkness are getting to be too scarce—too scarce and too shy. And such darkness as is now left is really of but an indifferent quality, and not dark enough for the game. The most of those People that Sit in Darkness have been furnished with more light than was good for them or profitable for us. We have been injudicious.

The Blessings-of-Civilization Trust, wisely and cautiously administered, is a Daisy. There is more money in it, more territory, more sovereignty and other kinds of emolument than there is in any other game that is played. But Christendom has been playing it badly of late years, and must certainly suffer by it, in my opinion. She has been so eager to get every stake that appeared on the green cloth, that the People who Sit in Darkness have noticed it—they have noticed it, and have begun to show alarm. They have become suspicious of the Blessings of Civilization. More—they have begun to examine them. This is not well. The Blessings of Civilization are all right, and a good commercial property; there could not be a better, in a dim light. In the right kind of a light, and at a proper distance, with the goods a little out of focus, they fur-

nish this desirable exhibit to the Gentlemen who Sit
in Darkness:

LOVE,	LAW AND ORDER,
JUSTICE,	LIBERTY,
GENTLENESS,	EQUALITY,
CHRISTIANITY,	HONORABLE DEALING,
PROTECTION TO THE	MERCY,
WEAK,	EDUCATION,
TEMPERANCE,	—and so on.

There. Is it good? Sir, it is pie. It will bring
into camp any idiot that sits in darkness anywhere.
But not if we adulterate it. It is proper to be emphatic upon that point. This brand is strictly for
Export—apparently. *Apparently.* Privately and
confidentially, it is nothing of the kind. Privately
and confidentially, it is merely an outside cover,
gay and pretty and attractive, displaying the special
patterns of our Civilization which we reserve for
Home Consumption, while *inside* the bale is the
Actual Thing that the Customer Sitting in Darkness buys with his blood and tears and land and
liberty. That Actual Thing is, indeed, Civilization,
but it is only for Export. Is there a difference between the two brands? In some of the details, yes.

We all know that the Business is being ruined.
The reason is not far to seek. It is because our
Mr. McKinley, and Mr. Chamberlain, and the
Kaiser, and the Czar and the French have been
exporting the Actual Thing *with the outside cover
left off.* This is bad for the Game. It shows that
these new players of it are not sufficiently acquainted with it.

It is a distress to look on and note the mismoves,
they are so strange and so awkward. Mr. Chamberlain manufactures a war out of materials so inadequate and so fanciful that they make the boxes
grieve and the gallery laugh, and he tries hard to
persuade himself that it isn't purely a private raid
for cash, but has a sort of dim, vague respectability
about it somewhere, if he could only find the spot;

and that, by and by, he can scour the flag clean again after he has finished dragging it through the mud, and make it shine and flash in the vault of heaven once more as it had shone and flashed there a thousand years in the world's respect until he laid his unfaithful hand upon it. It is bad play— bad. For it exposes the Actual Thing to Them that Sit in Darkness, and they say: "What! Christian against Christian? And only for money? Is *this* a case of magnanimity, forbearance, love, gentleness, mercy, protection of the weak—this strange and over-showy onslaught of an elephant upon a nest of field-mice, on the pretext that the mice had squeaked an insolence at him—conduct which 'no self-respecting government could allow to pass unavenged?' as Mr. Chamberlain said. Was that a good pretext in a small case, when it had not been a good pretext in a large one?—for only recently Russia had affronted the elephant three times and survived alive and unsmitten. Is this Civilization and Progress? Is it something better than we already possess? These harryings and burnings and desert-makings in the Transvaal—is this an improvement on our darkness? Is it, perhaps, possible that there are two kinds of Civilization—one for home consumption and one for the heathen market?"

Then They that Sit in Darkness are troubled, and shake their heads; and they read this extract from a letter of a British private, recounting his exploits in one of Methuen's victories, some days before the affair of Magersfontein, and they are troubled again:

"We tore up the hill and into the intrenchments, and the Boers saw we had them; so they dropped their guns and went down on their knees and put up their hands clasped, and begged for mercy. And we gave it them—*with the long spoon.*"

The long spoon is the bayonet. See *Lloyd's*

Weekly, London, of those days. The same number —and the same column—contains some quite unconscious satire in the form of shocked and bitter upbraidings of the Boers for their brutalities and inhumanities!

Next to our heavy damage, the Kaiser went to playing the game without first mastering it. He lost a couple of missionaries in a riot in Shantung, and in his account he made an overcharge for them. China had to pay a hundred thousand dollars apiece for them, in money; twelve miles of territory, containing several millions of inhabitants and worth twenty million dollars, and to build a monument and also a Christian Church; whereas the people of China could have been depended upon to remember the missionaries without the help of these expensive memorials. This was all bad play. Bad, because it would not, and could not, and will not now or ever, deceive the Person Sitting in Darkness. He knows that it was an overcharge. He knows that a missionary is like any other man; he is worth merely what you can supply his place for, and no more. He is useful, but so is a doctor, so is a sheriff, so is an editor; but a just Emperor does not charge war-prices for such. A diligent, intelligent, but obscure missionary, and a diligent, intelligent country editor are worth much, and we know it; but they are not worth the earth. We esteem such an editor, and we are sorry to see him go; but, when he goes, we should consider twelve miles of territory, and a church, and a fortune, over-compensation for his loss. I mean, if he was a Chinese editor, and we had to settle for him. It is no proper figure for an editor or a missionary; one can get shop-worn kings for less. It was bad play on the Kaiser's part. It got this property, true; but it *produced the Chinese revolt*, the indignant uprising of China's traduced patriots, the Boxers. The results have been expensive to Germany, and to the other

Disseminators of Progress and the Blessings of Civilization.

The Kaiser's claim was paid, yet it was bad play, for it could not fail to have an evil effect upon Persons Sitting in Darkness in China. They would muse upon the event, and be likely to say: "Civilization is gracious and beautiful, for such is its reputation; but can we afford it? There are rich Chinamen, perhaps they could afford it; but this tax is not laid upon them, it is laid upon the peasants of Shantung; it is they that must pay this mighty sum, and their wages are but four cents a day. Is this a better civilization than ours, and holier and higher and nobler? Is not this rapacity? Is not this extortion? Would Germany charge America two hundred thousand dollars for two missionaries, and shake the mailed fist in her face, and send warships, and send soldiers, and say: 'Seize twelve miles of territory, worth twenty millions of dollars, as additional pay for the missionaries; and make those peasants build a monument to the missionaries, and a costly Christian church to remember them by?' And later would Germany say to her soldiers: 'March through America and slay, *giving no quarter*; make the German face there, as has been our Hun-face here, a terror for a thousand years; march through the Great Republic and slay, slay, slay, carving a road for our offended religion through its heart and bowels?' Would Germany do like this to America, to England, to France, to Russia? Or only to China the helpless—imitating the elephant's assault upon the field-mice? Had we better invest in this Civilization—this Civilization which called Napoleon a buccaneer for carrying off Venice's bronze horses, but which steals our ancient astronomical instruments from our walls, and goes looting like common bandits—that is, all the alien soldiers except America's; and (Americans again excepted) storms frightened villages and cables the result to glad journals at home every day: 'Chinese losses, 450 killed; ours, *one officer and two men wounded*. Shall proceed against neighboring village to-morrow, where a *massacre* is reported.' Can we afford Civilization?'

And, next, Russia must go and play the game injudiciously. She affronts England once or twice —with the Person Sitting in Darkness observing and noting; by moral assistance of France and Germany, she robs Japan of her hard-earned spoil, all swimming in Chinese blood—Port Arthur—with the Person again observing and noting; then she seizes Manchuria, raids its villages, and chokes its great rivers with the swollen corpses of countless massacred peasants—that astonished Person still observing and noting. And perhaps he is saying to himself: "It is yet *another* Civilized Power, with its banner of the Prince of Peace in one hand and its loot-basket and its butcher-knife in the other. Is there no salvation for us but to adopt Civilization and lift ourselves down to its level?

And by and by comes America, and our Master of the Game plays it badly—plays it as Mr. Chamberlain was playing it in South Africa. It was a mistake to do that; also, it was one which was quite unlooked for in a Master who was playing it so well in Cuba. In Cuba, he was playing the usual and regular *American* game, and it was winning, for there is no way to beat it. The Master, contemplating Cuba, said: "Here is an oppressed and friendless little nation which is willing to fight to be free; we go partners, and put up the strength of seventy million sympathizers, and the resources of the United States: play!" Nothing but Europe combined could call that hand: and Europe cannot combine on anything. There, in Cuba, he was following our great traditions in a way which made us very proud of him, and proud of the deep dissatisfaction which his play was provoking in Continental Europe. Moved by a high inspiration, he threw out those stirring words which proclaimed that forcible annexation would be "criminal aggression;" and in that utterance fired another "shot heard round the world." The memory of that fine saying will be outlived by the remembrance of no act of his but one—that he forgot it within the twelvemonth, and its honorable gospel along with it.

For, presently, came the Philippine temptation. It was strong; it was too strong, and he made that bad mistake: he played the European game, the Chamberlain game. It was a pity; it was a great pity, that error; that one grievous error, that irrevocable error. For it was the very place and time to play the American game again. And at no cost. Rich winnings to be gathered in, too; rich and permanent; indestructible; a fortune transmissible forever to the children of the flag. Not land, not money, not dominion—no, something worth many times more than that dross: our share, the spectacle of a nation of long harassed and persecuted slaves set free through our influence; our posterity's share, the golden memory of that fair deed. The game was in our hands. If it had been played according to the American rules, Dewey would have sailed away from Manila as soon as he had destroyed the Spanish fleet—after putting up a sign on shore guaranteeing foreign property and life against damage by the Filipinos, and warning the Powers that interference with the emancipated patriots would be regarded as an act unfriendly to the United States. The Powers cannot combine, in even a bad cause, and the sign would not have been molested.

Dewey could have gone about his affairs elsewhere, and left the competent Filipino army to starve out the little Spanish garrison and send it home, and the Filipino citizens to set up the form of government they might prefer, and deal with the friars and their doubtful acquisitions according to

Filipino ideas of fairness and justice—ideas which have since been tested and found to be of as high an order as any that prevail in Europe or America.

But we played the Chamberlain game, and lost the chance to add another Cuba and another honorable deed to our good record.

The more we examine the mistake, the more clearly we perceive that it is going to be bad for the Business. The Person Sitting in Darkness is almost sure to say: "There is something curious about this—curious and unaccountable. There must be two Americas: one that sets the captive free, and one that takes a once-captive's new freedom away from him, and picks a quarrel with him with nothing to found it on; then kills him to get his land."

The truth is, the Person Sitting in Darkness *is* saying things like that; and for the sake of the Business we must persuade him to look at the Philippine matter in another and healthier way. We must arrange his opinions for him. I believe it can be done; for Mr. Chamberlain has arranged England's opinion of the South African matter, and done it most cleverly and successfully. He presented the facts—some of the facts—and showed those confiding people what the facts meant. He did it statistically, which is a good way. He used the formula: "Twice 2 are 14, and 2 from 9 leaves 35." Figures are effective; figures will convince the elect.

Now, my plan is a still bolder one than Mr. Chamberlain's, though apparently a copy of it. Let us be franker than Mr. Chamberlain; let us audaciously present the whole of the facts, shirking none, then explain them according to Mr. Chamberlain's formula. This daring truthfulness will astonish and dazzle the Person Sitting in Darkness, and he will take the Explanation down before his mental vision has had time to get back into focus. Let us say to him:

9

"Our case is simple. On the 1st of May, Dewey destroyed the Spanish fleet. This left the Archipelago in the hands of its proper and rightful owners, the Filipino nation. Their army numbered 30,000 men, and they were competent to whip out or starve out the little Spanish garrison; then the people could set up a government of their own devising. Our traditions required that Dewey should now set up his warning sign, and go away. But the Master of the Game happened to think of another plan—the European plan. He acted upon it. This was, to send out an army—ostensibly to help the native patriots put the finishing touch upon their long and plucky struggle for independence, but really to take their land away from them and keep it. That is, in the interest of Progress and Civilization. The plan developed, stage by stage, and quite satisfactorily. We entered into a military alliance with the trusting Filipinos, and they hemmed in Manila on the land side, and by their valuable help the place, with its garrison of 8,000 or 10,000 Spaniards, was captured—a thing which we could not have accomplished unaided at that time. We got their help by—by ingenuity. We knew they were fighting for their independence, and that they had been at it for two years. We knew they supposed that we also were fighting in their worthy cause—just as we had helped the Cubans fight for Cuban independence—and we allowed them to go on thinking so. *Until Manila was ours and we could get along without them.* Then we showed our hand. Of course, they were surprised —that was natural; surprised and disappointed; disappointed and grieved. To them it looked un-American; un-characteristic; foreign to our established traditions. And this was natural, too; for we were only playing the American Game in public—in private it was the European. It was neatly done, very neatly, and it bewildered them so they could not understand it; for we had been so friendly—so affectionate, even—with those simple-minded patriots! We, our own selves, had brought back out of exile their leader, their hero, their hope, their Washington —Aguinaldo; brought him in a warship, in high honor, under the sacred shelter and hospitality of the flag; brought him back and restored him to his people, and got their moving and eloquent gratitude for it. Yes, we had been so friendly to them, and had heartened them up in so many ways! We had lent them guns and ammunition; advised with them; exchanged pleasant courtesies with them; placed our sick and wounded in their kindly care; entrusted our Spanish prisoners to their humane and honest hands; fought shoulder to shoulder with them against "the common enemy" (our own phrase); praised their courage, praised their gallantry, praised their mercifulness, praised their fine and honorable conduct; borrowed their trenches, borrowed strong positions which they had previously captured from the Spaniards; petted them, lied to them—officially proclaiming that our land and naval forces came to give them their freedom and displace the bad Spanish Government—fooled them, used them until we needed them no longer; then derided the sucked orange and threw it away. We kept the positions which we had beguiled them of; by and by, we moved a force forward and overlapped patriot ground—a clever thought, for we needed trouble, and this would produce it. A Filipino soldier, crossing the ground, where no one had a right to forbid him, was shot by our sentry. The badgered patriots resented this with arms, without waiting to know whether Aguinaldo, who was absent, would approve or not. Aguinaldo did not approve; but that availed nothing. What we wanted, in the interest of Progress and Civilization, was the Archipelago, unencumbered by patriots struggling for independence; and the War was what we needed. We clinched our opportunity. It is Mr. Chamberlain's case over again—at least in its motive and intention; and we played the game as adroitly as he played it himself."

At this point in our frank statement of fact to the Person Sitting in Darkness, we should throw in a little trade-taffy about the Blessings of Civilization —for a change, and for the refreshment of his spirit —then go on with our tale:

We and the patriots having captured Manila, Spain's ownership of the Archipelago and her sovereignity over it were at an end—obliterated—annihilated—not a rag or shred of either remaining behind. It was then that we conceived the divinely humorous idea of *buying* both of these spectres from Spain! [It is quite safe to confess this to the Person Sitting in Darkness, since neither he nor any other sane person will believe it.] In buying those ghosts for twenty millions, we also contracted to take care of the friars and their accumulations. I think we also agreed to propagate leprosy and smallpox, but as to this there is doubt. But it is not important; persons afflicted with the friars do not mind the other diseases.

"With our treaty ratified, Manila subdued, and our Ghosts secured, we had no further use for Aguinaldo and the owners of the Archipelago. We forced a war, and we have been hunting America's guest and ally through the woods and swamps ever since."

At this point in the tale, it will be well to boast a little of our war-work and our heroisms in the field, so as to make our performance look as fine as England's in South Africa; but I believe it will not be best to emphasize this too much. We must be cautious. Of course, we must read the war-telegrams to the Person, in order to keep up our frankness; but we can throw an air of humorousness over them, and that will modify their grim eloquence a

12

little, and their rather indiscreet exhibitions of gory exultation. Before reading to him the following display heads of the dispatches of November 18, 1900, it will be well to practice on them in private first, so as to get the right tang of lightness and gaiety into them:

"ADMINISTRATION WEARY OF PROTRACTED HOSTILITIES !"

"REAL WAR AHEAD FOR FILIPINO REBELS!"*

"WILL SHOW NO MERCY!"

"KITCHENER'S PLAN ADOPTED!"

Kitchener knows how to handle disagreeable people who are fighting for their homes and their liberties, and we must let on that we are merely imitating Kitchener, and have no national interest in the matter, further than to get ourselves admired by the Great Family of Nations, in which august company our Master of the Game has bought a place for us in the back row.

Of course, we must not venture to ignore our General MacArthur's reports—oh, why do they keep on printing those embarrassing things?—we must drop them trippingly from the tongue and take the chances:

"During the last ten months our losses have been 268 killed and 750 wounded; Filipino loss, *three thousand two hundred and twenty-seven killed*, and 694 wounded."

We must stand ready to grab the Person Sitting in Darkness, for he will swoon away at this confession, saying: "Good God, those 'niggers' spare their wounded, and the Americans massacre theirs !"

We must bring him to, and coax him and coddle him, and assure him that the ways of Providence are best, and that it would not become us to find fault with them; and then, to show him that we are only imitators, not originators, we must read the following passage from the letter of an American soldier-lad in the Philippines to his mother, pub-

*"Rebels !" Mumble that funny word—don't let the Person catch it distinctly.

lished in *Public Opinion,* of Decorah, Iowa, describing the finish of a victorious battle:

"WE NEVER LEFT ONE ALIVE. IF ONE WAS WOUNDED, WE WOULD RUN OUR BAYONETS THROUGH HIM."

Having now laid all the historical facts before the Person Sitting in Darkness, we should bring him to again, and explain them to him. We should say to him:

"They look doubtful, but in reality they are not. There have been lies; yes, but they were told in a good cause. We have been treacherous; but that was only in order that real good might come out of apparent evil. True, we have crushed a deceived and confiding people; we have turned against the weak and the friendless who trusted us; we have stamped out a just and intelligent and well-ordered republic; we have stabbed an ally in the back and slapped the face of a guest; we have bought a Shadow from an enemy that hadn't it to sell; we have robbed a trusting friend of his land and his liberty; we have invited our clean young men to shoulder a discredited musket and do bandit's work under a flag which bandits have been accustomed to fear, not to follow; we have debauched America's honor and blackened her face before the world; but each detail was for the best. We know this. The Head of every State and Sovereignty in Christendom and ninety per cent. of every legislative body in Christendom, including our Congress and our fifty State Legislatures, are members not only of the church, but also of the Blessings-of-Civilization Trust. This world-girdling accumulation of trained morals, high principles, and justice, cannot do an unright thing, an unfair thing, an ungenerous thing, an unclean thing. It knows what it is about. Give yourself no uneasiness; it is all right."

Now then, that will convince the Person. You will see. It will restore the Business. Also, it will elect the Master of the Game to the vacant place in the Trinity of our national gods; and there on their high thrones the Three will sit, age after age, in the people's sight, each bearing the Emblem of his service: Washington, the Sword of the Liberator; Lincoln, the Slave's Broken Chains; the Master, the Chains Repaired.

It will give the Business a splendid new start. You will see.

Everything is prosperous, now; everything is just as we should wish it. We have got the Archipelago, and we shall never give it up. Also, we have every reason to hope that we shall have an opportunity before very long to slip out of our Congressional contract with Cuba and give her something better in the place of it. It is a rich country, and many of us are already beginning to see that the contract was a sentimental mistake. But now—right now—is the best time to do some profitable rehabilitating work—work that will set us up and make us comfortable, and discourage gossip. We cannot conceal from ourselves that, privately, we are a little troubled about our uniform. It is one of our prides; it is acquainted with honor; it is familiar with great deeds and noble; we love it, we revere it; and so this errand it is on makes us uneasy. And our flag—another pride of ours, our chiefest! We have worshipped it so; and when we have seen it in far lands—glimpsing it unexpectedly in that strange sky, waving its welcome and benediction to us—we have caught our breath and uncovered our heads, and couldn't speak, for a moment, for the thought of what it was to us and the great ideals it stood for. Indeed, we *must* do something about these things; we must not have the flag out there, and the uniform. They are not needed there; we can manage in some other way. England manages, as regards the uniform, and so can we. We have to send soldiers—we can't get out of that—but we can disguise them. It is the way England does in South Africa. Even Mr. Chamberlain himself takes pride in England's honorable uniform, and makes the army down there wear an ugly and odious and appropriate disguise, of yellow stuff such as quarantine flags are made of, and which are hoisted to warn the healthy away from unclean disease and repulsive death. This cloth is called khaki. We could adopt it. It is light, comfortable, grotesque, and deceives the enemy, for he cannot conceive of a soldier being concealed in it.

And as for a flag for the Philippine Province, it is easily managed. We can have a special one—our States do it: we can have just our usual flag, with the white stripes painted black and the stars replaced by the skull and cross-bones.

And we do not need that Civil Commission out there. Having no powers, it has to invent them, and that kind of work cannot be effectively done by just anybody; an expert is required. Mr. Croker can be spared. We do not want the United States represented there, but only the Game.

By help of these suggested amendments, Progress and Civilization in that country can have a boom, and it will take in the Persons who are Sitting in Darkness, and we can resume Business at the old stand.

MARK TWAIN.

KING LEOPOLD'S

SOLILOQUY:

A DEFENSE OF

HIS CONGO RULE

PRICE TWENTY-FIVE CENTS

KING LEOPOLD'S SOLILOQUY

BY
MARK TWAIN

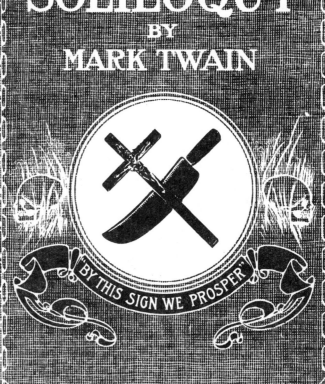

"BY THIS SIGN WE PROSPER"

KING LEOPOLD'S SOLILOQUY

"IT IS I"

"Leopold II is the absolute Master of the whole of the internal and external activity of the Independent State of the Congo. The organization of justice, the army, the industrial and commercial regimes are established freely by himself. He would say, and with greater accuracy than did Louis XIV., ' The State, it is I.'" *Prof. F. Cattier, Brussels University.* " Let us repeat after so many others what has become a platitude, the success of the African work is the work of a sole directing will, without being hampered by the hesitation of timorous politicians, carried out under his sole responsibility,— intelligent, thoughtful, conscious of the perils and the advantages, discounting with an admirable prescience the great results of a near future." *M. Alfred Poskine in " Bilans Congolais."*

" A memorial for the perpetuation of my name."— *Page 25*.

KING LEOPOLD'S

SOLILOQUY

A DEFENSE OF HIS CONGO RULE

BY

MARK TWAIN

————

THE P. R. WARREN CO.

BOSTON, MASS.

1905

King Leopold's Soliloquy

[*Throws down pamphlets which he has been reading. Excitedly combs his flowing spread of whiskers with his fingers; pounds the table with his fists; lets off brisk volleys of unsanctified language at brief intervals, repentantly drooping his head, between volleys, and kissing the Louis XI crucifix hanging from his neck, accompanying the kisses with mumbled apologies; presently rises, flushed and perspiring, and walks the floor, gesticulating*]

—— —— !! —— —— !! If I had them by the throat! [*Hastily kisses the crucifix, and mumbles*] In these twenty years I have spent millions to keep the press of the two hemispheres quiet, and still these leaks keep on occurring. I have spent other millions on religion and art, and what do I get for it? Nothing. Not a compliment. These generosities are studiedly ignored, in print. In print I get nothing but slanders—and slanders again—and still slanders, and slanders on top of slanders! Grant them true, what of it? They are slanders all the same, when uttered against a king.

Miscreants—they are telling *everything!* Oh, everything: how I went pilgriming among the Powers in tears, with my mouth full of Bible and my pelt oozing piety at every pore, and implored them to place the vast and rich and populous Congo Free State in trust in my hands as their agent, so that I might root out slavery and stop the slave raids, and lift up those twenty-five millions of gentle and harmless blacks out of darkness into light, the light of our blessed Redeemer, the light that streams from his holy Word, the light that makes glorious our noble civilization—lift them up and dry their tears and fill their bruised hearts with joy and gratitude—lift them up and make them comprehend that they were no longer outcasts and forsaken, but our very brothers in Christ; how America and thirteen great European states wept in sympathy with me, and were persuaded; how their representatives met in convention in Berlin and made me Head Foreman and Superintendent of the Congo State, and drafted out my powers and limitations, carefully guarding the persons and liberties and properties of the natives against hurt and harm; forbidding whisky traffic and gun traffic; providing courts of justice; making commerce free and fetterless to the merchants and traders of all nations, and welcoming and safe-guarding all missionaries of all creeds and denominations. They have told how I planned and prepared my establishment and selected my horde of officials—"pals" and "pimps" of mine,

4

"unspeakable Belgians" every one—and hoisted my flag, and "took in" a President of the United States, and got him to be the first to recognize it and salute it. Oh, well, let them blackguard me if they like; it is a deep satisfaction to me to remember that I was a shade too smart for that nation that thinks itself so smart. Yes, I certainly did bunco a Yankee—as those people phrase it. Pirate flag? Let them call it so— perhaps it is. All the same, *they were the first to salute it.*

These meddlesome American m i s s i o n - aries! these frank Brit- ish consuls! these blab-

"They were the first to salute it."

bing Belgian-born traitor officials!—those tiresome parrots are always talking, always telling. They have told how for twenty years I have ruled the Congo State not as a trustee of the Powers, an agent, a subordinate, a fore- man, but as a sovereign—sovereign over a fruitful domain four times as large as the German Empire—sovereign absolute, irres- ponsible, above all law; trampling the Berlin-made Congo charter under foot; bar- ring out all foreign traders but myself; re- stricting commerce to myself, through concessionaires who are my creatures and confederates; seizing and holding the State as my personal property, the whole of its vast revenues as my private "swag"—mine, solely mine—claiming and holding its millions of people as my private property, my serfs, my slaves; their labor mine, with or without wage; the food they raise not their property but mine; the rubber, the ivory and all the other riches of the land mine—mine solely—and gathered for me by the men, the women and the little children under compulsion of lash and bullet, fire, starvation, mutilation and the halter.

These pests!—it is as I say, they have kept back nothing! They have revealed these and yet other details which shame should have kept them silent about, since they were exposures of a king, a sacred personage and immune from reproach, by right of his selection and appointment to his great office by God himself; a king whose acts cannot be criticized without blasphemy, since God has observed them from the beginning and has manifested no dissatisfaction with them, nor shown disapproval of them, nor hampered nor interrupted them in any way. By this sign I recognize his approval of what I have done; his cordial and glad approval, I am sure I may say. Blest, crowned, beatified with this great reward, this golden reward, this unspeakably precious reward, why should I care for men's cursings and revilings of me? [*With a sudden outburst of feeling*] May they roast a million æons in—[*Catches his breath and effusively kisses the crucifix; sorrowfully murmurs, "I shall get myself damned yet, with these indiscretions of speech."*]

Yes, they go on telling everything, these chatterers! They tell how I levy incredibly burdensome taxes upon the natives—taxes which are a pure theft; taxes which they must satisfy by gathering rubber under hard and constantly harder conditions, and by raising and furnishing food supplies gratis—and it all comes out that, when they fall short of their tasks through hunger, sickness, despair, and ceaseless and exhausting labor without rest, and forsake their homes and flee to the woods to escape punishment, my black soldiers, drawn from unfriendly tribes, and instigated and directed by my Belgians, hunt them down and butcher them and burn their villages—reserving some of the girls. They tell it all: how I am wiping a nation of friendless creatures out of existence by every form of murder, for my private pocket's sake. But they never say, although they know it, that I have labored in the cause of religion at the same time and all the time, and have sent missionaries there (of a "convenient stripe," as they phrase it), to teach them the error of their ways and bring them to Him who is all mercy and love, and who is the sleepless guardian and friend of all who suffer. They tell only what is against me, they will not tell what is in my favor.

They tell how England required of me a Commission of Inquiry into Congo atrocities, and how, to quiet that meddling country, with its disagreeable Congo Reform Association, made up of earls and bishops and John Morleys and university grandees and other dudes, more interested in other people's business than in their own, I appointed it. Did it stop their mouths? No, they merely pointed out that it was a commission composed wholly of my "Congo butchers," "the very men whose acts were to be inquired into." They said it was equivalent to appointing a commission of wolves to inquire into depredations committed upon a sheepfold. *Nothing* can satisfy a cursed Englishman!*

And are the fault-finders frank with my pri-

---

*Recent information is to the effect that the resident missionaries found the commission as a whole apparently interested to promote reforms. One of its members was a leading Congo official, another an official of the government in Belgium, the third a Swiss jurist. The commission's report will reach the public only through the king, and will be whatever he consents to make it; it is not yet forthcoming, though six months have passed since the investigation was made. There is, however, abundant evidence that horrible abuses were found and conceded, the testimony of missionaries, which had been scouted by the king's defenders, being amply vindicated. One who was present at one hearing of the commission writes: "Men of stone would be moved by the stories that are being unfolded as the commission probes into the awful history of rubber collection." Certain reforms were ordered in the one section visited, but the latest word is that after the commission's departure, conditions soon became worse than before its coming. Very well, then, the king has investigated himself. One stage is achieved. The next one in order is the investigation of conditions in the Congo State *by the Powers responsible for the creation of the Congo State.* The United States is one of these. Such an investigation is advocated by Lyman Abbott, Henry Van Dyke, David Starr Jordan and other prominent citizens in a petition to Congress.—M. T.

"They tell only what is against me."—*Page 8.*

vate character? They could not be more so if
I were a plebeian, a peasant, a mechanic. They
remind the world that from the earliest days my
house has been chapel and brothel combined,
and both industries working full time; that I
practised cruelties upon my queen and my
daughters, and supplemented them with daily
shame and humiliations; that, when my queen
lay in the happy refuge of her coffin, and a
daughter implored me on her knees to let her
look for the last time upon her mother's face, I
refused; and that, three years ago, not being
satisfied with the stolen spoils of a whole alien
nation, I robbed my own child of her property
and appeared by proxy in court, a spectacle to
the civilized world, to defend the act and complete the crime. It is as I have said: they are unfair, unjust; they will resurrect and give new
currency to such things as those, or to any other
things that count against me, but they will not
mention any act of mine that is in my favor. I
have spent more money on art than any other
monarch of my time, and they know it. Do
they speak of it, do they tell about it? No,
they do not. They prefer to work up what they
call "ghastly statistics" into offensive kindergar-
ten object lessons, whose purpose is to make sentimental people shudder, and prejudice them
against me. They remark that "if the innocent
blood shed in the Congo State by King Leopold

were put in buckets and the buckets placed side by side, the line would stretch 2,000 miles; if the skeletons of his ten millions of starved and butchered dead could rise up and march in single file, it would take them seven months and four days to pass a given point; if compacted together in a body, they would occupy more ground than St. Louis covers, World's Fair and all; if they should all clap their bony hands at once, the grisly crash would be heard at a distance of—" Damnation, it makes me tired! And they do similar miracles with the money I have distilled from that blood and put into my pocket. They pile it into Egyptian pyramids; they carpet Saharas with it; they spread it across the sky, and the shadow it casts makes twilight in the earth. And the tears I have caused, the hearts I have broken—oh, nothing can persuade them to let *them* alone!

[*Meditative pause*] Well . . . no matter, I *did* beat the Yankees, anyway! there's comfort in that. [*Reads with mocking smile, the President's Order of Recognition of April 22, 1884*]

" . . . the government of the United States announces its sympathy with and approval of the humane and benevolent purposes of (my Congo scheme), and will order the officers of the United States, both on land and sea, to recognize its flag as the flag of a friendly government."

Possibly the Yankees would like to take that back, now, but they will find that my agents are not over there in America for nothing. But there is no danger; neither nations nor governments can afford to confess a blunder. [*With a contented smile, begins to read from "Report by Rev. W. M. Morrison, American missionary in the Congo Free State"*]

"I furnish herewith some of the many atrocious incidents which have come under my own personal observation; they reveal the *organized system* of plunder and outrage which has been perpetrated and is now being carried on in that unfortunate country by King Leopold of Belgium. I say King Leopold, because he and he *alone* is now responsible, since he is the *absolute sovereign*. *He styles himself such.* When our government in 1884 laid the foundation of the Congo Free State, by recognizing its flag, little did it know that this concern, parading under the guise of philanthropy, was really King Leopold of Belgium, one of the shrewdest, most heartless and most conscienceless rulers that ever sat on a throne. This is apart from his known corrupt morals, which have made his name and his family a byword in two continents. Our government would most certainly not have recognized that flag had it known that it was really King Leopold individually who was asking for recognition ; had it known that it was setting up in the heart of Africa an *absolute monarchy;* had it known that, having put down African slavery in our own country at great cost of blood and money, it was *establishing a worse form of slavery right in Africa.*"

[*With evil joy*] Yes, I certainly was a shade too clever for the Yankees. It hurts; it gravels them. They can't get over it! Puts a shame upon them in another way, too, and a graver way; for they never can rid their records of the reproachful fact that their vain Republic, self-appointed Champion and Promoter of the Liberties of the World, is the only democracy in history that has lent its power and influence to the establishing of an *absolute monarchy!*

[*Contemplating, with an unfriendly eye, a stately pile of pamphlets*] Blister the meddlesome missionaries! They write tons of these things. They seem to be always around, always

"They go to them with their sorrows"

spying, always eye-witnessing the happenings; and everything they see they commit to paper. They are always prowling from place to place; the natives consider them their only friends; they go to them with their sorrows; they show them their scars and their wounds, inflicted by my soldier police; they hold up the stumps of their arms and lament because their hands have been chopped off, as punishment for not bringing in enough rubber, and as proof to be laid before my officers that the required punishment was well and truly carried out. One of these missionaries saw eighty-one of these hands drying over a fire for transmission to my officials— and of course he must go and set it down and print it. They travel and travel, they spy and spy! And nothing is too trivial for them to print. [*Takes up a pamphlet. Reads a passage from Report of a "Journey made in July, August and September, 1903, by Rev. A. E. Scrivener, a British missionary"*]

" . . . . Soon we began talking, and without any encouragement on my part the natives began the tales I had become so accustomed to. They were living in peace and quietness when the white men came in from the lake with all sorts of requests to do this and that, and they thought it meant slavery. So they attempted to keep the white men out of their country but without avail. The rifles were too much for them. So they submitted and made up their minds to do the best they could under the altered circumstances. First came the command to build houses for the soldiers, and this was done without a murmur. Then they had to feed the soldiers and all the men and women—hangers on—who accompanied them. Then they were told to bring in rubber. This was quite a new thing for them to do. There was rubber in the forest several days away from their home, but that it was worth anything was news to them. A small reward was offered and a rush was made for the rubber. ' What strange white men, to give us cloth and beads for the sap of a wild vine.' They rejoiced in what they thought their good fortune. But soon the reward was reduced until at last they were told to bring in the rubber for nothing. To this they tried to demur ; but to their great surprise several were shot by the soldiers, and the rest were told, with many curses and blows, to go at once or more would be killed. Terrified, they began to prepare their food for the fortnight's absence from the village which the collection of rubber en-

13

tailed. The soldiers discovered them sitting about. 'What, not gone yet ?' Bang ! bang ! bang ! and down fell one and another, dead, in the midst of wives and companions. There is a terrible wail and an attempt made to prepare the dead for burial, but this is not allowed. All must go at once to the forest. Without food? Yes, without food. And off the poor wretches had to go without even their tinder boxes to make fires. Many died in the forests of hunger and exposure, and still more from the rifles of the ferocious soldiers in charge of the post. In spite of all their efforts the amount fell off and more and more were killed. I was shown around the place, and the sites of former big chiefs' settlements were pointed out. A careful estimate made the population of, say, seven years ago, to be 2,000 people in and about the post, within a radius of, say, a quarter of a mile. All told, they would not muster 200 now, and there is so much sadness and gloom about them that they are fast decreasing.''

'' We stayed there all day on Monday and had many talks with the people. On the Sunday some of the boys had told me of some bones which they had seen, so on the Monday I asked to be shown these bones. Lying about on the grass, within a few yards of the house I was occupying, were numbers of human skulls, bones, in some cases complete skeletons. I counted thirty-six skulls, and saw many sets of bones from which the skulls were missing. I called one of the men and asked the meaning of it. 'When the rubber palaver began,' said he, ' the soldiers shot so many we grew tired of burying, and very often we were not allowed to bury ; and so just dragged the bodies out into the grass and left them. There are hundreds all around if you would like to see them.' But I had seen more than enough, and was sickened by the stories that came from men and

"Some bones which they had seen"

women alike of the awful time they had passed through. The Bulgarian atrocities might be considered as mildness itself when compared with what was done here. How the people submitted I don't know, and even now I wonder as I think of their patience. That some of them managed to run away is some cause for thankfulness. I stayed there two days and the one thing that impressed itself upon me was the collection of rubber. I saw long files of men come in, as at Bongo, with their little baskets under their arms ; saw them paid their milk tin full of salt, and the two yards of calico flung to the head-men ; saw their trembling timidity, and in fact a great deal that all went to prove the state of terrorism that exists and the virtual slavery in which the people are held."

That is their way; they spy and spy, and run into print with every foolish trifle. And that British consul, Mr. Casement, is just like them. He gets hold of a *diary which had been kept by*

15

one of my government officers, and, although it is a private diary and intended for no eye but its owner's, Mr. Casement is so lacking in delicacy and refinement as to print passages from it. [*Reads a passage from the diary*]

"Each time the corporal goes out to get rubber, cartridges are given him. He must bring back all not used, and for every one used he must bring back a right hand. M. P. told me that sometimes they shot a cartridge at an animal in hunting; they then cut off a hand from a living man. As to the extent to which this is carried on, he informed me that in six months the State on the Mambogo River had used 6,000 cartridges, which means that 6,000 people are killed or mutilated. It means more than 6,000, for the people have told me repeatedly that the soldiers kill the children with the butt of their guns."

When the subtle consul thinks silence will be more effective than words, he employs it. Here he leaves it to be recognized that a thousand killings and mutilations a month is a large output for so small a region as the Mambogo River concession, silently indicating the dimensions of it by accompanying his report with a map of the prodigious Congo State, in which there is not room for so small an object as that river. That silence is intended to say, "If it is a thousand a month in this little corner, imagine the output of the whole vast State!" A gentleman would not descend to these furtivenesses.

Now as to the mutilations. You can't head

FOOT AND HAND OF CHILD DISMEMBERED BY SOLDIERS, BROUGHT
TO MISSIONARIES BY DAZED FATHER. FROM PHOTOGRAPH TAKEN AT
BARINGA, CONGO STATE, MAY 15, 1904. SEE MEMORIAL TO
CONGRESS, JAN., 1905

"Imagine the output of the whole vast State!"—*Page 16.*

off a Congo critic and make him stay headed-
off; he dodges, and straightway comes back at
you from another direction.   They are full of
slippery arts.   When the mutilations (severing
hands, unsexing men, etc.) began to stir Europe,
we hit upon the idea of excusing them with a
retort which we judged would knock them dizzy
on that subject for good and all, and leave them
nothing more to say; to wit, we boldly laid the
custom on the natives, and said we did not invent
it, but only followed it.   Did it knock them
dizzy? did it shut their mouths?   Not for an
hour.   They dodged, and came straight back at
us with the remark that "if a Christian king can
perceive a saving moral difference between inventing bloody barbarities, and *imitating them
from savages,* for charity's sake let him get what
comfort he can out of his confession!"

It is most amazing, the way that that consul
acts—that spy, that busy-body.   [*Takes up
pamphlet "Treatment of Women and Children
in the Congo State; what Mr. Casement Saw
in 1903"*] *Hardly two years ago!   Intruding* that date upon the public was a piece of
cold malice.   It was intended to weaken the
force of my press syndicate's assurances to the
public that my severities in the Congo *ceased,*
and ceased utterly, *years and years ago.*   This
man is fond of trifles—revels in them, gloats
over them, pets them, fondles them, sets them

all down. One doesn't need to drowse through his monotonous report to see that; the mere subheadings of its chapters prove it. [*Reads*]

"Two hundred and forty persons, *men, women and children*, compelled to supply government with *one ton* of carefully prepared foodstuffs *per week*, receiving in remuneration, all told, the princely sum of 15s. 10d!"

Very well, it was liberal. It was not much short of a penny a week for each nigger. It suits this consul to belittle it, yet he knows very well that I could have had both the food and the labor for nothing. I can prove it by a thousand instances. [*Reads*]

"Expedition against a village behindhand in its (compulsory) supplies; result, slaughter of sixteen persons; among them three women and a boy of five years. Ten carried off, to be prisoners till ransomed; among them a child, who died during the march."

But he is careful not to explain that we are *obliged* to resort to ransom to collect debts, where the people have nothing to pay with. Families that escape to the woods sell some of their members into slavery and thus provide the ransom. He knows that I would stop this if I could find a less objectionable way to collect their debts. . . . Mm—here is some more of the consul's delicacy! He reports a conversation he had with some natives:

Q. "How do you know it was the *white* men them-

selves who ordered these cruel things to be done to you?
These things must have been done without the white man's
knowledge by the black soldiers."

A.   " The white men told their soldiers : ' You only kill
women; you cannot kill men. You must prove that you
kill men.'   So then the soldiers when they killed us" (here
he stopped and hesitated and then pointing to . . . he said:)
"then they . . . and took them to the white men, who
said : ' It is true, you have killed *men.*' "

Q.   "You say this is true ? Were many of you so treated
after being shot ? "

All [*shouting out*] : "*Nkoto* ! *Nkoto* !" ("Very many !
Very many !")

There was no doubt that these people were not inventing.
Their vehemence, their flashing eyes, their excitement, were
not simulated."

Of course the critic had to divulge that; he
has no self-respect. All his kind reproach me,
although they know quite well that I took no
pleasure in punishing the men in that particular
way, but only did it as a warning to other delin-
quents. Ordinary punishments are no good with
ignorant savages; they make no impression.
[*Reads more sub-heads*]

" Devasted region ; population reduced from 40,000 to
8,000."

He does not take the trouble to say how it
happened. He is fertile in concealments. He
hopes his readers and his Congo reformers,
of the Lord-Aberdeen-Norbury-John-Morley-Sir

Gilbert-Parker stripe, will think they were all killed. They were not. The great majority of them escaped. They fled to the bush with their families because of the rubber raids, and it was there they died of hunger. Could we help that?

One of my sorrowing critics observes: "Other Christian rulers tax their people, but furnish schools, courts of law, roads, light, water and protection to life and limb in return; King Leopold taxes his stolen nation, but provides *nothing in return but hunger, terror, grief, shame, captivity, mutilation and massacre.*" That is their style! I furnish "nothing"! I send the gospel to the survivors; these censure-mongers know it, but they would rather have their tongues cut out than mention it. I have several times required my raiders to give the dying an opportunity to kiss the sacred emblem; and if they obeyed me I have without doubt been the humble ·means of saving many souls. None of my traducers have had the fairness to mention this; but let it pass; there is One who has not overlooked

On Dress Parade

20

it, and that is my solace, that is my consolation.

[Puts down the Report, takes up a pamphlet, glances along the middle of it]

This is where the "death-trap" comes in. Meddlesome missionary spying around—Rev. W. H. Sheppard. Talks with a black raider of mine after a raid; cozens him into giving away some particulars. The raider remarks:

" I demanded 30 slaves from this side of the stream and 30 from the other side; 2 points of ivory, 2,500 balls of rubber, 13 goats, 10 fowls and 6 dogs, some corn chumy, etc.

' How did the fight come up ? ' I asked.

' I sent for all their chiefs, sub-chiefs, men and women, to come on a certain day, saying that I was going to finish all the palaver. When they entered these small gates (the walls being made of fences brought from other villages, the high native ones) I demanded all my pay or I would kill them ; so they refused to pay me, and I ordered the fence to be closed so they couldn't run away ; then we killed them here inside the fence. The panels of the fence fell down and some escaped.'

' How many did you kill ? ' I asked.

' We killed plenty, will you see some of them ? '

That was just what I wanted.

He said: ' I think we have killed between eighty and ninety, and those in the other villages I don't know, I did not go out but sent my people.'

He and I walked out on the plain just near the camp. There were three dead bodies with the flesh carved off from the waist down.

' Why are they carved so, only leaving the bones ? ' I asked.

' My people ate them,' he answered promptly. He then

explained, 'The men who have young children do not eat people, but all the rest ate them.' On the left was a big man, shot in the back and without a head. (All these corpses were nude.)

' Where is the man's head ? ' I asked.

' Oh, they made a bowl of the forehead to rub up tobacco and diamba in.'

We continued to walk and examine until late in the afternoon, and counted forty-one bodies. The rest had been eaten up by the people.

On returning to the camp, we crossed a young woman, shot in the back of the head, one hand was cut away. I asked why, and Mulunba N'Cusa explained that they always cut off the right hand to give to the State on their return.

' Can you not show me some of the hands ? ' I asked.

So he conducted us to a framework of sticks, under which was burning a slow fire, and there they were, the right hands — I counted them, eighty-one in all.

There were not less than sixty women (Bena Pianga) prisoners. I saw them.

We all say that we have as fully as possible investigated the whole outrage, and find it was a plan previously made to get all the stuff possible and to catch and kill the poor people in the ' death-trap.' ' ''

Another detail, as we see!—cannibalism. They report cases of it with a most offensive frequency. My traducers do not forget to remark that, inasmuch as I am absolute and with a word can prevent in the Congo anything I choose to prevent, then whatsoever is done there by my permission is my act, my *personal* act; that *I* do it; that the hand of my agent is as truly

MULUNBA CHIEF OF CANNIBAL TRIBE NEAR LUEBO, CONGO STATE
FIGURE REPRODUCED FROM PHOTOGRAPH

"Mulunba N'Cusa explained."—*Page 22.*

my hand as if it were attached to my own arm;
and so they picture me in my robes of state, with
my crown on my head, munching human flesh,
saying grace, mumbling thanks to Him from
whom all good things come. Dear, dear, when
the soft-hearts get hold of a thing like that missionary's contribution they quite lose their tranquility over it They speak out profanely and
reproach Heaven for allowing such a fiend to
live. Meaning me. They think it irregular.
They go shuddering around, brooding over the
reduction of that Congo population from 25,-
000,000 to 15,000,000 in the twenty years of
my administration; then they burst out and call
me "the King with Ten Million Murders on his
Soul." They call me a "reccrd." The most of
them do not stop with charging merely the 10,-
000,000 against me. No, they reflect that but
for me the population, by natural increase, would
now be 30,000,000, so they charge another 5,-
000,000 against me and make my total death-
harvest 15,000,000. They remark that the man
who killed the goose that laid the golden egg was
responsible for the eggs she would subsequently
have laid if she had been let alone. Oh, yes,
they call me a "record." They remark that
twice in a generation, in India, the Great Famine
destroys 2,000,000 out of a population of
320,000,000, and the whole world holds up its
hands in pity and horror; then they fall to won-

dering where the world would find room for its emotions if I had a chance to trade places with the Great Famine for twenty years! The idea fires their fancy, and they go on and imagine the Famine coming in state at the end of the twenty years and prostrating itself before me, saying: "Teach me, Lord, I perceive that I am but an apprentice." And next they imagine Death coming, with his scythe and hour-glass, and begging me to marry his daughter and reorganize his plant and run the business. For the whole world, you see! By this time their diseased minds are under full steam, and they get down their books and expand their labors, with me for text. They hunt through all biography for my match, working Attila, Torquemada, Ghengis Khan, Ivan the Terrible, and the rest of that crowd for all they are worth, and evilly exulting when they cannot find it. Then they examine the historical earthquakes and cyclones and blizzards and cataclysms and volcanic eruptions: verdict, none of them "in it" with me. At last they do really hit it (as they think), and they close their labors with conceding—reluctantly— that I have *one* match in history, but only one— the *Flood*. This is intemperate.

But they are always that, when they think of me. They can no more keep quiet when my name is mentioned than can a glass of water control its feelings with a seidlitz powder in its bowels. The bizarre things they can imagine, with me for an inspiration! One Englishman offers to give me the odds of three to one and bet me anything I like, up to 20,000 guineas, that for 2,000,000 years I am going to be the most conspicuous foreigner in hell. The man is so beside himself with anger that he does not perceive that the idea is foolish. Foolish and unbusinesslike: you see, there could be no winner; both of us would be losers, on account of the loss of interest on the stakes; at four or five per cent. compounded, this would amount to— I do not know how much, exactly, but, by the time the term was up and the bet payable, a person could buy hell itself with the accumulation.

Another madman wants to construct a memorial for the perpetuation of my name, out of my 15,000,000 skulls and skeletons, and is full of vindictive enthusiasm over his strange project. He has it all ciphered out and drawn to scale. Out of the skulls he will build a combined monument and mausoleum to me which shall exactly duplicate the Great Pyramid of Cheops, whose base covers thirteen acres, and whose apex is 451 feet above ground. He desires to stuff me and stand me up in the sky on that apex, robed and crowned, with my "pirate flag" in one hand and a butcher-knife and pendant handcuffs in the other. He will build the pyramid in the centre of a depopulated tract, a brooding solitude covered with weeds and the mouldering ruins of burned villages, where the spirits of the starved and murdered dead will voice their laments forever in the whispers of the wandering winds. Radiating from the pyramid, like the spokes of a wheel, there are to be forty grand avenues of approach, each thirty-five miles long, and each fenced on both sides by skulless skeletons standing a yard and a half apart and festooned together in line by short chains stretching from wrist to wrist and attached to tried and true old handcuffs stamped with my private trade-mark, a crucifix and butcher-knife crossed, with motto, "By this sign we prosper;" each osseous fence to consist of 200,000 skeletons on a side, which is 400,000 to each avenue. It is remarked with satisfaction that it aggregates three or four thousand miles (single-ranked) of skeletons,— 15,000,000 all told—and would stretch across America from New York to San Francisco. It is remarked further, in the hopeful tone of a railroad company forecasting showy extensions of its mileage, that my output is 500,000 corpses a year when my plant is running full time, and that therefore if I am spared ten years longer there will be fresh skulls enough to add 175 feet to the pyramid, making it by a long way the loftiest architectural construction on the earth, and fresh skeletons enough to continue the transcontinental file (on piles) a thousand miles into

" A memorial for the perpetuation of my name." — *Page 25*.

the Pacific. The cost of gathering the materials from my "widely scattered and innumerable private graveyards," and transporting them, and building the monument and the radiating grand avenues, is duly ciphered out, running into an aggregate of millions of guineas, and then— why then, (—— ——!! —— ——!!) this idiot asks me *to furnish the money!* [*Sudden and effusive application of the crucifix*] He reminds me that my yearly income from the Congo is millions of guineas, and that "*only*" 5,000,000 would be required for his enterprise. Every day wild attempts are made upon my purse; they do not affect me, they cost me not a thought. But *this one*—this one troubles me, makes me nervous; for there is no telling what an unhinged creature like this may think of next. . · . *If he should think of Carnegie*—but I must banish that thought out of my mind! it worries my days; it troubles my sleep. That way lies madness. [*After a pause*] There is no other way —I have got to buy Carnegie.

[*Harassed and muttering, walks the floor a while, then takes to the Consul's chapter-headings again. Reads*]

"Government starved a woman's children to death and killed her sons."

"Butchery of women and children."

"*The native has been converted into a being without ambition because without hope.*"

"Women chained by the neck by rubber sentries."

"Women refuse to bear children because, with a baby to carry, they cannot well run away and hide from the soldiers."

"Statement of a child. 'I, my mother, my grandmother and my sister, we ran away into the bush. A great number of our people were killed by the soldiers. . . .

"Women chained by the neck"

After that they saw a little bit of my mother's head, and the soldiers ran quickly to where we were and caught my grandmother, my mother, my sister and another little one younger than us. Each wanted my mother for a wife, and argued about it, so they finally decided to kill her. They shot her through the stomach with a gun and she fell, and when I saw that I cried very much, because they killed my grandmother and mother and I was left alone. I saw it all done!'"

It has a sort of pitiful sound, although they are only blacks. It carries me back and back into the past, to when my children were little, and would fly—to the bush, so to speak—when they saw me coming. . . . [*Resumes the reading of chapter-headings*]

28

" My yearly income from the Congo is millions of guineas."
—*Page 27.*

"They put a knife through a child's stomach."

"They cut off the hands and brought them to C. D. (white officer) and spread them out in a row for him to see."

"Captured children left in the bush to die, by the soldiers."

"Friends came to ransom a captured girl; but sentry refused, saying the white man wanted her because she was young."

"Extract from a native girl's testimony. 'On our way the soldiers saw a little child, and when they went to kill it the child laughed, so the soldier took the butt of his gun and struck the child with it and then cut off its head. One day they killed my half-sister and cut off her head, hands and feet, because she had bangles on. Then they caught another sister, and sold her to the W. W. people, and now she is a slave there.'"

FROM PHOTOGRAPH, IKOKO, CONGO STATE.

"Somehow—I wish it had not laughed."

The little child laughed! [*A long pause. Musing*] That innocent creature. Somehow —I wish it had not laughed. [*Reads*]

"Mutilated children."

"Government encouragement of inter-tribal slave-traffic. The monstrous fines levied upon villages tardy in their supplies of foodstuffs compel the natives to sell their fellows— and children—to other tribes in order to meet the fine."

"A father and mother forced to sell their little boy."

"Widow forced to sell her little girl."

[*Irritated*] Hang the monotonous grumbler, what would he have me do! Let a widow off merely because she is a widow? He knows quite well that there is nothing much left, now, *but* widows. I have nothing against widows, as a class, but business is business, and I've got to live, haven't I, even if it does cause inconvenience to somebody here and there? [*Reads*]

"Men intimidated by the torture of their wives and daughters. (To make the men furnish rubber and supplies and so get their captured women released from chains and detention.) The sentry explained to me that he caught the women and brought them in (chained together neck to neck) by direction of his employer."

"An agent explained that he was forced to catch women in preference to men, as then the men brought in supplies quicker ; but he did not explain how the children deprived of their parents obtained their own food supplies."

"A file of 15 (captured) women."

"Allowing women and children to die of starvation in prison."

[*Musing*] Death from *hunger*. A lingering, long misery that must be. Days and days, and still days and days, the forces of the body failing, dribbling away, little by little—yes, it must be the hardest death of all. And to see food carried by, every day, and you can have none of it! Of course the little children cry for it, and that wrings the mother's heart. . . . [*A sigh*] Ah, well, it cannot be helped; circumstances make this discipline necessary. [*Reads*]

" The crucifying of sixty women ! ' "

How stupid, how tactless! Christendom's
goose flesh will rise with horror at the news.
"Profanation of the sacred emblem!" That is
what Christendom will shout. Yes, Christendom will buzz. It can hear me charged with
half a million murders a year for twenty years
and keep its composure, but to profane the Symbol is quite another matter. It will regard this
as serious. It will wake up and want to look
into my record. Buzz? Indeed it will; I seem
to hear the distant hum already. . . . It was
wrong to crucify the women, clearly wrong, manifestly wrong, I can see it now, myself, and am
sorry it happened, sincerely sorry. I believe it
would have answered just as well to skin them.
. . . [*With a sigh*] But none of us thought
of that; one cannot think of everything; and
after all it is but human to err.

It will make a stir, it surely will, these cruci-
fixions. Persons will begin to ask again, as now
and then in times past, how I can hope to win
and keep the respect of the human race if I continue to give up my life to murder and pillage.
[*Scornfully*] When have they heard me say
I wanted the respect of the human race? Do
they confuse me with the common herd? do they
forget that I am a king? What king has valued
the respect of the human race? I mean deep
down in his private heart. If they would reflect,

they would know that it is impossible that a king should value the respect of the human race. He stands upon an eminence and looks out over the world and sees multitudes of m e e k h u m a n things worshiping the persons, and submitting to the oppressions and exactions, o f a d o z e n human things who are in no way better or finer than themselves— made on just their own pattern, in fact, and

"He stands upon an eminence."
"Made on just their own pattern."

out of the same quality of mud. When it *talks,* it is a race of whales; but a king knows it for a race of tadpoles. Its history gives it away. If men were really *men,* how could a Czar be possible? and how could I be possible? But we *are* possible; we are quite safe; and with God's help we shall continue the business at the old stand. It will be found that the race will put up with us, in its docile immemorial way. It may pull a wry face now and then, and make large talk, but it will stay on its knees all the same.

Making large talk is one of its specialties. It works itself up, and froths at the mouth, and just when you think it is going to throw a brick,—it heaves a poem! Lord, what a race it is!

[*Reads*]   A CZAR — 1905

"A pasteboard autocrat ; a despot out of date ;
    A fading planet in the glare of day ;
    A flickering candle in the bright sun's ray,
Burnt to the socket ; fruit left too late,
    High on a blighted bough, ripe till it's rotten.

By God forsaken and by time forgotten,
Watching the crumbling edges of his lands,
    A spineless god to whom dumb millions pray,
    From Finland in the West to far Cathay,
Lord of a frost-bound continent he stands,
    Her seeming ruin his dim mind appalls,
And in the frozen stupor of his sleep
    He hears dull thunders, pealing as she falls,
And mighty fragments dropping in the deep."*

It is fine, one is obliged to concede it; it is a great picture, and impressive. The mongrel handles his pen well. Still, with opportunity, I would cruci—flay him. . . . "A spineless god." It is the Czar to a dot—a god, and spineless; a royal invertebrate, poor lad; soft-hearted and out of place. "A spineless god *to whom dumb millions pray.*" Remorselessly correct; concise, too, and compact—the soul and spirit of the human race compressed into half a sentence. On their knees—140,000,000. On their knees to a little tin deity. Massed together, they would stretch away, and away, and away, across the plains, fading and dimming and failing in a measureless perspective—why, even

*B. H. Nadal, in *New York Times.*

the telescope's vision could not reach to the final
frontier of that continental spread of human servility. Now *why* should a king value the respect
of the human race? It is quite unreasonable to
expect it. A curious race, certainly! It finds
fault with me and with my occupations, and for-
gets that neither of us could exist an hour without its sanction. It is our confederate and all-
powerful protector. It is our bulwark, our
friend, our fortress. For this it has our gratitude, our deep and honest gratitude—but not
our respect. Let it snivel and fret and grumble
if it likes; that is all right; we do not mind that.

[*Turns over leaves of a scrapbook, pausing
now and then to read a clipping and make a
comment*] The poets—how they do hunt that
poor Czar! French, Germans, English, Americans—they all have a bark at him. The finest
and capablest of the pack, and the fiercest, are
Swilburne (English, I think), and a pair of
Americans, Thomas Bailey Eldridge and Colonel Richard Waterson Gilder, of the sentimental
periodical called *Century Magazine and Louisville Courier-Journal.* They certainly have uttered some very strong yelps. I can't seem to
find them—I must have mislaid them. . . .
If a poet's bite were as terrible as his bark, why
dear me—but it isn't. A wise king minds
neither of them; but the poet doesn't know it.
It's a case of little dog and lightning express.

34

When the Czar goes thundering by, the poet
skips out and rages alongside for a little distance, then returns to his kennel wagging his
head with satisfaction, and thinks he has inflicted
a memorable scare, whereas nothing has really
happened—the Czar didn't know he was around.
They never bark at me; I wonder why that is.
I suppose my Corruption-Department buys
them. That must be it, for certainly I ought to
inspire a bark or two; I'm rather choice material, I should say. Why—here *is* a yelp at me.
[*Mumbling a poem*]

"... What gives thee holy right to murder hope
And water ignorance with human blood?

.   .   .   .   .   .

From what high universe-dividing power
Draws't thou thy wondrous, ripe brutality?

.   .   .   .   .   .

O horrible ... Thou God who seest these things
Help us to blot this terror from the earth."

... No, I see it is "To the Czar," * after
all. But there are those who would say it fits
me—and rather snugly, too. "Ripe brutality."
They would say the Czar's isn't ripe yet, but that
mine is; and not merely *ripe* but rotten. Nothing could keep them from saying that; they
would think it smart. "This terror." Let the
Czar keep that name; I am supplied. This long
time I have been "the monster"; that was their

*Louise Morgan Sill, in *Harper's Weekly.*

favorite—the monster of crime. But now I
have a new one. They have found a fossil
Dinosaur fifty-seven feet long and sixteen feet
high, and set it up in the museum in New
York and labeled it "Leopold II." But it
is no matter, one does not look for manners
in a republic. Um . . . that reminds me;
I have never been caricatured. Could it be
that the corsairs of the pencil could not
find an offensive symbol that was big enough
and ugly enough to do my reputation justice?
[*After reflection*] There is no other way—I
will buy the Dinosaur. And suppress it. [*Rests
himself with some more chapter-headings.
Reads*]

"More mutilation of children." (Hands cut off.)

"Testimony of American Missionaries."

"Evidence of British Missionaries."

It is all the same old thing—tedious repetitions and duplications of shop-worn episodes;
mutilations, murders, massacres, and so on, and
so on, till one gets drowsy over it. Mr. Morel
intrudes at this point, and contributes a comment
which he could just as well have kept to himself
—and throws in some italics, of course; these
people can never get along without italics:

"It is one heartrending story of human misery from
beginning to end,. and *it is all recent*."

Meaning 1904 and 1905. I do not see how a

person can act so. This Morel is a king's subject, and reverence for monarchy should have
restrained him from reflecting upon me with that
exposure. This Morel is a reformer; a Congo
reformer. That sizes *him* up. He publishes a
sheet in Liverpool called "The West African
Mail," which is supported by the voluntary contributions of the sap-headed and the
soft-hearted; and every week it
steams and reeks and festers with
up-to-date "Congo atrocities" of
the sort detailed in this pile of
pamphlets here. I will suppress
it. I suppressed a Congo atrocity
book there, after it was actually
in print; it should not be
difficult for me to suppress a
newspaper.
[*Studies some photographs of
mutilated negroes—throws them
down. Sighs*] The kodak has
been a sore calamity to us. The
most powerful enemy that has
confronted us, indeed. In the early years
we had no trouble in getting the press to
"expose" the tales of the mutilations as
slanders, lies, inventions of busy-body American
missionaries and exasperated foreigners who had
found the "open door" of the Berlin-Congo
charter closed against them when they inno-

"The only witness
I couldn't bribe"

37

cently went out there to trade; and by the press's help we got the Christian nations everywhere to turn an irritated and unbelieving ear to those tales and say hard things about the tellers of them. Yes, all things went harmoniously and pleasantly in those good days, and I was looked up to as the benefactor of a down-trodden and friendless people. Then all of a sudden came the crash! That is to say, the incorruptible *kodak*—and all the harmony went to hell! The only witness I have encountered in my long experience that I couldn't bribe. Every Yankee missionary and every interrupted trader sent home and got one; and now—oh, well, the pictures get sneaked around everywhere, in spite of all we can do to ferret them out and suppress them. Ten thousand pulpits and ten thousand presses are saying the good word for me all the time and placidly and convincingly denying the mutilations. Then that trivial little kodak, that a child can carry in its pocket, gets up, uttering never a word, and knocks them dumb!

. . . . What is this fragment? [*Reads*]

" But enough of trying to tally off his crimes ! His list is interminable, we should never get to the end of it. His awful shadow lies across his Congo Free State, and under it an unoffending nation of 15,000,000 is withering away and swiftly succumbing to their miseries. It is a land of graves ; it is *The* Land of Graves ; it is the Congo Free

"The pictures get sneaked around everywhere."—*Page 38.*

Graveyard. It is a majestic thought : that is, this ghastliest episode in all human history is the work of *one man alone ;* one solitary man; just a single individual—Leopold, King of the Belgians. He is personally and solely responsible for all the myriad crimes that have blackened the history of the Congo State. He is *sole* master there; he is absolute. He could have prevented the crimes by his mere command ; he could stop them today with a word. He withholds the word. For his pocket's sake.

It seems strange to see a king destroying a nation and laying waste a country for mere sordid money's sake, and solely and only for that. Lust of conquest is royal; kings have always exercised that stately vice ; we are used to it, by old habit we condone it, perceiving a certain dignity in it ; but *lust of money—lust of shillings—lust of nickels—lust of dirty coin,* not for the nation's enrichment but for *the king's alone*—this is new. It distinctly revolts us, we cannot seem to reconcile ourselves to it, we resent it, we despise it, we say it is shabby, unkingly, out of character. Being democrats we ought to jeer and jest, we ought to rejoice to see the purple dragged in the dirt, but—well, account for it as we may, we don't. We see this awful king, this pitiless and blood-drenched king, this money-crazy king towering toward the sky in a world-solitude of sordid crime, unfellowed and apart from the human race, sole butcher for personal gain findable in all his caste, ancient or modern, pagan or Christian, proper and legitimate target for the scorn of the lowest and the highest, and the execrations of all who hold in cold esteem the oppressor and the coward; and—well, it is a mystery, but *we do not wish to look* ; for he is a king, and it hurts us, it troubles us, by ancient and inherited instinct it shames us to see a king degraded to this aspect, and we shrink from hearing the particulars of how it happened. *We shudder* and *turn away* when we come upon them in print.''

Why, certainly—*that* is my protection. And you will continue to do it. I know the human race.

FROM PHOTOGRAPH, IKOKO, CONGO STATE.

'' To THEM it must appear
very awful and
mysterious.''
Joseph Con-
rad.

AN ORIGINAL MISTAKE

" This work of ' civilization' is
an enormous and continual
butchery."   " All the facts we
brought forward in this chamber were denied at first most
energetically ; but later, little
by little, they were proved by documents and by official texts."
" The practice of cutting off hands is said to be contrary
to instructions ; but you are content to say that indulgence
must be shown and that this bad habit must be corrected
' little by little ' and you plead, moreover, that only the hands
of *fallen* enemies are cut off, and that if hands are cut off
' enemies ' not quite dead, and who, after recovery, have
had the bad taste to come to
the missionaries and show
them their stumps, it was
due to an original mistake in
thinking that they were
dead." *From Debate in Belgian Parliament, July, 1903.*

SUPPLEMENTARY

OUGHT KING LEOPOLD TO BE HANGED? *

INTERVIEW BY MR. W. T. STEAD WITH THE REV. JOHN H. HARRIS,
BARINGA, CONGO STATE, IN THE ENGLISH REVIEW OF
REVIEWS FOR SEPTEMBER, 1905.

For the somewhat startling suggestion in the heading of
this interview, the missionary interviewed is in no way
responsible. The credit of it, or, if you like, the discredit,
belongs entirely to the editor of the *Review*, who, without
dogmatism, wishes to pose the question as a matter for serious
discussion. Since Charles I's head was cut off, opposite
Whitehall, nearly two hundred and fifty years ago, the sanc-
tity which doth hedge about a king has been held in slight
and scant regard by the Puritans and their descendants.
Hence there is nothing antecedently shocking or outrageous
in the discussion of the question whether the acts of any
Sovereign are such as to justify the calling in of the services
of the public executioner. It is not, of course, for a journalist to pronounce judgment, but no function of the public
writer is so imperative as that of calling attention to great
wrongs, and no duty is more imperious than that of insisting
that no rank or station should be allowed to shield from
justice the real criminal when he is once discovered.

The controversy between the Congo Reform Association
and the Emperor of the Congo has now arrived at a stage in
which it is necessary to take a further step towards the

---

*The above article which comes to hand as the foregoing is in press
is commended to the king and to readers of his Soliloquy.— M. T.

45

redress of unspeakable wrongs and the punishment of no less
unspeakable criminals. The Rev. J. H. Harris, an English
missionary, has lived for the last seven years in that region
of Central Africa — the Upper Congo — which King Leopold has made over to one of his vampire groups of financial
associates (known as the A.B.I.R. Society) on the strictly
business basis of a half share in the profits wrung from the
blood and misery of the natives. He has now returned to
England, and last month he called at Mowbray House to
tell me the latest from the Congo. Mr. Harris is a young
man in a dangerous state of volcanic fury, and no wonder.
After living for seven years face to face with the devastations
of the vampire State, it is impossible to deny that he does
well to be angry. When he began, as is the wont of those
who have emerged from the depths, to detail horrifying
stories of murder, the outrage and torture of women, the
mutilation of children, and the whole infernal category of
horrors, served up with the background of cannibalism,
sometimes voluntary and sometimes, incredible though it
seems, enforced by the orders of the officers, I cut him short,
and said : —

" Dear Mr. Harris, as in Oriental despatches the India
Office translator abbreviates the first page of the letter into
two words ' after compliments,' or ' a.c.,' so let us abbre-
viate our conversation about the Congo by the two words
' after atrocities,' or ' a.a.' They are so invariable and so
monotonous, as Lord Percy remarked in the House the other
day, that it is unnecessary to insist upon them. There is no
longer any dispute in the mind of any reasonable person as to
what is going on in the Congo. It is the economical
exploitation of half a continent carried on by the use of
armed force wielded by officials the aim-all and be-all
of whose existence is to extort the maximum amount of

rubber in the shortest possible time in order to pay the largest possible dividend to the holders of shares in the concessions.''

''Well,'' said Mr. Harris reluctantly, for he is so accustomed to speaking to persons who require to be told the whole dismal tale from A to Z, ''what is it you want to know?''

'' I want to know,'' I said, ''whether you consider the time is ripe for summoning King Leopold before the bar of an international tribunal to answer for the crimes perpetrated under his orders and in his interest in the Congo State.''

Mr. Harris paused for a moment, and then said : — '' That depends upon the action which the king takes upon the report of the Commission, which is now in his hands.''

'' Is that report published ?''

''No,'' said Mr. Harris ; '' and it is a question whether it will ever be published. Greatly to our surprise, the Commission, which every one expected would be a mere blind whose appointment was intended to throw dust in the eyes of the public, turned out to be composed of highly respectable persons who heard the evidence most impartially, refused no *bona fide* testimony produced by trustworthy witnesses, and were overwhelmed by the multitudinous horrors brought before them, and who, we feel, *must* have arrived at conclusions which necessitate an entire revolution in the administration of the Congo.''

'' Are you quite sure, Mr. Harris,'' I said, ''that this is so ?''

''Yes,'' said Mr. Harris, '' quite sure. The Commission impressed us all in the Congo very favorably. Some of its members seemed to us admirable specimens of public-spirited, independent statesmen. They realized that they were acting in a judicial capacity; they knew that the eyes of Europe were upon them, and, instead of making their inquiry a farce, they made it a reality, and their conclusions must be, I feel sure, so damning to the State, that if King Leopold were to take no action but to allow the whole infernal business to proceed unchecked, any international tribunal which had powers of a crimal court, would upon the evidence of the Commission alone, send those responsible to the gallows."

"Unfortunately," I said, "at present the Hague Tribunal is not armed with the powers of an international assize court, nor is it qualified to place offenders, crowned or otherwise, in the dock. But don't you think that in the evolution of society the constitution of such a criminal court is a necessity?"

"It would be a great convenience at present," said Mr. Harris; "nor would you need one atom of evidence beyond the report of the Commission to justify the hanging of whoever is responsible for the existence and continuance of such abominations."

"Has anybody seen the text of the report?" I asked.

"As the Commission returned to Brussels in March, some of the contents of that report are an open secret. A great deal of the evidence has been published by the Congo Reform Association. In the Congo the Commissioners admitted two things: first, that the evidence was overwhelming as to the existence of the evils which had hitherto been denied, and secondly, that they vindicated the character of the missionaries. They discovered, as anyone will who goes out to that country, that it is the missionaries, and the missionaries alone, who constitute the permanent European element. The Congo State officials come out ignorant of the language, knowing nothing of the country, and with no other sense of their duties beyond that of supporting the concession companies in extorting rubber. They are like men who are dumb and deaf and blind, nor do they wish to be otherwise. In two or three years they vanish, giving place to other migrants as ignorant as themselves, whereas the missionaries remain on the spot year after year; they are in personal touch with the people, whose language they speak, whose customs they respect, and whose lives they endeavor to defend to the best of their ability.''

'' But, Mr. Harris,'' I remarked, '' was there not a certain Mr. Grenfell, a Baptist Missionary, who has been all these years a convinced upholder of the Congo State?''

'''Twas true,'' said Mr. Harris, '' and pity 'tis 'twas true; but 'tis no longer true. Mr. Grenfell has had his eyes opened at last, and he has now taken his place among those who are convinced. He could no longer resist the overwhelming evidence that has been brought against the Congo Administration.''*

''Was the nature of the Commissioners' report,'' I resumed, ''made known to the officials of the State before they left the Congo?''

''To the head officials—yes,'' said Mr. Harris.

''With what result?''

''In the case of the highest official in the Congo, the man who corresponds in Africa to Lord Curzon in India, no sooner was he placed in possession of the conclusions of the Commission than the appalling significance of their indictment convinced him that the game was up, and he went into his room and cut his throat. I was amazed on returning to Europe to find how little the significance of this suicide was appreciated. A paragraph in the newspaper announced the suicide of a Congo official. None of those who read that

---

* Mr. Grenfell's station is in the Lower Congo, a section remote from the vast rubber areas of the interior.

paragraph could realize the fact that that suicide had the same significance to the Congo that the suicide, let us say, of Lord Milner would have had if it had taken place immediately on receiving the conclusions of a Royal Commission sent out to report upon his administration in South Africa.''

"Well, if that be so, Mr. Harris," I said, "and the Governor-General cuts his throat rather than face the ordeal and disgrace of the exposure, I am almost beginning to hope that we may see King Leopold in the dock at the Hague, after all.''

"I will comment upon that," Mr. Harris said, " by quoting you Mrs. Sheldon's remark made before myself and my colleagues, Messrs. Bond, Ellery, Ruskin, Walbaum and Whiteside, on May 19th last year, when, in answer to our question, 'Why should King Leopold be afraid of submitting his case to the Hague tribunal ?' Mrs. Sheldon answered, 'Men do not go to the gallows and put their heads in a noose if they can avoid it.' ''

AFTERWORD

Fred Kaplan

When Mark Twain set off in July 1895 on a round-the-world reading and lecture tour, his eye was mostly on the poor state of his bank account. There had been a series of business failures, the most devastating of which were the collapse of his publishing house and the failure of the Paige typesetting machine. For over ten years he had been investing all his own and almost all his wife's capital and credit in these enterprises. With the help of a new friend, Henry Huttleston Rogers, a wealthy and powerful vice president of Standard Oil and a successful stock market investor, Twain attempted to put his house in order. His house meant a great deal to him. With his usual sense of perverse rectitude, for which other people sometimes paid the price, he insisted on paying Webster and Company debtors in full. Rogers thought this a good idea precisely because he expected to help Twain capitalize on the reputation it would earn him. America's most famous writer would be an even more valuable market property if he epitomized the lie that nineteenth-century Americans liked to believe about their great men: they always paid their debts in full. In such matters Rogers was even shrewder than Twain, and each by his own reasoning reached the same conclusion. Consequently, already exhausted and demoralized by the hard work ahead of him, Twain began the voyage one result of which was to be the travel book *Following the Equator*.

During the three years *before* his oriental voyage Twain had been doing a great deal of hard traveling. He had gone back and forth across the Atlantic about ten times, in each instance leaving his family in Europe, usually in Paris. He hated to be without them, but his desperate business affairs demanded his occasional presence in America. Since the early 1890s, his native land had become a preoccupation that he preferred to view from abroad. With less reluctance than his wife and three daughters, he had declared that they could no longer afford America. The cost of keeping up his grand house in Hartford was too great for a man who had invested every bit of his ample spare cash in a revolutionary typesetter that had not repaid him a penny. Also, he had been both generous and casual with his publishing house; the account books of what had seemed a sound business now demanded more capital. Twain did not have any. And much of the money at risk was his wife's, which made him even more frantic and desperate. He fought hard to keep the company afloat. Each time he returned to New York to wrestle with the financial demons, he lost some vital outward and inward stamina. After one session with the creditors, he remarked, "I have lost 84 pounds in the last 48 hours. Not meat, but moral fat."[1] He stayed abroad partly because he had convinced himself that he could not afford to live at home in the style to which he had become accustomed. But he also stayed away because he felt an only partly articulated dissatisfaction with some of the corruptions of the American Gilded Age and his own ambivalent relationship with them. Life was physically and morally simpler for Twain in Europe.

Meanwhile, he kept working, partly out of habitual obsessiveness, partly to keep from worrying about business affairs and about the health of his wife and two of his daughters. Livy Clemens had always been physically frail; there were now signs of heart disease as well as neurasthenia, that indeterminate nineteenth-century illness, part headache, part lassitude, part nervous breakdown. In Livy's case it also involved a strong toxin: a doting mother's worry about her children. Her eldest daughter, Susy, brilliant and beautiful, was frequently ill; she was also dangerously high-strung. The youngest, Jean, did not seem quite right; the illness was later diagnosed as epilepsy, and treated disastrously. Only the middle daughter, Clara, seemed fit to cope with the trials of life. Twain's women also suffered, among other things, from the restlessness and the occasional imperiousness of their dearly loved and only man of the house, who knew how to keep doing one thing no matter what turmoil surrounded him — to keep writing brilliantly.

In 1894, he had written in Italy and published in Hartford his trenchant but underdeveloped novel *The Tragedy of Pudd'nhead Wilson* (with the appended farce *Those Extraordinary Twins*). Despite its weaknesses, it was not a commercial success. Still obsessed with and recapitalizing the matter of Hannibal, he also wrote the brief *Tom Sawyer, Detective*, which he published in New York in 1896. He had been writing in fits and starts one of his longest books, *Personal Recollections of Joan of Arc*, which he finished in France in early 1895 and first published anonymously in installments in *Harper's* magazine. Still haunted by the fear that he could never escape his reputation as a humorist, he thought that if *Joan* appeared without his name the critics and the reading public would take it more seriously. They did and they did not. Soon it became widely known that he was either its author or editor. All in all, the sales were excellent. Still, excellence and the likelihood of some better business arrangements for his publishing house debts and for the reissuing of his books did not solve his major problem. He had made himself personally liable for the Webster bankruptcy. And he could neither borrow nor write his way into solvency.

Twain had grown to hate lecturing, and for almost ten years he had kept the vow that he would never do it again. Now a more extensive reading and lecturing tour than ever before seemed the pill he had to swallow. He saw no alternative. Because neither he nor Livy could stand the idea of a long separation, they decided that she would accompany him. Perhaps, they hoped, it would even do her health some good. But fifteen-year-old Jean and twenty-three-year-old Susy would stay with their aunt at one of the family's favorite places, Quarry Farm, near Elmira, New York. Susy, who desperately missed their lost Hartford home, felt that she particularly had suffered from her father's world-traveling restlessness. Only twenty-one-year-old Clara, who was physically and emotionally sturdier than her sisters and liked the idea of having her parents to herself, would go with them.

The three left from Quarry Farm in July 1895. After Midwestern and Western lectures, they sailed from Vancouver, British Columbia, in late August. Eventually, they circumnavigated the globe. Twain read and lectured about one hundred times, in Australia, New Zealand, Ceylon, India, and South Africa, before sailing northward to Britain in July 1896. The trip was both exhausting and refreshing. He brought himself and his lectures to a part of the world he had not visited before. Sometimes both the traveling and the lecturing were physically wearing, but often enough he glowed happily in applause that rose to the level of veneration in these Anglo-colonial countries: the British colonial structure embraced and supported him; the well-to-do and socially superior natives frequently revered him as a literary deity. All the while, he kept sending his profits back to Rogers in New York, where they were credited against his losses and obligations. And all the while he did one of the things he did best, for which he had a genius that he had transformed into literary gold from the beginning of his career: he observed. In the Jamesian sense, "nothing was lost" on him. And for the last time in his life he had sufficient physical vigor to engage in the kind of cultural sightseeing that had helped create *The Innocents Abroad* (1869), *Roughing It* (1872), *A Tramp Abroad* (1880), and *Life on the Mississippi* (1883).

But between the observation and the writing act fell a dark shadow. In the Orient, Twain had taken sensual pleasure in the uninhibited colorfulness of what was for him an exotic world. Most of all, he found India brightly riveting. "The colors were often dazzling, yellows and greens in shades one never finds in America anywhere," Clara wrote to her aunt about their voyage to Bombay.[2] Twain was dazzled. "It is all color, bewitching color, enchanting color — everywhere — all around. . ." (347). The variegated brilliance seemed to him a vivid reminder of the integrity of other cultures and of the drabness of Victorian Anglo-American provincialism. Under the right circumstances, he would have happily "gone native," as he had done in a limited way when he spent four months in Hawaii as a young man. In South Africa, railroading from Johannesburg to Bloemfontein, he found the landscape colorfully Edenic: "*I* think the veld is just as beautiful as Paradise — rolling, & swelling, & rising & subsiding, & sweeping on, & on, & on, like an ocean, toward the remote horizon, & changing its pale brown by delicate shades, to rich orange and finally to purple & crimson where it washes against the hills at the base of the sky."[3] To his surprise, everyone seemed refreshed by the exotic landscapes and the social flattery. Only his tightly rehearsed and exactly timed readings tired him enough to remind him of his disinclination to have made this journey at all. Sporadic bronchitis and bothersome skin eruptions were with him constantly, and he later complained that he "was never very well, from the first night in Cleveland to the last one in Cape Town and I found it pretty hard work on that account. I did a good deal of talking when I ought to have been in bed."[4]

True enough, but far from the whole truth. Mostly, he flourished during the voyage. Clara expressed the fear and the temptation, since "we are enjoying the traveling, every minute of it, and even the hardships seem to us much less hard," that "life will be monotonous without them. That's always the danger of traveling. I hope we shan't get the regular wandering mania, for we couldn't satisfy it, I suppose."[5] They arrived in England at the end of July 1896. Rather than return to America, Twain had decided to write *Following the Equator* abroad. He had friends and admirers in England. Susy and Jean would join them there shortly. Why go back to the land of his bankruptcy? By mid-August 1896, they had found a temporary house an hour from London where it seemed, on the whole, that "the weather has been ordered from hell."[6] Under gray English skies, the bright colors were fading.

Gradually, everything went black. Susy and Jean did not come, and he was never to see Susy again. Troubling news came from America about a delay. They "are not on their way hither, we do not yet know why." Then came word that Susy had been taken ill. Then silence. The silence seemed charged with anxiety and foreboding — in itself it seemed "disastrous news."[7] In Hartford, where she had been visiting friends and regretting her lost life and the wonderful house that she might never live in again, Susy lay ill in the midst of a blistering August heat wave. It was clear to the family in England that her condition was serious enough to warrant Livy's immediate departure for New York. While Livy and Clara were at sea, Twain had a cable from America with hopeful news. "The illness has moderated and . . . a sea-voyage may presently be possible. So I shall not go . . . as I was expecting to do, but will wait a little in the hope that a cablegram will soon tell me that the family are on their way to England."[8] The next day a new telegram made the previous day's optimism seem excruciatingly cruel. He opened it, expecting to read "something pleasant."[9] Instead, he learned that Susy had died of spinal meningitis. "Burn letter. Blot it from your memory. Susie is dead."[10] With remarkable writerly self-control, he composed an obituary notice for the London papers and told a friend that Susy "was the prodigy of our flock, in intellectuality, in the gift of speech, & in music — not instrumental but vocal. Will you hand the enclosed half-sheet to any newspaper you please — no, *copy* it and give them the copy, so that they will not know it came from me. I have many personal friends in England, & they should know of my disaster."[11] He cabled Rogers to meet the fearful but uninformed Livy, who would arrive at the dock three days too late and "to have Dr. Rice at the ship and keep all other friends prudently out of sight."[12]

In London, alone, Twain absorbed his disaster. "But though my heart *break* I will still say she was fortunate; and I would not call her back if I could. I eat," he wrote to Livy, "because you wish it; I go on living — because you wish it. I play billiards, and billiards, and billiards, till I am ready to drop — to keep from going mad with grief and with resentful thinkings."[13] All his life a golden sun had shone upon him. And all his life he had been expecting misfortune. His Presbyterian mother had prepared him for stoicism. Deep in his heart and personality were intimations of unworthiness, tremors of life's precariousness. This was the damn human race that he belonged to. He had long been steeling himself and protecting himself with philosophical resignation. But resignation coexisted with resentment, even fury. This was indeed, eventually for everyone, a vale of tears, but there was no sense and no justice in the pain that human beings had to suffer. "Not being dead, I knew that calamities, & again calamities, & still other calamities were in store for me — but I had not thought of this one. This beggars imagination: it is not calamity, it is martyrdom. I bear it as I bear all heavy hardships that befal me — with a heart bursting with rebellion."[14] Susy was buried at Elmira without her father to watch the coffin lowered into the grave.

2

As he often did in times of misery as well as happiness, Twain turned to his work. He had the challenge of writing an interesting and entertaining travel book while nursing a broken heart. When Livy and Clara returned to England in September, they found a house in Chelsea, "not to live in public there, but to hide ... for a time and let the wounds heal."[15] For the rest of 1896 the family secluded themselves in their rented house in Tedworth Square, and Twain hid himself from everyone all day. "He goes to his study directly after breakfast & works until seven o'clock in the evening," Livy reported. By Christmas he seemed actually to be taking some interest in what he was writing.[16]

Considering the circumstances, or perhaps because of them, once he got started he wrote with remarkable rapidity. Soon he was "hard at work on a book, and the proceeds will go to the payment of debts contracted for me by others."[17] He had made notes during his voyage, partly in the form of a journal, which he now drew on. Like all Twain's travel books, the work in progress embodied his loose definition of the genre as a kind of potpourri entertainment that had room for substantial quotations from his notebooks and from other source materials, including previously published works. His agreement with Frank Bliss that the American Publishing Company would publish *Following the Equator* for subscription rather than bookstore sales meant that it would be about the same length as *Roughing It* and *Life on the Mississippi*; his British publisher, Chatto and Windus, did not argue with these terms. The length would be partly satisfied by borrowed materials; the apparent structure would be determined by the linear coordinates of the actual voyage itself, and the book would have two major vectors of emotional intensity: Twain's encounters with third-world cultures and his encounters with himself, those moments in which the lifelong interaction between personal memory and present circumstance would create another autobiographical act, a point of self-definition.

Not that Twain himself determined purposely to make the book these things. It was simply a question of being himself, of being the writer he had become. And he knew that he was always most noticeably himself, in style and tone and subject, when he allowed himself to write with the loosest constraints. At its intermittent best, *Following the Equator* contains some of Twain's most characteristic and revealing writing, some passages of exquisite beauty and significance; they shine even more brightly because of the long patches of heavy quotations or tired, workmanlike prose that occasionally surround them. But this is no less the case with his better-known travel books. *Following the Equator* deserves more attention than it has received, partly because it has unexpected and mostly unreceived pleasures to give. Equally importantly, like Twain's other travel books, it has something significant to *say*, especially about what had become by the late 1890s a crucial topic for Twain: the recapitulation on the larger world stage of the painful and discriminatory distinction between Anglo-American whiteness and people of color about which America had fought, and in Twain's view was still fighting (with lynchings rather than with cannons), a civil war. For Twain, American and European imperialism were the international side of the coin of white America's treatment of blacks. The American eagle and empire were poised on the verge of high flight. A year or so after the publication of *Following the Equator*, Teddy Roosevelt's troops were to charge up San Juan Hill and Admiral Dewey's fleet was to steam into Manila Bay.

Following the Equator focuses on the mother of all modern empires, Queen Victoria's far-flung British imperium. It hardly needed professional historical analysis for Twain to find his target. The American eagle was only about to fly, while Britannia ruled the waves and also much of the shore. As Twain traveled westward, he left from the west coast of North America, two former British colonies, to sail into a sunset whose sun also rose on New Zealand, Australia, Singapore, India, and South Africa, not to speak of some lands along the way that England did not rule but dominated. South Africa — that land of Kimberly diamond mines, Cecil Rhodes, and nationalistic Boers — was in political turmoil when Twain arrived. He found the struggle fascinating and appalling. Four years later warfare erupted between the English and Dutch South Africans, and the troops of the empire came to South Africa where Twain had noticed that the native South Africans played no role in the disposition of the country and its wealth. Undoubtedly "the white man's burden" was a heavy one, and Twain was not an unambivalent anti-imperialist. He was an Anglophile who revered British culture and loved living in England, where he was warmly embraced and widely read. He supported social and political organization and hierarchy. But he also detested inhumanity and exploitation; he carried with him his own personal scars as an American Southerner who had embraced the North only to find the North in its own way as prejudiced as the South; and he found the whole human race invariably worthy of Swiftian satirical condemnation. He also vigorously detested nationalistic self-centeredness, the complacent and self-serving assumption that other people are inferior and ought to be made to see the light for their own good. He did not believe that the Christian God or western culture shone so brightly that other cultures should abandon their primitive ways and follow the holy beacon.

Actually, he had become, at the beginning of his writing career, a mildly satirical commentator on ethnic, racial, and cultural differences. In his early essays, stereotypical American Irishmen, Chinese, and Indians appear, expressive of a struggle within Twain to define the prejudices and the problems of the melting pot. About the Irish and the Indians (as they were then known) Twain rarely if ever transcended mainstream Anglo-American cultural caricatures. The Chinese in California and Nevada were another matter. They had qualities that appealed to Twain, including their work ethic, and they were visibly different enough in their looks and in their gentle manners that Twain could use them as an object lesson about cultural oppression and the cruel inhumanity of racial prejudice. In Hawaii, in 1866, he had his first encounter with colonialism and with a native culture still sufficiently in possession of its patrimony to provide a large-scale comparison with the invading culture. America and Britain competed for dominance in Hawaii. There seemed only two types of white people there: missionaries and sea captains. Nobody would believe Twain that he was not one or the other. In his letters from Hawaii and in the Hawaii section of *Roughing It* he deliciously, intelligently, and satirically highlights the ludicrousness and the destructiveness of colonial exploitation, without ever maintaining that native Hawaiian culture is in itself especially virtuous. The issue for Twain was never better or worse: human nature was essentially the same everyplace, and the difference that cultural difference made was part of the balance to be weighed on a larger scale. He had come to the conclusion as early as his Hawaii visit that in general national cultures should be in charge of themselves.

At the same time, he admired Yankee initiative and inventiveness. He thought thrift, investment, shrewdness, competence, and accomplishment positive values. He approved of a growing gross national product to whose ascent of the upward graph international trade made an important contribution. But his notion of the men who made this happen resonated with a pre–Civil War America's emphasis on the small entrepreneur, the individual investor, the Jacksonian idea of the frontier American who did it with his own brains and his own hand. Much as he admired and grew to love Henry Rogers, Twain was not entirely at home in corporate America. His dis-ease never allowed him to accept one of the historical corollaries of Yankee values: the connection between international trade and international political dominance.

But Twain made adjustments. He construed grand schemes of capitalistic grandeur. He dreamed sometimes of cornering markets. Still, they were usually domestic markets. In *A Connecticut Yankee in King Arthur's Court* (1889) the international subject was much on his mind. The shrewd technological Yankee Hank Morgan finds himself in a colonialist's paradise, a third-world country of the Dark Ages. To his delight, he is "just as much at home in that century [the sixth] as I could have been in any other; and as for preference, I wouldn't have traded it for the twentieth. Look at the opportunities here. . . . The grandest field that ever was; and all my own; not a competitor; not a man who wasn't a baby to me in acquirements and capacities. . . . What a jump I had made! I couldn't keep from thinking about it, and contemplating it, just as one does who has struck oil" ("The Boss," chapter 8). Diamonds in South Africa, gold in the Belgian Congo, oil in Arabia, pineapples in the Philippines — Twain could feel and dramatize both sides of the moral equation.

Twain's "Boss," like many British and American colonists, believes that enlightened colonialism brings significant benefits to the local population. Deeply attracted to progressive values, Twain sometimes felt the temptation of the argument that there is virtue in doing for others what they cannot do for themselves. In the end, though, the objects of Hank Morgan's reforms become electrocuted corpses. Very few readers feel confident that the gory triumph of Dark Age chaos at the end of *A Connecticut Yankee* can be understood as a victory for either side. It seems like the damn human race again, and perhaps Twain's own ambivalence about the issues. But without question Hank Morgan is on the right side of Twain's moral ledger when he expresses his horror at institutionalized slavery in Camelot. And, in *Following the Equator*, Twain dramatically and beautifully makes the connection between the slavery (and allied forms of dominance) that he has grown to abhor and the colonial experience. Not surprisingly, he conceptualizes the connection in strong personal terms. A German hotel manager in India, settling the Clemenses in their rooms with the assistance of three natives, sharply hits one of them in the face "without *explaining* what was wrong."

The native took it with meekness, saying nothing, and not showing in his face or manner any resentment. I had not seen the like of this for fifty years. It carried me back to my boyhood, and flashed upon me the forgotten fact that this was the *usual* way of explaining one's desire to a slave. I was able to remember that the method seemed right and natural to me in those days, I being born to it . . . but I was also able to remember that those unresented cuffings made me sorry for the victim and ashamed for the punisher. My father was a refined and kindly gentleman. . . . He laid his hand upon me in punishment only twice in his life . . . and never any other member of the family at all; yet every now and then he cuffed our harmless slave boy, Lewis, for trifling little blunders and awkwardnesses. My father had passed his life among the slaves from his cradle up, and his cuffings proceeded from the custom of the time, not from his nature. When I was ten years old I saw a man fling a lump of iron-ore at a slave-man in anger, for merely doing something awkwardly — as if that were a crime. It bounded from the man's skull, and the man fell and never spoke again. He was dead in an hour. I knew that the man had the right to kill his slave if he wanted to, and yet it seemed a pitiful thing and somehow wrong, though why wrong I was not deep enough to explain if I had been asked to do it. Nobody in the village approved of that murder, but of course no one said much about it. . . . For just one second, all that goes to make the *me* in me was in a Missourian village . . . vividly seeing again these forgotten pictures of fifty years ago . . . and in the next second I was back in Bombay. (351–52)

In Twain's mind and memory, the two were closely connected. The aging Twain at once understood, disapproved of, and sometimes felt at home in a colonial world that had much in common with the world of his childhood.

3

By mid-January 1897, Twain had written a substantial portion of the new book; he hoped to be finished by the beginning of March, which proved characteristically overoptimistic. His plan was to overwrite and then "scratch out as much as I want to." "I must do all this extraordinary revising because this book has to come into comparison with the Innocents, & so I must do my level best to bring it chock up to the mark." Early in March he spent a week "gutting" out a third of the manuscript and then sent what remained to a typist in ten-thousand-word increments. The business of producing a subscription book intensified: he marked passages for Frank Bliss to use in the canvassing advertisements and gave careful consideration to whether or not there would be illustrations and who would do them.[18] Now he was hoping to send a typescript to the American Publishing Company in June. By the end of March he felt that he had almost completed the revision, and he urged his typist to work more quickly. Suddenly he was buoyant about what he had done. The book seemed wonderful. "I wouldn't trade it for any book I have ever written — & I am not an easy person to please. Shall it be called *Imitating the Equator*? or shall it be called *Another Innocent Abroad*? The first may be the best, possibly. People would ask the canvasser 'What does it *mean?*' And so might the newspapers. And by & by somebody would guess it — 'the equator goes around the world.' "[19]

But the book in this version did not go round the world. It stopped at India, as he first intended. There would be "room for nothing more" if he were to keep to the agreed-on length, he told Bliss. "At a later day I can make a book about South Africa if there is material enough in that rather uninteresting country to make the job worth while."[20] Livy worked along with him; as soon as he finished revising a section, she took her turn. Between late March and mid-April 1897 she did two revisions, editing out passages, Twain explained, "on the ~~hypercritical~~ ground that the first part is not delicate & that the last part is *in*delicate. Now *there's* a nice distinction for you — & correctly stated, too, and perfectly true."[21] Livy's pen, though, was not too narrowly censorious, and some morally as well as visually colorful sections remained. And despite his extraordinary claim that South Africa was not an interesting country, Twain contradicted himself sometime between the end of March and the summer by writing a lengthy, fascinating concluding section on his visit there, giving a detailed account of his contacts, shortly after the disruptive Jameson raid, with the Anglo–South African ruling class and the Dutch colonists. Why he changed his mind is unclear. Perhaps he and Livy had cut so much that he needed more. Perhaps he realized that South Africa was more interesting to him than he had allowed himself to think in his eagerness to finish and would likely be of interest to his reading public, especially in Britain. He was still at work, presumably on the South African section, in May 1897.

Between late May and July, the British edition was set in type and returned to him for correction, which, as usual, drove him mad with impatience and exasperation at the typesetters' liberties: "These printers pay no attention to my punctuation. Nine-tenths of the labor and vexation put upon me by Messrs. Spothswoode and Co consists in annihilating their ignorant and purposeless punctuation and *restoring* my own. . . . [I know] more about punctuation in two minutes than any damned bastard of a proof-reader can learn in two centuries."[22] Actually, the Chatto and Windus edition, published under the title *More Tramps Abroad*, was more accurate than the American Publishing Company edition; the former had been set from the manuscript, the latter from a typescript. He had a lot more to say along these lines, and he kept saying it right through August. "I never imagined that the book was going to last until *this* time. But it did — just 10 months; I read the last chapter in proof yesterday — English edition. It was the only book I have ever confined myself to from title-page to Finis without the relief of shifting to other work meantime; & I would rather go hang myself than do the like again. It was a contract, & couldn't be helped. But that slavery is over."[23] He dedicated the book to Harry Rogers, the son of his friend Henry who was soon to tell him that he had been brought out of the Egypt of debt as well. It was almost exactly the first anniversary of Susy's death. The initial period of mourning was over. In Switzerland for the rest of the summer, and about to begin a long residence in Vienna, Mark Twain felt alive again.

Following the Equator was the least commercially successful of Twain's travel books. It completed the downward trend in sales, which had begun smashingly with *Innocents Abroad* and gradually descended from *Roughing It* to *A Tramp Abroad* to *Life on the Mississippi*. If the travel genre was not played out, Twain *was* in regard to such books. And subscription publishing, which the American Publishing Company and Twain had pioneered, had already proved itself unsuited to the changing world of book marketing. Both Twain and Frank Bliss had hoped for a success that would rival *Innocents Abroad*. For the American Publishing Company, however, *Following the Equator* was a last moment of reflected glory in its inevitable and imminent extinction; and Twain never wrote another travel book.

But one of the dominant subjects of *Following the Equator* was increasingly on his mind, though the American imperium rather than the British Empire became the focus. Not long after the publication of *Following the Equator*, America burst militarily onto the world stage. Twain supported the Spanish-American War initially; it seemed a good thing to free Cuba from Spanish tyranny and to let the Cubans decide their own future. So too the Philippines. But when it soon became clear that neither the Cubans nor the Filipinos were going to be allowed to rule themselves, Twain denounced

American imperialism and, along with William James and others, joined the Anti-Imperialist League. When the Boxer rebellion dramatized European and American commercial colonization of China, Twain's opposition rose to bitter fury, especially at American Christian missionaries who brought both Christ and the money changers, as if they were necessary partners, into the Orient. In one of numbers of letters and essays, some of which he published, some of which he did not, Twain addressed the "person sitting in darkness"; this essay, which appeared in the *North American Review* in February 1901, was among his most scathing condemnations of western smugness, exploitativeness, and greed. When the European powers and America, negotiating spheres of influence, turned the Congo over to King Leopold of Belgium, Twain bristled, then was sickened. The sadistic sovereign mutilated both the Congo and its people. In a savage monologue, "King Leopold's Soliloquy," accompanied with a photographic montage of examples of Leopold's depredations and published as a pamphlet in September 1905 on behalf of the Congo Reform Association, Twain condemned the unspeakable. Both essays infuriated the defenders of Christian imperialism; they showed the white man's burden to be a heavy one indeed. For Mark Twain, they were the moral and literary culmination of a journey that had begun in his Missouri childhood.

NOTES

1. MT to Henry Robinson, April 20, 1894. Mark Twain's previously unpublished words as well as the words of his wife, Olivia L. Clemens, and his daughter, Clara Clemens, are © 1996 by Chemical Bank as Trustee of the Mark Twain Foundation, which reserves all reproduction or dramatization rights in every medium. Quotation is made with the permission of the University of California Press and Robert H. Hirst, General Editor, Mark Twain Project. Each of these quotations is identified by a dagger (†) in its citation.

2. Clara Clemens to Pamela Moffett, February 1896.†

3. MT to Olivia Clemens, June 1, 1896.†

4. MT to James B. Pond, August 10, 1896.†

5. Clara Clemens to Pamela Moffett, February 1896.†

6. MT to Andrew Chatto, August 11, 1896.†

7. MT to Henry Huttleston Rogers, August 14, 1896, in *Mark Twain's Correspondence with Henry Huttleston Rogers, 1893–1909*, ed. Lewis Leary (Berkeley: University of California Press, 1969), p. 232.

8. MT to Mrs. Armstrong, August 18, 1896.†

9. MT to Orion Clemens, September 14, 1896.†

10. MT to Henry Watterson, August 18, 1896.†

11. MT to Andrew Chatto, August 19, 1896.†

12. MT to Olivia Clemens, August 21, 1896, in *The Love Letters of Mark Twain*, ed. Dixon Wecter (New York: Harper, 1949), p. 323.

13. Ibid.

14. MT to Franklin G. Whitmore, September 10, 1896.†

15. Ibid.

16. Olivia Clemens to Mary Fairbanks, December 28, 1896.†

17. MT to James Ross Clemens, March 5, 1897.†

18. MT to Frank Bliss, January 19, 1897, and March 19, 1897.†

19. MT to Frank Bliss, March 26, 1897.†

20. Ibid.

21. MT to Orion Clemens, March 28, 1897 †; MT to James MacAlister, April 14, 1897.†

22. MT to Chatto and Windus, July 25, 1897.†

23. MT to Wayne MacVeagh, August 22, 1897.†

Fred Kaplan

Mark Twain as a traveler has been written about briefly but brilliantly in Richard Bridgman, *Traveling in Mark Twain* (Berkeley: University of California Press, 1987). A number of general biographical works deal with Twain's round-the-world voyage and with his anti-imperialism, particularly Justin Kaplan, *Mr. Clemens and Mark Twain* (New York: Simon and Schuster, 1966). Hamlin Hill, *Mark Twain: God's Fool* (New York: Harper and Row, 1973), focuses on the last ten years of Twain's life. Twain's own words, in letters and in his autobiography, are riveting, especially in the *Mark Twain-Howells Letters*, ed. Henry Nash Smith and William M. Gibson (Cambridge: Harvard University Press, 1960), in *Mark Twain's Correspondence with Henry Huttleston Rogers*, ed. Lewis Leary (Berkeley: University of California Press, 1969), and in *The Autobiography of Mark Twain*, ed. Charles Neider (New York, Harper and Row, 1969).

His talents as a lecturer and his lectures themselves are the subjects of Paul Fatout, *Mark Twain on the Lecture Circuit* (Bloomington: Indiana University Press, 1960), and *Mark Twain Speaking*, ed. Paul Fatout (Iowa City: University of Iowa Press, 1976). Mark Twain's travels to Australia and New Zealand are well documented in Miriam Jones Schillingsburg, *At Home Abroad: Mark Twain in Australasia* (Jackson: University Press of Mississippi, 1988), and in her "From Ballarat to Bendigo with Mark Twain," *Australian Literary Studies* 12 (May 1985): 116–19. There are scattered articles on the subject here and there, including Sarah Searight, "Mark Twain in New Zealand," *New Zealand Heritage* 13 (1972): 1703–5, and Coleman O. Parsons' four articles: "Mark Twain in Adelaide," *Mark Twain Journal* 21 (Spring 1983): 51–55; "Mark Twain in Australia," *Antioch Review* 21 (Winter 1961–62): 455–68; "Mark Twain in Melbourne," *Mark Twain Journal* (Spring 1984): 41–42; and "Mark Twain in New Zealand," *South Atlantic Quarterly* 61 (1962): 51–76. Twain's activities in England and his relationship with the British are the subject of Dennis Welland, *Mark Twain in England* (London: Chatto and Windus, 1978).

On the subject of Mark Twain and imperialism, Twain's own words, collected in *Mark Twain's Weapons of Satire: Anti-imperialist Writings on the Philippine-American War*, ed. Jim Zwick (Syracuse: Syracuse University Press, 1992), which has a full bibliography, and Twain's autobiographical fulminations against Theodore Roosevelt and others in *Mark Twain in Eruption*, ed. Bernard DeVoto (New York: Harper and Brothers, 1940), are an excellent starting point. William M. Gibson's *Theodore Roosevelt Among the Humorists: W. D. Howells, Mark Twain, and Mr. Dooley* (Knoxville: University of Tennessee Pres, 1960) and Philip S. Foner's *Mark Twain: Social Critic* (New York: International Publishers, 1958) are useful works. William R. Macnaughton, *Mark Twain's Last Years as a Writer* (Columbia: University of Missouri Press, 1979), has some general discussion of the anti-imperialist essays, and there is some focused discussion in William M. Gibson, "Mark Twain and Howells: Anti-imperialists," *New England Quarterly* 20 (1947): 435–70; Fred Harrington, "The Anti-imperialist Movement in the United States," *Mississippi Valley Historical Review* 22 (September 1935): 211–30; and Fred Harrington, "Literary Aspects of American Anti-imperialism, 1898–1902," *New England Quarterly* 10 (December 1937): 650–67. On Twain and the Belgian Congo, see Robert Wuliger, "Mark Twain on *King Leopold's Soliloquy*," *American Literature* 25 (March 1953): 234–37; Hunt Hawkins, "Mark Twain's Involvement with the Congo Reform Movement," *New England Quarterly* 51 (June 1978): 147–75; and Robert Giddings, "Mark Twain and King Leopold of the Belgians," in *Mark Twain: A Sumptuous Variety*, ed. Robert Giddings (Totowa, N.J.: Barnes and Noble, 1955).

ILLUSTRATORS AND ILLUSTRATIONS
IN MARK TWAIN'S FIRST AMERICAN EDITIONS

Beverly R. David & Ray Sapirstein

From the "gorgeous gold frog" stamped into the cover of *The Celebrated Jumping Frog of Calaveras County* in 1867 to the comet-riding captain on the frontispiece of *Extract from Captain Stormfield's Visit to Heaven* in 1909, illustrators and illustrations were an integral part of Mark Twain's first editions.

Twain marketed most of his major works by subscription, and illustration functioned as an important sales tool. Subscription books were packed with pictures of every type and size and were bound in brassy gold-stamped covers. The books were sold by agents who flipped through a prospectus filled with lively illustrations, selected text, and binding samples. Illustrations quickly conveyed a sense of the story, condensing the proverbial "thousand words" and outlining the scope and tone of the work, making an impression on the potential purchaser even before the full text had been printed. Book canvassers were rewarded with up to 50 percent of the selling price, which started at $3.50 and ranged as high as $7.00 for more ornate bindings. The books themselves were seldom produced until a substantial number of customers had placed orders. To justify the relatively high price and to reassure buyers that they were getting their money's worth, books published by subscription had to offer sensational volume and apparent substance. As Frank Bliss of the American Publishing Company observed, these consumers "would not pay for blank paper and wide margins. They wanted everything filled up with type or pictures." While authors of trade books generally tolerated lighter sales, gratified by attracting a "better class of readers," as Hamlin Hill put it, authors of subscription books sacrificed literary respectability for popular appeal and considerable profit.[1]

The humorist George Ade remembered Twain's books vividly, offering us a child's-eye view of the nineteenth-century subscription book market.

Just when front-room literature seemed at its lowest ebb, so far as the American boy was concerned, along came Mark Twain. His books looked at a distance, just like the other distended, diluted, and altogether tasteless volumes that had been used for several decades to balance the ends of the center table . . . so thick and heavy and emblazoned with gold that [they] could keep company with the bulky and high-priced Bible. . . . The publisher knew his public, so he gave a pound of book for every fifty cents, and crowded in plenty of wood-cuts and stamped the outside with golden bouquets and put in a steel engraving of the author, with a tissue paper veil over it, and "sicked" his multitude of broken-down clergymen, maiden ladies, grass widows, and college students on the great American public.

Can you see the boy, Sunday morning prisoner, approach the book with a dull sense of foreboding, expecting a dose of Tupper's *Proverbial Philosophy*? Can you see him a few minutes later when he finds himself linked arm-in-arm with Mulberry Sellers or Buck Fanshaw or the convulsing idiot who wanted to know if Christopher Columbus was sure-enough dead? No wonder he curled up on the hair-cloth sofa and hugged the thing to his bosom and lost all interest in Sunday school. *Innocents Abroad* was the most enthralling book ever printed until *Roughing It* appeared. Then along came *The Gilded Age*, *Life on the Mississippi*, and *Tom Sawyer*. . . . While waiting for a new one we read the old ones all over again.[2]

Publishers, editors, and Twain himself spent a good deal of time on design — choosing the most talented artists, directing their interpretations of text, selecting from the final prints, and at times removing material they deemed unfit for illustration.[3]

With the exception of *Following the Equator* (1897), books released in the twilight of Twain's career were not sold by subscription. Twain's later books, published for the trade market by Harper and Brothers, seldom contained more than a frontispiece and a dozen or so tasteful illustrations, rather than the hundreds of illustrations per volume that subscription publishing demanded. Illustration, however, remained a major component of Twain's later work in two important cases: *Extracts from Adam's Diary*, illustrated by Fred

Strothmann in 1904, and *Eve's Diary*, illustrated by Lester Ralph in 1906.

The stories behind the illustrators and illustrations of Mark Twain's first editions abound in back-room intrigue. The besotted or negligent lapses of some of the artists and the procrastinations of the engravers are legendary. The consequent production delays, mistimed releases, and copyright infringements all implied a lack of competent supervision that frequently infuriated Twain and ultimately encouraged him to launch his own publishing company.

In many cases, Twain took illustrations into account as he wrote and edited his text, using them as counterpoint and accompaniment to his words, often allowing them to inform his general narrative strategy and to influence the amount of detail he felt necessary to include in his written descriptions. In the most artful and carefully considered illustrated works, an analysis of the relationships between author and illustrator and between text and pictures illuminates key dimensions of Twain's writings and the responses they have elicited from readers. Examinations of even the most straightforward examples of decorative imagery yield insights into the publishing history of Twain's books and his attitudes toward the production process.

The original illustrations in Twain's works have often been replaced in the twentieth century by subsequent visual interpretations. But while Norman Rockwell's well-known nostalgic renderings of *Tom Sawyer* and *Huckleberry Finn* may tell us much about 1930s sensibilities, we would do well to reacquaint ourselves with the first American editions and the artwork they contained if we want to understand the books Twain wrote and the world they affected.

Illustrated books, like the illustrated weekly magazines that first appeared in the 1860s, were a significant source of visual images entering nineteenth-century homes. Because of their widespread popularity and the relative paucity of other sources of visual information, Twain's books helped to define America's perceptions of remote people, exotic scenes, and historic events. In addition to being an essential element of Mark Twain's body of work, illustrations are a documentary source in their own right, a window into Twain's world and our own.

NOTES

1. For background on subscription book publishing, see Hamlin Hill, *Mark Twain and Elisha Bliss* (Columbia: University of Missouri Press, 1964), chapter 1. See also R. Kent Rasmussen, "Subscription-book publishing" entry, *Mark Twain A to Z: The Essential Reference to His Life and Writings* (New York: Facts on File, 1995), p. 448.

2. George Ade, "Mark Twain and the Old-Time Subscription Book," *Review of Reviews* 61 (June 10, 1910): 703–4; reprinted in Frederick Anderson, ed., *Mark Twain: The Critical Heritage* (London: Routledge and Kegan Paul, 1971), pp. 337–39.

3. Beverly R. David, *Mark Twain and His Illustrators, Volume 1 (1869–1875)* (Troy, N.Y.: Whitston Publishing Company, 1986), discusses in detail Twain's involvement in the production of his early books.

READING THE ILLUSTRATIONS IN *FOLLOWING THE EQUATOR* AND *KING LEOPOLD'S SOLILOQUY*

Beverly R. David & Ray Sapirstein

The book cover of *Following the Equator* (1897) promised readers a vicarious trip around the world. A graphically elegant African elephant and Indian temple appeared in full color on a pastel pink background flanked by richly decorated gold arabesques; the cursive, gilt-lettered title on a deep blue background suggested Eastern exoticism and romantic fantasy. In scope and lavishness, the book outstripped any of Mark Twain's previous travel narratives. The creator of the cover image, F. Berkeley Smith, was but one of a large team of illustrators and photographers who contributed to the account of Mark Twain's monumental odyssey.

Twain was probably responsible for enlisting Dan Beard as principal illustrator for the book. Beard (1850–1941) had illustrated *A Connecticut Yankee in King Arthur's Court* in 1889, and Twain, enthusiastic about the artist's abilities as a collaborator and kindred spirit, used him regularly in the 1890s, on *The American Claimant*, *The £1,000,000 Bank-Note*, and *Tom Sawyer Abroad*. Many of the other *Equator* illustrators were equally well known. A. B. Frost (1851–1928), of *Uncle Remus* fame, had previously illustrated *Tom Sawyer, Detective*, Frederick Dielman (1847–1935) was the illustrator of the deluxe editions of Longfellow and Hawthorne, and the drawings of A. G. Reinhart (1854–1926) filled popular magazines. Several of the contributors were up-and-coming artists, some of whom would rank as masters of the craft in the next decade: Smith (1868–1931), B. W. Clinedinst (1859–1931), Peter Newell (1862–1924), Thomas Fogarty (1873–1938), and C. Allan Gilbert (1873–1929). C. H. Warren and F. M. Senior (the latter's name misspelled in the list of illustrations) had done the drawings for the first American edition of *The Tragedy of Pudd'nhead Wilson and the Comedy Those Extraordinary Twins*.[1] Most of the artists were friends and had studied at the Art Students League in New York.

To augment the number of images in the book, Twain sought the assistance of three photographers who had accompanied him on several legs of the journey. (Introduced in the late 1880s, Kodak cameras and roll film had made photography popularly accessible for the first time.) Walter Chase of Boston, F. R. Reynolds of England, and Major James B. Pond, Twain's American lecture agent, took snapshots that covered various segments of the trip, as did the Clemenses themselves. Among the scenes of relaxed shipboard camaraderie, Olivia Clemens appears in the photo "Watching for the Blue Ribbon," seated demurely on a bench behind her bantering husband (66).[2] Many other candid travel snapshots found their way into the pages of *Following the Equator*, a feat made possible by the technical innovations of the 1880s and 1890s. The recent development of the half-tone printing process allowed the reproduction of photographic images in ink, as well as the direct reproduction of painted illustrations. Still a novelty, such images could be printed directly from the artist's hand without the intermediate engraving process, which was necessarily the secondhand interpretation of a craftsman trained to scratch the image into wood blocks and steel plates.

For the first time, one of Mark Twain's books had a good many halftone photographs to supplement the usual multitude of engravings, though many of these halftones were heavily altered and retouched by hand. More images of Twain himself appear in *Following the Equator* than in any of his books since *Sketches, New and Old*, published in 1875. The frontispiece, a photogravure of the author aboard ship, is from a photograph by Walter Chase. The first full-page illustration, "They Passed in Review" (24), effectively a second frontispiece, shows a playfully stunned Twain encountering a parade of otherworldly creatures, most of them half human, half beast. A product of Dan Beard's incisive creativity, the drawing suggests Twain's revelry in the diversity of the world's inhabitants while poking fun at the author's peculiar tendency to play fast and loose with the truth.

With twenty-six signed illustrations and three others probably attributable to him, Beard contributed more prints than any other artist, and his spade logo is found on most of them. Best known as a founder of the Boy Scouts of America, he later transformed the symbol into the trademark emblem of scouting, the familiar fleur de lis. Beard shared Twain's sense of humor and was a staunch ally in his fierce hatred of cruelty and oppression. He firmly endorsed Twain's outspoken denunciation of Anglo-Australian outrages against Aborigines. In a full-page allegory placed arrestingly without direct relation to the text (187), Beard shows a heavily armed white hunter coolly wading in gore, his trophies including several human bodies. Heedless of the biblical injunction "Thou Shalt Not Kill," which is inscribed above his head, the hunter is crowned with laurels by the Grim Reaper. The colonial adventurer appears in the service of death, offering sacrificial violence and brutality to appease a voracious master. The drawing entitled "The Political Pot" (655) offers a similarly direct indictment of colonialism in Africa. Beard's sharp images in *Following the Equator* are as powerful and ambitious as any he drew for Twain.

F. M. Senior's illustrations, scattered randomly throughout the work in disparate episodes, rank second in number to Beard's. Senior's highlights include cartoons in the Queensland chapters, numerous chapter tailpieces, the full-page drawing of Cecil Rhodes and the shark that made him a millionaire by delivering the newspaper (147), and a charging mad elephant (409). Clinedinst, whose specialty was portraits, made several memorable contributions: a full-page image of President Grant with Twain (39), Twain in pajamas on his way out of Bombay (458), and "I Was Embarrassed," in which Twain appears in a pith helmet in conversation with a young native woman in Ceylon (601).

Frederick Dielman's eight illustrations include an inventive triptych detailing the undersea exploration of a ghastly shipwreck in Hawaii (56), a crowded Indian train station (402), and several reiterations of popular stereotypes of non-western peoples. The images by Warren, Fogarty, and Gilbert appear in the book sporadically, without any discernible pattern. The work of the remainder of the illustrators is generally concentrated in short bursts, as with Newell's five contributions to chapter 64 and Reinhardt's three contributions to chapter 68. A. B. Frost, the most renowned illustrator of the group, contributed all the drawings to chapters 62 and 63. "The Mate's shadow froze fast to the deck" (613) is comical in its matter-of-fact delineation of Twain's anecdote, and Frost's caricatures and scenes typically exult in the absurd: the androgynous "A Female Uncle" (610) and the loincloth-clad native carrying an umbrella in "The Wettest Place on Earth" (619) border on the surreal. For the most part, the last chapters of the book are illustrated photographically, with dry news photos and stock images of noteworthy locations.

The variation in illustrative style and sensibility among this diverse group of collaborators heightens the episodic nature of the narrative. Where a single illustrator might have brought a polished artistic unity to the book, the jumble of styles and the irregular distribution of the illustrators' work instead contribute to an offhand, improvisatory mood reminiscent of the breezy illustrated magazines of the 1890s. This innocuous look was a convenient cover for Twain's anti-imperial polemics and helped *Following the Equator* retain its commercial appeal. As in many of his books, Twain relied on the illustrators' humor to tone down his most bitter and challenging criticism.

The illustrations in *King Leopold's Soliloquy* (1905) operate in a very different manner. Without commercial motive, inspired purely by a moral imperative, they were calculated to heighten the impact of Twain's words by providing incontrovertible proofs of brutality and genocide. The documentary-style images of savage crimes in the Congo sought to make readers witnesses, shock them into moral outrage, and catalyze popular action to depose Leopold and strip him of his colonial holdings. Although viewers today may be inured to the steady stream of images of depredations in remote places, in Twain's time direct confrontation with vast human suffering was novel and had great potential to prick readers' consciences and inspire them to active condemnation.

King Leopold's Soliloquy and *Following the Equator* are the only works by Twain substantially illustrated with photographs. In *Following the Equator*, the informal and often awkward snapshots of Twain lend the narrative a feeling of candor and intimacy; and the stark, unembellished photographs of scenes and characters gave contemporary readers doses of pure information and the sense that they were encountering actuality itself. In *King Leopold's Soliloquy*, photography brought readers to the sites of atrocities, with little apparent mediation or manipulation of the information. Twain, through

Leopold's resentful words, presented the camera as an "incorruptible" witness and "the most powerful enemy" of injustice because it captured the terrible crimes committed in secrecy and opened them to public scrutiny. Publishing the images both informed the world and let the perpetrators know they were being watched. According to Philip Foner, Twain conceived the pamphlet as an illustrated work and specified that it was to depict the mutilations of the victims explicitly.[3]

The drawings in *King Leopold's Soliloquy* were made by an artist who evidently wished to remain anonymous. While the photographs, too, are unattributed, they were supplied by Edmund Dene Morel and the Congo Reform Association, and were taken on site by missionaries seeking to expose human rights abuses. Several of the images in *King Leopold's Soliloquy* had been published the previous year in Morel's *King Leopold's Rule in Africa*. The drawing of the shockingly placid man on page 18 is a severely cropped version of a photograph in Morel's book that included several other figures; a hand-drawn rendering of the photograph appeared in *King Leopold's Soliloquy*, a generation removed from the original, it mitigated the gruesome immediacy of the scene. The caption in Morel's book is also more explicit, adding that rubber plantation overseers had "killed [the man's] wife, his daughter, and a son, cutting up the bodies, cooking and eating them." In addition, Morel's caption identifies the photographer, John H. Harris, and the victim, "Nsala of Wala in the Nsongo District (Abir Concession)."[4] Although not in this edition, a photograph of Harris, a missionary and reformer, appears in the appendix of the second edition of *King Leopold's Soliloquy* (1906), accompanying the interview "Ought Leopold to Be Hanged?" Other photographs in Morel's book were credited to Herbert Frost, who documented the abuses of the rubber trade in Baringa in 1903. Frost was probably responsible for some of the photographs used in *King Leopold's Soliloquy* as well.

Three of the photographs that appear in the collage of mutilation victims on page 40 were also reproduced in the Morel book, each within a separate frame and captioned with the victim's name, under the joint title "Children Mutilated by Congo Soldiery"; elsewhere in the book, Morel related their individual stories.[5] As the caption for the collage in *King Leopold's Soliloquy* suggests, the images were in fact "sneaked around everywhere": the Congo Reform Association evidently kept files of stock photos on hand to be used for publicity purposes. In the service of an undeniably noble effort in this case, the images of the Congo, photographic and otherwise, were manipulated in a sophisticated manner to produce a calculated response, despite Twain's portrayal of the Kodak as an incorruptible instrument of disinterested reportage.

NOTES

1. See entries on Beard, Frost, Newell, Senior, Smith, Warren, and *Following the Equator* in R. Kent Rasmussen, *Mark Twain A to Z: The Essential Reference to His Life and Writings* (New York: Facts on File, 1995).

2. Thanks to Kent Rasmussen for pointing out this detail.

3. Philip Foner, *Mark Twain, Social Critic* (New York: International Publishers, 1958), p. 388.

4. Edmund Dene Morel, *King Leopold's Rule in Africa* (London: William Heinemann, 1904; rpt., Westport, Conn.: Negro Universities Press, 1970), p. 145.

5. Ibid., p. 112.

A NOTE ON THE TEXT

Robert H. Hirst

The text of *Following the Equator: A Journey around the World* is a photographic facsimile of a copy of the first American edition dated 1897 on the title page. The first edition was published in November 1897; two copies were deposited with the Copyright Office on November 13 (*BAL* 3451). §The text of *To the Person Sitting in Darkness* is a photographic facsimile of a copy of the first American edition dated 1901 on the title page (*BAL* 3470). The first edition reprinted the serial text from the February 1901 *North American Review* and was published in March 1901 (Jim Zwick, *Mark Twain's Weapons of Satire,* Syracuse University Press, 1992, p. xxiii). §The text of *King Leopold's Soliloquy; a Defense of His Congo Rule* is a photographic facsimile of a copy of the first American edition dated 1905 on the title page. The first edition was published in September 1905; two copies were deposited with the Copyright Office on September 28. The copy reproduced here is an example of Jacob Blanck's second issue (*BAL* 3485). All copies of the first editions of *To the Person Sitting in Darkness* and *King Leopold's Soliloquy* bear the year of first publication on the title page. §The first two original volumes reproduced here are in the collection of the Mark Twain House in Hartford, Connecticut (810/C625fo/1897/c. 10 and +810/C625to/1901/pam). The third is in the First Editions Collection of the American Antiquarian Society in Worcester, Massachusetts.

THE MARK TWAIN HOUSE

The Mark Twain House is a museum and research center dedicated to the study of Mark Twain, his works, and his times. The museum is located in the nineteen-room mansion in Hartford, Connecticut, built for and lived in by Samuel L. Clemens, his wife, and their three children, from 1874 to 1891. The Picturesque Gothic-style residence, with interior design by the firm of Louis Comfort Tiffany and Associated Artists, is one of the premier examples of domestic Victorian architecture in America. Clemens wrote *Adventures of Huckleberry Finn*, *The Adventures of Tom Sawyer*, *A Connecticut Yankee in King Arthur's Court*, *The Prince and the Pauper*, and *Life on the Mississippi* while living in Hartford.

The Mark Twain House is open year-round. In addition to tours of the house, the educational programs of the Mark Twain House include symposia, lectures, and teacher training seminars that focus on the contemporary relevance of Twain's legacy. Past programs have featured discussions of literary censorship with playwright Arthur Miller and writer William Styron; of the power of language with journalist Clarence Page, comedian Dick Gregory, and writer Gloria Naylor; and of the challenges of teaching *Adventures of Huckleberry Finn* amidst charges of racism.

CONTRIBUTORS

Beverly R. David is professor emerita of humanities and theater at Western Michigan University in Kalamazoo. She is currently working on volume 2 of *Mark Twain and His Illustrators*, and on a Mark Twain mystery entitled *Murder at the Matterhorn*. She has written a number of sections on illustration for the *Mark Twain Encyclopedia* and her *Mark Twain and His Illustrators, Volume 1 (1869–1875)* was published in 1989. Dr. David resides in Allegan, Michigan, in the summer and Green Valley, Arizona, in the winter.

Shelley Fisher Fishkin, professor of American Studies and English at the University of Texas at Austin, is the author of the award-winning books *Was Huck Black? Mark Twain and African-American Voices* (1993) and *From Fact to Fiction: Journalism and Imaginative Writing in America* (1985). Her most recent book is *Lighting Out for the Territory: Reflections on Mark Twain and American Culture* (1996). She holds a Ph.D. in American Studies from Yale University, has lectured on Mark Twain in Belgium, England, France, Israel, Italy, Mexico, the Netherlands, and Turkey, as well as throughout the United States, and is president-elect of the Mark Twain Circle of America.

Robert H. Hirst is the General Editor of the Mark Twain Project at The Bancroft Library, University of California at Berkeley. Apart from that, he has no other known eccentricities.

Fred Kaplan, Distinguished Professor at Queens College and the Graduate Center of the City University of New York, is the author of *Miracles of Rare Device: The Poet's Sense of Self in Nineteenth-Century Poetry* (1972), *Dickens and Mesmerism: The Hidden Springs of Fiction* (1975), *Sacred Tears: Sentimentality in Victorian Literature* (1987), *Dickens: A Biography* (1988), and *Henry James, The Imagination of Genius: A Biography* (1992). His *Thomas Carlyle: A Biography* (1983) was nominated for the National Book Critics' Circle Award and was a jury-nominated finalist for the

Pulitzer Prize. The editor of *Dickens' Book of Memoranda* (1983) and *Traveling in Italy with Henry James* (1994), he has held Guggenheim and National Endowment for the Humanities Fellowships and has been a fellow of the National Humanities Center, the Rockefeller Foundation, and the Huntington Library.

Ray Sapirstein is a doctoral student in the American Civilization Program at the University of Texas at Austin. He curated the 1993 exhibition *Another Side of Huckleberry Finn: Mark Twain and Images of African Americans* at the Harry Ransom Humanities Research Center at the University of Texas at Austin. He is currently completing a dissertation on the photographic illustrations in several volumes of Paul Laurence Dunbar's poetry.

Gabriel Garcia Márquez praised "*Gore Vidal*'s magnificent series of historical novels or novelized histories" that deal with American life as viewed by one family from the Revolution to the present: *Burr, Lincoln, 1876, Empire, Hollywood, and Washington, D.C.* Vidal's interest in politics has not been limited to commentary; he ran for Congress in New York in 1960, and in 1982 came in second in the California Democratic senatorial primary. *Myron* and *Myra Breckinridge*, mine of the vein of fanciful, sometimes apocalyptic humor that inform *Kalki, Duluth*, and *Live from Golgotha*, works described by Italo Calvino as "the hyper-novel or the novel elevated to the square or to the cube." After *Julian*, Vidal continued to explore the ancient world in the wide-ranging *Creation*. The *Boston Globe* noted, "He is our greatest living man of letters." In 1993 *United States: Essays 1952–1992* won the National Book Award.

ACKNOWLEDGMENTS

There are a number of people without whom The Oxford Mark Twain would not have happened. I am indebted to Laura Brown, senior vice president and trade publisher, Oxford University Press, for suggesting that I edit an "Oxford Mark Twain," and for being so enthusiastic when I proposed that it take the present form. Her guidance and vision have informed the entire undertaking.

Crucial as well, from the earliest to the final stages, was the help of John Boyer, executive director of the Mark Twain House, who recognized the importance of the project and gave it his wholehearted support.

My father, Milton Fisher, believed in this project from the start and helped nurture it every step of the way, as did my stepmother, Carol Plaine Fisher. Their encouragement and support made it all possible. The memory of my mother, Renée B. Fisher, sustained me throughout.

I am enormously grateful to all the contributors to The Oxford Mark Twain for the effort they put into their essays, and for having been such fine, collegial collaborators. Each came through, just as I'd hoped, with fresh insights and lively prose. It was a privilege and a pleasure to work with them, and I value the friendships that we forged in the process.

In addition to writing his fine afterword, Louis J. Budd provided invaluable advice and support, even going so far as to read each of the essays for accuracy. All of us involved in this project are greatly in his debt. Both his knowledge of Mark Twain's work and his generosity as a colleague are legendary and unsurpassed.

Elizabeth Maguire's commitment to The Oxford Mark Twain during her time as senior editor at Oxford was exemplary. When the project proved to be more ambitious and complicated than any of us had expected, Liz helped make it not only manageable, but fun. Assistant editor Elda Rotor's wonderful help in coordinating all aspects of The Oxford Mark Twain, along with literature editor T. Susan Chang's enthusiastic involvement with the project in its final stages, helped bring it all to fruition.

I am extremely grateful to Joy Johannessen for her astute and sensitive copyediting, and for having been such a pleasure to work with. And I appreciate the conscientiousness and good humor with which Kathy Kuhtz Campbell heroically supervised all aspects of the set's production. Oxford president Edward Barry, vice president and editorial director Helen McInnis, marketing director Amy Roberts, publicity director Susan Rotermund, art director David Tran, trade editorial, design and production manager Adam Bohannon, trade advertising and promotion manager Woody Gilmartin, director of manufacturing Benjamin Lee, and the entire staff at Oxford were as supportive a team as any editor could desire.

The staff of the Mark Twain House provided superb assistance as well. I would like to thank Marianne Curling, curator, Debra Petke, education director, Beverly Zell, curator of photography, Britt Gustafson, assistant director of education, Beth Ann McPherson, assistant curator, and Pam Collins, administrative assistant, for all their generous help, and for allowing us to reproduce books and photographs from the Mark Twain House collection. One could not ask for more congenial or helpful partners in publishing.

G. Thomas Tanselle, vice president of the John Simon Guggenheim Memorial Foundation, and an expert on the history of the book, offered essential advice about how to create as responsible a facsimile edition as possible. I appreciate his very knowledgeable counsel.

I am deeply indebted to Robert H. Hirst, general editor of the Mark Twain Project at The Bancroft Library in Berkeley, for bringing his outstanding knowledge of Twain editions to bear on the selection of the books photographed for the facsimiles, for giving generous assistance all along the way, and for providing his meticulous notes on the text. The set is the richer for his advice. I would also like to express my gratitude to the Mark Twain Project, not only for making texts and photographs from their collection available to us, but also for nurturing Mark Twain studies with a steady infusion of matchless, important publications.

I would like to thank Jeffrey Kaimowitz, curator of the Watkinson Library at Trinity College, Hartford (where the Mark Twain House collection is kept), along with his colleagues Peter Knapp and Alesandra M. Schmidt, for having been instrumental in Robert Hirst's search for first editions that could be safely reproduced. Victor Fischer, Harriet Elinor Smith, and especially Kenneth M. Sanderson, associate editors with the Mark Twain Project, reviewed the note on the text in each volume with cheerful vigilance. Thanks are also due to Mark Twain Project associate editor Michael Frank and administrative assistant Brenda J. Bailey for their help at various stages.

I am grateful to Helen K. Copley for granting permission to publish photographs in the Mark Twain Collection of the James S. Copley Library in La Jolla, California, and to Carol Beales and Ron Vanderhye of the Copley Library for making my research trip to their institution so productive and enjoyable.

Several contributors — David Bradley, Louis J. Budd, Beverly R. David, Robert Hirst, Fred Kaplan, James S. Leonard, Toni Morrison, Lillian S. Robinson, Jeffrey Rubin-Dorsky, Ray Sapirstein, and David L. Smith — were particularly helpful in the early stages of the project, brainstorming about the cast of writers and scholars who could make it work. Others who participated in that process were John Boyer, James Cox, Robert Crunden, Joel Dinerstein, William Goetzmann, Calvin and Maria Johnson, Jim Magnuson, Arnold Rampersad, Siva Vaidhyanathan, Steve and Louise Weinberg, and Richard Yarborough.

Kevin Bochynski, famous among Twain scholars as an "angel" who is gifted at finding methods of making their research run more smoothly, was helpful in more ways than I can count. He did an outstanding job in his official capacity as production consultant to The Oxford Mark Twain, supervising the photography of the facsimiles. I am also grateful to him for having put me in touch via e-mail with Kent Rasmussen, author of the magisterial *Mark Twain A to Z*, who was tremendously helpful as the project proceeded, sharing insights on obscure illustrators and other points, and generously being "on call" for all sorts of unforeseen contingencies.

I am indebted to Siva Vaidhyanathan of the American Studies Program of the University of Texas at Austin for having been such a superb research assistant. It would be hard to imagine The Oxford Mark Twain without the benefit of his insights and energy. A fine scholar and writer in his own right, he was crucial to making this project happen.

Georgia Barnhill, the Andrew W. Mellon Curator of Graphic Arts at the American Antiquarian Society in Worcester, Massachusetts, Tom Staley, director of the Harry Ransom Humanities Research Center at the University of Texas at Austin, and Joan Grant, director of collection services at the Elmer Holmes Bobst Library of New York University, granted us access to their collections and assisted us in the reproduction of several volumes of The Oxford Mark Twain. I would also like to thank Kenneth Craven, Sally Leach, and Richard Oram of the Harry Ransom Humanities Research Center for their help in making HRC materials available, and Jay and John Crowley, of Jay's Publishers Services in Rockland, Massachusetts, for their efforts to photograph the books carefully and attentively.

I would like to express my gratitude for the grant I was awarded by the University Research Institute of the University of Texas at Austin to defray some of the costs of researching The Oxford Mark Twain. I am also grateful to American Studies director Robert Abzug and the University of Texas for the computer that facilitated my work on this project (and to UT systems analyst Steve Alemán, who tried his best to repair the damage when it crashed). Thanks also to American Studies administrative assistant Janice Bradley and graduate coordinator Melanie Livingston for their always generous and thoughtful help.

The Oxford Mark Twain would not have happened without the unstinting, wholehearted support of my husband, Jim Fishkin, who went way beyond the proverbial call of duty more times than I'm sure he cares to remember as he shared me unselfishly with that other man in my life, Mark Twain. I am also grateful to my family — to my sons Joey and Bobby, who cheered me on all along the way, as did Fannie Fishkin, David Fishkin, Gennie Gordon, Mildred Hope Witkin, and Leonard, Gillis, and Moss

Plaine — and to honorary family member Margaret Osborne, who did the same.

My greatest debt is to the man who set all this in motion. Only a figure as rich and complicated as Mark Twain could have sustained such energy and interest on the part of so many people for so long. Never boring, never dull, Mark Twain repays our attention again and again and again. It is a privilege to be able to honor his memory with The Oxford Mark Twain.

Shelley Fisher Fishkin
Austin, Texas
April 1996